Sustainable Solutions for Modern Economies

RSC Green Chemistry

Series Editors:
James H Clark, *Department of Chemistry, University of York, York, UK*
George A Kraus, *Department of Chemistry, Iowa State University, Iowa, USA*

Titles in the Series:
1: The Future of Glycerol: New Uses of a Versatile Raw Material
2: Alternative Solvents for Green Chemistry
3: Eco-Friendly Synthesis of Fine Chemicals
4: Sustainable Solutions for Modern Economies

How to obtain future titles on publication:
A standing-order plan is available for this series. A standing order will bring delivery of each new volume immediately on publication.

For further information please contact:
Sales and Customer Care, Royal Society of Chemistry, Thomas Graham House, Science Park, Milton Road, Cambridge, CB4 0WF, UK
Telephone: +44 (0)1223 432360, Fax: +44 (0)1223 420247, Email: sales@rsc.org
Visit our website at http://www.rsc.org/Shop/Books/

Sustainable Solutions for Modern Economies

Edited by

Rainer Höfer
Cognis GmbH, Monheim, Germany

RSCPublishing

The front cover image has been taken from the website of EFPRA, the European Fat Processors and Renderers Association, Rijswijk, Netherlands, http://www.efpra.eu. The picture shows SARIA Bio-Industries' SIFDDA SAS site in Benet, France. Reproduction with kind permission of EFPRA and SARIA Bio-Industries, Selm, Germany.

RSC Green Chemistry No. 4

ISBN: 978-1-84755-905-0
ISSN: 1757-7039

A catalogue record for this book is available from the British Library

© Royal Society of Chemistry 2009
Reprinted 2012

Published by The Royal Society of Chemistry,
Thomas Graham House, Science Park, Milton Road,
Cambridge CB4 0WF, UK

Registered Charity Number 207890

For further information see our website at www.rsc.org

Foreword

There's a funny thing about design. You can't do design by accident. If you wind up with a wonderful new product through serendipity, you can say all kinds of things about it but you can never claim that it was designed. This is important because what we face today is the greatest design challenge of all time. How do we design the products and processes that are the basis of our society and our economy so that they are benign to humans and the environment and are sustainable? It is a difficult challenge for many reasons.

First, we have designed things so wrong for so long, we have many old, bad habits to break. As we look across the Twelve Principles of Green Chemistry, one could view them as common sense. However, common sense is not common in chemical design. The amount of waste generated per kilogram of product is often of higher magnitude than the production volume. Our feedstocks are usually depleting finite resources, our reagents are often toxic and our products persistent and bioaccumulating. The good news is that many of the best practitioners in the world have recognized the shortcomings of our chemical design and their work is featured in this book.

Second, we don't view hazard as a design flaw. We are very good at designing for performance. The past 150 years of chemistry can be viewed as nothing short of a technological miracle in the development of new medicines, dyes, materials and catalysts. However, adverse consequence to humans or the environment was never considered as a design criterion. In part, this was due to the fact that we didn't have the molecular basis of understanding hazard in a way that would inform design. However, with the advancement of the science, we have insights that allow us to design intrinsically less hazardous products and processes as can be seen in this volume.

Third, we don't think in terms of systems. Even when we approach some of the big sustainability challenges, climate change, renewable energy, pure water, food supply, toxics, *etc.*, we approach these challenges in a fragmented manner. We often forget that climate change is inextricably linked to energy, and energy to water purification, and water to food, *etc.* We often wind up doing the "right things, wrong". We purify water with acutely lethal substances. We make energy-efficient bulbs with neurotoxins, and solar energy with scarce, depleting and toxic metals. The Twelve Principles of Green Chemistry have supplied a framework by which to recognize how to do the "right things, right". In other words, to know when your solutions to sustainability challenges are themselves sustainable.

This book is a collection of work by thoughtful designers who have approached their work with sustainability in mind; who recognize the errors of our past and are designing new systems that reduce or eliminate intrinsic hazard wherever possible. It is one of the great scientific challenges that we face and we need to face it with creative, rigorous design. We cannot count on accident or serendipity to get us off the unsustainable trajectory that we are on currently.

The achievements of the field of Green Chemistry and sustainable design in its short life are truly amazing. They span every molecular sub-discipline. The achievements can be found across virtually every industry sector that chemistry touches from electronics to aerospace, to chemicals, pesticides and medicines, to paints, plastics and cosmetics. However, the most remarkable thing about the accomplishments of the field of Green Chemistry thus far is that collectively they are just a small fraction of the power and the potential of the achievements yet to be realized. The achievements in this book are yet another glimmer of how thoughtful design can lead us towards a sustainable civilization.

Paul T. Anastas
Teresa and H. John Heinz III Professor
In the Practice of Chemistry for the Environment
Yale University
USA

Preface

Apocalypse now? Was the financial crisis which erupted in 2008 the "writing on the wall", the Menetekel for the Industrial Age? Is mankind approaching the impasse of Easter Island, Anasazi and Maya societies shortly before collapse – "which followed swiftly upon the society's reaching its peak of population, monument construction and environmental impact"? Or will mankind be capable of a new global common sense? After 200 years of industrial development largely based on easily available, abundant, and hence cheap fossil raw materials, will there be a concept and an agreed-upon action plan to preserve these more and more precious materials, because they are finite, fossil resources and substitute them with renewable raw materials, enforcing sustainable development on a global basis and bringing global warming to a halt?

This introduction to *Sustainable Solutions for Modern Economies* has been written in the first week of April 2009, after the G20, NATO and EU-USA summits in London, Kehl-Strasbourg and Prague, which have created hope that such a vision might become a reality. There is no doubt, however, that concepts for energy savings on a global basis and a fair value for finite fossil resources need to be part of such reality.

Sustainable Solutions for Modern Economies is not meant as a political pamphlet. However, the very concept of sustainability and its social, economical and ecological aspects have been established and accepted at the Earth Summit in Rio de Janeiro as a political initiative obligating the signatory states to implement Agenda 21, the wide-ranging blueprint for action to achieve sustainable development worldwide. *Sustainable Solutions for Modern Economies* is meant as an essay to reflect the aspects of sustainability in the different sectors of national and global economies, to draft a roadmap for public and corporate sustainability strategies, and to outline the current status of markets, applications, use and research and development for renewable resources.

RSC Green Chemistry No. 4
Sustainable Solutions for Modern Economies
Edited by Rainer Höfer
© The Royal Society of Chemistry 2009
Published by the Royal Society of Chemistry, www.rsc.org

Besides history of the sustainability concept, Chapter 1 brings up philosophical aspects of the relationship between man and nature and highlights the key sustainability initiatives of the chemical industry, *i.e.* *The Responsible Care®️ Global Charter* and the *24 Principles of Green Chemistry and Green Engineering*.

Chapter 2 depicts the position and the systemic role of the financial market in the economic circuit on the one hand and, on the other, recently developed key performance indicators for the sustainability rating of companies used as criteria for socially responsible investments and asset management, and to analyze and measure the non-financial enterprise value on a normative basis. A normative basis necessary to comparatively measure sustainability in industrial products, processes and applications is provided by the eco-efficiency analysis. Chapter 3 describes the eco-efficiency analysis as a management tool incorporating economic and environmental aspects for the comprehensive evaluation of products over their entire life-cycle from concept development, design, implementation and marketing to end-of life issues like recycling or disposal. For the first time, Chapter 4 describes a holistic approach to define sustainability as a guiding principle for modern logistics, *i.e.* throughout the process that plans, implements and controls the effective, efficient, forward and reverse flow and storage of goods, services, finance and/or information between the point of origin and the point of consumption in order to meet customers' requirements.

Consumer behaviour and expectations, indeed, are crucial aspects to be considered when dealing with further development of the sustainability concept. This is done in Chapter 5 for consumer goods, taking detergents as an example with the life-cycle of the washing process acting as indicator, while Chapter 6 specifies the achievement of food security at a global level as a key element of sustainable development and details the importance of, and attention attributed to, the food and nutrition industries to consumer expectations throughout the value chain starting with green agriculture, animal husbandry and fishing followed by sustainable food production and processing, packaging, retailing and service.

Key challenges for society at the beginning of the twenty-first century are energy economy and alternative energies. Tens of millions of years ago, biomass provided the basis for what we actually call fossil resources and biomass again is by far the most important resource for renewable energies today. The actual status and the potential of biomass as well as biomass conversion technologies to provide green energy in the form of heat and/or power are detailed in Chapter 7, while Chapter 8 summarizes the manufacturing and usage of first-generation biofuels and gives an outlook to biomass-based second- and third-generation transportation fuels.

Together with the increasingly efficient utilization of fossil resources for heat and power generation and as fuel for transportation of people and goods, the chemical industry has established the basis for more or less all modern industries. Machinery wouldn't work and cars and trucks wouldn't move without synthetic lubricants. The chemical industry provides dyes and pigments which make our world bright and colourful. Hunger has been a problem throughout history until chemical fertilizers and pesticides made efficient agriculture and plant protection possible. Lightweight and shock resistant plastics guarantee the safe transport and storage of goods. Modern communication and information storage systems depend on liquid crystals, printed circuit boards or ultrapure silicon wafers. Human population growth, increased life expectancy and reduced risk of physical infirmity (as well as voluntary birth control) only became possible when the chemical industry emanated into the pharmaceutical industry, and when synthetic detergents ensured hygiene in personal care, laundry care and institutional cleaning. It needs to be noted, however, that organic molecules are composed of small molecular building blocks predominantly derived from coal, natural gas and crude oil. The efficient complementation and eventual substitution of these raw fossil materials by biomass is the subject matter of green chemistry and is comprehensively described in Chapter 9, which comprises lipid-based biomass (natural fats and oils, Chapter 9.1), industrial applications of carbohydrate-based biomass (starch, Chapter 9.2, and sucrose, Chapter 9.3), wood (Chapter 9.4), natural rubbers (Chapter 9.5), natural fibres (Chapter 9.6) and plant-based biologically active ingredients for cosmetics (Chapter 9.7).

The notion of sustainability in highly specialized markets where specifications and performance are key requirements is discussed in Chapter 10 (green solvent alternatives for fine chemicals, for metal treatment, for coatings and for crop protection formulations) and in Chapter 11 (sustainable solutions for adhesives and sealants).

Last but not least, White Biotechnology (Chapter 12) is largely regarded as a particularly promising gateway to a sustainable future. Reduction in greenhouse gas emissions, energy and water usage are examples of the benefits brought about by greener, cleaner and simpler biotechnology processes. White biotechnology can contribute to the reduction in the dependency on fossil resources through the utilization of renewable raw materials. An especially notable feature of white biotechnology, though, is the ability to perform specific biochemical reactions without by-product formation or waste generation, which synthetic chemistry is not able to provide.

As an employee of Henkel and Cognis I have had the chance to follow, observe and contribute to the successful implementation of sustainability as a guiding principle and business model for the company and for relations with our customers. I would like to thank my colleagues Benoît Abribat, Carsten Baumann, Manfred Biermann, Joaquim Bigorra, Paul Birnbrich, Christoph Breucker, Wolfgang H. Breuer, Stefan Busch, Dieter Feustel, Matthias Fies, Roland Grützmacher, Bernhard Gutsche, Jochen Heidrich, Uwe Held, Karlheinz Hill, Klaus Hinrichs, Ronald Klagge, Alfred Meffert, Harald Rößler, Thorsten Roloff, Setsuo Sato, Harald Sauthoff, Jörg Schmitz, Ulrich Schörken, Markus Scherer, Heinz-Günther Schulte, Alfred Westfechtel, Andreas Willing and Guido Willems, who have accompanied this enterprise and, in one way or another, have framed the concept and the content of this book.

I would like to thank all the authors for their commitment and for bringing in their knowledge, their professional experience and their expertise.

I would also like to thank the Management Board of Cognis GmbH, particularly Paul Allen, Helmut Heymann and Antonio Trius, for their support of this project.

Rainer Höfer
Düsseldorf

Contents

Abbrevations   **xxi**

Chapter 1   **History of the Sustainability Concept – Renaissance of**
                Renewable Resources   **1**
                Rainer Höfer

                1.1   From Evolution to Apocalypses   2
                1.2   Our Common Future   3
                1.3   Sustainable Chemistry   6
                1.4   Renaissance of Renewable Raw Materials   7
                References   9

Chapter 2   **Sustainability in Finance – Banking on the Planet**   **12**
                Philippe Spicher, Juliane Cramer von Clausbruch and
                Pablo von Waldenfels

                2.1   Introduction   12
                2.2   Sustainability and Asset Value   13
                2.3   Socially Responsible Investment, SRI   15
                      2.3.1   Exclusion   17
                      2.3.2   Best-in-class   18
                      2.3.3   Engagement   19
                2.4   Responsible Investment: the Mainstreaming of SRI   19
                2.5   Conclusion   22
                References   23

Chapter 3   **Metrics for Sustainability**   **25**
                Peter Saling

RSC Green Chemistry No. 4
Sustainable Solutions for Modern Economies
Edited by Rainer Höfer
© The Royal Society of Chemistry 2009
Published by the Royal Society of Chemistry, www.rsc.org

3.1   Introduction   25
3.2   The Eco-Efficiency Analysis as an Approach for the
       Checking of Sustainability in Industrial Products and
       Applications   26
       3.2.1   Conducting an Eco-Efficiency Analysis   27
3.3   Industrial Examples for Using Sustainability
       Metrics   29
       3.3.1   Eco-Efficiency Study of Curing Alternatives
               for Wooden Substrates   29
       3.3.2   Vitamin B₂ Case Study   30
       3.3.3   Eco-Efficiency Analysis Confirms: Ionic
               Liquids Provide Benefits   33
3.4   Beneficial Uses of Eco-Efficiency Analysis and
       Metrics for Sustainability   34
3.5   Outlook   35
References   35

Chapter 4   **Sustainable Logistics as a Part of Modern Economies**   **37**
                  Thierry Jouenne

4.1   Introduction   37
4.2   Definition and Role of Logistics   38
4.3   Current Situation   40
4.4   The Four Logistic Drivers   41
       4.4.1   Logistic Reliability   42
       4.4.2   Logistic Efficiency   43
       4.4.3   Logistic Agility   45
       4.4.4   Eco-logistics   46
4.5   Towards Sustainable Logistics in the Service of
       Sustainable Development   48
4.6   Conclusion   50
References   51

Chapter 5   **Sustainable Solutions for Consumer Products**   **53**
                  Frank Roland Schroeder

5.1   Introduction   53
5.2   Demographic Dynamics and Global Megatrends   54
5.3   Life-cycle of the Washing Process – an Example for
       Sustainability in Consumer Goods   57
       5.3.1   Raw Materials   59
       5.3.2   Logistics   61
       5.3.3   Production   61
       5.3.4   Use Phase   61
       5.3.5   Disposal Phase   62

5.4 Sustainability Profiles of Detergent Formulations 62
5.5 Conclusion 65
References 65

Chapter 6 Sustainable Solutions for Nutrition: A Consumer Expectation 68
Sven Thormahlen

6.1 Introduction 68
6.2 Sustainability in Food and Nutrition 69
  6.2.1 Sustainable Milk Procurement in Rural Turkey 71
  6.2.2 Sustainable Cow Feed in France 74
  6.2.3 Sustainable Exploitation of the Evian Mineral Water Source 80
6.3 Conclusion 84
References 85

Chapter 7 Biomass-based Green Energy Generation 86
Martin Kaltschmitt and Daniela Thrän

7.1 Introduction 86
7.2 Biomass Sources 90
  7.2.1 Properties 90
  7.2.2 Biomass Potential 93
7.3 Biomass Conversion 95
  7.3.1 Thermo-chemical Conversion 95
  7.3.2 Physico-chemical Conversion 107
  7.3.3 Bio-chemical Conversion 109
7.4 Biomass Use 113
7.5 Final Considerations 115
  7.5.1 Competition Areas 116
  7.5.2 Effects on Competition 118
  7.5.3 Configuration Approaches 119
  7.5.4 Conclusions and Recommendations 122
References 123

Chapter 8 Green Fuels – Sustainable Solutions for Transportation 125
Eckhard Dinjus, Ulrich Arnold, Nicolaus Dahmen, Rainer Höfer and Wolfgang Wach

8.1 Introduction 125
8.2 First-generation Biofuels 126
  8.2.1 Bioethanol 127
8.3 Lipid-based Biofuels 137
  8.3.1 Vegetable Oils as Transportation Fuels 137

8.3.2   Vegetable Oils as Biodiesel Feedstock   138
8.3.3   Fats and Oils as BTL Raw Material   140
8.3.4   Lipid-based Jet Fuels   141
8.3.5   Conclusions for Lipid-based Biofuels   142
8.4   Methane *via* Anaerobic Digestion   142
8.5   Second-generation Biofuels   142
8.5.1   Hydrogen *via* Biomass Gasification   144
8.5.2   Synthetic Natural Gas *via* Biomass
        Gasification   144
8.5.3   Biobutanol   144
8.5.4   HTU Diesel   145
8.5.5   Pyrolysis Oil   145
8.5.6   Syngas-based Biofuels   145
8.6   Third-generation Biofuels and Beyond   157
References   158

Chapter 9   Biomass for Green Chemistry   164
            Karlheinz Hill and Rainer Höfer

            References   166

Chapter 9.1 Natural Fats and Oils   167
            Karlheinz Hill and Rainer Höfer

9.1.1   Introduction   167
9.1.2   Paradigm Changes in Global Fats and Oils
        Production, Use and Trade   168
9.1.3   Production of Oils and Fats   174
        9.1.3.1   Production of Vegetable Oils and Fats   174
        9.1.3.2   Production of Animal Oils and Fats 178
9.1.4   Chemical Composition of Fats and Oils   179
        9.1.4.1   Animal Fats and Oils   182
        9.1.4.2   Vegetable Fats and Oils   182
9.1.5   The Value Chain of Fats and Oils – Industrial
        Non-food Uses   213
        9.1.5.1   Fats and Oils as Precursors for
                  Biopolymers   214
        9.1.5.2   Fatty Acids – Keystones of
                  Oleochemistry   216
        9.1.5.3   Fatty Acid Esters   218
        9.1.5.4   Green Lubricants and Carrier Oils   218
        9.1.5.5   Glycerine as C3-Building Block   222
        9.1.5.6   Fatty Alcohols   222
        9.1.5.7   Green Surfactants   224
9.1.6   Perspectives   228
References   228

Chapter 9.2 Starch: A Versatile Product from Renewable Resources for Industrial Applications **238**
Andrea Gozzo and Detlev Glittenberg

9.2.1   Markets   238
9.2.2   Starch and Derivatives   239
9.2.3   Food Applications   245
9.2.4   Pharmaceutical and Chemical Applications   247
9.2.5   Industrial Binder Applications   256
9.2.6   Paper and Board Applications   257
9.2.7   Outlook   259
References   262

Chapter 9.3 Industrial Sucrose **264**
Stefan Frenzel, Siegfried Peters, Thomas Rose and Markwart Kunz

9.3.1   Industrial Production of Sucrose   266
9.3.2   Chemistry of the Sucrose Molecule   270
        9.3.2.1   Basic Organic Chemicals by Sucrose
                  Degradation   272
        9.3.2.2   Sucrose-derived Products of Industrial
                  Relevance Maintaining the Sugar
                  Skeleton   276
        9.3.2.3   Sugar Derivatives While Maintaining
                  Carbohydrate Structure   279
9.3.3   Outlook   290
References   291

Chapter 9.4 Wood **300**
Elisabeth Windeisen and Gerd Wegener

9.4.1   Introduction   300
        9.4.1.1   Perspectives of Sustainability   302
        9.4.1.2   Forest as Ecosystem and Resource   303
        9.4.1.3   From Wood Resources to Wood
                  Products   303
9.4.2   Chemistry of Wood   306
        9.4.2.1   Survey   306
        9.4.2.2   Cellulose   307
        9.4.2.3   Polyoses (Hemicelluloses)   309
        9.4.2.4   Lignin   311
        9.4.2.5   Extractives   313
        9.4.2.6   Inorganic Components (Ash)   318
9.4.3   Pulp and Paper   319

9.4.3.1   Production and Environmental Aspects   319
                 9.4.3.2   Products   324
        9.4.4   Wood-based Composites   325
                 9.4.4.1   Conventional Concepts and Products   325
                 9.4.4.2   New Concepts and Products   326
        9.4.5   Modified Solid Wood Products   329
                 9.4.5.1   Chemical Modification   330
                 9.4.5.2   Thermal Modification   331
        9.4.6   Outlook   334
        References   335

Chapter 9.5 Natural Rubber   339
Laurent Vaysse, Frédéric Bonfils, Philippe Thaler and
Jérôme Sainte-Beuve

        9.5.1   Introduction   339
        9.5.2   Challenges Facing the Supply Chain   340
        9.5.3   Water and Carbon Budget of the Rubber Tree   343
                 9.5.3.1   Carbon and Water in Plants   343
                 9.5.3.2   Photosynthesis and Water in the Rubber
                           Tree   343
                 9.5.3.3   Tapping, Latex Yield and Carbon Budget
                           of the Rubber Tree   344
                 9.5.3.4   Tapping and Water Budget of the Rubber
                           Tree   345
        9.5.4   Biosynthesis of poly(*cis*-1,4-isoprene)   347
                 9.5.4.1   Polyisoprenoids   347
                 9.5.4.2   Biosynthetic Pathway   347
                 9.5.4.3   Localization of Rubber Biosynthesis   351
                 9.5.4.4   Conclusion   351
        9.5.5   Natural Rubber Structure   351
                 9.5.5.1   Introduction   351
                 9.5.5.2   Microstructure   352
                 9.5.5.3   Mesostructure   352
        9.5.6   Non-isoprene Components of Natural Rubber   355
                 9.5.6.1   Non-isoprene in the Different
                           Compartments of *Hevea brasiliensis*
                           Latex   355
                 9.5.6.2   Non-isoprene Families   356
                 9.5.6.3   Conclusion   359
        9.5.7   Specific Properties *versus* Synthetic Counterparts   359
                 9.5.7.1   Elasticity   360
                 9.5.7.2   Strain-induced Crystallization   360
                 9.5.7.3   Heat Build-up   360
                 9.5.7.4   Tack and Green Strength   361

9.5.7.5   Vulcanization 361
9.5.8   Conclusion 361
Acknowledgement 362
References 362

Chapter 9.6   Natural Fibres **368**
Martin Möller and Crisan Popescu

9.6.1   Generalities 368
9.6.2   Demands and Restraints for Sustainable Fibres 369
9.6.3   Fibre Structure 371
9.6.3.1   Chemistry and Structure of the Cellulose Fibres 372
9.6.3.2   Chemistry and Structure of the Protein Fibres 374
9.6.4   Fibre Sourcing 380
9.6.4.1   Cotton 380
9.6.4.2   Bast Fibres (Flax, Hemp) 381
9.6.4.3   Animal Fibres 383
9.6.4.4   Silk 384
9.6.5   Summary of the Properties of Natural Fibres 385
9.6.6   Processing of Natural Fibres 385
9.6.6.1   Operations which Transform Fibres into Fabric 388
9.6.6.2   The Cleaning Operations 388
9.6.6.3   Stabilizing the Dimensions 389
9.6.6.4   Coating and Infiltrating 389
9.6.6.5   Surface Treatments 391
9.6.7   Conclusions 391
References 391

Chapter 9.7   Plant-based Biologically Active Ingredients for Cosmetics **394**
Charlotte d'Erceville, Florence Henry, Patrice Lago and Andreas Rathjens

9.7.1   Introduction 394
9.7.2   Active Ingredients and their Functionality in Cosmetic Applications 395
9.7.3   Plant-based Raw Materials 396
9.7.4   Sustainability Concept and Corporate Social Responsibility (CSR) 397
9.7.5   From the Botanical Raw Material Towards the Final Product 398

9.7.6   Sustainable Development and CSR for the
        Supply of Natural Products Derived from the
        Argan Tree   399
        9.7.6.1   Targanine Network   401
        9.7.6.2   Partnership Between EIG Targanine
                  and Cognis   402
9.7.7   Conclusion   404
References   405

Chapter 10   Sustainable Solutions – Green Solvents for Chemistry   407
             Carles Estévez

10.1   Introduction   407
10.2   The Design of Safer Chemicals and Solvent
       Innovation   408
10.3   SOLVSAFE: A Roadmap for the Design and
       Application of Safer Functional Organic
       Solvents   411
       10.3.1   Background and Sustainability Goals   411
       10.3.2   Design Strategy   412
10.4   Industrial Application of SOLVSAFE Solvents:
       Results and Perspectives   416
       10.4.1   Fine Chemicals   416
       10.4.2   Metal Degreasing   418
       10.4.3   Paints and Varnishes   419
       10.4.4   Crop Protection Formulations   420
10.5   Conclusions   422
References   423

Chapter 11   Sustainable Solutions for Adhesives and Sealants   425
             Jürgen O. Wegner

11.1   Introduction   425
11.2   Features and Requirements of Adhesives and
       Sealants   426
11.3   Chemical Composition of Adhesives and
       Sealants over Time   428
11.4   Ongoing Sustainability Evolution   429
11.5   Quality Features and Gaps with Natural-based
       Adhesives and Sealants   432
11.6   Current Use of Renewable Raw Materials in
       Adhesives and Sealants   432

11.7 Major Use Areas for Adhesives Based on
Natural Resources 433
11.8 Outlook and Conclusion 433
References 434

Chapter 12 White Biotechnology **436**
Thomas Haas, Manfred Kircher, Tim Köhler,
Günter Wich, Ulrich Schörken and Rainer Hagen

12.1 The Status of White or Industrial Biotechnology 436
    12.1.1 Introduction 436
    12.1.2 Relevant Market Segments 437
    12.1.3 The Drivers of White Biotechnology 447
12.2 Recent Examples 449
    12.2.1 Sphingolipids 449
    12.2.2 L-Cysteine 457
    12.2.3 Lipid Biotechnology 462
    12.2.4 PLA (Polylactic Acid) 466
12.3 Outlook of White Biotechnology 472
References 473

Subject Index **479**

Abbreviations

AcTR	acetyl-CoA:long chain base acetyltransferase
A.I.S.E.	Association Internationale de la Savonnerie, de la Détergence et des Produits d'Entretien
ALA	α-linolenic acid
AMF	atomic force microscopy
AMT	accepted modern technology
API	active pharma ingredients
APME	Association of Plastics Manufacturers in Europe, now PlasticsEurope
ARA	arachidonic acid
BAT	best available technology
BSE	Bovine spongiforme Enzephalopathie, also known as 'Mad Cow Disease'
CAP	Common Agricultural Policy (of the European Community)
CBOT	Chicago Board of Trade
CDM	Clean Development Mechanism, an arrangement under the Kyoto Protocol allowing industrialised countries with a greenhouse gas reduction commitment to invest in projects that reduce emissions in developing countries as an alternative to more expensive emission reductions in their own countries
CEFIC	Conseil Européen des Fédérations de l'Industrie Chimique
CEN	European Committee for Standardization
CEPEA	Research Center of the University of São Paulo/BR, focused on agribusiness issues, market analysis and price discovery, international trade, entrepreneurship and family farming
CIS	Commonwealth of Independent States: Russia, Ukrainia, Kazakhstan, Belarus, Moldova, Armenia, Aserbaijan, Georgia, Kyrgystan, Tajikistan, Turkmenistan, Uzbekistan
CLA	conjugated linoleic acid
CMAI	Chemical Market Associates

CMC	carboxymethyl cellulose
CR	Corporate Responsibility
DHA	docosahaxaenoic acid
DJSI	Dow Jones Sustainability Index
dl	decilitre
DLS	Dynamic Light Scattering
DMDHEU	dimethylol dihydroxy ethylene urea
DOE	Department of Energy (USA)
DP	degree of polymerization
DPNR	Deproteinized Natural Rubber
DP-s	direct polycondensation in high boiling solvents
DVFA	Deutsche Vereinigung für Finanzanalyse und Asset Management
EBFM	Ecosystem-based fisheries management
ECF	elemental chlorine-free
EFA	essential fatty acid[s]
EFFAS	European Federation of Financial Analysts Societies
EDI	Electronic Data Interchange
ERP	Enterprise Resource Planning
EU-27	European Union: Austria, Belgium, Bulgaria, Cyprus, Czech Republik, Denmark, Estonia, Finland, France, Germany, Greece, Hungary, Ireland, Italy, Latvia, Lithuania, Luxemburg, Malta, Netherlands, Poland, Portugal, Romania, Slovakia, Slovenia, Spain, Sweden, United Kingdom
EPA	eicosapentanoic acid
EPD	Environmental Product Declaration
EPS	expanded polystyrene
ESG	Environmental sustainability, Social responsibility and corporate Governance (often referred to as "extra-financial criteria")
FAME	fatty acid methyl ester
FAO	Food and Agriculture Organization of the United Nations
GFSI	Global Food Safety Initiative
GJ	gigajoule, equal to 10^9 joules
GM	genetically modified
GTA	Glyphosate tolerant soybeans, also referred to as Roundup Ready® soybeans
Ha	hectar
HDR	Human Development Report
HDT	heat distortion temperature
HEC	hydroxyethyl cellulose
HERA	human and environmental risk assessment
HRT	Hevea Rubber Transferase

ICE	Intercontinental Exchange, https://www.theice.com/homepage.jhtml
ICIS	Chemical & Oil Industry business information system, part of Reed Business Information (RBI), a division of Reed Business and a member of Reed Elsevier plc.
IL	ionic liquid
INRA	Institut Scientifique de Recherche Agronomique
IP	intellectual property, biotech.: inositol phosphoryl
IPP	Isopentenyl diphosphate
IR	isoprene rubber (polyisoprene rubber)
IRSG	International Rubber Study Group
LA	linoleic acid
LCA	Life-Cycle Assessment
LCI	Life-Cycle Inventory
LCIA	Life-Cycle Impact Assessment
LOHAS	Lifestyle of Health and Sustainability
MC	methyl cellulose
MDF	medium-density fibreboard
Mercosur or Mercosul Spanish	*Mercado Común del Sur*, Portuguese: *Mercado Comum do Sul*, English: *Southern Common Market* is a Regional Trade Agreement (RTA) among Brazil, Argentina, Uruguay and Paraguay, founded in 1991 by the Treaty of Asunción, which was later amended and updated by the 1994 Treaty of Ouro Preto. Its purpose is to promote free trade and the fluid movement of goods, people, and currency
Mio	million
MPa	megapascal = 1 Million Pa = 1 N/mm
Mto	metric tons
MSC	Marine Stewardship Council
MSCI	Morgan Stanley Capital International
MMD	Molar Mass Distribution
MWL	milled-wood-lignin
NAS	N-acetylserine
NMMO	4-methylmorpholine-4-oxide
NMR	nuclear magnetic resonance
NOW	new option wood
NR	natural rubber
NREL	National Renewable Energy Laboratory (USA)
NTFP	non-timber-forest-products
OAS	O-acetylserine
OHT	oil-heat-treatment
OSB	oriented strand board
PA	polyamide

PBT	polybutylene terephthalate
PDO	propanediol
PE	polyethylene
PET	polyethylene terephthalate
PETA	People for the Ethical Treatment of Animals
PJ	Peta joule $= 10^{15}\,\mu$J
PLA	polylactic acid
PLATO	provided lasting advanced timber option
PL/DLA	poly-rac.-lactide
PLLA	poly-L-lactide, the product resulting from polymerization of L,L-lactide
PP	polypropylene
PPO	processed palm oil ("processed" stands for neutralized, bleached, deodorized and/or fractionated)
PS	polystyrene
PSR	Product Specific Requirements
PVC	polyvinyl chloride
PUFA	polyunsaturated fatty acid
PUR	polyurethane
REACH	Registration, Evaluation and Authorisation of Chemicals
REF	Rubber Elongation Factor
RFID	Radio Frequency Identification
ROP	ring opening polymerisation
RSS	ribbed smoked sheets: rubber quality, prepared from intentionally coagulated latex
SARS	Severe Acute Respiratory Syndrome, a respiratory illness caused by a virus. SARS was first reported in Asia in 2003. It spread worldwide over several months before the outbreak ended
SBR	styrene- butadiene-rubber
SC	stratum corneum
SCF	Scientific Committee for Food
SEC	Size Exclusion Chromatography
SME	small & medium enterprises
SMR	Standard Malaysian Rubber
SR	synthetic rubber
SRI	Socially Responsible Investment
TAPS	tetraacetyl phytosphingosine
TCF	totally chlorine-free
TFA	trans-fatty acid[s]
T_g	glass transition temperature
TLV-TWA	Threshold Limit Value-Time Weighted Average
TMP	thermo mechanical pulp
TMT	thermally treated timber
TOE	Tons of Oil Equivalent

TPS	thermoplastic starch
TriASa	triacetyl sphinganine
TriASo	triacetyl sphingosine
TS	technical specification
TSR	Technically specified natural rubber
TÜV	Technischer Überwachungsverein
UNEP	United Nations Environment Program
VOC	volatile organic compound[s]
WPC	wood plastic composites
WUE	Water use efficiency

CHAPTER 1

History of the Sustainability Concept – Renaissance of Renewable Resources

RAINER HÖFER

Cognis GmbH, Rheinpromenade 1, D-40789 Monheim, Germany

"One World One Dream"

Slogan of the Beijing 2008 Olympic Games

"One World One Dream" fully reflects the essence and the universal values of the Olympic spirit – Unity, Friendship, Progress, Harmony, Participation and Dream. It expresses the common wishes of people all over the world, inspired by the Olympic ideals, to strive for a bright future for Mankind. In spite of the differences in colors, languages and races, we share the charm and joy of the Olympic Games, and together we seek the ideal of Mankind for peace. We belong to the same world and we share the same aspirations and dreams." (*The Official Website of the Beijing 2008 Olympic Games.*)
"In the middle of the 20th century, we saw our planet from space for the first time . . . . From space, we see a small and fragile ball dominated not by human activity and edifice but by a pattern of clouds, oceans, greenery, and soils. Humanity's inability to fit its activities into that pattern is changing planetary systems, fundamentally. Many such changes are accompanied by life-threatening hazards. This new reality, from which there is no escape, must be recognized – and managed." (Brundtland-Report "Our Common Future" – Introduction: *From one Earth to one world.*)[1]

RSC Green Chemistry No. 4
Sustainable Solutions for Modern Economies
Edited by Rainer Höfer
© The Royal Society of Chemistry 2009
Published by the Royal Society of Chemistry, www.rsc.org

"Warming of the climate system is unequivocal, as is now evident from observations of increases in global average air and ocean temperatures, widespread melting of snow and ice, and rising global average sea level." (Summary for Policymakers of the *Synthesis Report of the IPCC Fourth Assessment Report*.)[2]

1.1   From Evolution to Apocalypses

The twentieth century has seen a phenomenal growth of the global economy, a continuous improvement of the standard of living in industrialized countries and the transformation of "underdeveloped" or "Third World" nations into emerging economies. Since the Iron Curtain disappeared at the beginning of the 1990s mankind has become aware that the Earth is not "endless", that there are no longer insuperable borders setting limits to migration, that there is no unknown territory left to be discovered, conquered, cultivated or exploited.[3] The motto of the Beijing Olympic Games 2008 was aimed at expressing this awareness and the common wishes of people all over the world to strive for a bright future of mankind. Anticipating the development towards a global community, Hans Küng developed programmatically the idea of a Global Ethic, a fundamental consensus on binding values, irrevocable standards and personal attitudes common to all religions emanating into the Global Ethic project and the "Declaration towards a Global Ethic", which was endorsed by the Parliament of the World's Religions in Chicago in 1993.[4]

Globalization, global growth of economies and increases in living standards, though, have had their price: exploitation of natural resources to their limits and an ever-increasing contamination of the environment.[5]

History of mankind to a large extent means history of the relations between human beings and nature.[6] Evolution of the human species in prehistoric times was driven by nature, by genetic selection and by geological and climatic changes. In biblical times mankind received the divine message to "replenish the earth and subdue it" (Genesis 1:28). This has been read not only by Jews but also by Christians and Muslims and by the western hemisphere, at least, as man's charter, granting him the right to "have dominion over the fish of the sea, and over the fowl of the air, and over the cattle, and over all the earth, and over every creeping thing that creepeth upon the earth" (Genesis 1:26). However, "*heaven and earth, . . . and every thing that creepeth upon the earth*" has been made by God and "after his kind: and God saw that *it was good*" (Genesis 1:1–1:25), a clear statement that the entire Creation is divine, although *creation care* (*Bewahrung der Schöpfung* in German) does not directly appear as a divine commandment. Nature virtually exists as a resource. It is man's relationship with God, however, that really matters.[7]

Nature in early times was largely regarded as uncontrollable and threatening. Early man feared the night and the wilderness.[8] Natural phenomena, like floods, thunderstorms, famine and plagues of locusts, were perceived as the wrath of God. Outside the Abrahamic religions, Nature was regarded as God or Gods, arguably because of such threats. The theorem of a direct link between the biblical subordination of nature to man and devastation of the environment by mankind.[9,10] though, is doubtful.[6] Indeed, the outcry *"spectant victores ruinam naturae …"* (English: *triumphant they contemplate their victory over Nature …*) goes back to the Roman officer and encyclopedist Plinius the Elder (AD 23–79).[11] Collapses of ancient civilizations, such as the Maya, Easter Island and Anasazi,[12] generally happened outside the Hebrew–Christian sphere and it was the Greek philosopher Plato, living around 400 BC, who gave us the oldest description of a natural disaster, the deforestation of Attika.[13] Although the antique Greek culture and later the Roman Empire before converging in Christianity apparently had a close relation to Nature, venerated river gods, Nereids and Naiads, they are claimed to be at the origin of one of the first ecological disasters, the complete deforestation of all areas around the Mediterranean Sea and the resulting erosion, loss of humus soil and formation of karsts landscape in the plains followed by climatic changes which last until our modern times. Cedar trees, which once flourished throughout the Levant, the Mountains of southern Lebanon and Syria, much heralded in the times of antiquity for their beauty and fragrance, were nearly extinct when the Phoenicians ruled the Mediterranean Sea between 1200 and 900 BC. Cedar trees simply were too good a raw material for ship-building and construction.

The perception that Earth is a fragile body and easily thrown out of balance by the hand of man may have surfaced for the first time after the Trinity nuclear test on July 16, 1945, when Robert Oppenheimer quoted the Sanskrit Bhagavad Gita (English: "Song of God") text *I am become Death, the shatterer of Worlds.*[14]

1.2 Our Common Future

In 1972, the UN Stockholm Conference on the Human Environment marked the first great international meeting on how human activities were harming the environment and putting humans at risk. In the same year, the Club of Rome's report *Limits to Growth*[15] was published, which, together with the first oil crisis in 1973, had an enormous impact on public opinion worldwide and started a political debate and thinking process. Based on a mathematical simulation technique known as *systems dynamics*[16] and with input factors such as population growth, food production and fertilizer demand, energy consumption, availability of non-renewable raw materials, the report predicted that, within a time span of less than 100 years, with no major change in physical, economic or social relationships, society will run out of the non-renewable resources on which the industrial base depends. The economic system will consume successively larger amounts of depletable resources until they are completely used up. The characteristic behaviour of the system is overrun and collapse.

In autumn 1983 the General Assembly of the United Nations asked the Secretary General to appoint a World Commission on Environment and Development. The idea was to forecast, on a global scale, how man-made activities would affect the environment of the Earth encompassing the industrial as well as the social and economic aspects. The Secretary General of the UN entrusted the chair of this committee to Mrs Gro Harlem Brundtland, who was then Prime Minister of Norway. Work on the report was completed in March 1987 and it was published later that year under the title *Our Common Future*.[1] Working on that subject, the commission faced a double problem: on the one hand it was obviously human activities that were behind the deterioration of the environment, especially in developed nations. On the other hand, it was inconceivable for the UN to create difficulties for developing nations, for people who had no access to decent living conditions but who, by catching up *en masse*, would significantly add to the deleterious effects of pollution and degradation of the environment. One of the ways in which the Brundtland commission sought to overcome this dilemma consisted in the creation of the "sustainable development" concept. This concept was meant to provide a long-term balance between the environment, the economy and the social well-being of humanity – *i.e*: whereas in prehistoric times of evolution the human race was driven by nature, since biblical times humans were filling and subduing the Earth, now, at the doorstep of the third millennium, the concept of balance was born – the conclusion was drawn that the nature and action of humans between themselves and towards nature need to be in balance.[3]

In fact, the term and the concept of *sustainability* (*Nachhaltigkeit* in German) dates back to the eighteenth century. Historically, depletion of natural resources (*Raubbau* in German) is not new and hit renewable resources first. Scarcity of wood was the concern of feudal Europe and led to the introduction of sustainability principles in forest management. It was in order to preserve wood supplies for the important silver mines in the Erzgebirge, the economic backbone of the Kingdom of Saxony and its famous capital Dresden, when Hannß Carl von Carlowitz (Oberberghauptmann and Chief Executive of the Royal Saxon Mining Department) in 1713 for the first time clearly formulated the concept that forestry had to be "*sustainable*", which meant that logging and reforestation had to be in balance.[17] Similar concepts arose in France (*La grande réformation des forêts*, Colbert[18]) and in Japan.[12,19] The home of von Carlowitz, a Renaissance-style townhouse built in 1542, can still be visited at the *Obermarkt*, the market square of the Saxony mining burg of Freiberg (Figure 1.1). The principles of sustainability since then have emanated into the worlds of finance[20] (see Chapter 3), education[21] and administration.[22]

Our Common Future was the third in a series of important UN initiatives;[23] the first being the Brandt Commission's *North-South: A Program for Survival* and *Common Crisis – North-South Co-operation for World Recovery*, which spelt out the extent of the mutual interests between north and south and appealed for a program to avert disaster for the poorest countries, the need for

Hannß Carl von Carlowitz

Obermarkt Freiberg

Skyline Dresden

Figure 1.1 Hannß Carl von Carlowitz (copper engraving, Johann Martin Bernigeroth (1711), with the permission of Stadt- und Bergbaumuseum, Freiberg), Obermarkt Freiberg (Aquarell, Carl August Müller (1883), with the permission of Stadt- und Bergbaumuseum, Freiberg) and Skyline Dresden (with the permission of D. Berthold, B&V Verlag, Dresden).

a longer-term reorganization of the global economic system and methods to deal with worsening economic conditions and the lack of global cooperation.[24] The Brandt reports were followed by the Palme Commission's work on security and disarmament, *Common Security*.[25] Appearing on the public scene at the end of the cold war, *Our Common Future* cannot be separated from these earlier UN initiatives. Coinciding with the upcoming public awareness of the environment, the World Commission on Environment and Development emphasized the fact that sustainable development should be employed to safeguard the Earth's resources thereby improving social well-being and creating a better quality of life for future generations, in other words:

"Development that meets the needs of the present without compromising the ability of future generations to meet their own needs" (Brundtland, 1987).

This way, the report became the catalyst for global thinking processes about the relationship between man and nature[9,26] and about future prospects of mankind in the potentially conflicting contexts of ethics, state policies and social, ecological and economical interests.[4,23,27] In 1992, the UN Conference on Environment and Development (UNCED), more commonly known as the

Rio Earth Summit, established a number of initiatives to promote the uptake of sustainable development worldwide:

1. The Convention on Biological Diversity with three main goals:
   - conservation of biodiversity;
   - sustainable use of its components; and
   - fair and equitable sharing of benefits arising from genetic resources.

   The convention recognized for the first time in international law that the conservation of biological diversity is "a common concern of humankind" and is an integral part of the development process. The agreement covers all ecosystems, species and genetic resources. It links traditional conservation efforts to the economic goal of using biological resources sustainably. It sets principles for the fair and equitable sharing of the benefits arising from the use of genetic resources, notably those destined for commercial use.

2. The Rio Declaration on Environment and Development, reaffirming the Declaration of the UN Conference adopted at Stockholm, 1972.

3. A Statement of Principles for the Sustainable Management of Forests.

4. The Agenda 21, a comprehensive and dynamic plan of action for the twenty-first century addressing the UNCED goals and initiatives and identifying means and resources for their implementation.

1.3   Sustainable Chemistry

Chemistry has laid the foundations for many essential materials that have shaped our modern world. Earlier than many other sectors of the economy, the chemical industry has faced an imbalance in the utilization of its products (excessive use of fertilizers and pesticides in agriculture, *Silent Spring*, Rachel Carson[28]), malpractice when applied as weapons (Napalm, Agent Orange) and a series of critical safety issues during manufacturing (ICMESA, Seveso; UCC, Bhopal; Sandoz, Basel). Consequently a set of guiding principles for pollution prevention, employee health and safety, product stewardship, process safety and distribution codes originally developed since 1977 by the Canadian Chemical Producers' Association (CCPA) merged into *Responsible Care*, "the chemical industry's own, unique initiative".[29] Under *Responsible Care*, the worldwide chemical industry is committed to continual improvement in all aspects of health, safety and environmental performance and to open communication about its activities and achievements. Broken down to the level of individual companies, *Responsible Care* became a mission statement for sustainable operations (Figure 1.2).

In February 2006, The International Council of Chemical Associations (ICCA) launched the *Responsible Care Global Charter* and the *Global Product Strategy* (GPS), marking a renewal of the chemical industry's former commitment.

We do not own the world, we borrowed it from our children

Responsible Care®

Cognis' guiding principle following a saying by Antoine de Saint Exupéry

Figure 1.2 Responsible Care®.

The *Responsible Care*® initiative achieved a kind of normative enforcement, when Paul Anastas, then of the Environmental Protection Agency (EPA), in 1991 coined the phrase *Green Chemistry* and together with John C. Warner[30] and J. B. Zimmerman[31] developed the *24 Principles Of Green Chemistry & Green Engineering* (Figure 1.3). Green chemistry, also termed sustainable chemistry, is an umbrella concept that seeks to unite government, academic and industrial communities by placing more focus on environmental impacts at the earliest stage of innovation and invention. Paul Anastas and John Warner also provided the first definition of green chemistry: "Applying fundamental knowledge of chemical processes and products to achieve elegant solutions with the ultimate goal of hazard-free, waste-free, energy efficient synthesis of non-toxic products without sacrificing efficacy of function."[30] This approach represents a significant departure from the traditional principles. Instead of trying to minimize exposure to chemicals, green chemistry emphasizes the design and creation of chemicals that are not hazardous to people or the environment, in other words a kind of molecular-level pollution prevention, applying the principle that it is better to consider waste prevention options during the design and development phase, rather than disposing or treating waste after a process or material has been developed.

1.4 Renaissance of Renewable Raw Materials

Use of renewable rather than depletable feedstock is one of the Green Chemistry principles.

Renewable raw materials are already making an extensive contribution to positioning the concept of sustainability even more firmly in the public mind

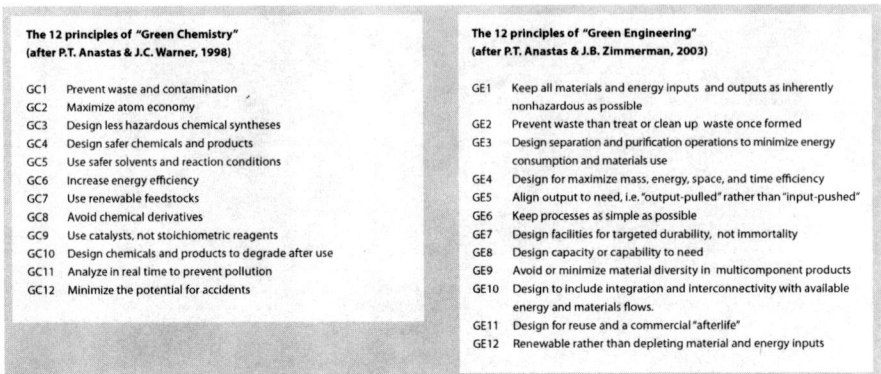

The 12 principles of "Green Chemistry" (after P.T. Anastas & J.C. Warner, 1998)	The 12 principles of "Green Engineering" (after P.T. Anastas & J.B. Zimmerman, 2003)
GC1 Prevent waste and contamination	GE1 Keep all materials and energy inputs and outputs as inherently nonhazardous as possible
GC2 Maximize atom economy	GE2 Prevent waste than treat or clean up waste once formed
GC3 Design less hazardous chemical syntheses	GE3 Design separation and purification operations to minimize energy consumption and materials use
GC4 Design safer chemicals and products	GE4 Design for maximize mass, energy, space, and time efficiency
GC5 Use safer solvents and reaction conditions	GE5 Align output to need, i.e. "output-pulled" rather than "input-pushed"
GC6 Increase energy efficiency	GE6 Keep processes as simple as possible
GC7 Use renewable feedstocks	GE7 Design facilities for targeted durability, not immortality
GC8 Avoid chemical derivatives	GE8 Design capacity or capability to need
GC9 Use catalysts, not stoichiometric reagents	GE9 Avoid or minimize material diversity in multicomponent products
GC10 Design chemicals and products to degrade after use	GE10 Design to include integration and interconnectivity with available energy and materials flows.
GC11 Analyze in real time to prevent pollution	GE11 Design for reuse and a commercial "afterlife"
GC12 Minimize the potential for accidents	GE12 Renewable rather than depleting material and energy inputs

Figure 1.3   24 Principles of Green Chemistry & Green Engineering.

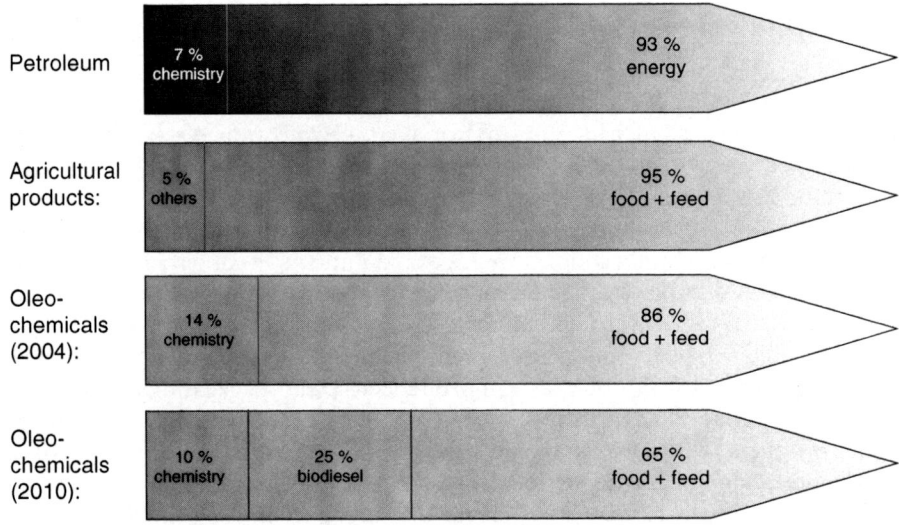

Figure 1.4   Utilization of carbonic raw material sources.

worldwide. Renewable raw materials may in the future also be the only solution when fossil resources are gradually exhausted.

Currently, 93% of crude petroleum is used for energy generation, while only 7% is used for chemistry. Similarly, 95% of agricultural products are used for food and feed, with only 5% being used for the chemical industry (Figure 1.4). 45 Mio mto of annual worldwide biodiesel capacity (forecast for 2010 based on published projects;[32] a more conservative forecast envisages 32 Mio mto by 2012[33]) would already represent 25% of total renewable fats and oils production and would reduce the share of food and feed from 86% to 65% and the percentage of chemical uses from 14% to 10%. Finally, 30% of the global arable land (450 Mio ha of 1.4 billion ha) is needed for biomass cultivation to meet 10% of the 2030 annual world oil demand[34] (41 billion barrels forecast; forecast 2009: 32 billion barrels[35]).

These figures demonstrate the rift that must be overcome before renewable resources can completely substitute fossil supplies of raw materials.[36] If both the emerging energy gap and the fuel crisis are to be solved while at the same time providing sufficient feedstock for the chemical industry and foodstuffs for humans and animals by cultivation of renewable raw materials, then a well-organized, anticipatory, cross-border cultivation policy including water management is required. This entails concrete measures to increase yields per area unit; that means concentrated advancement of green agricultural research and green crop protection, including white biotechnology and green genetic engineering.[37] A great deal could be gained by giving worldwide priority to sparing fossil resources to permit development of constructive and long-term successful solutions based on renewable raw materials.

References

1. G. Brundtland, ed., *Our Common Future*, The World Commission on Environment and Development, Oxford University Press, Oxford, 1987, http://www.worldinbalance.net/agreements/1987-brundtland.html; http://www.un-documents.net/ocf-ov.htm (retrieved 31.07.2008).
2. Climate Change 2007: Synthesis Report – Summary for Policymakers, *An Assessment of the Intergovernmental Panel on Climate Change*, Valencia, 2007, http://www.ipcc.ch/pdf/assessment-report/ar4/syr/ar4_syr_spm.pdf (retrieved 05.08.2008).
3. R. Höfer and J. Bigorra, *Green Chem.*, 2007, **9**, 203.
4. Global Ethic Foundation, http://www.weltethos.org/dat-english/index.htm; H. Küng, *Projekt Weltethos*, Piper, München, 1990.
5. C. Brandt, *Chemie in unserer Zeit*, 2002, **36**(4), 224.
6. G. Zirnstein, *Ökologie und Umwelt in der Geschichte*, Ökologie und Wirtschaftsforschung, Bd. 14, Metropolis, Marburg, 1996.
7. J. Passmore, *Man's Responsibility for Nature*, Duckworth, London, 2nd edn, 1980.
8. R. Nash, *Wilderness and the American Mind*, revised edn. Yale University Press, New Haven, London, 1973.
9. L. White Jr, *Science*, 1967, **155**(3767), 1203.
10. W. Haedecke, *Neue Rundschau*, 1975, **86**(2), 290.
11. Gaius Plinius Secundus, *Historia Naturalis*, Pliny the Elder, *The Natural History*, ed. and translation: English J. Boston, Book XXXIII, http://www.perseus.tufts.edu/cgi-bin/ptext?lookup=Plin. + Nat. + toc (retrieved 08.08.2008).
12. J. Diamond, *Collapse*, Penguin Books, London, 2006.
13. K. -W. Weeber, *Smog über Attika: Umweltverhalten im Altertum*, Rowohlt, Reinbek, 1993.

14. http://en.wikipedia.org/wiki/Trinity_test (retrieved 08.08.2008); quotation is taken from Robert Jungk, *Brighter Than a Thousand Suns*, Harcourt, Brace, 1958.

15. D. L. Meadows, *et al.*, *The Limits to Growth*, Universe Books, New York, 1972.

16. J. W. Forrester, *Industrial Dynamics*, Pegasus Communications, Waltham, 1961; J. W. Forrester, *World Dynamics*, Pegasus Communications, Waltham, 1973.

17. H. C. von Carlowitz, *Sylvicultura oeconomica – Anweisung zur wilden Baum-Zucht*, reprint der Ausg. Leipzig, Braun, 1713/bearb. von K. Irmer, A. Kießling, TU Bergakademie Freiberg u. Akad. Buchh., Freiberg, 2000; U. Grober, DIE ZEIT, 25.11.1999, **48**, 98 (nur in der Print-Ausgabe); U. Grober, *Von Kursachsen nach Rio – Ein Lebensbild über den Erfinder der Nachhaltigkeit Hannß Carl Edler von Carlowitz und die Wegbeschreibung eines Konzeptes – aus der Silberstadt Freiberg*, http://www.forschung-sheim.de/fachstelle/arb_carl.htm (retrieved 04.11.2007).

18. B. Boutefeu, *VertigO – La Revue en sciences de l'environnement*, Sept. 2005, **6**(2), 1.

19. G. Bachmann, in *Kollaps von Gesellschaften*, GAIA, 2006, **15**(4), 260M.

20. M. Jeucken, *Sustainability in Finance: Banking on the Planet*, Eburon, Delft, 2005.

21. S. Sterling, *Sustainable Education: Re-visioning Learning and Change*, Schumacher Briefings 6, Green Books for the Schumacher Society, Bristol, 2001.

22. M. Kopatz, *Nachhaltigkeit und Verwaltungsmodernisierung: eine theoretische und empirische Analyse am Beispiel nordrhein-westfälischer Kommunalverwaltungen*, Diss., Univ. Oldenburg, 2006.

23. D. Pearce, E. Barbier and A. Markandya, *Sustainable Development*, Edward Elgar Publ., Hants, Brookfield, 1990.

24. W. Brandt, *North-South: A Programme for Survival*, MIT Press, Cambridge, 1980; The Brandt Commission, *Common Crisis – North-South Co-operation for World Recovery*, MIT Press, Cambridge, 1983.

25. Palme Commission, *Common Security*, Pan Books, London, 1982.

26. G. D. Bennett, *Perspectives of Science and Christian Faith*, March 2008, **60**(1), 16.

27. K. E. Goodpaster and K. M. Sayre, *Ethics and Problems of the 21st Century*, University of Notre Dame Press, Notre Dame, London, 1979; H.-J. Harborth, *Dauerhafte Entwicklung statt Selbstzerstörung*, Ed. Sigma, Berlin, 1991; C. V. Kidd, *J. Agr. Environ. Ethics,* 1992, **5**(1), 1; W. Beckerman, *Environ. Val.*, 1994, **3**, 191; H. E. Daly, *Environ. Val.*, 1995, **4**, 49; W. Beckerman, *Environ. Val.*, 1995, **4**, 169; P. Fritz, J. Huber and H. W. Levi, Hrsg., *Nachhaltigkeit in naturwissenschaftlicher und sozialwissenschaftlicher Perspektive*, S. Hirzel, Wiss. Verlagsges., Stuttgart, 1995; K. Rennings, K. L. Brockmann, H. Koschel, H. Bergmann and I. Kühn, *Nachhaltigkeit, Ordnungspolitik und freiwillige Selbstverpflichtung*, Physica-Verl., Heidelberg, 1996; H. G. Kastenholz, K.-H. Erdmann and M. Wolff,

Hrsg., Nachhaltige Entwicklung: Zukunftschancen für Mensch und Umwelt, Springer, Berlin, Heidelberg, New York, 1996; K.-W. Brand, *Hrsg., Nachhaltige Entwicklung: Eine Herausforderung an die Soziologie*, Leske + Budrich, Opladen, 1997; BUND/Misereor, Hrsg., *Zukunftsfähiges Deutschland – Ein Beitrag zu einer global nachhaltigen Entwicklung*, [R. Loske ...unter Mitarb. von T. Böhmer ...] 5. Aufl., Birkhäuser Verl., Basel, 1998.

28. R. L. Carson, *Silent Spring*, REPRINT, Houghton Mifflin, 1994.
29. http://www.cefic.be/Templates/shwStory.asp?NID=471&HID=8.
30. P. T. Anastas and J. C. Warner, *Green Chemistry: Theory and Practice*, Oxford University Press, New York, 1998.
31. P. T. Anastas and J. B. Zimmerman, *Env. Sci. Tech.*, 2003 **37**(5), 95A–101A.
32. R. Gubler, *Biodiesel*, SRI Consulting, ed., CEH Marketing Research Report, Nov. 2006.
33. S. Hansen, *Rabobank's View on the Global Biofuel Market*, 6th European Motor BioFuels Forum, Rotterdam, 2008.
34. S. Marcinowski, *Renewable Raw Materials – a Novel Approach in Polymers*, Bio & Polymers, Biannual Meeting of the GdCh-Division of "Macromolecular Chemistry", Aachen, 2008.
35. IEA, *Oil Market Report*, http://omrpublic.iea.org/ (retrieved 30.09.2008).
36. R. Höfer and J. Bigorra, *Green Chemistry Letters and Reviews*, June 2008, **1**(2), 79.
37. R. Höfer in F. Brickwedde, R. Erb, M. Hempel and M. Schwake, Hrsg., *Nachhaltigkeit in der Chemie-13. Internat. Sommerakademie St. Marienthal, Initiativen zum Umweltschutz* Bd. 70, Erich Schmidt Verlag, Berlin, 2008, p. 150–166.

CHAPTER 2
Sustainability in Finance – Banking on the Planet

PHILIPPE SPICHER,[a] JULIANE CRAMER VON CLAUSBRUCH[b] AND PABLO VON WALDENFELS[b]

[a] Centre Info SA, Sustainable Investment Consulting, Rue de Romont 2, CH-1700 Fribourg, Switzerland; [b] PricewaterhouseCoopers AG Wirtschaftsprüfungsgesellschaft, Germany

2.1 Introduction

Markets are a product of society and must reflect society's concerns; society simply could not exist without financial services (Figure 2.1: economic circuit, *Wirtschaftskreislauf* in German). In the production industry material flow is paralleled by "cash flow"; from buying a house to insuring a car or saving for a pension, financial services also touch every aspect of private households; central banks are responsible for achieving price stability and maintaining the overall financial stability of a country or a community of countries.

As an integral part of the economy, financial systems *ex ante* produce and screen information about potential investments which enable the market participants to make investment decisions; they mobilize investment capital and *ex post* monitor the investment. Monitoring includes supervising investments to ensure they are on-course and on-schedule in meeting the objectives and performance targets in order to effect value enhancement and preclude value destruction. With regard to Corporate Responsibility (CR) and the contribution of enterprises to sustainable development, screening operations of financial systems fix capital expenditure and the marginal productivity of invested

RSC Green Chemistry No. 4
Sustainable Solutions for Modern Economies
Edited by Rainer Höfer
© The Royal Society of Chemistry 2009
Published by the Royal Society of Chemistry, www.rsc.org

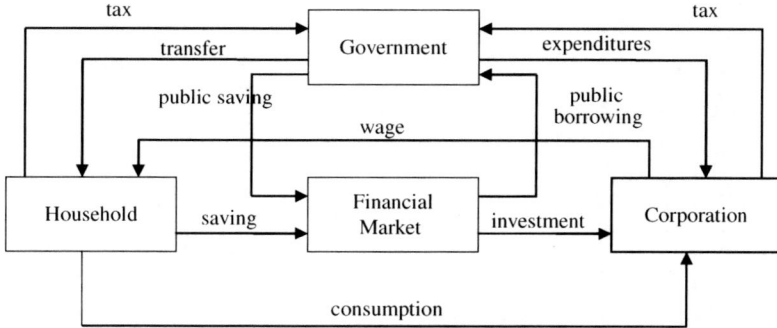

Figure 2.1 Economic circuit (adapted from http://de.wikibooks.org/wiki/Betriebs-wirtschaft/_Grundlagen/_Wirtschaftsteilnehmer).

capital (*via* allocation of capital to selected projects or assets) while monitoring and risk management control the sustainability of investments.[1]

2.2 Sustainability and Asset Value

Whereas the book value of enterprises is calculated retrospectively based on financial data, the market value contains a strategic prospect and depends on its sustainability rating (in the sense of "fitness for the future" (*Zukunftsfähigkeit* in German). The analysis comprises elements like:

- Climate change;
- Energy efficiency;
- Social aspects in the supply chain;
- Corruption handling and compliance with laws and regulations;
- Health and safety of products and services.

It values soft factors like reputation, customer satisfaction, intellectual capital, R&D-expenses, innovation pipeline, talent attraction, staff turnover, risk management and deployment of renewable energies. Since March 2008 the *Deutsche Vereinigung für Finanzanalyse & Asset Management* (DVFA) has for the first time published a terminology for ESG (Environmental, Social, Governance) aspects, and defined performance requirements for "non-financial" CR-data, which are key performance indicators (KPIs) for analysis and rating of companies (Figure 2.2). In May 2008 the DVFA framework received the endorsement of the European Federation of Financial Analysts Societies (EFFAS) and gained the status of an official EFFAS standard as a guideline for corporates on how to report on ESG and as a benchmark for investment professionals on how to integrate ESG into financial analysis.[2]

	E Environmental	S Social	G Governance	V Longterm Viability
General: KPIs which apply to all industry-groups	E1 Energy Efficiency E2 Deployment of Renewable Energy Sources	S1 Staff Turnover S2 Training & Qualification S3 Maturity of Workforce S4 Absenteeism S5 Restructuring-related Relocation of Jobs	G1 Contributions to Political Parties G2 Anti-competitive Behaviour, Monopoly G3 Corruption	V1 Customer Satisfaction V2 Revenues from New Products
Sector-Specific: KPIs which apply to select sectors	E4 NO,SO Emissions E3 CO2 Emissions E5 Waste E6 Environmental Compatibility E7 End-of-Lifecycle Impact	S6 Diversity S7 % of Credit Loans, Undergone ESG Screening S8 % of Funds Managed in Accordance to ESG Criteria S9 Financial Instruments held in Accordance to ESG Criteria S10 Investments in Accordance with ESG S11 Supplier Agreements in Accordance with ESG S12 Health & Safety of Products	G4 Litigation Payments G5 Dimension of Pending Legal Proceedings	V3 R&D Expenses V4 Number of Patents V5 Investments in Research on New Risk V6 Customer Retention

Figure 2.2   ESG grid and KPIs for the analysis of "non-financials".[2]

2.3 Socially Responsible Investment, SRI

Nevertheless, should banks use financial instruments to allow sustainable development in their own dealings, *i.e.* base their credit and investment policy on sustainability ratios instead of exclusively financial ratios?[3] The question has turned out to be rhetorical. In the past, in many businesses, protection

---

Main Socially Responsible Indices

Domini 400 Social Index (DS400)

In 1990, Domini Social Investments and Kinder, Lyndenberg, Domini & Co. (KLD) created the Domini 400 Social Index (DS400) – the first socially responsible investment benchmark – to measure how social and environmental screens affect investment performance.

Over the years, KLD has created a family of indexes to provide investors with a variety of ways to integrate environmental, social and governance factors into their investment decisions; among them KLD Broad Market Social Index series, KLD Global Climate 100 Index, KLD Global Sustainability Index.

For more information, visit www.kld.com.

Dow Jones Sustainability Indexes

Launched in 1999, the Dow Jones Sustainability Indexes are the first global indexes tracking the financial performance of the leading sustainability-driven companies worldwide. Based on the cooperation of Dow Jones Indexes, STOXX Limited and SAM, they provide asset managers with benchmarks to manage sustainability portfolios.

For more information, visit www.sustainability-index.com.

FTSE4Good Index Series

Launched in 2001, the FTSE4Good Index Series (acronym for *F*inancial *T*imes & *S*tock *E*xchange *for Good*) is a series of benchmark and tradable indices for responsible investors. The index series is derived from the FTSE Global Equity Index Series.

The FTSE4Good criteria are applied to the FTSE Developed Index Series, which covers 23 markets and over 2000 potential constituents. In the UK, the universe of eligible constituents is drawn from the FTSE All-Share Index. The series consists of five benchmark indices covering the Global and European regions, the US, Japan and the UK.

For more information, visit www.ftse.com/Indices/FTSE4Good_Index_Series/index.jsp.

of the environment has been regarded as a burden, leading to an increase in manufacturing and other business costs. However, since Sarokin and Schulkin's paper "Environmental Concerns and the Business of Banking" was published in the *Journal of Commercial Bank Lending*[4] there is mounting recognition of the importance of environmental and social issues to industry clients across sectors and markets, and to their financial sector partners, from banks to asset managers to private equity investors. In the public sector as a result of the Rio Conference UNEP FI a global partnership between the United Nations Environment Program (UNEP) and the financial sector was initiated in 1992. The partnership comprises over 160 institutions, including banks, insurers and fund managers, working with UNEP to understand the impacts of environmental and social considerations on financial performance. Sustainability definitely entered the financial scene, when in 1999 the Dow Jones Sustainability Indexes (DJSI) were launched.

This chapter aims to explore the role the financial sector can play in pursuing sustainable development. This problematic area is generally referred to as "sustainability in finance" and has two different meanings:

1. financing sustainable development (clean technologies, renewable energies, micro-credit, *etc.*)
2. financial activities generating sustainable development benefits.

There are several ways or channels through which financial activities (or transactions) have the potential to impact – positively and negatively – upon sustainable development (project finance, credit activities, retail banking, *etc.*). We will focus in this chapter on asset management (in traditional asset classes, *e.g.* stock quoted companies and corporate bonds).

Under the generic name of socially responsible investment, there are actually a multitude of different philosophies: ethics, sustainable development, materiality of extra-financial factors, *etc.* These various approaches differ as regards both the criteria and analytical tools used to assess corporate behaviour and performance, and the techniques used to build portfolios.

Even though the term "Socially Responsible Investment" covers many different approaches, it is useful to adopt a general definition:

Socially Responsible Investment (SRI) can be defined as the process by which factors other than standard financial ones are affecting investment decision-making and ownership practices.

Those factors are generally referred to as "extra-financial criteria" and encompass ethical considerations, **environmental** sustainability, **social** responsibility and corporate **governance** (ESG) issues. The abbreviation ESG in newer terminology is often used synonymously with SRI.

With the first fund created in 1928 and documented practices of making investment decisions based on moral grounds before that, the practice of taking into account extra-financial criteria when making investment decisions is not new. This was initially the initiative of mostly faith-based organizations. However, it is generally admitted that the modern area of socially responsible investment started in 1971 when the Episcopal Church presented at the General Motors annual meeting a shareholder resolution calling for the company to withdraw from segregated South Africa.

Since then assets managed under the SRI label have increased dramatically, particularly over the past decade. According to Eurosif (European Social Investment Forum),[5] "the growth of the broad SRI market between 2003 and the amount identified in 2006 is estimated at 106%; adjusted to the progression of the Morgan Stanley Capital International (MSCI) Europe Index, the real growth of European Broad SRI is 36%".

A key differentiating factor of the different approaches hidden behind the general and broad definition given above is how extra-financial criteria are taken into account alongside financial ones for the construction of the portfolio. As a matter of fact, the different approaches used in integrating ethical, environmental, social and governance factors into investment processes fall into the following three broad categories:

- Exclusion;
- Best-in-class;
- Engagement.

There is a continuum between these approaches and two of them, or even all three, are quite commonly applied together. It is, however, true that many investors focus only on one. It is worth noting that exclusion and best-in-class techniques affect the selection of the stocks which are held in the portfolio (*ex ante* screening), while the engagement technique does not (*ex post* monitoring).

In the following pages we will review each of these approaches by focusing on the underlying research needed to implement them, the kind of results this research delivers and to what extent and how investors are actually using these results.

2.3.1 Exclusion

Exclusion involves the creation of an investable universe. In other words, it means defining some companies as eligible for an investment and some others as not. The companies that are not eligible – the excluded companies – are those that do not meet certain criteria, called the exclusion criteria or screen. These criteria can be of three types:

1. activity-based
2. policy-based
3. practice-based

Activity-based criteria focus on the products and services a company produces and brings to the market. Some common activity-based criteria include weapons and military contracting, tobacco, alcohol, nuclear power, gambling, pornography and genetically modified organisms (GMO).

Policy-based criteria seek to identify companies that have not adopted formal policies on one or several environmental, social or governance (ESG) related issues. In that context, the adoption of a formal policy by a company is considered as a necessary condition for investment. Although any kind of issue can fit into this criteria, in practice the most commonly used issues specifically relate to human rights and to the environment.

Practice-based criteria are intended to capture the actual corporate behaviour and not just the corporate commitments. As for policy-based criteria, the practice-based ones are generally aimed at capturing human rights violations by companies. This specific type of exclusion is also sometimes called norm-based screening.

According to Eurosif, 266,000 Mio € are invested in Europe based on simple exclusion screens,[6] with the arms trade (138,000 Mio €) and human rights (61,000 Mio €) the most commonly used screens. In the United States, the Social Investment Forum[7] reports that in 2005 a total of 1,679,000 Mio US$ was invested using social screens, among these approximately 17% included a human rights screen. It has to be noted that while the data reported by Eurosif makes a clear distinction between exclusion screen and best-in-class strategy (see also the next section), reflecting current practice in Europe, it is not the case for the data reported by the US Social Investment Forum. This makes a direct comparison difficult, as part of the 1,679,000 Mio US$ reported as using social screens may also include some best-in-class strategies.

2.3.2   Best-in-class

A best-in-class approach involves the selection or the overweighting of stocks of companies that perform particularly well compared to their peers. A best-in-class approach thus involves the realization of an assessment of corporate ESG performance and behaviour. Such a sustainability rating can be based on a sector assessment (chemical industry, energy industry, *etc.*) and on the sustainability assessment of individual companies and their peers in different industry sectors.[8]

This kind of assessment is conducted by dedicated teams within asset management firms or banks and, more generally, by specialized research organizations and rating agencies. More recently, brokerage houses, encouraged by the Enhanced Analytics Initiative (EAI, see below) growing demand, have started to deliver interesting pieces of research in this area.

According to Eurosif, a total of 29,000 Mio € is invested in Europe using best-in-class strategies. This figure may be slightly underestimated as a part of the 641,000 Mio € assets managed using an "integration" approach[9] is most probably using the best-in-class technique as defined above.

2.3.3 Engagement

Engagement is used by investors to apply direct pressure to corporations to improve their social and environmental performance. It encompasses various means to gain dialogue with management and influence corporate behaviour, including letter writing, meetings with top management and board of directors, proxy voting and ultimately filling in shareholder resolutions. In 2005, investment through engagement strategies accounted for 703,000 Mio $ in the United States[7] and 730,000 Mio € in Europe.[5]

Shareholder advocacy is a particular form of engagement that is used by stakeholders other than investors, mainly Non-Governmental Organizations (NGOs, *e.g.* Amnesty International and Reporters Without Borders) and unions. Shareholder advocacy is more popular in countries where shareholders' rights require little capital to bring resolutions to an annual general meeting such as in the United States.

Table 2.1 below gives an overview of the importance of engagement strategies in the US market in 2005 as reported by the US Social Investment Forum. Among a total of 348 resolutions, 98 were withdrawn, which is an indication of the responsiveness of companies to pre-vote dialogues and negotiations. On average, the shareholders' resolutions that have been voted on received 10% of the votes. This low figure does not necessarily mean, however, that these resolutions were not successful since quite often companies take these votes as a signal and undertake some of the requested changes.

2.4 Responsible Investment: the Mainstreaming of SRI

The evolution of assets under management in socially responsible investment presented in the previous chapter is accompanied by and probably also – at least to some extent – supported by an institutionalization and a mainstreaming of such practices.

This institutionalization and mainstreaming phenomenon is characterized by changes in the legal environment pushing pension funds into SRI (see box on p. 20) as well as by several financial industry initiatives encouraging major players to embrace the SRI concept.

Table 2.1 Shareholder advocacy: Resolutions to Annual Meetings, USA 2005.[7]

Number of resolutions	Issue
42	Political contributions
35	Climate change
25	Global labour standards
11	Human rights

Changing legal environment pushes pension funds into SRI

United Kingdom

Since July 2000, the "Trustee Act" requires all pension fund trustees to disclose their policies on socially responsible investment.

Germany

Since January 2002, the new pension law requires private and corporate pension schemes to disclose their policies on socially responsible investment.

France

Since February 2001, the French law on Employee Saving Plans requires mutual funds, which collect money from the Employee Saving Plans, the Inter-companies Saving Plans and the Voluntary Partnership Employee Saving Plans, to report on their policies on socially responsible investing.

Sweden

Since January 2001, the five biggest public pension funds have to consider the environmental and social aspects in their environmental policy.

The UNEP-FI/Freshfields Bruckhaus Deringer study[10]

UNEP and lawyers at Freshfields Bruckhaus Deringer in 2005 released a study entitled "*A Legal Framework for the Integration of Environmental, Social and Governance Issues into Institutional Investment*", surveying the law and legal trends in the jurisdictions of the world's largest capital markets, including Australia, Canada, France, Germany, Italy, Japan, Spain, the United Kingdom and the United States. The study states that ESG considerations must be integrated into an investment decision where a consensus (express or in certain circumstances implied) amongst the beneficiaries mandates a particular investment strategy and may be integrated into an investment decision where a decision-maker is required to decide between a number of value-neutral alternatives. Particularly institutional investors have a far greater opportunity – and in some cases a legal obligation – to incorporate environmental, social and governance issues into their investment decision-making than is traditionally believed.

Two important financial industry initiatives were launched in 2004:

The *Who Cares Wins* initiative was launched by mainstream investment houses representing 6,000,000 Mio US$ in assets following an invitation by Kofi

Annan to "develop guidelines and recommendations on how to better integrate environmental, social and corporate governance issues in asset management, securities brokerage services and associated research functions".[11]

The *Enhanced Analytics Initiative* (EAI) was launched by fund managers representing 1,000,000 Mio US$ in assets to encourage ESG research by brokers. This initiative is designed to give research providers a commercial incentive to produce research that captures the value of ESG issues on corporate performance. Twice a year, the EAI evaluates the research submitted by research providers and the EAI members then allocate a minimum of 5% of brokerage commissions to the producers of the selected extra-financial research. The EAI has definitely been a key driver in the growing interest of mainstream financial analysts for ESG issues; in turn this gives investors access to more and more research into these areas.

Later, in April 2006, the Principles for Responsible Investment (see box below) were launched by the United Nations and signed by asset owners and investment managers representing more than 2,000,000 Mio US$.

One year later, more than 180 institutional investors managing assets worth 8,000,000 Mio US$ have signed up to these principles. In June 2008, signatories of the Principles for Responsible Investment represented more than 13,000,000 Mio US$ of assets.

The Principles for Responsible Investment

As institutional investors, we have a duty to act in the best long-term interests of our beneficiaries. In this fiduciary role, we believe that environmental, social and corporate governance (ESG) issues can affect the performance of investment portfolios (to varying degrees across companies, sectors, regions and asset classes and through time). We also recognize that applying these Principles may better align investors with broader objectives of society. Therefore, where consistent with our fiduciary responsibilities, we commit to the following:

1. We will incorporate ESG issues into investment analysis and decision-making processes.
2. We will be active owners and incorporate ESG issues into our ownership policies and practices.
3. We will seek appropriate disclosure on ESG issues by the entities in which we invest.
4. We will promote acceptance and implementation of the Principles within the investment industry.
5. We will work together to enhance our effectiveness in implementing the Principles.
6. We will each report on our activities and progress towards implementing the Principles.

As shown in the second annual report on PRI implementation,[12] signatories – asset owners and investment managers – have made considerable progress in implementing the six principles.

From a niche market socially responsible investment as an integrative approach (integration of ESG issues with industrial analysis and valuation on a sector-by-sector basis to identify [mid- and long-term] investment opportunities) has now definitely entered the mainstream arena,[13] with a particular focus on the link between ESG issues and financial performance.

2.5 Conclusion

As we have seen, the same words can have different meanings, depending on who is using them and in what context. More particularly *Sustainability in Finance* has two meanings which are overlapping but not identical: on one hand we have a sustainable management of companies rated by financial and sector specific strategic non-financial ESG aspects; on the other hand we have sustainable investment strategies directing capital according to SRI/ESG principles. Whatever the meaning, and behind the differences in approaches and in the objectives followed by the investors (investing according to own values, seeking higher performance by capturing the materiality of ESG factors), in the end, from a sustainable development perspective, the key question is:

Are these investment practices having a tangible impact on sustainable development?

If SRI/ESG is able to push companies to rethink their activities and business models, it certainly already makes a contribution to sustainable development. The question that remains is: does SRI turn out to be the key driver for sustainable development? To be able to answer this question, one would need to have metrics actually measuring this impact. Research in this area is still in its infancy and a lot remains to be done. In this regard, however, it is worth mentioning two recent initiatives going in this direction.

In a paper[14] published in May 2008, Christoph Butz and Olivier Pictet argue that "Financial performance is not a sufficient condition for the success of an SRI investment strategy. Social investors' objectives are not one-dimensional but multi-dimensional. Their utility function also explicitly includes a social and an environmental performance dimension. Hence the need for a credible and transparent extra-financial reporting." They present in their paper how such extra-financial reporting might look and "show that the companies in the sustainable portfolio emit less CO_2 and create more jobs than their peers and thus provide the sustainable investor with a measurable social and environmental added value".

In a similar attempt to inform investors, the Caisse d'Epargne – a French banking group – introduced in June 2008 the sustainability label for savings products. This labelling allows consumers to compare passbook savings accounts, mutual funds and life insurance offerings on the basis of three criteria: financial risk (Security Criterion), the use of social and environmental criteria in managing the product (Responsibility Criterion) and the impact on the climate of the activities financed with the products (Climate Criterion).[15]

These two initiatives will most certainly be followed by others, which will hopefully lead to a comprehensive system of reporting the extra-financial performance – or sustainable added value – of investment.

In the same way as the chemical industry, following growing ecological concerns over waste as well as catastrophic events like Seveso and Bhopal, had to reconsider the way it was doing business, the current financial crisis will also lead to a redefinition and reconsideration of the role and functioning of the financial system. In order to make a sustainable contribution to the solution of the current crisis the financial markets must return to their supportive function (Figure 2.1). In the last years they have taken the goal setting position within the economy which in turn has driven short termism within the real economy. The concepts and current practices related to sustainability in finance presented in this chapter are now more relevant than ever before.

References

1. R. Levine, *Finance and Growth, Theory and Evidence*, NBER Working Paper 10766, NBER, Cambridge, 2004, http://www.econ.brown.edu/fac/ Ross_Levine/Publication/Forthcoming/Forth_- Book_Durlauf_FinNGrowth.pdf (retrieved 17.03.2009); B. Scholtens, *J. Bus. Ethics*, 2006, **68**(1), 19–33.
2. H. Garz and F. Schnella, *KPIs for ESG*, DVFA Financial Papers, No. 8/ 08_e, Version 1.1 – Draft, DVFA, Dreieich, 2008.
3. J. Bouma, M. H. A. Jeucken and L. Klinkers, *Sustainable Banking – The Greening of Finance*, Greenleaf Publ., Sheffield, 2001.
4. D. Sarokin and J. Schulkin, *Journal of Commercial Bank Lending*, Feb. 1991, **73**, 6–19.
5. *European SRI Study*, Eurosif, 2006, http://www.eurosif.org/publications/ sri_studies_2006_2003 (retrieved 14.10.2008).
6. In its "European SRI Study 2006", Eurosif defines simple screens as "an approach that excludes a single given sector from a fund such as arms manufacture, publication of pornography, tobacco, animal testing, *etc.* Simple screens also include simple human rights screens (such as excluding companies for activities in Sudan or Myanmar) and Norms-based screening".
7. Social Investment Forum, *2005 Report on Socially Responsible Investing Trends in the United States*, Washington, 2006, http://www.socialinvest.org/ pdf/research/Trends/2005%20Trends%20Report.pdf (retrieved 15.10.2008).
8. E. Plinke, in *Nachhaltigkeit in der Chemie – 13. Internat. Sommerakademie St. Marienthal, Initiativen zum Umweltschutz.* ed. F. Brickwedde, R. Erb, M. Hempel, M. Schwake, Bd. 70, Erich Schmidt Verlag, Berlin, 2008, p. 329.
9. Eurosif defines integration as "the explicit inclusion by asset managers of [corporate governance]/[social, environmental, ethical] risk into traditional financial analysis".

10. UNEP FI/Freshfields Bruckhaus Deringer, *A Legal Framework for the Integration of Environmental, Social and Governance Issues into Institutional Investment*, Genève, October 2005.
11. The Global Compact, *Who Cares Wins – Connecting Financial Markets to a Changing World*, Financial Sector Initiative, December 2004, http://www.unglobalcompact.org/docs/news_events/8.1/WhoCaresWins.pdf (retrieved 15.10.2008).
12. UNEP Finance Initiative, PRI Report on Progress 2008, New York, 2008, http://www.unpri.org/files/2008PRI_Report_on_Progress.pdf   (retrieved 15.10.2008).
13. W. Baue, *Social Funds*, Oct. 05, 2005, http://www.socialfunds.com/news/article.cgi/1826.html (retrieved 14.03.2009); S. Forrest, *Goldman Sachs ESG: Integrating ESG into Investment Research*, June 2006, www.ahcgroup.com/powerpoint/GoldmanSachsPresentation.ppt (retrieved 14.03.2009).
14. C. Butz and O. Pictet, *The SRI Performance Paradox – How to Gauge and Measure the Extra-financial Performance of Socially Responsible Investment*, Pictet & Cie, Geneva, May 2008, www.pictet.com.
15. Groupe Caisse d'Epargne & Utopies, *Etiquetage Développement Durable des produits bancaires Première approche méthodologique*, M.-C. Korniloff, S. Dupré (Coordination), Juin 2008, http://programme.beneficesfutur.fr/themes/perso/pdf/Methodologie_Generale_Juin_2008.pdf (retrieved 15.10.2008).

CHAPTER 3
Metrics for Sustainability

PETER SALING

BASF SE, D-67056 Ludwigshafen, Germany

3.1 Introduction

Meeting society's needs without damaging the environment requires new ways of thinking. In spite of a strong record of innovation – in products that meet customers' needs, in manufacturing processes that protect the environment and human health and in solutions that directly address environmental problems[1] – the greater public often has a critical relationship with industrial chemistry. Society perceives a need that industry acts more responsibly and governments all around the world introduce increasingly strict legislation to ensure it does so.[2] Quantifying the sustainability benefits of industrial chemistry is therefore an important issue for the further development of this field of opportunities in the future. A realistic and validated estimate of innovative potentials of industrial chemistry by using quantitative methods is essential for the development of new products and processes. Therefore, a link between industrial chemistry and metrics for sustainability will be one of the key factors of sustainable development within the chemicals industry. New approaches and developments of "Green Chemistry", "Green Engineering", *etc.* in combination with the development of industrial processes should be evaluated with a powerful and extensive methodology.[3]

The chemical industry has played a key role in the development of new tools and techniques as well as in impact assessment research, consistently allocating a large portion of its resources to research and development.[4] To promote communication and understanding between the academic community and the

RSC Green Chemistry No. 4
Sustainable Solutions for Modern Economies
Edited by Rainer Höfer
© The Royal Society of Chemistry 2009
Published by the Royal Society of Chemistry, www.rsc.org

chemical industry is also important. The best way to bridge the gap between the ideas that academia generate and the ideas that the product development industry seeks should also be evaluated.[5]

3.2 The Eco-Efficiency Analysis as an Approach for the Checking of Sustainability in Industrial Products and Applications

A particularly important issue is the investigation and valuation of the environmental impacts of a given product or service caused or necessitated by its existence. Different approaches have been made to analyze and standardize such impacts throughout a product's life-cycle:

- **Life-Cycle Assessment** (LCA, also known as life-cycle analysis, ecobalance and cradle-to-grave analysis) is the method to evaluate the mass balance of inputs and outputs of systems and to organize and convert those inputs and outputs into environmental impacts relative to different ecological areas. The procedures of LCA are part of the ISO 14000 environmental management standards: in ISO 14040:2006 and 14044:2006. One of the first steps before starting an LCA is to define the "functional unit", which is related to the function that a product or service will deliver. LCA also allows comparisons between equivalent stages of life-cycles (*i.e.* the consumer stage of product A and the consumer stage of product B), provided that the LCIs as basic data rely on the same databases and the same assumptions. LCA normally follows the "cradle-to-grave" approach and means the full Life-Cycle Assessment from manufacture ("cradle") to use phase and disposal phase ("grave").
- **Life-Cycle Inventory** (LCI) pursues the compilation and quantification of inputs and outputs for a given product system throughout its life-cycle on a volume or mass basis (*e.g.* kg of CO_2, kg of cadmium, cubic metre of solid waste).
- **Product-Specific Requirements** (PSR) is the phase of life-cycle assessment involving the compilation and quantification of inputs and outputs, for a given product system throughout its life-cycle. Additionally, specific technical data information will be a part of PSR or EPD (Environmental Product Declaration).
- **Life-Cycle Inventory analysis result** (LCI result) provides information about all inputs and outputs in the form of elementary flow to and from the environment from all the unit processes involved in the study. An outcome of a life-cycle inventory analysis result includes the flows crossing the system boundary and provides the starting point for life-cycle impact assessment.
- **Life-Cycle Impact Assessment** (LCIA) is the phase of life-cycle assessment aimed at understanding and evaluating the magnitude and significance of the potential environmental impacts of a product system.

- **Life-Cycle Interpretation** is the phase of life-cycle assessment in which the findings of either the inventory analysis or the impact assessment, or both, are combined in consistence with the defined goal and scope in order to reach conclusions and recommendations.

Following these general methodology approaches, a new approach was developed in the late 1990s by BASF in cooperation with Roland Berger Consulting, Munich. This methodology, the **Eco-Efficiency Analysis**, among the other tools mentioned above, is seen as a life-cycle management tool and can be involved in assessments of the entire product life-cycle, from concept development to design and implementation, further to marketing and, finally, end-of-life issues. The analysis method may incorporate both economic and environmental aspects and lead to a comprehensive evaluation of products and processes over their entire life-cycle.

In the method of Eco-Efficiency Analysis, results are presented as aggregated information on costs and environmental impact and show the strengths and weaknesses of a particular product or process. The ecological calculations of the single results in each category follow the ISO-rules 14040 and 14044.

The quantitative weighting steps to obtain the ecological fingerprint and the portfolio are additional features.

3.2.1   Conducting an Eco-Efficiency Analysis

Every eco-efficiency analysis passes through several key stages. This ensures consistent quality and the comparability of different studies. Environmental impacts are determined by LCA and economic data is calculated using the usual business or, in some instances, national economical models.

The basic preconditions in eco-efficiency analysis are:

- Products or processes studied have to meet the same defined customer benefit;
- The entire life-cycle is considered;
- Both an environmental and an economic assessment are carried out.

The eco-efficiency analysis is worked out by following specific and defined calculations:

- Calculation of total cost from the customer viewpoint;
- Preparation of a specific life-cycle analysis for all investigated products or processes according to the rules of ISO 14040 ff;
- Determination of impacts on the health, safety and risk to people, assessing use of area over the whole life-cycle;
- Calculation of relevance and calculation factors for specific weighting;
- Weighting of life-cycle analysis factors with societal factors;
- Determination of relative importance of ecology *versus* economy;
- Creation of an eco-efficiency portfolio;

- Analyses of weaknesses, scenarios, sensitivities and business options;
- Optionally: inclusion of social aspects.

The methodology has been approved by the German TÜV ("Technischer Überwachungsverein", English: "Technical Inspection Association") and was also used by the Öko-Institut ("Institute for Applied Ecology") in Freiburg, Germany, in different APME-studies (APME, now "Plastics Europe", "Association of Plastics Manufacturers in Europe"). TNO in the Netherlands is using the BASF standard method with a different weighting system (Figure 3.1).

For the calculation and comparison of the environmental position of each alternative, data from the different production methods is assimilated and analyzed to provide a value for energy consumption, raw-material consumption, emissions and use of area, risk potential and toxicity potential. All raw-materials required in the process and how these are derived are factored into the study, as are the steps required to bring the product to the end-user.

In the same manner economical data from the life-cycle chain of a product application or process evaluation may also be calculated and summarized. In the end, this analysis can lead to better decisions with regard to product design, material utilization and capital investment. The rationale behind this assessment tool has been described by Saling *et al.* and by Landsiedel and Saling.[6,7]

Practical examples can show how the metrics for sustainability can support decision-making processes answering different questions.

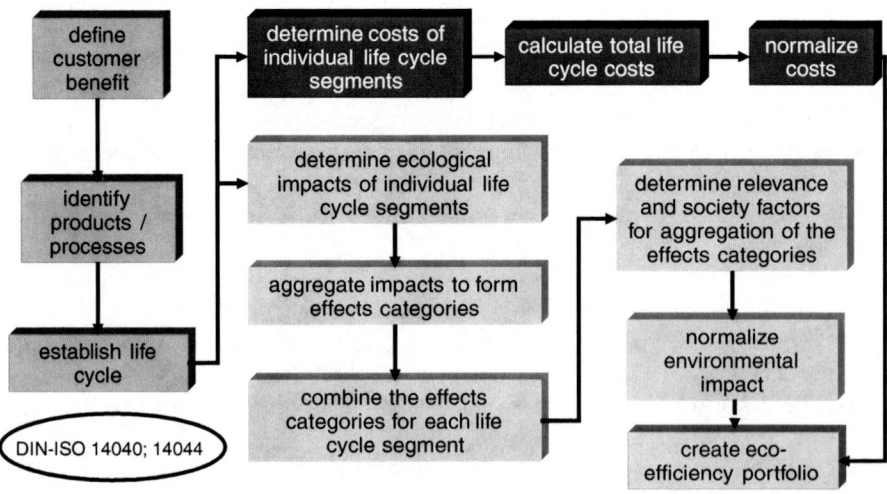

Figure 3.1 Process steps of eco-efficiency analysis.

3.3 Industrial Examples for Using Sustainability Metrics

3.3.1 Eco-Efficiency Study of Curing Alternatives for Wooden Substrates

In this example radiation curing is compared with several other coating technologies, such as solvent-based two-component polyurethane, acid curing, nitrocellulose and water-based coatings. In the application discussed here, a specific customer use was defined as: performance application of 1,000 wooden front doors.

The most important differences between the processes are the curing mechanisms of each alternative. With radiation curing technology, UV-light is used for curing. This allows the skipping of the energy-intensive thermal-oven drying process steps, which are required for all other alternatives.

The objective is to express the properties of the UV-curable coating technologies with the method of eco-efficiency. The eco-efficiency fingerprint (Figure 3.2) shows that the UV-roller coating has the best results in all environmental categories. The low film weight and the absence of a need for thermal drying induce high productivity and the lowest production costs, as shown in Figure 3.3.

This overall result is shown in the eco-efficiency portfolio in Figure 3.4. The alternative with the highest eco-efficiency is in this case the UV-curing alternative. It might be different in other applications, so a "case-by-case" decision is necessary. A lot of scenarios are also evaluated in this study. One of these shows the influence of the radiation energy of the UV-curing system (Figure 3.5). Even if the energy is doubled, the UV system is still the most eco-efficient alternative.[8]

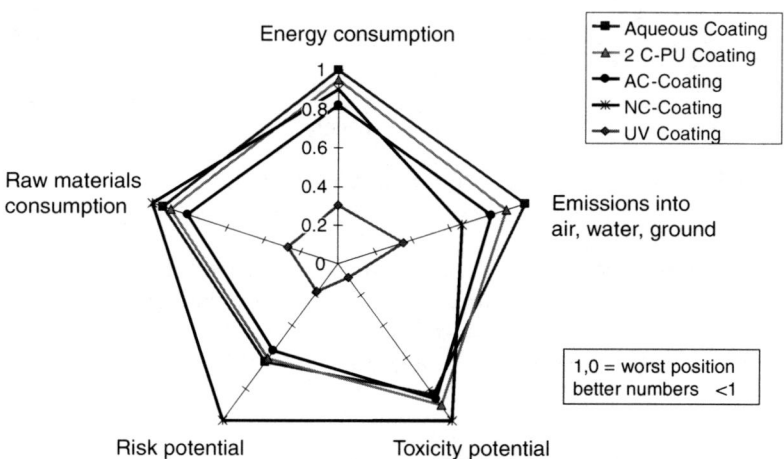

Figure 3.2 Ecological fingerprint for alternative curing systems.

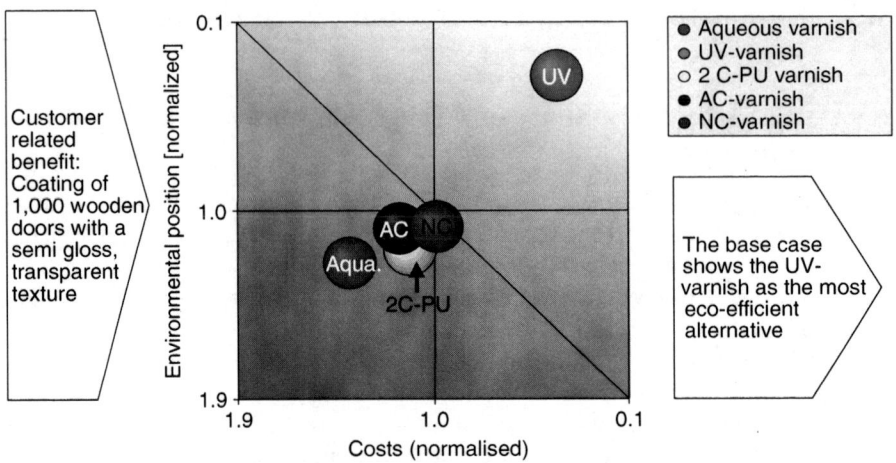

Figure 3.3 Life-cycle cost calculation of alternative curing systems.

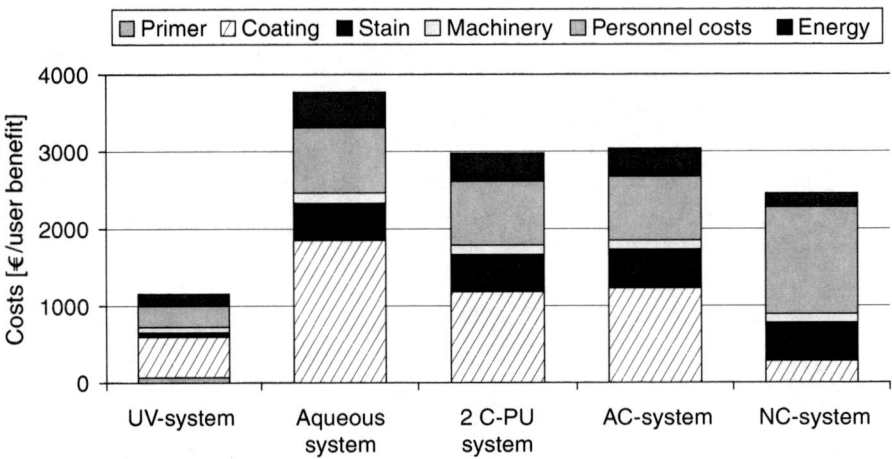

Figure 3.4 Eco-efficiency portfolio for alternative curing systems.

Manufacturers need to take the results of an eco-efficiency analysis into consideration for future planning. It could support the process of switching over to the more eco-efficient UV-technology. Eco-efficiency analysis provides guidance for internal planning and decisions.

3.3.2 Vitamin B₂ Case Study

Vitamin B_2 is produced by BASF's Agricultural Products & Nutrition segment for use as a vitamin for human and animal nutrition. As a component of animal

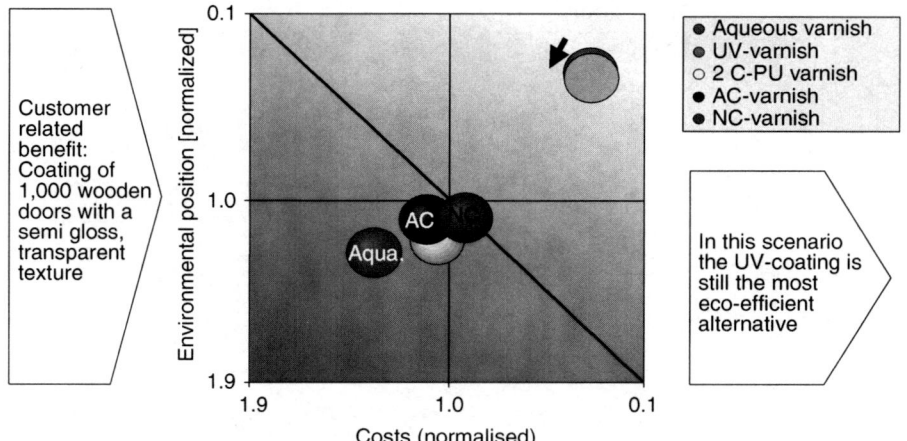

Figure 3.5 Scenario of the eco-efficiency analysis for alternative curing systems.
2 CPU = two-component polyurethane, AC = acid curing, NC = nitro-
cellulose.

feed, it is vital to ensure the animals' health and fitness; vitamin B_2 deficiency
leads to slower growth and poor feed conversion.[9]

Eco-efficiency demonstrated which vitamin B_2 production process is the most
eco-efficient. Three "bio-technological" processes and one "chemical" process
were evaluated for the production of 100 kg of vitamin B_2 for use in animal feed
pre-mix. All of the processes include renewable resources such as plant oil or
glucose as a raw material (Figure 3.6). The bio-technological processes use
fermentation, while the chemical process starts with a bio-technological precursor like glucose or soybean oil and afterwards uses traditional chemistry to
produce the vitamin B_2.

As Figure 3.7 shows, Biotech process 1 was the most eco-efficient. It had
the least overall environmental impact, and was one of the lowest cost alternatives. Biotech process 3 had noticeably higher environmental impact and
higher costs. In this case the chemical process alternative had the highest
cost and a greater environmental impact than Biotech process 1, resulting in the
lowest eco-efficiency. BASF produces vitamin B_2 *via* one-step fermentation
from vegetable oil by using the fungus *Ashbya gossypii*. BASF pioneered the
shift from chemical to biotechnological vitamin B_2 production on an industrial
scale and runs a production facility in Korea. It is an excellent example of
industrial-scale production using the most eco-efficient technology currently
available.

The eco-efficiency analysis in this case study was able to outline and to
describe new goals of research activities. It was able to highlight the most
important factors that influence the system.

The results can be used for decision-making processes in "white biotechnology", which can be assessed and compared to the actually established
chemical processes. The eco-efficiency study is able to support strategic
decision-making processes with a different life-cycle-based view on the different

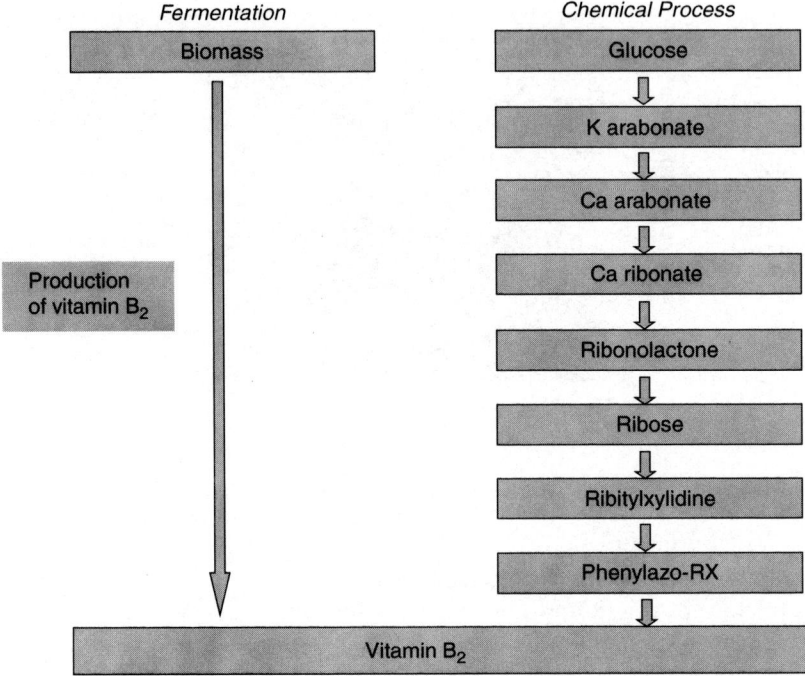

Figure 3.6 Chemical and biotechnological production pathways for vitamin B_2.

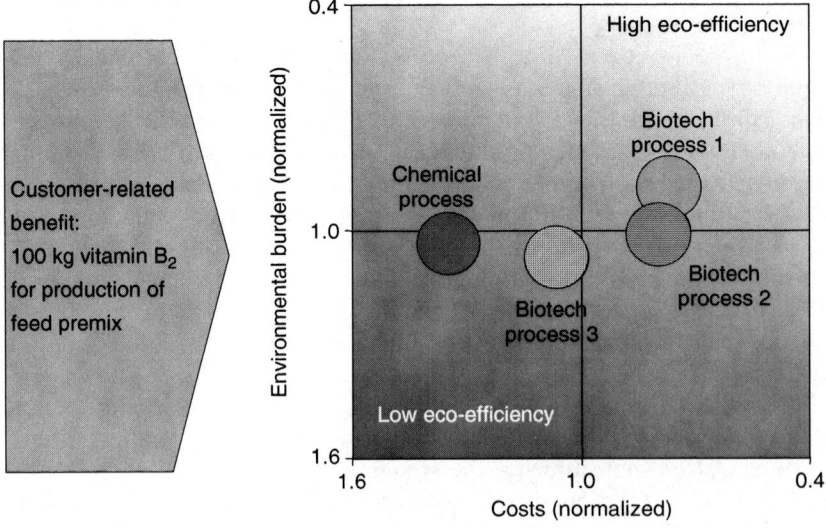

Figure 3.7 Portfolio of vitamin B_2 production for the feed segment.

technologies. Without any prejudices and preferences for a certain technology, the most sustainable process can be selected and realized.[9,10]

3.3.3 Eco-Efficiency Analysis Confirms: Ionic Liquids Provide Benefits

On account of their unique set of characteristics, ionic liquids are potentially suited for numerous applications and are thus attracting increasing attention. Today, they are viewed as promising alternatives in chemical reactions and separation processes, as well as in processing metals and polymers – especially biopolymers such as cellulose.[11]

Using the BASIL™ process (the acronym stands for **B**iphasic **A**cid **S**cavenging utilizing **I**onic **L**iquids), BASF was the first company to transfer ionic liquids from laboratory to commercial dimensions, and the BASIL™ process became the first large-scale industrial process worldwide that uses ionic liquids.

The use of BASF's BASIL™ process for scavenging hydrochloric acids in the chemical synthesis of phosphorus compounds like $\emptyset_2POR$ or $\emptyset P (OR)_2$ by reaction of $\emptyset_2PCl$ (or $\emptyset PCl_2$) and ROH[12] offers significant advantages over the conventional system (Figure 3.8).

Compared to amines, which have traditionally been used as HCl-scavengers in this type of reaction, the BASF process based on 1-methylimidazole circumvents the problematic, time-consuming and expensive filtration of a solid ammonium hydrochloride cake by yielding an ionic liquid, the 1-methylimidazolinium

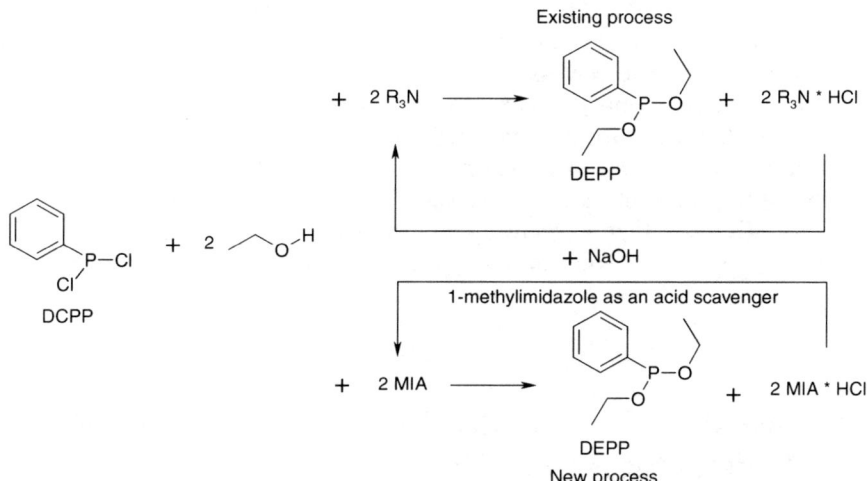

Figure 3.8 BASIL™: a process alternative for the synthesis of alkoxyphenylphosphines. DCPP = dichlorphenylphosphine; MIA = 1-methylimidazome; DEPP = diethoxyphenylphospine.

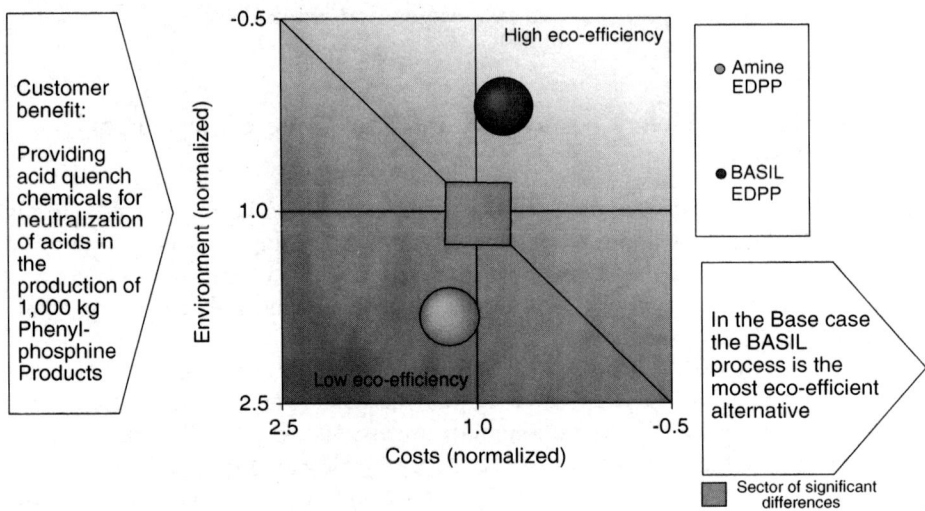

Figure 3.9 Portfolio of the use of BASIL™ in EDPP (ethoxydiphenyl phosphine)-production alternatives.

hydrochloride salt. The BASIL™ process is less cost-intensive and at the same time better for the environment (Figure 3.9). The new process for synthesizing phosphorus compounds, which are used as chemical building blocks to produce photo initiators in UV-curable coatings, reliably avoids a number of problems encountered to date. Stability and product yield improve, and the process is less laborious. The ionic liquids can be easily separated from the desired products, like oil from water, and can also be recycled.

The eco-efficiency analysis has demonstrated that BASIL™ is much more eco-efficient than the conventional method.[13]

This tool can be used for further improvement, even for chemical synthesis in the laboratory phase to realize more sustainable chemical processes.

The BASIL™ developers received the "Innovation Award" of the "European Chemical News" trade journal and the BASF Innovation Award in October 2004.

3.4 Beneficial Uses of Eco-Efficiency Analysis and Metrics for Sustainability

Since 1996 more than 350 analyses have been carried out in key fields at BASF (paints and dyes, plastics, life science, oil and gas, chemicals) but also externally with customers, retailers, associations and authorities by eco-efficiency experts. In doing so, eco-efficiency analysis has been employed in four major fields of application; strategic decisions, research and product development, communication with policy makers and marketing.

In strategic decisions it is possible for the application investigated to identify solutions with a promising future. Even in investment decisions eco-efficiency analysis provides valuable perspectives.

The second field of application relates to research and product development. Promising products are identified at an early stage, thus facilitating decision-making about the prime thrust of the development.

The third field of application is the drawing up of position papers for discussions with opinion and policy makers. Eco-efficiency analysis makes it possible to present the complex, holistic interconnections in industrial production and product use in a graphic and readily communicable form. By these means, it is possible to conduct quantitative discussions, with politicians for instance, about the effects of planned legislation.

Eco-efficiency analysis is even used in marketing, the fourth main application area. Since the entire life-cycle of a product is considered, the effects for customers are integrated into the analysis. As a result the total vision inherent in products can be communicated to customers.

Determination of the sustainability of products and processes should be done by employing an accepted scientific method such as the presented eco-efficiency analysis.

Results of an evaluation with metrics of sustainability can be expressed by a label that is designed for b-2-b or b-2-c purposes. BASF labelled about 15 products with such a label; examples can be found on the internet (www.ecoefficiency.basf.com).

Sustainability will be an important key factor for the development of new products, processes and services.

3.5 Outlook

In order to arrive at a comprehensive assessment of products or processes, it is in general necessary to include all three dimensions of sustainability. The objective was to develop a tool, comparable to the BASF eco-efficiency analysis, which is simple to use by LCA-experts and easy to understand for people without any prior experience in the field. Moreover, the results of even the most complex studies should be understandable at a glance.

This new method is named SEEBALANCE. SEEBALANCE targets the adoption of life-cycle social data to the results of eco-efficiency analysis for a complete sustainability assessment methodology. It is a helpful tool in different fields for the evaluation of product or process alternatives.[14]

References

1. P. T. Anastas and J. B. Zimmerman, *Environ. Sci. Technol.*, 2003, **37**(5), 94A.

2. J. Bigorra and R. Höfer, *Comunicaciones, Jorn. Com. Esp. Deterg.*, Barcelona, 2008, **38**, 179.
3. D. R. Shonnard, A. Kicherer and P. Saling, *Environ. Sci. Technol.*, 2003, **37**(23), 5340.
4. Cefic, the European Chemical Industry Council: One Vision, One Voice http://www.cefic.be.
5. H. Gavaghan, *Nature*, 2000, **406**(6797), 809.
6. P. Saling, A. Kicherer, B. Dittrich-Krämer, R. Wittlinger, W. Zombik, I. Schmidt, W. Schrott and S. Schmidt, *Int. J. LCA*, 2002, **7**(4), 203.
7. R. Landsiedel and P. Saling, *Int. J. LCA*, 2002, **7**(5), 261.
8. P. Saling, *Kunststoff-Trends*, GIT Verlag, 2007, **4**, 8.
9. BASF Group, ed., Press release, *A big step forward in the extension of BASF's vitamins business*, 10 November 2003, P-03-495.
10. P. Saling, *Appl. Microbiol. Biotechnol.*, 2005, **68**, 1.
11. R. P. Swatloski, R. D. Rogers and J. D. Holbrey, US 2003/0157351, 2002 (Univ. of Alabama); J. Holbrey, R. P. Swatloski, J. Chen, D. Daly and R. D. Rogers, US 2005/0288484, 2005 (Univ. of Alabama); D. A. Fort, R. C. Rensing, R. P. Swatloski, P. Moyna, G. Moyna and R. D. Rogers, *Green Chem.*, 2007, **9**, 63.
12. M. Maase, *C & E News*, 2003, **81**(13), 9.
13. P. Saling, M. Maase and O. Huttenloch, ACHEMA, Frankfurt, 2006; P. Saling, *Green Innovation in the Chemical Industry*, E.N.G.'s 2 senior executive summit, Amsterdam, 2007; http://corporate.basf.com/de/sustainability/oekoeffizienz/label.htm.
14. I. Schmidt, M. Meurer, P. Saling, A. Kicherer, W. Reuter and C. Gensch, in *Greener Management International*, ABI/INFORM Global, ed. S. Seuring, Greenleaf Publishing Ltd., Sheffield, Spring 2004, **45**, p. 79.

CHAPTER 4

Sustainable Logistics as a Part of Modern Economies

THIERRY JOUENNE

Conservatoire National des Arts et Métiers (CNAM), 292, rue Saint Martin, F-75003 Paris, France

4.1 Introduction

Given the close relationship with transportation, logistics clearly lies at the heart of sustainable development issues because of the environmental pollution generated in supplying factories, warehouses, sales outlets and consumers. But logistics also makes a major contribution to social equity in terms of employment, regional development and fair trade. At the same time, it contributes to sustainable growth by the enormous saving potentials still to be raised by quality improvement and efficient supply chain management. Transport consumes 70% of the oil used in Europe and represents the primary generator of greenhouse gas emissions (in 2004, inland transport accounted for 21% of total greenhouse gas emissions in the EU-15). Other factors on the downside are traffic congestion in urban areas, energy consumption, toxic exhaust emissions, noise and the imprint left by logistic infrastructure on the environment. The situation is all the more alarming if we consider that a 60% increase in trade is expected between 2005 and 2015 within the European Union. Conversely, under the Kyoto Protocol greenhouse gases are due to be reduced by 5.2% with respect to the 1990 levels between 2008 and 2012. While it is the subject of much criticism regarding the negative impact of transport on the environment, the logistic function gives rise to a great deal of expectation in

RSC Green Chemistry No. 4
Sustainable Solutions for Modern Economies
Edited by Rainer Höfer
© The Royal Society of Chemistry 2009
Published by the Royal Society of Chemistry, www.rsc.org

meeting the challenges of sustainable development, as in the broadest sense this management discipline plays a role in controlling and optimizing production and distribution systems. This feature makes it one of the key players capable of contributing across the board to sustainable development, operating not only on reducing greenhouse gas emissions but also on business competitiveness and the economic vitality of society. This less well-known aspect of logistics as a performance driver connects it *de facto* with the other two pillars of sustainable development – the economic and social – once it is able to influence economic growth, employment and purchasing power while reducing the impact of its activities on the environment. While acknowledging that there is still a lot of headway to be made, the purpose of this chapter is to show how sustainable logistics can reconcile the three economic, social and environmental aspects by developing eco-technological solutions and adopting "responsible practices" between suppliers, manufacturers, retailers and logistic service providers. We shall see that this global project calls for the same vision of the value chain to be shared and for collaboration in achieving common objectives for the benefit of business, society and the environment.

4.2   Definition and Role of Logistics

Several definitions have been put forward over the last 40 years but what is known as modern logistics is generally viewed as a business planning framework for the management of material, service, information and capital flows. It includes the increasingly complex information, communication and control systems required in today's business environment. It fits into the supply chain as defined by Mentzer *et al. "as a set of three or more entities (organizations or individuals) directly involved in the upstream and downstream flows of products, services, finances and/or information from a source to a customer"* (see Figure 4.1).[12]

The logistics approach is global and applies to all the players in the logistic chain from the design, manufacture and distribution phase until the products are withdrawn from the market. The logistic strategy is in line with corporate and marketing strategies.

The NF X 50-600 standard specifies that the purpose of the logistic function is to *"meet expressed or latent needs, internal or external, at the best economic conditions for a given level of service".*[1] This definition calls up the many dimensions of logistics, the boundaries of which extend beyond the particular company to reach a global optimum and meet the expectations of all the stakeholders in the supply chain. In order to fully appreciate developments over the last few years, we shall recall that logistics, until the 1970s, focused on the techniques tied to the physical processing of flows of goods (inventory management, transportation optimization, job shop scheduling). For the last 20 years or so, the assignments have been considerably extended and the associated information flows have linked up with physical flows. The logistic function extends henceforth from the design of the industrial and logistic

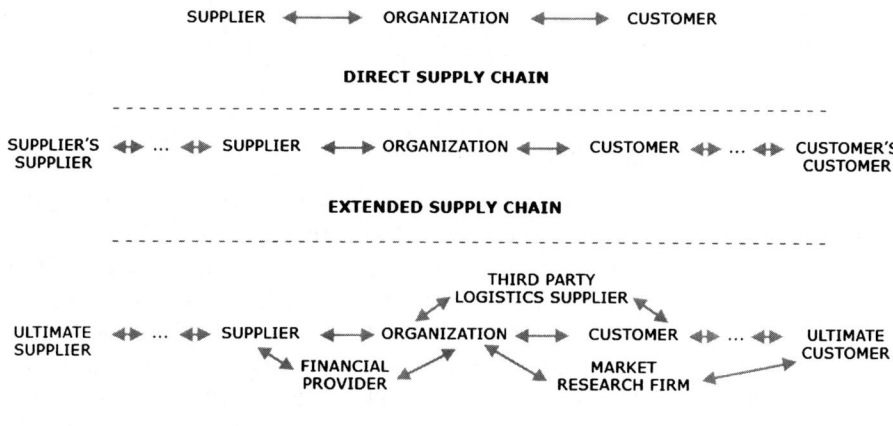

Figure 4.1 Types of channel relationships. Source: Mentzer *et al.*[12]

system – necessary for the manufacture, distribution and support of products throughout their life-cycle – to the management of the logistic system dedicated to controlling flows and inventories up to the point of consumption, including the management of returns.

In the same way, we may note that performance indicators have moved from a management philosophy focused on the achievement of local performance to a philosophy of management of the supply chain as a driver of value creation for all stakeholders.

Logistics is a cross-cutting, multi-disciplinary function covering, according to the AFNOR reference model, no less than 25 trades and 600 activities spread along the value chain. It plays a primary role from day-to-day business to major corporate projects. Its field of operation has widened, its missions have diversified – becoming more complex than before – and its competencies have developed, just as the range of its methods has become more elaborate. This transformation has occurred over the last 15 years and has not yet ended.

Under the impetus of Supply Chain Management (SCM), a concept that appeared in the early 1990s and is defined as the integrated management of the logistic process in a coordinated flow between companies linked in the same value chain, logistics is increasing its scope of operation. It benefits nowadays from all the headway brought by developments in the computing world and, in particular, since it deals with a chain of interfaces, the progress made in information and communication technology.

The logistic function runs the supply chain in terms of customer service objectives under the best economic conditions. This is an exercise calling on planning, coordination and management capabilities. It requires also educative and mediation abilities so as to get all internal and external players singing from the same hymn sheet despite conflicts of interest. This approach is global and targets the general interest of business in a spirit of intra- and inter-company collaboration.

4.3   Current Situation

Industry is faced today with endemic problems that remain difficult to solve despite the progress in technology. For instance, through lack of flow control and proper synchronization, shelf out-of-stocks amount on average to 10% of items, which translates into a loss of revenue for 1449 hypermarkets in France alone estimated at 200 million euros per quarter.[2] The capacity utilization of the trucks – assessed at 60–70% on average – is also problematic. It is reckoned that there are today some 20–25% empty miles, a rate which goes up to 35% for private carriers.[3] It should be recalled here that the Just-in-Time requirement when taken to the extreme is not irrelevant to deliveries being split up and the increase in empty returns. Other individualistic practices generate situations that are counterproductive in terms of sustainable development. For example, the search for the lowest production costs in the so-called *low-cost* countries – dictated by short-term profitability considerations – entails an explosion of logistic costs and CO_2 emissions and a loss of responsiveness, which is nonetheless essential to meet the fluctuations in demand.

Let us also mention the practices of some contractors in terms of customer service rates with penalties – which may reach 80% of the order in the event of delayed delivery – that deter any attempt at optimising the logistic chain overall. E-commerce is also costly and polluting, particularly in the last mile. These practices certainly have an impact on product price, company profitability margins and of course on the environment, exacerbated by the fact that not all companies have the same logistic maturity or the same capacity for optimization. For example, small and medium enterprises (SMEs) do not have the critical size, the organization or the facilities to optimize their flows. In such cases, frequent deliveries of small batches (fragmented orders) to distant customers weigh heavily on the transport bill and the carbon footprint.

At the same time, urban areas are saturated and do not permit effective distribution of goods in the absence of logistic hubs set up in sufficient numbers. Frequent imbalances between transport supply and demand are seen leading to shortages of road transportation capacity on certain routes at certain times of year, for which alternatives are difficult to find because of the lack of availability and responsiveness of rail freight. Lastly let us mention the lack of maritime and rail motorways and the lack of harmonization for legislation in goods transport Europe-wide to appreciate the structural difficulties that players on the supply chain have to contend with.

This alarming observation in terms of the survival of businesses and respect for the environment calls for a breakthrough transformation in doing business at all levels in the value chain in the direction of better balance, transparency, responsibility and collaboration between stakeholders. Vigorous step changes are needed to significantly reduce CO_2 emissions, traffic congestion, infrastructure footprint and supply-chain costs, while at the same time reducing out-of-stocks. The Grenelle Environment meeting in France in October 2007 highlighted the economic, social and environmental impacts of human activities in general, and transportation in particular. It already advocates an assessment of greenhouse gas emissions from transport services. The consequences for the organization of production, distribution networks, warehousing and circulation of goods are considerable. To assist businesses in meeting the challenges of sustainable development and translate their strategic objectives into concrete realities, logistics offers four essential drivers in the service of economic growth, social development and the protection of the environment.

4.4   The Four Logistic Drivers

Logistics extends from end to end of the value chain where its role consists of connecting the two poles of the economy in synchronizing effectively and on the best economic terms the supply chain with real consumer demand, however complex, uncertain and fluctuating it may be. In line with company strategy, well orchestrated and integrated, it makes it possible, according to Christopher Martin's dictum to do "better, quicker, cheaper and nearer", to obtain competitive advantage:[4]

- Better, by delivering perfect orders and achieving customer satisfaction;
- Quicker, by cutting down lead-times and cutting out unnecessary activities that hamper process flow;
- Cheaper, by reducing stock, operating costs and structural costs that bear down on the logistic chain;
- Nearer, through establishing customer loyalty by offering added-value services like customized products, responsiveness to demand, flow traceability, order tracking *via* the internet, *etc.*

This attractive vision is nonetheless difficult to achieve. It calls for a gradual, long-haul approach and depends on the ability of companies to modernise their working methods, integrate the logistic process, cut down costs and lead-times, measure performance, automate information exchange (Electronic Data Interchange, EDI and interfacing the Enterprise Resource Planning, ERP), coordinate activities and share information, resources and facilities between partners.

Incorporating sustainable development brings in two further dimensions – social and environmental. These add to the economic challenge for the logisticians who, let us note, have not yet – specifically in small and medium-sized businesses – taken it on board that they have to incorporate the other two. Fortunately, the three features to be reconciled are not mutually contradictory – as the economic feature lies in the same direction as the environmental and social ones – although this statement remains to be demonstrated.

We introduce here the concept of logistic drivers making the link between the company's strategic objectives and their operational translation in the supply chain. These drivers are in line with the aim of logistics as defined by James Hesket (1977) in these terms: "*Respond to demand at a given level of service at the lowest cost*" and also cross-check with Martin Christopher's dictum. Four in total, they incorporate in addition the new environmental component. We quote:

- logistic reliability;
- logistic efficiency;
- logistic agility;
- eco-logistics.

Far from being limited in action, logistics consequently offers several drivers in the service of sustainable development. In the following sections we shall be attempting to show how logistics can reconcile the expectations of shareholders, customers, the staff, society and the environment by being more reliable, efficient, agile and mindful of the preservation of social and environmental balances.

4.4.1   Logistic Reliability

An organization is called reliable when the probability of it fulfilling its mission within a given time corresponds to what is laid down in the contract or specifications. In the case of logistics, reliability of operations consists of the ability to meet customer demand according to a service rate fixed on the best economic, social and environmental terms. Portrayed as a lever in balance, logistic reliability covers the concepts of respect for the commitment of input and outcome in accordance with the specifications and pre-defined expectations (see Figure 4.2). It requires reliable and appropriate resources, skills and knowledge all the way along the supply chain with regard to the agreed-upon competencies. Information must be in phase with products. For example, the product information must correspond to the products, and stocks on the computer screen must reflect physical inventories.

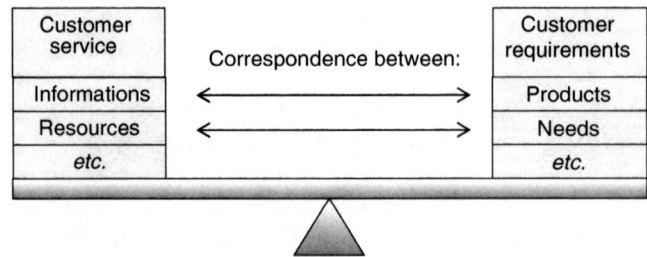

Figure 4.2   "Logistic reliability" drives.

The application of procedures and the use of equipment and packaging in accordance with the regulations and/or best practices for secure handling and transport of products contribute equally to respecting the quality and integrity of merchandize and limiting its impact on the environment. The global language used by players in the supply chain, *i.e.* the international coding and marking standards for products and logistic units, just like EDI message standards, is also a means of producing and exchanging reliable, precise and complete information.

To counter risks of incidents or crises, input errors, errors in receiving, labelling, picking, shipping and billing but also delivery delays, possible wrongdoing, *etc.*, programmes to make information and flows secure, technologies based on barcodes, radiofrequency (Radio Frequency IDentification, RFID) and geo-location for tracking products and vehicles, Supply Chain Event Management (SCEM) and risk pooling systems are already at work.

Logistic reliability measures at the various levels of the supply chain mainly concern customer service, logistic service and supplier service ratios, as well as the quality of product information, the record accuracy, the supply chain inventory visibility, the sales forecast accuracy, the shrinkage rate, the scrap rate, the incident rate, the procedure compliance rate, the absentee rate, staff training, the certification of skills, the number of insurance contracts, *etc.* The issues for the stakeholders are found in savings in terms of finance, time, natural resources and image quality. The proper application of procedures and regulations reduces the risk of breakdowns that may be prejudicial particularly in the case of storage, handling, transport and the use of perishable and/or hazardous materials.

"First time round, deliver the right product to the right place at the right time under the specifications" also means reducing on-costs and pollution connected to delays and the need for re-deliveries. Besides company-internal ratings, no published study quantifying the cost of non-reliability in the supply chain is available to date, but we can safely say that it is not negligible. Lastly, reliability of operation is not just a driver to reduce costs and pollution; it also represents a business driver tied to satisfaction and customer loyalty. This point is particularly true in an economic context overshadowed by hyper-competition.

4.4.2 Logistic Efficiency

Efficiency is the "effectiveness/cost" ratio. It specifies the achievement of an objective with minimum engagement of resources. It is not to be confused with effectiveness, which measures only the achievement of an objective without reference to the outlay involved. The principles of industrial efficiency call upon economies of scale, the standardization of products and processes, automation, improvement of visibility of demand, organization into process flows, demand-driven systems, the optimization of resources, the sharing of logistic facilities and IT applications, the pooling of company functions (order processing management, planning, monitoring, optimization of process flows, *etc.*) and

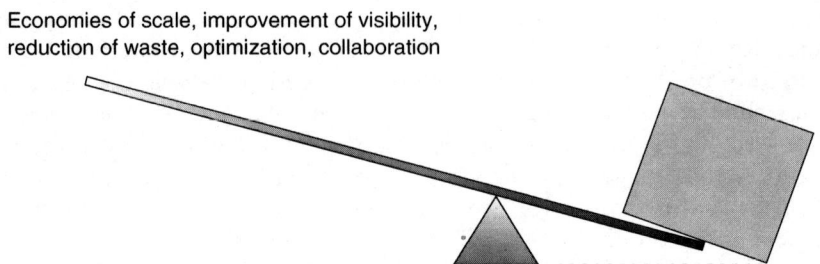

Economies of scale, improvement of visibility,
reduction of waste, optimization, collaboration

Figure 4.3   "Logistic efficiency" driver.

intercompany collaboration. They also involve Total Quality Management techniques for the streamlining of products and processes, the reduction of costs and the systematic elimination of waste under a continuous improvement approach. Logistic efficiency is portrayed as a lever reducing the effort required to achieve an optimum outcome. (Figure 4.3)

Being efficient is being effective in making proper use of resources (human, informational, material, financial, *etc.*) with a positive impact upon company profitability and cash-flow and on the environment, once consumption of resources is minimized. Developing leverage for internal optimization is the first stage. This involves, for example, defining inventory and service policy per segment of products/market and determining key variables like the order quantity, process batch, safety stock, delivery frequency and delivery lead-time in terms of customer requirements and economic and environmental trade-offs.

The repercussions on the fall of inventories and costs of transport with equal service quality are generally spectacular. We recall here that logistic costs lie between 5 and 30% of product cost depending on the sector of activity and company supply-chain maturity. From a systemic viewpoint, "the whole being greater than the sum of its parts", the participants in the supply chain tend to go further in search of overall optimization greater than the sum of local optimizations. This requires collaboration between economic partners. First applied to product development in the automotive and aeronautical industries (simultaneous engineering), the practise of intercompany collaboration made its appearance in the middle of the 1990s with Vendor-Managed Inventory (VMI) in the fast-moving consumer goods industry.

This revolutionary management model was a genuine intercompany collaborative process and made it possible for stocks to be cut down by 50% in retail warehouses and significantly reduced transport costs by maximizing truck fill rate while ensuring a service level close to 99%. It was supplemented in 1999 by CPFR (Collaborative Planning, Forecasting and Replenishment),[5] a customer–supplier collaboration based on plans, promotions and sales forecasts. Henceforth pitched at extending throughout the industrial sector, collaborative supply-chain management techniques bring in an increasing number of players such as logistic service providers and small and medium-sized industries through joint implementation of collaborative warehousing and transportation programmes.

These approaches are new to most companies which need time to catch up in their operational modes and technology investments. However, they represent global efficiency drivers capable of reducing stocks, handling costs, transport costs, total truck kilometres travelled and CO_2 emissions to an unheard-of degree with no negative impact on customer service. We should recall that stocks weigh heavily on cash-flow and the growth of business. According to economists, it is a strategic issue. Thanks to a global vision and optimum inventory management in the supply chain, companies are able to unlock tied-up capital so that they increase their cash-flow and finance growth while at the same time avoiding applying to the banks. The same is true of warehousing and transport optimized by consolidating different delivery streams (different products based on different orders from different ordering facilities, all for the same customer). Efficiency measures rely on different performance indicators, mainly based on costs. We find, for example, ordering cost, carrying cost, storage cost, picking cost, transportation cost, price list, hourly rate, *etc.*, the variables connected with productivity like utilization, yield factor, turnover ratio, vehicle fill rate, *etc.* and other variables like demand visibility, order quantity, lead-time, delivery frequency, rate of standard pallets, *etc.*

4.4.3 Logistic Agility

Agility is at the heart of reactive organizations. A reactive company enjoys flexible facilities, which, if they are lightweight enough, enable it to be agile. Reactivity is the ability to rapidly adapt production volumes and product variety to fluctuations of demand and speed up getting a new product to market. From an agility viewpoint, it is the flexibility and adaptability of processes, resources, organizations and logistic chains that are sought after to cope with unstable, turbulent, uncertain and risky environments, and to seize market opportunities. One of the keys of agility is the systematic reduction of lead-times in design, sourcing, manufacturing, arranging setup (SMED: Single-Minute Exchange of Die) and adapting logistic services to fluctuations of demand.

For hybrid products (half-generic, half-customized), postponement is another technique that achieves mass customization by offering the customer greater variety at a lower total cost. This strategy makes use of the advantages of standardization in terms of reducing costs (low cost production of components and generic modules, more flexible generic stocks, more reliable generic forecasting), while maximizing the commercial offering by customizing products to the local market. By way of example, agility indicators refer typically to time-to-market, time-to-volume, inventory turnover, flow velocity, cycle time, run time, tact time, transit time, wait time, down time, lot size, order-to-cash cycle, cash-to-cash cycle, *etc.*

Figure 4.4   "Logistic agility" driver.

The various time-based strategies also offer prospects for sustainable development by enabling local industries to be more reactive while holding less stock to meet a demand that is more difficult to forecast. In responding to demand better, quicker and more cheaply, turnover, profitability and working capital are all up, while the impact on the environment is down, particularly as regards excess inventory and obsolete products to recycle. To illustrate logistic agility, we use a flexible lever consisting of quickly serving products and services in terms of fluctuations in demand, as in Figure 4.4.

4.4.4   Eco-logistics

According to Martinet and Reynaud, "*companies must internalise a part of the environmental costs and the social costs they would [earlier] have rejected. Managing sustainable development becomes an element of differentiation*".[6] Today, the application of several sustainable development programmes is possible, such as ISO14001 certification concerning environmental management, the use of renewable energies, reduction of water consumption, sorting and recycling packaging (the Eco-packaging program) and regional development, thanks in particular to the development of local products, the development of fair trade, the integration of social workers, *etc*. But these projects have to be balanced against the achievement of economic and financial performance if they are to be sustained and develop.

In terms of logistics, sustainable programmes concern more specifically fuel-efficient driving, the use of hybrid modes of propulsion or a shift towards environmentally friendly transport modes. Freight initiatives concern multi-modal transport combining road, rail, waterway, air freight and shipping to reduce energy consumption, greenhouse gas emission and traffic congestion. Limitations on packaging and the increased recyclability of products also represent practical measures to reduce the ecological footprint of goods. In the same way, reverse logistics provides the collection, sorting, dismantling and recovery of value from used products.

Other avenues concern the certification of logistic platforms and buildings according to the HEQ® (High Environmental Quality) method promoted by the AFILOG association in France. This standard exists in other forms in

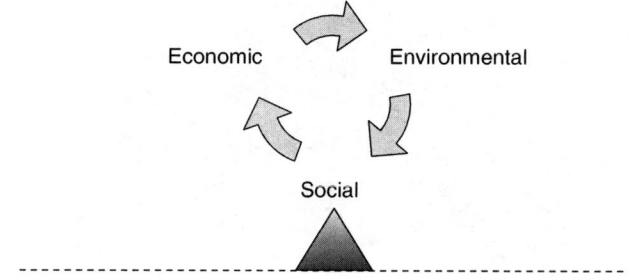

Figure 4.5   "Eco-logistic" driver.

other countries in Europe and reviews various criteria such as flow impact on the immediate environment, the use of combined transport, energy consumption in offices and warehouses, water management (reduced waterproofing on the grounds, landscaping of catchment area, saving water for fire extinguishing purposes, *etc.*), the treatment of hazardous materials, health quality of air and of working conditions.

Finally, the transport and logistic trades offer a considerable source of employment with the development of trade and commerce. According to a study by the Centre d'Analyse Stratégique,[7] it should offer about 700,000 new positions to be filled between 2005 and 2015, out of a total of 2 million jobs dedicated to logistics in a country like France.

Because of its approach directed towards social and environmental objectives allied to economic performance, the eco-logistic driver is represented as a virtuous loop combining the three pillars of sustainable development applied to logistics (see Figure 4.5).

The measurement of eco-logistic performance potentially uses several indicators such as energy consumption, number of tonnes of CO_2 emitted by logistics platforms and transport (according to weight carried, mode used and distance travelled), traffic congestion rate, infrastructure simplification rate, hours of staff training, the number of social disparities, *etc.*

Combined with the three leverage drivers of reliability, efficiency and logistic agility – sources of social, economic and environmental benefits – the eco-logistic lever strengthens the contribution of the supply chain to the social and environmental aspects of sustainable development.

Mastery of these four logistic drivers ensures optimum customer service without out-of-stocks or overstocks under the best economic, social and environmental terms. It requires a structure, an organization and a process to be set up plus an integrated cross-cutting information system as demonstrated in the Supply Chain Wheel[8] (Figure 4.6).

This figure symbolizes all the logistic supply chain components drawn from the Supply Chain Master® reference model. To improve their logistic performance, companies have these drivers available dedicated to reliability,

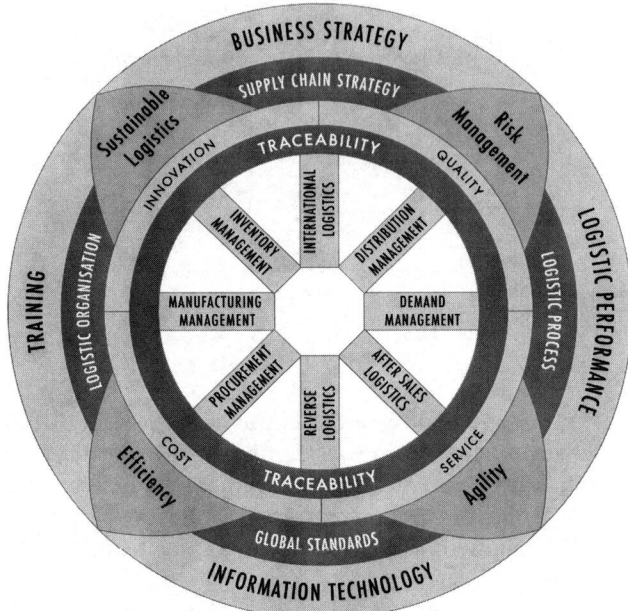

Figure 4.6   The Supply Chain Wheel.[8]

efficiency, agility and eco-logistics which clearly have a positive effect on economic growth, social development and the protection of the environment.

The whole issue now hangs on the ability of the players in the supply chain, in particular SMEs, to operate these drivers in accord with all the partners to get significant leverage on sustainable development.

4.5   Towards Sustainable Logistics in the Service of Sustainable Development

Following the revision of the European Sustainable Development Strategy in 2006, Member States reaffirmed their determination to promote sustainable transport with a series of measures concerning:

- behaviour;
- policy on inter-modality development;
- infrastructure developments;
- pricing for the use and management of infrastructure;
- technological improvements.

The guide published in 2007 highlights the Marco Polo programme[9] aimed at encouraging the use of combined transport and developing more efficient and sustainable transport chains. The idea is to combat congestion, environmental degradation, accidents and the danger of loss of competitiveness of European industry, which needs to be able to rely on cost-effective and reliable transport systems to manage its supply chains. However, forecasts on the increase of road freight in 2013 exceed 60% in the European Union and it is expected to double in the ten new states by 2020. In this context, greater use of inter-modality is required. It contributes to better use of existing infrastructure and service resources by integrating short journey shipping, rail and waterway transport in the supply chain. The sea highways are an illustration of this with the launch in 2009 of the sea link between the French Atlantic coast (Nantes) and the north of Spain (Gijón – Asturias) short-cutting and relieving thousands of kilometres of international state roads.

Other initiatives for logistics directed towards the requirements of sustainable development are under way. For example, in the fast-moving consumer goods industry, the Global Commerce Initiative (GCI) defined the key changes required for the future supply chain in 2016.[10] Seven areas for improvement are identified in:

- in-store logistics: includes in-store visibility, shelf-ready products, shopper interaction;
- collaborative physical logistics: shared transport, shared warehouse, shared infrastructure;
- reverse logistics: product recycling, packaging recycling, returnable assets;
- demand fluctuation management: joint planning, execution and monitoring;
- identification and labelling;
- efficient assets: alternative forms of energy, efficient/aerodynamic vehicles, switching modes, green buildings;
- joint scorecard and business plan.

Thus, the players in the supply chain are called on to play their part in bringing about the changes that partially cover existing challenges but also fresh challenges involving sustainable development. According to the GCI estimates, "*the total impact of this supply chain redesign could potentially reduce transport costs per pallet to the order of more than 30%, cut handling costs per pallet to the order of 20%, reduce lead time by 40% and lower CO_2 emissions per pallet to the order of 25%, while also improving on-shelf availability. This does not include additional energy cost savings stemming from more efficient assets such as green buildings and fuel-efficient/aerodynamic and jumbo trucks*".

These improvements hinge on maximizing loads, cutting down empty journeys, training drivers in eco-driving, the renewal of truck fleets and the transfer from road to combined freight. Apart from eco-technological innovations, collaboration between the players in the supply chain is the *sine qua non* for the programme's success. While customer–supplier collaboration has been developing successfully for fifteen years driven jointly by ECR (Efficient Consumer Response), VICS (Voluntary Inter-industry Commerce Solutions) and GS1 organizations in different models (Vendor-Managed Inventory,

Collaborative Planning, Forecasting and Replenishment, stream consolidation, cross-docking, better traceability, *etc*.), it must be backed up and henceforth take in the transporters, warehouse operators and the small and medium-sized industries hitherto not generally incorporated in this approach.

To do this, real-time information sharing and particularly point-of-sales data sharing in standardized formats must become general practice between retailers, manufacturers, logistic service providers and suppliers. This development is designed to enhance visibility of demand and events on the supply chain with a view to limiting out-of-stocks and further optimizing the use of resources. To reduce stocks, transport costs and CO_2 emissions, the emergence of collaborative warehouses and hubs – where flows from different industries and different retailers can be bulked and cross-docked before being distributed optimally to consumers (to city stores, local stores and home delivery *via* shared facilities) – will revolutionize the way the supply chain works today.

According to the Global Commerce Initiative, the challenges ahead will force the players in the supply chain to change their operation to a future model leading to sustainability and new business opportunities. It is actually under way. Companies are by no means helpless. Several technologies and support tools are available today such as global communications standards (product identification, barcodes, RFID, EDI), Supply Chain Management software, tracing and tracking tools, the optimization of shipping units and of secondary and tertiary packaging, *etc*. The real challenge lies elsewhere: it involves adopting in unison "*new ways of working together in the physical supply chain (including management of required investments, capabilities, organisational resources and design, incentives and measures, social regulations like working hours, etc.*).[11]

To achieve this, the key success factor is to convince tens of thousands of small and medium-sized businesses to join the future supply chain so as to reach the critical mass necessary for the reduction by 20 to 40% of the impact of supply chain activities on the environment. Of course, this objective has to be reconciled with the needs of economic growth and the sharing of the gains between the players while supporting purchasing power and regional development.

4.6 Conclusion

Representing an essential function to trade, logistics is a complex system made up of a multitude of players and interrelationships between suppliers, manufacturers, logistic service providers and the retailers whose job it is to deliver to the end user the product or service required under the best economic, social and environmental conditions. Meanwhile, its impact on the environment is important because of the many forms of pollution generated by road transport – the major delivery mode used today. The transport sector consumes 70% of all petrol consumed in the EU and produces 21% of all greenhouse gas emissions. At the same time, it is well recognized that transport is the very life blood of a modern economy and its effects go far beyond the direct economic effects of employment and the added value of the transport sector in an economy.

In this chapter, we have reported the principal initiatives launched in Europe to ensure that transport systems meet society's economic, social and environmental needs while minimizing their undesirable impacts on the economy, society and the environment. But the biggest challenge seems to come from the difficulty of getting the supply chain players to work together on the three aspects of sustainable development and to prepare all the people involved for the new world of collaboration. To assist them in working together, we have outlined the purpose of "modern" logistics and highlighted the existence of the four logistic drivers of individual, joint and global performance. Taken separately, these aim at improving reliability, efficiency, agility and relations with the environment. Operating together, they prove to be powerful performance activators in the service of economic growth, social development and protection of the environment. This includes shared success measures across financial, operational, environmental and consumer-based dimensions that drive outcomes throughout the value chain and greater overall value for all stakeholders.

Finally, we note that the economic, social and environmental aspects of logistics are indissociable and all three participate in the emergence of this strategic function in the service of sustainable development, alongside the sustainable solutions of eco-design, eco-industrialization and eco-production used in industry.

References

1. AFNOR, NF X 50-600, *Logistique – Fonction et démarche logistiques*, AFNOR 2005, **12**, 4.
2. Cf. Barometre ECR France IRI based on the analysis of out-of-stocks (OOS) of 61 categories of products in 1449 hypermarkets in France, 2008.
3. S. Pasi, *Chargements moyens, distances et parcours à vide dans le transport routier de marchandises,* Eurostat, 2007, **8**, 6.
4. M. Christopher, *Supply Chain Management, Create networks with high added value, Créer des réseaux à forte valeur ajoutée*, Village Mondial, 2005, **318**, 288.
5. *Collaborative Planning, Forecasting and Replenishment Voluntary Guidelines; Roadmap to CPFR*, VICS, 1999.
6. E. Reynaud, *Développement durable et entreprise: vers une relation symbiotique?* CNRS, 2003, **15**, 9.
7. Rapport du groupe, Prospectives des métiers et qualifications, *Les métiers en 2015*, Centre d'Analyse Strategique, DARES, 2007, **179**, 123.
8. T. Jouenne, *How to contribute to the competitiveness and development of SMEs by using APICS concepts,* Revue Française de Gestion Industrielle, 2008, **27**(3), 3, 7.
9. European Commission, *A sustainable future in our hands*, European Communities, Brussels (2008), http://bookshop.europa.eu/eubookshop/download.

action?fileName = KA7007020ENC_002.pdf&eubphfUid = 602686&catalog Nbr = KA-70-07-020-EN-C (retrieved 21.06.2009).

10. *Future Supply Chain 2016, Serving consumers in a sustainable way, Global Commerce Initiative*, Cap Gemini, 2008, **52**, 24.
11. *Future Supply Chain 2016, Serving consumers in a sustainable way, Global Commerce Initiative*, Cap Gemini, 2008, **52**, 46.
12. J. T. Mentzer *et al.*, *Defining Supply Chain Management, Journal of Business Logistics*, **52**(2), 2001, 5, 41.

CHAPTER 5

Sustainable Solutions for Consumer Products

FRANK ROLAND SCHROEDER

Henkel AG & Co. KGaA, Henkelstrasse 67, D-40589 Düsseldorf, Germany

5.1 Introduction

"Everybody in the consumer industry, from fast-food services to big-box retailers, have put out their green agenda to assure customers, regulators, shareholders and Non Governmental Organiations (NGOs) that they are walking toward the right, ethical direction. Most of the key initiatives made by global consumer businesses last year deal with reducing their carbon footprint, such as using or selling more environmentally responsible products, reducing carbon dioxide emissions in stores or supply chains, phasing out ingredients and raw materials that they deem hazardous, reducing waste and pushing their supply chains to improve their own sustainability, among others."[1] This statement reflects the current worldwide concern about climate change and the state of the environment and consequently consumers are asking for more sustainable products. Consumer goods manufacturers have to take these concerns seriously and develop new types of product concepts, which combine superior performance with improved environmental profile and greater social benefits. Since there is no clear cut definition on how a sustainable product might look, this chapter describes those elements in a fast-moving consumer goods life-cycle that might contribute to the overall sustainability profile. Detergents are chosen as an example since a lot of relevant information is available for this product category.

RSC Green Chemistry No. 4
Sustainable Solutions for Modern Economies
Edited by Rainer Höfer
© The Royal Society of Chemistry 2009
Published by the Royal Society of Chemistry, www.rsc.org

5.2   Demographic Dynamics and Global Megatrends

The human population has been increasing constantly since the beginning of mankind. Between the years 1800 and 2006 population has increased more than six-fold (see Figure 5.1). In parallel to this demographic development *per capita* consumption and consequently total production volume have been growing proportionally (see Figure 5.2). This process has been accelerated even further by globalization.

However, with respect to population development there are striking differences between the northern and the southern hemisphere. In the developed countries located in the northern hemisphere, population growth is generally slow or even stagnant in some countries, resulting in a fundamental change in societal structure. The development in Spain between 1995 and 2005 may serve as a typical example:[3]

- 0.7% p.a. population growth but 20% more households in 10 years;
- 56% more smaller households (1 and 2 persons) in 10 years;
- 30% more senior households (>65 years old) in 10 years;
- 50% decrease in women's unemployment rate in 10 years.

All these factors, in addition to an increase in consumers' income, result in a profound change in consumer purchasing behaviour. In parallel to the growing purchasing power, consumers feel that they are "*poor in time*". Convenience is becoming a driver of the market, with slogans like "*open and serve*" or "*boil and eat*". In the detergent industry this results in an increase of single dose products. Moreover there is an increasing concern about health and environmental matters. Consumers that are seen as environmentally aware, socially attuned and with a view of the world that takes into account personal, community and planetary outcomes are often characterized as LOHAS (Lifestyle **Of** **H**ealth **A**nd **S**ustainability). LOHAS has become a megatrend, a term made popular by John Naisbitt,[4] since the increased desire for healthier and more sustainable lives is leveraging the sociological developments described above. The LOHAS construct has been adopted and embraced around the world over the past

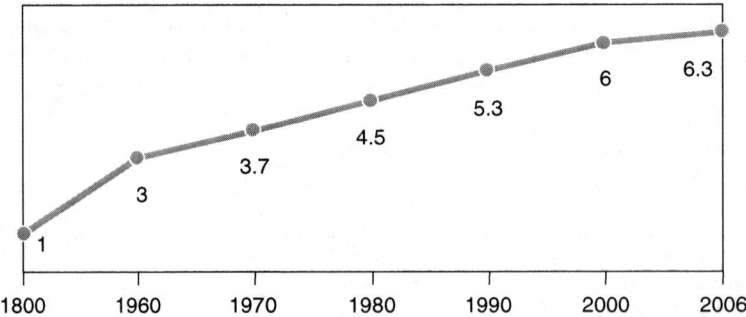

Figure 5.1   World population in billions.

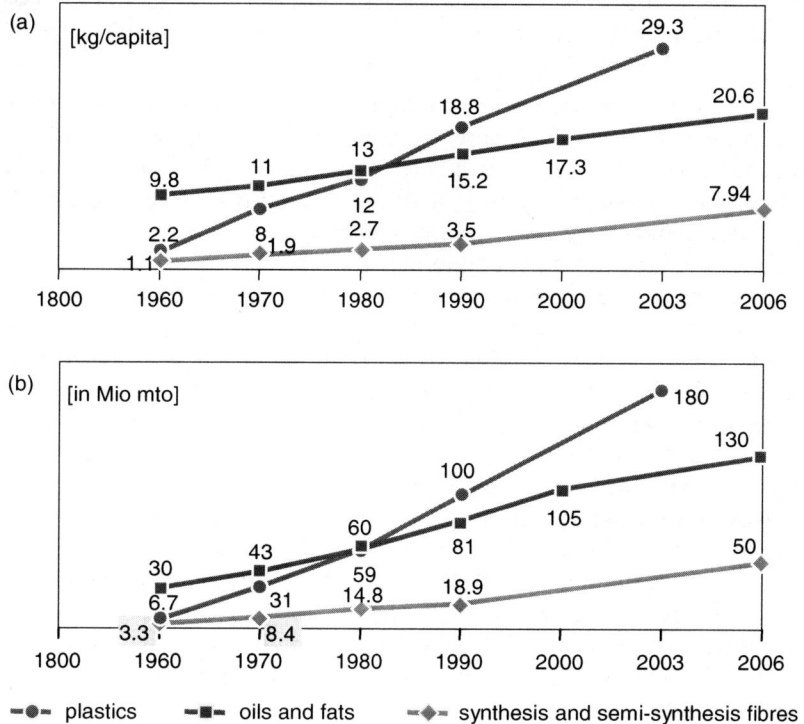

Figure 5.2 *Per capita* consumption (a) and total production (b) for major chemical raw materials.

decade. In western countries, such as the USA and Europe, LOHAS is an industry term while in Japan and East Asia LOHAS is a widely recognized consumer term. International research estimates the LOHAS marketplace has a global value in excess of 500,000 Mio $.

According to an analysis of the consumer market in Australia[5]

- nearly 4 million adult Australians (26% of the adult population) are LOHAS aligned;
- individuals with a LOHAS outlook are drawn from all parts of society; their values and world view are not strongly tied to income, geography or gender.

LOHAS-aligned consumers seek to integrate healthier, more sustainable options into all aspects of their lives by making consumption decisions for products and services in key market segments including:

- food and nutrition – *e.g.* organic and natural food, vitamin supplements;
- home life – *e.g.* natural cleaning products, efficient appliances, recycled paper;

- buildings and energy – *e.g.* water tanks, solar hot water, "green" energy;
- work and money – *e.g.* socially responsible investing, "green" loans.

Contrary to the demographic development in the developed world, many regions in the southern hemisphere are experiencing strong population growth, so that the population of the Earth as a whole is increasing (Figure 5.1). At the same time, the growing prosperity of emerging countries such as Brazil, China and India is matched by their greater consumption of resources and energy. Crucial factors include greater mobility and changes in eating habits. Many emerging countries are experiencing a growth in demand for "high-quality" foods with a high proportion of animal protein. Since the production of one kilogram of animal-based food needs about nine kilograms of grain there is a considerable increase in water, fertilizers and land needed to fulfil the demand. This change is especially dramatic in countries where vegetable protein has always played a key role. China may serve as an example where meat consumption has doubled over the last twenty years.[6] All these factors manifest themselves in a mounting scarcity of raw materials and in higher prices.

The change in eating behaviour not only reinforces pressure on resources but additionally there is a marked effect on the climate. Intensive livestock farming of ruminants results in methane (CH_4) as a waste product of digestion. Methane, weight for weight, is 21 times more potent as a greenhouse gas than CO_2. Overall, the agricultural sector's emissions account for 14% of all greenhouse gases.[7]

The emission of greenhouse gases and the associated rise in temperature have far-reaching consequences in several areas. Rising temperatures enable pests and their vectors (*i.e.* host organisms transmitting pathogens) to spread more easily, resulting in a greater risk of infectious diseases and even epidemics occurring. This applies especially to densely populated tropical regions, where the population in conurbations is expanding disproportionately rapidly. Hence, appropriate measures have to be taken to counter the possible spread of infectious diseases associated with the rise in temperature.

Rising temperatures also increase the pressure on freshwater resources. 96% of the Earth's freshwater is locked up in glaciers and geological strata. Of the remaining 4%, equivalent to 110,000 Mio m^3, humans use 4,000 Mio m^3 for their needs. There is a qualitative rather than a quantitative shortfall here, as 1 billion people do not have access to clean drinking water. The development of agriculture is putting additional stress on freshwater resources since agriculture is responsible for 70% of "blue water" consumption, *i.e.* liquid water in rivers, lakes and groundwater aquifers. Food production to satisfy a person's daily dietary needs takes about 3000 litres of water – a little more than one litre per calorie.[6]

Although it is virtually impossible to carry out a quantitative analysis of all these factors and their effects on future development, some of them can be illustrated very well in qualitative terms (arrow model). Given the increasing population of the southern hemisphere and the increase in living standards in these regions, raw materials will steadily become scarcer. At the same time there will be growing pressure on water resources. This, and a general rise in temperatures, will tend to promote the incidence of infectious diseases and impair hygiene. Scarcer resources, growing pressure on water resources and poorer hygiene are the three megatrends on which we will have to focus in future (see Figure 5.3).

5.3 Life-Cycle of the Washing Process – an Example for Sustainability in Consumer Goods

The washing process can be seen as an example of how an industrial sector can provide technology and innovation to improve the sustainability profile of a market segment that is relevant to virtually all consumers.

The global household detergent and cleanser market had a value in 2008 of 86,565 Mio €.

The washing process and the technology of laundry detergents it requires are an integral part of human cultural history.[8–10] Currently, washing is carried out in all regions of the Earth under widely varying environmental and economical conditions. Nonetheless, consumers have very clear expectations. A detergent used in a household has to deliver excellent performance in soil removal and general cleaning. The consumer wants to save energy and money through a low washing temperature and short washing cycle. The textile fibres and the colour must look like they did before washing, or even better. The consumer wants a detergent that is simple and convenient to use and has an individual appeal through a special product form or aesthetics such as perfume. Clustering the demands from consumers in three different claim dimensions results in:

- performance;
- convenience;
- aesthetics.[11]

The functions of detergents, however, go beyond laundry care and aesthetic considerations as they make an important contribution to hygiene and there-fore to the maintenance of good health. Obviously those megatrends described earlier, like scarcity of raw materials and freshwater as well as increasing demand for better hygiene, clearly impact the washing process.

For an assessment of the sustainability profile, it is not enough to look at laundry detergent surfactants in isolation; the whole life-cycle has to be taken into account.[12] In order to arrive at soundly based conclusions, the total washing process must be analyzed. To enable the findings for different systems to be compared, all results must be related to a functional unit *e.g.* the cleaning of 1 kg of laundry. A systematic analysis of the total process is then carried out in the context of a so-called life-cycle assessment (LCA), in which all the sys-tem's inputs and outputs are systematically recorded and assessed.[13]

Figure 5.4 gives a schematic overview of the different stages of the washing process. The most relevant ones are discussed below from the point of view of sustainability.

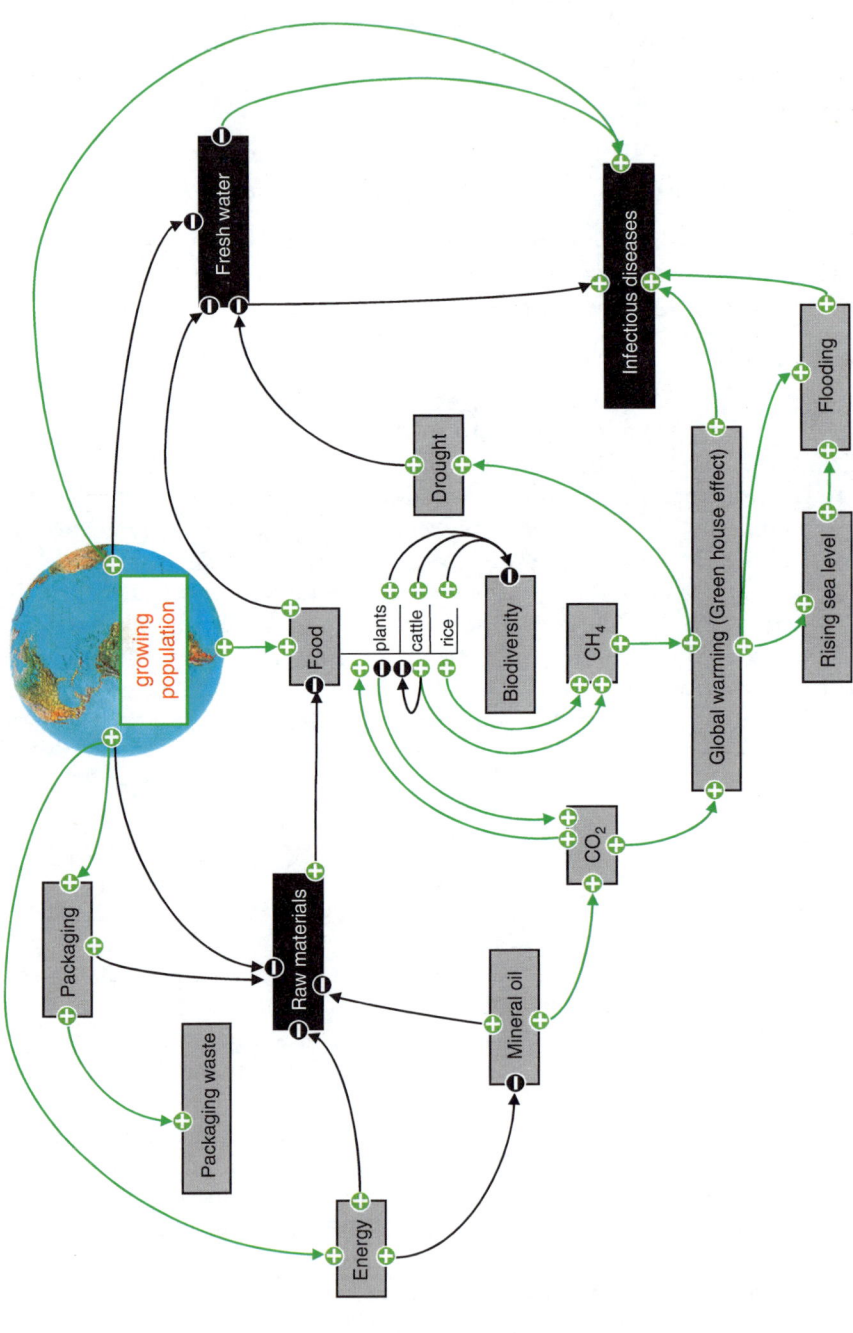

Figure 5.3   Arrow model to illustrate development megatrends.

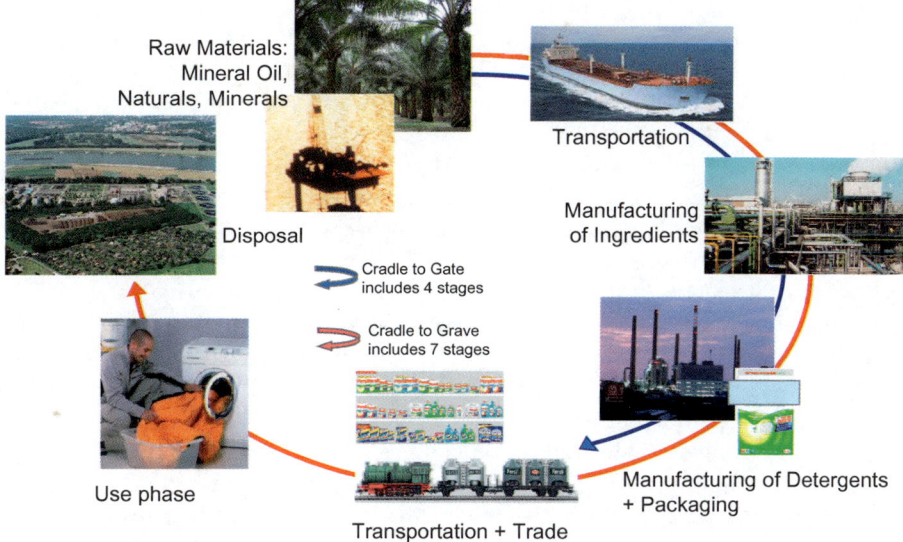

Raw Materials: Mineral Oil, Naturals, Minerals

Transportation

Manufacturing of Ingredients

Disposal

Cradle to Gate includes 4 stages

Cradle to Grave includes 7 stages

Use phase

Transportation + Trade

Manufacturing of Detergents + Packaging

Figure 5.4   Key phases of the life-cycle of the washing process.

5.3.1 Raw Materials

Household detergents (whether liquid or powder products) are mixtures of inorganic and organic chemicals. Detergents for household and institutional use can be categorized into:

- surfactants (organic);
- builders (typically inorganic);
- bleaching agents (typically inorganic);
- auxiliaries.

Inorganic chemicals are obtained either chemically or by mining of natural resources. In general, it is possible to recover inorganic chemicals from waste streams (recycling), but this usually requires considerable technical effort and the consumption of correspondingly large amounts of energy. Such processes have therefore not yet been used on an industrial scale.

Most organic chemicals are derived from mineral oils, and in earlier times they were also obtained to a large extent from coal. Since the increasing pressure on the Earth's resources is a megatrend, renewable raw materials have also been increasingly used as a source of organic chemicals in recent years[14] and especially in detergents.[15] The most important feedstock for chemicals based on renewable raw materials are oils derived from plants or materials generated by means of biotechnical methods.

Palm kernel oil and coconut oil are the most common plant oils for the production of surfactants. Through chemical reactions, the oils are transformed to yield glycerol, as well as surfactants with medium carbon chain lengths (C_{12-14}), which have a number of especially advantageous properties for washing and cleaning processes. Fatty alcohol sulfates, fatty alcohol ether sulfates and fatty alcohol ethoxylates are typical representatives (see Chapter 9.1). All these surfactants are usually made up of a combination of a chemical moiety deriving from renewable raw materials and a part based on inorganic or classical organic chemistry derived from mineral oils. Product concepts based on the greatest possible use of renewable resources have led to the development of sugar-based surfactants by reacting fatty alcohols with glucose yielding alkyl polyglycosides (APG®-surfactants).[16]

Fermentation is the conversion of biological feedstock to functional chemical substances on an industrial scale with the help of specialized micro-organisms. Organic acids like citric acid and lactic acid as well as enzymes are typical detergent ingredients manufactured by fermentation processes on an industrial scale. The biological feedstock is typically relatively cheap and can be either derived from agricultural crops (*e.g.* starch, which is obtained from corn or wheat) or biological waste materials (*e.g.* molasses, which is a residue in the production of sugar).

At first sight there are two striking advantages of chemicals based on renewable raw materials. Since plants use atmospheric CO_2 for the build-up of biomass they are, in theory, climate neutral. Moreover, plants are a nearly unlimited resource for renewable raw materials.[14] On the other hand energy for processing and transportation needs to be taken into account. Additionally, many crops used as sources for vegetable raw materials require fertilizers, which are one source of agricultural greenhouse gases. Therefore, the full effects of renewable raw materials on the generation of greenhouse gases can only be evaluated by LCA.

To explore the full picture with respect to sustainability, social factors have to be taken into account as well. At present, the effect of the increased use of renewable raw materials on the availability and price of food is a subject of intense debate. It is not yet clear which factors are responsible for the recent volatility in food prices. While the World Bank estimates the influence of biofuels at 75%, the FAO also points to other factors that contribute to the rise in prices. These include changed nutritional habits in many emerging countries, poor harvests and increased fuel prices, which account for about 30% of production costs.

In 2008, the total vegetable oil market amounted to 128.3 Mio mto (see Chapter 9.2). Palm oil, with 38 Mio mto, accounted for the largest share. 80% of palm oil comes from Southeast Asia (Malaysia and Indonesia). Palm kernel oil and coconut oil both provide medium carbon chain oils (C_{12-14}) often referred to as laurics. Together they account for 7.4 Mio mto. While palm oil is mainly used for nutritional purposes palm kernel oil, which can be seen as a side product of palm oil production, is preferred for the production of surfactants. Against the background of a growing demand for biofuels, there has been an increasing demand for vegetable oils as well that has increased the pressure on primary forests (rainforests). While only 12% of Malaysia's primary forest is still intact,

Indonesia has 25%. To further protect high conservation areas (*e.g.* primary forests, peat lands) and the animals that live in them, the Roundtable on Sustainable Palm Oil, an alliance of palm oil growers, down stream users, non-governmental organizations and financing institutes, has developed a certification system intended to ensure that vegetable oils – and above all palm oil and palm kernel oil – are produced under sustainable conditions (see Chapter 9.4.2.2).

5.3.2 Logistics

From an energy perspective (this also includes the generation of greenhouse gases) logistics plays a minor role in the life-cycle of a detergent. Nevertheless, there are clear differences between the various modes of transport. One metric ton of freight transported by ship generates about 6–11 kg CO_2 eq km^{-1}, while its transport by rail and road generates respectively 39 and 117 kg CO_2 eq km^{-1}.[17] An optimized transport mode mix can therefore be devised to take account of the geographical circumstances.

5.3.3 Production

In general, there are basic differences between the production of liquid and powder laundry detergents. In terms of energy, liquid detergents are more favourable, as no drying steps are required.[18] Other opportunities for optimizing production with respect to the use of primary energy include the recycling of process water, and combined heat and power generation.

5.3.4 Use Phase

Based on figures retrieved for Germany in 2006, laundry washing accounts for only 3% of households' total energy[19] and for 12% of households' water consumption,[20] respectively.

If the washing life-cycle is looked at in isolation, the use phase is the most critical with regard to energy consumption with the heating of the washing water playing the most important role. This is especially true for countries with a high prevalence for washing at elevated temperatures. If energy is generated from the combustion of lignite and coal, considerable amounts of greenhouse gases (mainly CO_2) and solid waste (slag and ash created by burning solid lignite and coal) are generated. If the main source of energy is hydroelectric or nuclear power plants, the contribution to the greenhouse effect is low, although the use of nuclear energy may entail other problems. Reducing the washing temperature is therefore a key means of saving energy and reducing greenhouse gases as well as waste material. The main challenge is to retain the same level of washing performance and hygiene at lower temperatures.

5.3.5 Disposal Phase

Relative to other anthropogenic activities, the disposal phase plays only a minor role as regards energy consumption and the generation of greenhouse gases. However, since the wash water is discharged to the sewer and, depending on the country, this wastewater is treated or it is discharged directly into the environment, the contribution of detergent ingredients to the loads in surface waters is considerable.[21] Surfactants in particular must be viewed critically. Their surface activity makes them relatively toxic to water organisms, and they show the potential to produce easily visible foam on surface waters. Aquatic toxicity is an inherent property of these substances, so their biodegradability is especially important, as it reduces the amount of surfactants in surface waters.

For this reason, the industry started to develop biodegradable surfactants in the 1950s. In parallel, screening tests were devised, which allow biodegradation potential to be verified in the laboratory. The OECD has officially recognized six of these test methods (OECD 301 A–F) that measure biodegradability in the presence of oxygen. They involve inoculating relatively high concentrations of surfactants with relatively small amounts of micro-organisms from public sewage treatment plants. The OECD test methods differ in how the biodegradation is monitored. If the limit values (60 or 70% biodegradation) are reached within the specified time period, the test substance can be described as "readily biodegradable". There are also test methods that determine biodegradability in the absence of oxygen (anaerobic biodegradability). The data obtained is relevant for, *e.g.*, sediments or the digesters of sewage treatment plants. Finally there is also the metabolites test. A positive result in the metabolites test indicates that no poorly biodegradable molecular groups are still present.

Subsequently, an efficient method of ecological risk assessment was developed. It is based on a comparison of the predicted environmental concentration (PEC) and the predicted no-effect concentration (PNEC). If PEC $\leq$ PNEC, the substance under consideration is expected to pose no ecological risk in the environment. A comprehensive review about the ecological profile was provided by J. Steber.[21]

In the HERA project,[22] a European voluntary initiative commissioned by A.I.S.E.[23] (representing the formulators and manufacturers of household and industrial and institutional cleaning products) and CEFIC[24] (representing the suppliers and manufacturers of the raw materials), the most important ingredients of laundry detergents are subject to a systematic risk assessment. This covers product safety from an ecological and human health perspective. The underlying reports are available to the public *via* the HERA website.

5.4 Sustainability Profiles of Detergent Formulations

At the beginning of the industrial revolution, the societal focus was clearly on increasing production and multiplying capital. Only in individual companies did additional considerations play a role. In the second half of the nineteenth century, as the power of the workers increased, social issues also moved into the foreground. Occupational health, working hours and pay now played an increasingly important role. Nature and environmental protection were still neglected, however, and it was not until the 1960s that environmental protection became a major social, political and industrial issue.

But how to improve sustainability in this context? Improving sustainability implies that all three elements, the economic, ecological and social factors, need to be developed in parallel to achieve an overall optimum for the whole system.

If a product or service is optimized with respect to their sustainability profile the optimization process can be best illustrated by using a graphic model (Sustainability Profile Circle – SPC, Sustainability Master®). The starting point of the optimization is a group of three equal circles, representing the economic, ecological and social dimensions, with a surface area that is chosen to be 100% by default. The circle's centres are connected, resulting in an equilateral triangle – the "sustainability zone". Changing relevant parameters will impact the surface area of the appropriate circle – and consequently the shape and area of the resulting triangle (Figure 5.5).

One example makes this clear. A household cleaner based on a sugar ether as a surfactant would certainly have an excellent ecological profile, given its ready biodegradability and the renewable raw materials that serve as its basis. The circle representing the ecological dimension would therefore grow. These surfactants do not have such good performance characteristics, however, so that the social dimension reflecting consumers' benefit would tend to shrink. In view of the higher price, the consumer would have to pay more (causing this circle to shrink still further) or the producer's profit margin would be reduced

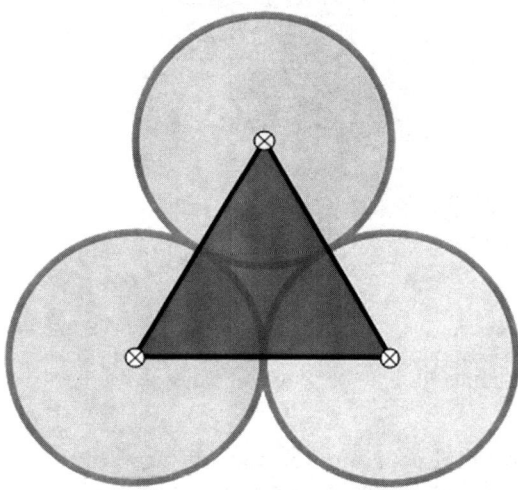

Figure 5.5 SPC model, Sustainability Master® to describe the sustainability profile of an ecological cleanser – reference scenario; all circles set to 100%.

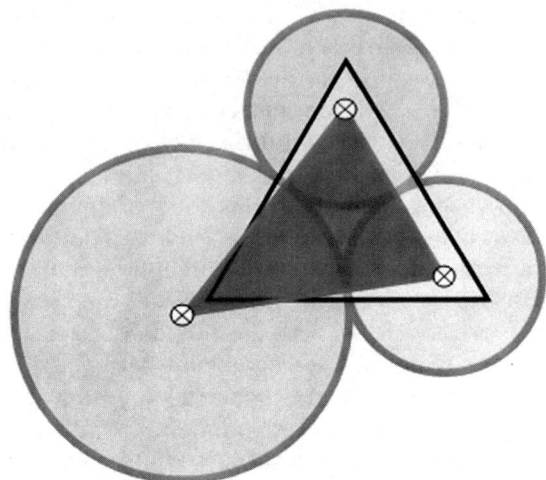

Figure 5.6   SPC model, Sustainability Master[R] to describe the sustainability profile of an ecological cleanser – environmental, social and economic circles set to 120, 70 and 70% of original area, respectively.

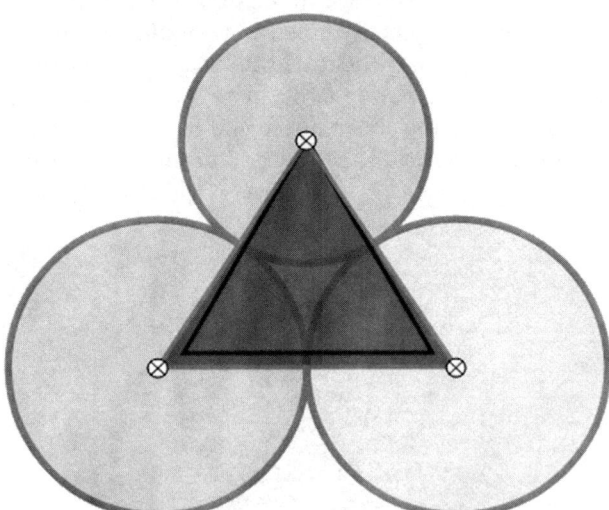

Figure 5.7   SPC model, Sustainability Master[R] to describe the impact of compaction on the sustainability profile – environmental, economic and social circles set to 120, 120 and 100% of original area, respectively.

resulting in a decreased economic profile. By using the SPC approach it can be clearly visualized that a growing ecological benefit alone will not increase the area of sustainability if there is a decrease in the economic and social profile (see Figure 5.6).

Another example: modern laundry detergents are made more compact[25] in order to reduce the environmental burden of disposal to receiving surface waters. Moreover, these compacted products show a positive environmental profile since they require less material for packaging and less energy for transportation. At the same time material and logistics costs decrease as well, resulting in an increase of the economic dimension. Since there is no impact on the performance, provided the detergents are added in the correct dosages, the area of the "social circle" would therefore remain constant. Overall this example would result in an increased sustainability zone (see Figure 5.7).

5.5 Conclusion

Against the background of a growing concern about the global climate and the general state of the world's environment, sustainability became a megatrend in recent years. Consumer goods manufacturers have to take public concerns and the demand of consumers into account and develop new product concepts. To fulfil all requirements, a holistic approach is needed that takes all single phases of a product's life-cycle into account. In the case of laundry detergents this encompasses the raw material extraction, the manufacture of the ingredients and the formulation of the end product as well as transportation, consumer use of the product, disposal and treatment in the wastewater treatment plant. To collate all this information, "Life-Cycle Assessment" provides useful information. To further consolidate this information and visualize the impact on the area of sustainability, the Sustainability Profile Circle (SPC) model, Sustainability Master® is introduced. Using the SPC approach the individual impact of ecological, economical and social parameters can be assessed separately and their contribution to a "sustainability zone", where these parameters converge, can be displayed.

References

1. D. De Guzman, *ICIS Chemical Business*, 30 Jan. 2008, http://www.icis.com/Articles/2008/02/11/9096925/consumer-industry-increases-sustainability-initiatives.html (retrieved 16.10.2008).
2. R. Höfer and J. Bigorra, *Green Chem. Lett. Rev.*, June 2008, **1**(2), 79.
3. C. Miñarro García, in *Comunicaciones*, 36 Jorn. CED, Barcelona, 2006, p. 219.
4. J. Naisbitt, *Megatrends*, Warner Books, Clayton, 1982.
5. Research and Markets, *Detailed Analysis of the Lifestyles of Health and Sustainability Consumer Market in Australia*, Mobium Group, Dublin, Aug 2007, http://www.researchandmarkets.com/reports/c86981 (retrieved 27.10.2008).
6. Anonymous, *Agricultural Ecosystems, Facts and Trends*, World Business Council for Sustainable Development/IUCN, Gland, Conches-Geneva, 2008.

7. Stern review on the economics of climate change, Annex 7g: *Emissions from the agriculture sector*, http://www.hm-treasury.gov.uk/stern_review_final_report.htm (retrieved 22.11.2008).
8. F. Bertrich, *Kulturgeschichte des Waschens*, Econ-Verl., Düsseldorf, Wien, 1966.
9. E. Smulders, *Laundry Detergents*, Wiley-VCH, Weinheim, 2002.
10. E. Smulders, W. von Rybinski, E. Sung, W. Rähse, J. Steber, F.Wiebel and A. Nordskog, in *Ullmann's Encyclopedia of Industrial Chemistry*, Wiley-VCH, Weinheim, online posting 2007.
11. U. Lehner, *Surfactants Surfing Around the World,* 6th World Surfactant Congress, CESIO, Berlin, 2004.
12. M. Stalmans, H. Berenbold, J. L. Berna, L. Cavalli, A. Dillarstone, M. Franke, F. Hirsinger, D. Janzen, K. Kosswig, D. Postlethwaite, T. Rappert, C. Renta, D. Scharere, K. P. Schick, W. Schul, H. Thomas and R. Van Sloten, *Tenside Surf. Det.*, 1995, **32**(2), 84–109.
13. C. Gutzschebauch, G. van Hoof, H. King, H. J. Klüppel, E. Saouter and T. Taylor, 2001, *Life Cycle Assessment of Washing Systems: Results from the AISE-LCA Working Group*, European Climate Change Program (ECCP), Renewable Raw Materials Working Group 5, AISE presentation; E. Saouter, G. van Hoof and P. White, in *Handbook of Detergents*, Part B: Environmental Impact, ed. U. Zoller, *Surfactant Science Series*, **121**, Marcel Dekker, New York, 2004, p. 195.
14. W. Umbach, in *Perspektiven nachwachsender Rohstoffe in der Chemie*, ed. H. Eierdanz, VCH, Weinheim, 1996.
15. F. R. Schroeder, in *Nachhaltige Chemie*, ed. M. Angrick, K. Kümmerer and L. Meinzer, Metropolis Verl., Marburg, 2006.
16. K. Hill, W. von Rybinski and G. Stoll, ed., *Alkyl Polyglycosides*, VCH, Weinheim, 1996.
17. Data Ecoinvent 2.1, www.ecoinvent.org.
18. M. Franke, H. J. Klüppel, K. Kichert and P. Olschewski, *Tenside Surf. Det.*, 1995, **32**, 508.
19. K. Schmitt, Ökoinstut, Freiburg, personal communication.
20. BDEW Bundesverband der Energie und Wasserwirtschaft, Berlin, 2008, http://www.bdew.de/bdew.nsf/id/DE_7DBKHD_Grafiken (retrieved 01.12.2008).
21. Miljøstyrelsen, *Environmental and Health Assessment of Substances in Household Detergents and Cosmetic Detergent Products*, Miljømini-steriet, Environmental Project 615, 2001, http://www2.mst.dk/common/Udgivramme/Frame.asp?http://www2.mst.dk/udgiv/publications/2001/87-7944-596-9/html/helepubl_eng.htm (retrieved 23.11.2008); E. Smulders, W. Rähse, W. von Rybinski, J. Steber, E. Sung and F. Wiebel, *Laundry Detergents*, Wiley-VCH, Weinheim, 2002.
22. http://www.heraproject.com.
23. http://www.aise.eu.
24. http://www.cefic.be.

25. *Concentrated Laundry Detergents Become Latest Trend in Green Retail Packaging*, http://www.sustainableisgood.com/blog/2007/09/concentrated-la.html; *Wal-Mart to Only Sell Concentrated Liquid Laundry Detergent by 2008*, http://www.sustainableisgood.com/blog/2007/09/wal-mart-to-onl.html (retrieved 23.11.2008); *Sustainability Report 2006*, Henkel KGaA, Düsseldorf, 2007, p. 20; M. McCoy, *C&EN*, January 29, 2007, **85**(5), 13–19.

CHAPTER 6

Sustainable Solutions for Nutrition: A Consumer Expectation

SVEN THORMAHLEN

Danone Research, Centre Daniel Carasso, Direction Générale, Route départementale 128, F-91767 PALAISEAU CEDEX, France

6.1 Introduction

Sustainable development has become a subject of animated public discussion in recent years. The term covers economic, social and environmental dimensions. The heightened interest in the subject can be explained by challenges and threats from already observed or merely anticipated changes in climate and environment. Innovative companies have identified the field of sustainable development as a source of differentiation and opportunity. Consumers have become very sensitive to the environmental impact of their consumption habits. They have begun to modify their behaviour and have come to expect from consumer brands the inclusion of sustainability aspects in their product design and marketing. At Danone, sustainable development has always been a point of great importance and attention. Over the years stable practices were developed inside the company: fundamental social principles; a business, environmental and health and nutrition charter; as well as a biodiversity charter; form the company's approach to sustainable growth. It is of importance for the company to build a trust relationship with consumers, to guarantee absolute product safety, to respect the environment and to be concerned with the social

RSC Green Chemistry No. 4
Sustainable Solutions for Modern Economies
Edited by Rainer Höfer
© The Royal Society of Chemistry 2009
Published by the Royal Society of Chemistry, www.rsc.org

and societal impact to the various markets where the business is present. Equally important is the solid foundation of mutually profitable relationships with suppliers. This chapter reports on recent progress in sustainable development efforts, which were undertaken by the company. In the area of agriculture Danone decided to become more involved with upstream processes in order to assure more sustainable processes for the production of key agricultural ingredients. For Danone's water business a comprehensive water source protection policy guides all businesses in all markets.

6.2 Sustainability in Food and Nutrition

Food and drink satisfy the elementary needs of humans. They provide liquids and calories to feed the human metabolism. Eating and drinking also contribute to social togetherness with family or friends. Achieving food security at a global level for this and future generations is a key element of sustainable development. The value chain of sustainability in the food industry[1] as a prerequisite for high-quality food products and nutrition starts with green agriculture, animal husbandry and fishing followed by sustainable food production and processing, packaging, retailing and service.

The world's most pressing problem is poverty and associated hunger (Figure 6.1). 600 million children live in absolute poverty (SCF, Beat Poverty 2003). Over 1 billion people live on less than $1 a day with nearly half the world's population (2.8 billion) living on less than $2 a day (UN HDR, 2003). Eradication of hunger and extreme poverty is now the first millennium goal of the United Nations.[2,3]

Agricultural intensification including growing of bioengineered crop varieties, an agro-industrial model that stresses uniformity and standardized technologies maximizing yields of commercial crops have fuelled a global food system during the last decades. As a result in North and Latin America, for example, agricultural production per agricultural worker has more than doubled between 1980 and 2003 and also crop production per hectare has seen a noticeable increase.[4] On the other hand, hunger is predominantly a problem of rural regions, where families who own smallholdings often possess too little fertile land of their own, live in agriculturally disadvantaged regions or are denied access to credits for equipment or seeds and have no alternative source of employment or income. Ensuring the income and the long-term productivity of the rural production in order to combat hunger[5] and to achieve sustainable agriculture and rural development is a challenge that has been entrusted to the FAO by the Rio Conference. More particularly for the introduction of the micro-credit instrument (giving loans to poor people without sufficient financial security) Muhammad Yunus and Grameen Bank have been awarded the Nobel Peace Price.[6]

Prerequisites for a pro-poor policy in a globalizing world are peace, democracy and good governance. "Good governance" comprises stimulation of the development process by national governments, banishing corruption,

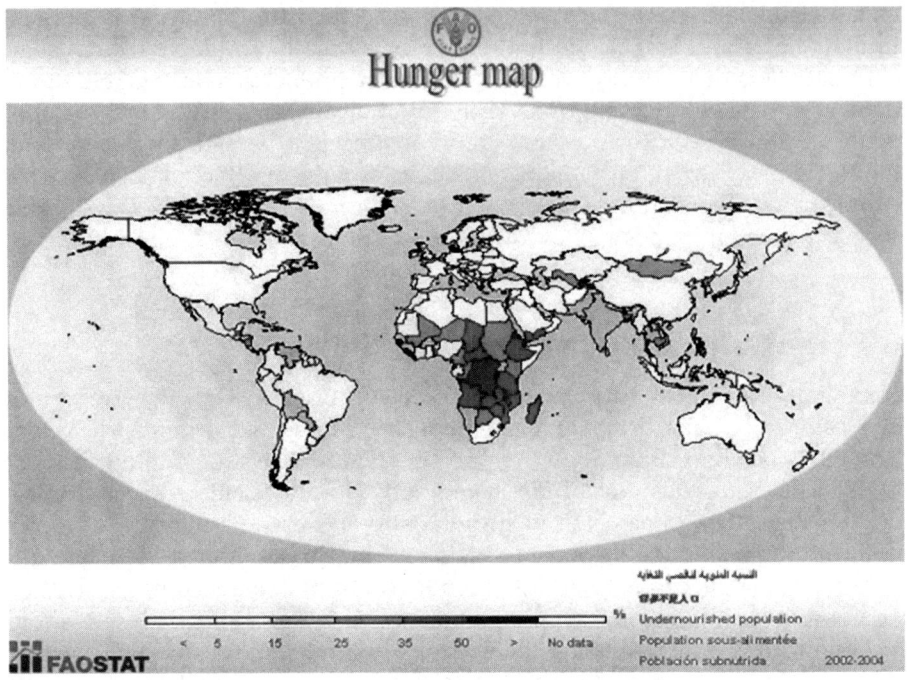

Figure 6.1 FAO hunger map. http://www.fao.org/es/ess/_faostat/foodsecurity/
FSMap/_map14.htm, published 2007.

ensuring civil order, providing access to credits, investing in human capital
(education, health, sanitary services), provision of physical infrastructure and
facilitating institutions (property rights, predictable dispute resolution, information and communication).[7]

Since the BSE crisis, consumer protection is regarded as an essential element of
sustainability in food production.[8] The need to enhance food safety, ensure
consumer protection, strengthen consumer confidence, by setting requirements
for food safety schemes and for improved cost efficiency throughout the food
supply chain, has led to the launch of the Global Food Safety Initiative (GFSI)
in May 2000. The purpose of the GFSI is to monitor issues around food safety
standards (The Global Standards Project). It has developed a Guidance Document, against which food safety standards for manufacturing can be bench-
marked. The benchmark requirements in the Guidance Document are made up
of three key elements: Food Safety Management Systems, Good Practices for
Agriculture, Manufacturing or Distribution and HACCP (Hazard Analysis and
Critical Control Point) and those for the delivery of food safety systems.

The science of nutrition is the knowledge repository for the understanding
of essential nutrients, their quantitative requirements at different life stages,
the mechanisms behind their essentiality and the pathological phenotypes

experienced by populations who fail to consume sufficient quantities of them.[9] But the contribution of diet to health is more than the sum of its chemical components and goes beyond molecular science; it is more than the intake of calories/kilojoules and nutrients recognized as essential. In 1995 FAO and WHO sponsored an expert consultation on the preparation and use of population-specific Food-Based Dietary Guidelines (FBDGs) advocating food cultural differences, population genetic differences, the broad non-nutritional role of food as well as physico-chemical properties and phytochemicals which confer human biological advantage.[10,11] Furthermore, food diversity, including food processing variety, besides food adequacy and quality confers advantages regarding health and survival. The traditional Mediterranean diet (notably Spanish, Italian and French cuisine), for example, has a protective myocardial effect and reduces the rate of recurrence after a first myocardial infarction as has been demonstrated by the Lyon Diet Heart Study.[12]

The following three concrete examples are chosen to give the reader an impression of the variety of activities a food company undertakes in the area of sustainable development. The first example shows efforts in the improvement of milk procurement in Turkey, the second example explains improvements in cow feed quality in France and the third example demonstrates efforts and progress in the responsible and sustainable exploitation of the mineral water source in Evian, France.

6.2.1 Sustainable Milk Procurement in Rural Turkey

Milk is a staple food but it is also a key agricultural raw material. Danone transforms milk into dairy products such as yoghurts, fresh cheeses and desserts and also manufactures probiotic milk-based products such as Actimel and Activia. For these products fermentation is done using specially selected probiotic cultures. These are living bacteria that, when consumed in sufficient quantity, convey concrete health benefits to the consumer. Approximately 4.5 billion litres of milk are transformed per year in Danone factories worldwide. Fresh dairy products require refrigeration, which is why most of them are manufactured in the country where they are being marketed. One fast-growing and dynamic market for dairy products is Turkey. After a successful launch of some of Danone's major brands in Turkey in 2002, the local team was soon confronted with problems in milk procurement. Between 2002 and 2008 milk quantities in Danone Turkey had already doubled and future volume growth had to be secured. At the same time, the local team had to ensure that Turkish milk complies with EU guidelines for milk quality, cow registration, recording of illnesses and other items. The major threat to future growth came from the large proportion of isolated small family farms in Turkey's milk supply. These farms are largely characterized by small numbers of cows, often just one single cow, and low productivity with less than 2,000 litres per cow per year – one of the lowest rates in Europe. For comparison, a German cow produces more than 6,000 litres of milk per year, a French cow produces 5,700 litres per year. Bacteriological quality in milk from small Turkish family dairies was also poor compared to other European countries. In order to improve this situation Danone Turkey had to get involved upstream in the national milk supply. The objectives of the milk development program were ambitious. Four projects were run from 2002 to 2007 under a program called "Project Village Farm", which tried to consolidate individual family farmers into small milk producing co-operatives, to improve professionalism in milk farming and to elevate quality. This program was composed of four projects. Between 2002 and 2004, during the first phase of the program, basic steps were undertaken by Danone Turkey in collaboration with Turkish milk farmers in order to improve milk quality. This was achieved by implementing a quality premium system which was designed to give incentives for good quality and to penalize poor quality. No such quality management system existed among Turkish milk farmers prior to this period. The microbiological burden of raw milk in Danone Turkey's milk collection was diminished seven-fold during this period. Regular antibiotic controls on milk trucks helped manage antibiotic use in milk farms. Cold chain improvement allowed extension of shelf life of finished products. Training programmes for dairy farming professionals, in combination with quality improvement in cow feed, helped to increase the milk's protein content. The second phase of this project, which started in 2004 and is still under way, has a double objective. On one side it aims to further improve milk quality in Turkey (reaching EU standards), and to secure volume growth and best cost. On the other side, this project creates better income for Turkish dairy farmers, increases their professionalism and improves social life among Turkish farmers and their families. In collaboration with Cargill, DeLaval and Ecolab, Danone Turkey implemented a program designed to provide farmers with cheaper meals. A credit support program was launched with the Turkish Deniz Bank in order to give farmers access to affordable credit and to improve livestock. Another project was designed to develop new agricultural land in Turkey. Finally, in collaboration with the Turkish Ministry of Education, Danone Turkey helped farmers to build schools in dairy farming villages.

Four years into the projects the number of small-scale dairy co-operatives, federated by Danone Turkey, has increased from 111 in 2005 to 172 in 2007 and will be greater than 200 by the end of 2009. Total milk from these co-operatives has risen from 36,912 mto in 2005 to 62,414 mto in 2007. The number of cows per farm has risen to 31 on average in 2007, and productivity has gone up to 6,850 l per cow per year with improved protein content of 3.31 g l^{-1} and greatly improved bacteriological quality.

The following pictures illustrate this project. The example taken is from the village of Yenikoy. The photos were taken on May 8, 2008. The Yenikoy project represents an investment of 110,000 € plus a 50% incentive for milking equipment from the Turkish government. A long-term loan for 75,000 € was given from Deniz Bank over 4 years. Payback of this loan is planned to come from Danone quality and productivity premiums, Figures 6.2–6.4.

Figure 6.2   Central milking parlour.

Figure 6.3   Central feed producing unit.

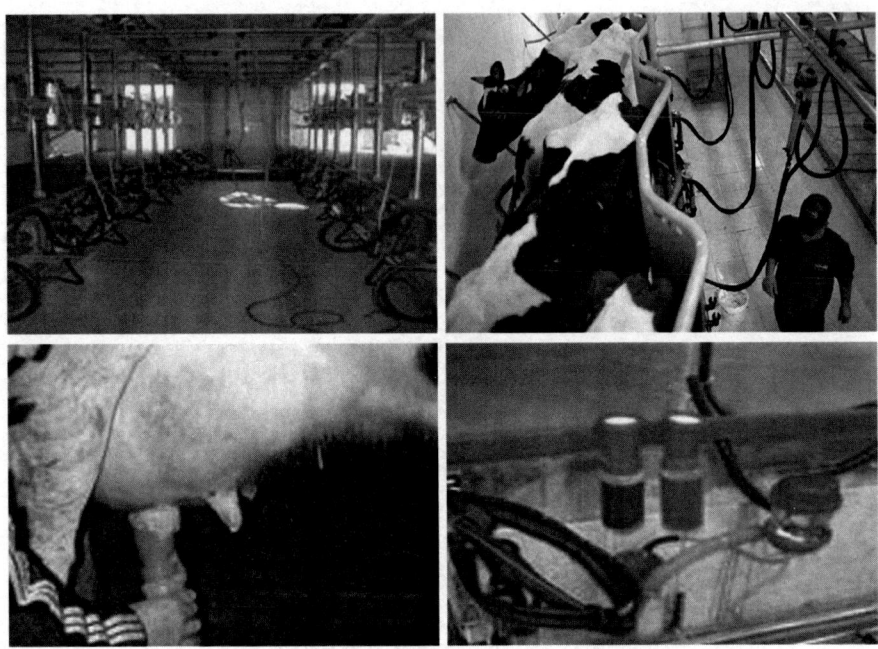

Figure 6.4   The Yenikoy milk co-operative in 2008.

6.2.2   Sustainable Cow Feed in France

Danone France, one of Danone's largest business units, decided to get involved with upstream dairy farming practices for different reasons from those described above. Danone France has strong traditional links with the French dairy farming community where the company has developed exclusive milk collectives, which work exclusively for Danone. France is Europe's leading milk producing country with a longstanding history of dairy farming. Traditional agriculture has in large parts given way to large-scale farming practices at very high productivity levels. In some parts of the country agricultural practices can be described as intense. Intensive farming practices can badly influence cows' health and welfare. Furthermore, significant changes in feeding practices today have largely modified the lipid composition of milk fat. This modified lipid composition is no longer in accordance with generally recommended dietary intakes. In cooperation with the French foundation Bleu-Blanc-Coeur, Danone France has started an initiative to introduce extruded flax into the cattle feed in order to improve the health of the animals and at the same time improve the lipid composition pattern. This project was undertaken with a double objective in mind: improvement of animal health and welfare, and improvement of milk fat composition.

Dairy farmers in France, like anywhere else, are concerned about productivity and profitability of their enterprises. Good breeding practices and good farm management practices must be accompanied by good feeding programs. Feeding for high milk production is highly efficient.[13] Nutrient requirement for maintenance comprises a smaller portion of the total requirement of high-producing cows. Feeding dairy cows for efficient production involves supplying energy, protein, minerals, vitamins and water in sufficient amounts. The forage program is selected by the farmer. Danone France provides best practices to their milk collective farmers in order to improve yields and quality in the framework of responsible farming practices (*agriculture raisonnée* in French). There is no single best forage program; all have advantages and disadvantages. Pasture, the traditional cow feed, has low harvesting costs and provides a convenient way of spreading manure. But pasture is not well adapted to large herds in confinement systems. Furthermore, quality varies through the season, which can influence the quality of finished dairy products. Hay is somewhat close to pasture. It can be high in protein content. It requires low equipment levels and also low facility costs. Some hay is desirable to maintain good rumen function. High labour requirement, waste problems and greater risk from weather at harvest make hay only the second choice behind crop silage. Crop silage offers the advantage of high tonnage; it is well adapted to automation and provides reduction of waste. There is, however, the requirement of high upfront investment. Crop silage can be low in protein and there is a considerable risk of nutrient loss due to spoilage. Grains commonly used in dairy feed include corn, barley, oats and wheat. By-product feeds include protein supplements, brewers' dried grains, corn gluten, whole cottonseed, soy hulls, wheat bran and others. A problem associated with intensive dairy farming is the control of acidity in the cow's rumen. Digestion upsets may occur if the cow feed is not sufficiently buffered. Sodium bicarbonate is often added at 1.5% to cow feed.

Flax (also known as lin) was a traditional ingredient in cow feed frequently used in earlier times. Flax is native to the region extending from the eastern Mediterranean to India. It was extensively cultivated for the first time in ancient Egypt. It is grown both for its seeds and for its fibres. Various parts of the plant have been used to make fabric, dye, paper, medicines, fishing nets and soap. For a long time flax was used as an important component in animal feed, but today it has almost entirely vanished from cattle feeding. Only a handful of horse or steer breeders, all fervent advocates of this century-old tradition, continue to feed linseed to their animals. They do so because they know about the particular nutritional value of flax, its natural richness in omega-3 fatty acids (see chapter 9.1.4.2.18), for which they are willing to put in the extra effort. Linseed cannot be used in its crude state. Horse breeders need to boil the flax seeds before feeding them. Linseed meal is used in steer feeding. Raw untreated linseeds are never used because of the difficulty animals have in digesting them. But farm-side boiling and milling are time and labour intensive, and so linseed has been replaced in general farming over time by soy and corn. Soy and corn are easier to cultivate and to harvest and their grains do not require boiling or milling. They are economically more favourable. At the beginning of the twenty-first century linseed has almost completely vanished from animal feeding. As a consequence of changed feeding practices the lipid composition of milk and animal fat has changed over time and today has become poor in omega-3 and overly rich in omega-6 fatty acids. The following paragraphs explain the importance of the ratio of ω-6 : ω-3 for human health and animal welfare.

Fats and lipids are multi-functional nutrients. They supply energy to the body. They are also structural building blocks for the formation of cellular membranes. In addition, their metabolites affect numerous important physiological systems such as the cardiovascular, the immune and the nervous systems. Official nutritional recommendations invariably stress the importance of lipid control as a very important factor in prevention of poor nutrition. Such recommendations primarily include quantitative aspects of lipid consumption in relation to total energy need (no more than one-third of the total energy contribution should come from lipids). Increasingly they also reflect on the qualitative aspects of nutritional lipids, in particular with regard to essential fatty acids (EFAs) and poly-unsaturated fatty acids (PUFAs). Insufficient PUFA supply can deregulate the metabolism.

Sources of lipids in the food of Europeans are mainly of animal origin. Approximately two-thirds of lipids come from livestock products. Dairy lipids and lipids from pork are the dominating animal sources in Europe, followed by lipids from eggs, beef, poultry and fish. The remaining third is from plant oils with great differences occurring between the individual European countries. Rapeseed (*Colza*, in French and Spanish) and sunflower oils, very different in composition from one another, are the most used oils in Europe. Olive oil is increasingly present in table oils, but it is quasi-absent from industrially prepared foods (margarines, sauces, biscuits and prepared dishes). Various European consumption surveys have shown that lipid consumption in Europe has exceeded recommendations for the past decades. According to these studies, total energy contributions from lipids today reach between 35% and 45%. These surveys also reveal considerable changes, both qualitative and quantitative, that have occurred in lipid consumption over the past 40 years. Table 6.1 presents data from lipid consumption in France. These data are approximate as they have been reconstituted from different published surveys.

The data presented here are representative of Western European consumption patterns. There is an important gap between nutritional recommendations and actual consumption. In addition to overall excessive lipid consumption in Europe there is over-consumption of saturated fatty acids, as well as an excessive ratio of ω-6 : ω-3 fatty acids. Based on many convergent data from different epidemiological studies, this imbalanced consumption pattern is likely to be at the origin of the steady increase of "lifestyle" related illnesses observed in Europe. The extent of the gap between consumption and recommendations is notable and calls for corrective measures such as nutritional education, a better choice of plant oils and a better control of animal lipids. Since animal lipids represent two thirds of all consumed lipids, it is of particular importance to improve their composition. For several decades different research teams have studied the link between animal feed and lipid composition of their products.

Table 6.1 Recommended and reported daily consumption of Essential Fatty Acids.

| | Daily consumption of major Essential Fatty Acids (EFA) and EFA ratios ($g\ day^{-1}$) | | | | | | | |
	$EFA\ g\ day^{-1}$	C16:0	C18:1	C18:2 n-6	C18:3 n-3	P4 FA n-3	LA/ALA	C16:0/ALA
TOTAL	104	24.6	30.6	17.6	1.0	0.74	17.6	24.6
Recommended daily ration (ANC)	81	/	/	10	2	/	<5	/
Reported daily ration over recommended dose (in percentage)	+28%	/	/	+76%	−50%	/	+252%	/

LA = linoleic acid; ALA = α-linolenic acid

Their studies have shown that animal fat composition depends almost exclusively on the animal feed, regardless of the particular species involved. Changes in cattle feeding practices directly influence the animal fat composition. For example, conventional beef contains a 4 : 1 ω-6 : ω-3 ratio while grass-only diets produce a 2 : 1 ω-6 : ω-3 ratio. As another example, progressive introduction of cereals into dairy cattle feed has increased the level of palmitic acid, from 20% in grass-fed dairy cows to 40% in cereal-fed cows.

In collaboration with the French foundation Bleu-Blanc-Coeur, and together with the company Valorex, Danone France succeeded in introducing 5% of a special linseed preparation into the daily ration for cows delivering milk for Danone's fresh cheese "Jockey". This product, popular in France, is manufactured in the Danone Ferrières-en-Bray factory in the north of France. Valorex has been chosen as a supplier for their particular variety of linseed, named *Tradi-Lin*®, which is extremely rich in Omega-3 fatty acid content. Approximately 60% of total fatty acids in this feed supplement are ALA Omega-3. A patented high-pressure cooking and extrusion method transforms raw linseed into extruded linseed pellets which are void of any toxic components while being at the same time easily digestible for cows (Figures 6.5–6.8).[14]

Figure 6.5   Milling of linseeds.

Figure 6.6   Vapour treatment and maturation.

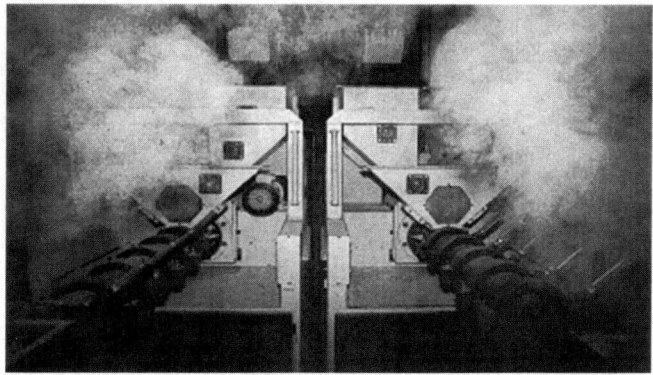

Figure 6.7   Compressing extrusion.

Figure 6.8   Extruded linseed pellets.

As already mentioned above, French dairy farmers are concerned about productivity and profitability of their enterprises. They needed to be convinced of the economic advantages of linseed fortification of cow feed. The initial increase in feed cost is off-set over time by higher productivity, better health of the cows and higher price obtained for omega-3 rich milk. The role of the foundation Bleu-Blanc-Coeur was to share experiences and best practices regarding *Tradi-Lin*® fortification with French farming communities. Comparative analysis between Danone's Jockey and similar products competing in the marketplace clearly demonstrate the improved lipid profile of milk from linseed-fed cows, Table 6.2.

Of all analyzed products, only Jockey does in fact correspond to the recommended lipid profile.

In addition to the improved lipid profiles indicated above, linseed supplementation of dairy feed offers another advantage. Farmers who participated in the program soon realized improvements in health and welfare in their animals.

Table 6.2   Saturated and unsaturated fatty acids in four fresh cheeses.

	Target	Jockey	Competitor 1	Competitor 2	Competitor 3
Saturated Fatty Acids	<68	65.6	72.4	70.3	70.9
Unsaturated Fatty Acids	>30	33.8	27.3	28.8	28.4
Omega 3	>0.6	0.7	0.3	0.3	0.5
Omega 6	>0.8	2	1.7	1.8	1.7
Omega 6 : Omega 3	<5	2.9	5.7	6	3.4

in % of total fatty acids

These improvements are due to the better digestibility of linseed enriched feed. Dairy cows ferment their food and produce fermentation gases, notably carbon dioxide and methane. One cow produces up to 900 litres of CO_2 per day along with 600 litres of CH_4. Both are greenhouse gases. Methane's contribution to atmospheric warming is 21 times that of CO_2 per molecule. Patrick Herpin and his group at the French National Institute of Agriculture (INRA) have shown that linseed feeding in cows reduces methane emission by 37%. In view of the actual debate about greenhouse gas emissions, this reduction appears to be highly significant. Danone Research collaborates with the INRA group to study possible gas reductions in the Danone milk collective. Extension of the linseed project to Danone's entire milk collective is under current investigation.

6.2.3   Sustainable Exploitation of the Evian Mineral Water Source

Evian is a mineral water from Danone coming from several sources near Evian-les-Bains, on the south shore of Lake Geneva. The virtues of Evian mineral water were discovered in 1789 by the Marquis de Lessert. Bottling from the spring started in 1826 and the first mineral water company was founded three years later. The French Ministry of Health reauthorized the bottling of Evian water on the recommendation of the Medicine academy in 1878. Evian water began to be sold in glass bottles in 1908. In 1969 the first PVC bottle was launched. Danone, at the time called BSN Group, took control of Evian in 1970. Evian switched to collapsible PET bottles in 1995 and to 50% recycled PET in 2007. Over the years Evian established its name as a natural and healthy beverage throughout the world. During the time of Evian's commercial development, Danone paid attention to sustainable growth and the relationship with all stakeholders.

Danone is historically a pioneer in plastic recycling with collection programs that were started in 1970 under the title "Vacances Propres". Today Danone is involved with more than 20 national and international partners in the domain of packaging recycling. At the same time continued efforts were made to reduce the weight of packaging. Today, one empty bottle of Evian weighs 30 grams compared to 43 grams in the early days. Evian has undertaken a series of other environmental initiatives that include: the creation of the Evian Water Protection Institute, the incorporation of recycled PET plastic into the bottle sizes that receive the most sales, and collaboration with the RecycleBank, an award-based company that gives participating households redeemable points according to the amount of materials they recycle.

But despite these continued efforts and improvements, consumers of bottled mineral water these days are accused of committing an environmentally irresponsible act. Recent newspaper headlines sent warning messages to its readers: "Give up bottled water to save the planet", "No to bottled H_2O", "Just turn on the tap and join the drive to stop drinking bottled water". At Danone we remain convinced that Evian, being a pure mineral water, unadulterated and free from chemicals such as nitrate, chlorine or heavy metals, presents a valuable choice to consumers who are concerned about their health and well-being. At the same time, now more than ever, Danone needs to explain actions and initiatives undertaken by the company to protect the source and the water quality of Evian and to make sure that the natural underground reservoirs are not depleted. Consumers have begun to take interest in the nature and origin of Evian mineral water. It is important to explain that the groundwaters that generate Evian natural mineral water are not fossil waters; they take part in the global water cycle. They find their origin in the infiltration of water coming from rainfall and snowmelt in a specific zone called their catchment area. Their main difference to surface waters is their residence time. While the surface run-off needs only a few days to a few months to reach the sea through rivers, the groundwaters usually stay several months to several decades underground before reaching the surface. This long transit allows them to self-purify through several processes (filtration, adsorption, buffering, bioremediation, *etc.*) and to catch minerals from their aquifer (hydrolysis because of their CO_2 content). Therefore these waters can naturally reach the drinking water standards (spring waters) or even get a specific constant mineral content (natural mineral waters) that can favour human health, as is the case for Evian. Nevertheless, as with all other natural resources, this one needs to be properly managed and protected to ensure sustainability. A company policy was put in place in 2004, followed by a comprehensive groundwater protection plan. These elements were created to ensure sustainable management of the natural and patrimonial resources. This policy is based on the following objectives: guarantee the purity and quality of the water in order to ensure the quality of its products and the safety of its consumers, guarantee the sustainability of its water resources, exercise social responsibility at the local level and protect and promote the natural heritage of the sites where we operate. To this end, the divisions, subsidiaries and production facilities of Danone apply the policy by implementing the following measures: understand the hydrogeology of water resources and the natural and human environment of the sites, respect the natural recovery capacity of the water resources, develop lasting relationships with local communities and contribute to local development, train designated managers and monitor and evaluate results regularly. The policy has been translated into five related action fields: knowledge of the resource and its environment, management of the resource, actions of protection, local development and organization. The target is to acquire a good knowledge of the groundwater system and its surface environment in order to carry out the next actions in a relevant way.

First, the focus is on the spring for which all characteristics must be established (technical, geological, hydraulic), plus a local mapping of the neighbouring activities abstracting groundwater; next, a geological report from a local hydrogeologist must have defined all possible vulnerable areas around the spring. In this area the land use and the human activities must be inventoried. Finally hydrogeological studies must provide comprehensive and congruent information of the groundwater system (catchment area, transit time, conceptual model, vulnerability mapping, *etc.*) and a socio-environmental assessment of the whole concerned area bringing understanding of the potential hazards, risks of conflicts and socio-economic levers.

The Evian company understood early the need of management and protection of its mineral water resource. Thanks also to its local economic weight it has been able to introduce, in 1992, a model of local cooperation which is still in operation.

6.2.3.1   Knowledge of the Resource

Good geological and hydrogeological knowledge are important. Where does water come from and how? Several studies including various methods

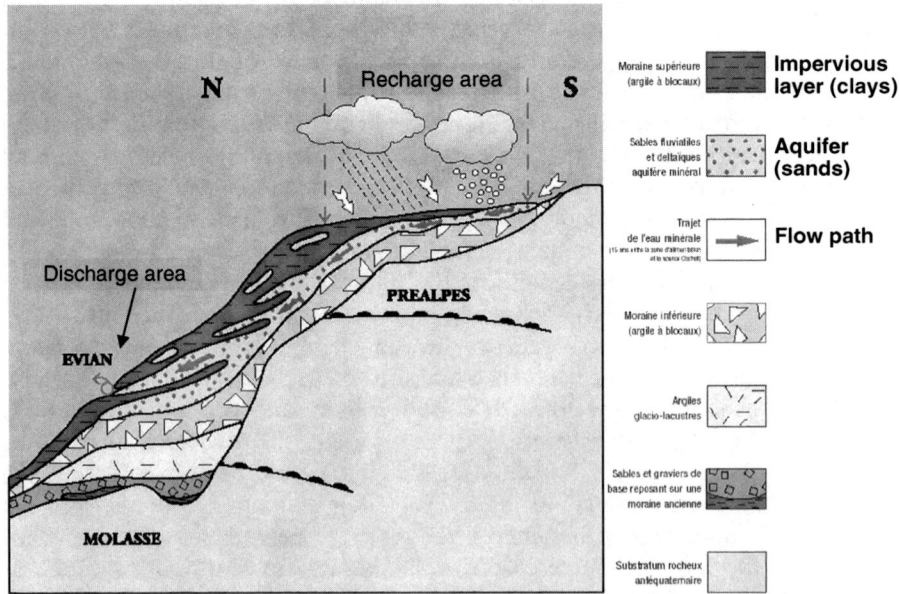

Figure 6.9   Knowledge of resource and environment: cornerstones for sustainable management.

(geohydrochemistry, isotopy) have been done in order to answer these questions. Geological maps of the aquifer were drawn up (Figure 6.9). A definition of the catchment area, a description of all water flows and a full description of the discharge area were established. The residence time of water in the Evian water system was calculated to be 50 years on average. All vulnerable areas were identified and drawn on a map. A complete environmental diagnosis, current land use and an inventory of risk activities allow constant monitoring of the water system.

6.2.3.2 Resource Management: Sustainable Exploitation

Based on flow rates, water levels and other parameters, a continuous hydrodynamical and hydrochemical assessment is made. Annual archives allow proper forecasting for future exploitation. Maximum flow rates, minimum water level and chemical and microbiological standard give the frame for sustainable use of the Evian source. Figure 6.10 shows the water balance in a simplified form.

6.2.3.3 Protection of the Resource

The natural geological protection, that is the integrity of the impermeable layers of the reservoirs with a thickness of tens of metres and an average transit

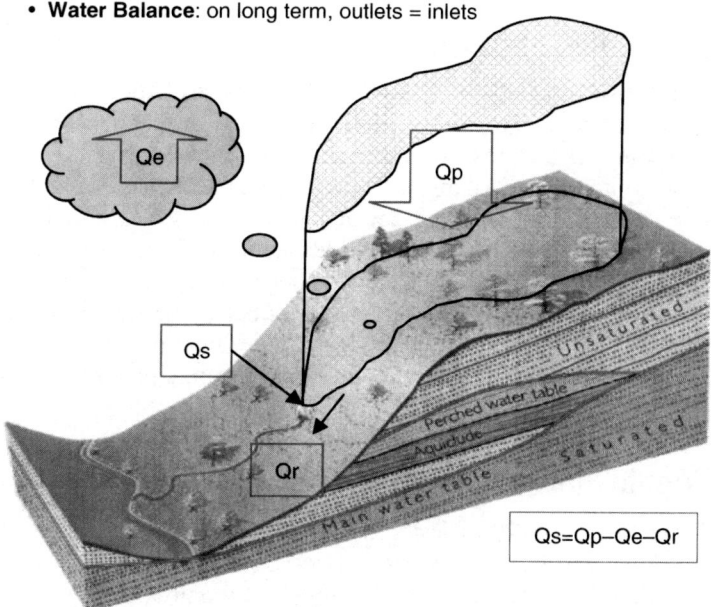

• **Water Balance**: on long term, outlets = inlets

$$Qs = Qp - Qe - Qr$$

Figure 6.10 Illustration for the Water Balance at the Source, $Qs =$ amount of surface water, $Qe =$ evaporation, $Qp =$ precipitation, $Qr =$ run-off.

time of 20 years, is constantly monitored. Particular attention is drawn to the installations at the catchment sites. This is an area of specific technical know-how. Quality management includes sanitation protocols and detailed hazard analysis of critical control points. A sanitary defence perimeter is described around the discharge area. This is a physically enclosed zone prohibiting all activity and entry of unauthorized persons. It corresponds to the size of the zone owned by the company. The recharge area, the zone where water infiltrates the reservoir, is protected in close collaboration with the local community. A special association in Evian works with local partners to implement and finance the protection of this sensitive zone. A specific convention has been signed with local farmers to stop the use of phyto-sanitary products. Buildings for livestock were constructed with the help of this association in order to protect the source from animal waste. A risk management system was established together with 12 villages in the vicinity of the Evian source. Risks related to human activity such as drainage works, waste treatment, fuel storage, wetland management, road maintenance and others are identified and controlled in this local network. Water exploitation limits have been established in conventions signed by four villages in order to promote sustainable management of the communal drinking water supply. Specific efforts were made to measure and influence local economic and social impacts. Evian contributes largely to the industrial employment of the region. Tourism is developed to introduce the Evian area to Evian consumers. Transportation methods are constantly optimized.

6.2.3.4   The "Observatory"

This institution was created to verify the effectiveness of the protection actions since there is, at Evian, an average delay of 20 years from the recharge area (where the protection actions are implemented) and the discharge area (where there are the sources and drillings). The Observatory is based on the measure of certain physical, economic and social parameters, which can highlight the evolution of risks. These parameters include chemical parameters of streams and sources of the area, agricultural parameters such as the quantities of fertilizers, phyto-sanitary products, de-icing salts and number of signed local conventions and agreements. This Observatory allows the continuation or re-adjustment of engaged actions in order to ensure proper water resource protection at all times.

6.3   Conclusion

Three different examples of projects for sustainable development inside a food company were illustrated. They have two important elements in common: first, they are strictly business oriented and firmly embedded with the overall commercial strategy. Second, they are mid-term to long-term projects, which require continued motivation, resources and energy from the entire organization. It is fair to say that in today's world sustainability aspects in business projects are no longer "nice-to-haves", but they have become central elements for successful business project management.

References

1. C. J. Baldwin, ed., *Sustainability in the Food Industry*, Institute of Food Technologists Series, Wiley-Blackwell, Oxford, 2009.
2. United Nations Development Programme, *Millennium Development Goals*, Millennium Declaration, UN Millennium Summit (Sept. 2000), http://www.undp.org/mdg/basics.shtml (retrieved 19.09.2008).
3. M. L. Wahlquist, *Publ. Health Nutr.*, Sept. 2005, **8**(6A), 766.
4. FAO, ed., *Summary of World Food and Agricultural Statistics*, Rome (2005), http://www.fao.org/ES/ESS/sumfas/sumfas_en_web.pdf (retrieved 22.09.2008).
5. SUSTAINET, Sustainable Agriculture Information Network, *Combating World Hunger through Sustainable Agriculture*, gtz, ed., http://www2.gtz.de/dokumente/bib/04-5888.pdf (retrieved 19.09.2008); M. A. Altieri, P. Rosset and L. A. Thrupp, *The Potential of Agroecology to Combat Hunger in the Developing World*, http://www.cnr.berkeley.edu/~agroeco3/the_potential_of_agroecology.html (retrieved 19.09.2008).
6. O. D. Mjøs, *The Nobel Peace Price 2006: Presentation Speech*, Oslo (2006), http://nobelprize.org/nobel_prizes/peace/laureates/2006/presentation-speech.html (retrieved 21.09.2008).
7. IFPRI (International Food Policy Research Institute), *Putting globalization to work for the poor*, Summary Note, R. S. Johnson, Panelist, in 2020 Vision: Sustainable food security for all by 2020, Bonn (2001), http://www.ifpri.org/2020conference/PDF/summary_johnsonrobbin.pdf; http://www.ifpri.org/pubs/books/2020conpro/ch11.pdf (retrieved 23.09.2008).
8. K. Bergmann, in K.-M. Brunner and G. U. Schönberger, Hrsg., *Nachhaltigkeit und Ernährung: Produktion – Handel – Konsum*, Campus Verl., Frankfurt, New York, 2005.
9. D. G. Lemay, C. J. Dillard and J. B. German, *Food Structure for Nutrition*, Special publication – Royal Society of Chemistry, 2006, **308**, 1.
10. World Health Organization/FAO, *Preparation and use of food-based dietary guidelines*, WHO Technical Report Series No. 880, Geneva (1998).
11. M. L. Wahlquist, *Requirements for healthy nutrition: Integrating food sustainability, food variety, and health*, World Congress of Food Science and Technology No. 12, Chicago, *J. Food Sci.*, 2004, **69**(1), CRH16, http://www.iuns.org/features/requirement_for_healthy_nutrition.pdf.
12. M. de Lorgeril, P. Salen, J.-L. Martin, I. Monjaud, J. Delaye and N. Mamelle, *Mediterranean diet, traditional risk factors, and the rate of cardiovascular complications after myocardial infarction: final report of the Lyon Diet Heart Study*, Circulation 99, 779–785.
13. M. J. VandeHaar, *J. Dairy Sci.*, 1998, **81**(1), 272; M. J. VandeHaar and N. St-Pierre, *J. Dairy Sci.*, 2006, **89**(4), 1280.
14. P. Weill, EP1155626, 2001 (Valorex).

CHAPTER 7

Biomass-based Green Energy Generation

MARTIN KALTSCHMITT[a,b] AND DANIELA THRÄN[a]

[a] German Biomass Research Centre, Torgauer Str. 116, D-04347 Leipzig, Germany; [b] Hamburg University of Technology, Institute for Environmetal Technology and Energy Economics, Eissendorfer Str. 40, D-21073 Hamburg, Germany

7.1 Introduction

Energy from biomass currently contributes on a world-wide basis 10 to 12% to the gross primary energy demand.

Due to geographical, economic and climatic differences, the share of biomass energy in relation to total energy consumption differs widely between different countries, ranging from less than 1% in some industrialized countries like the United Kingdom to significantly more than 50% in some developing countries in Africa and Asia. Biomass is by far the most important renewable energy source, being significantly larger in energetic terms than the second largest, hydropower. This becomes obvious in Figure 7.1, which shows the energy consumption exemplarily for Germany subdivided into the different energy carriers contributing to cover the given energy demand. Overall on a global scale the energy from the oldest fuels (*i.e.* biofuels) utilized by human beings is much larger in absolute terms than the energy from the newest, nuclear fuels. In terms of numbers of people, biofuels dominate the world picture, for it is probably true to say that most people in the world still rely on biomass fuels for most of their energy, a situation that has not changed

RSC Green Chemistry No. 4
Sustainable Solutions for Modern Economies
Edited by Rainer Höfer
© The Royal Society of Chemistry 2009
Published by the Royal Society of Chemistry, www.rsc.org

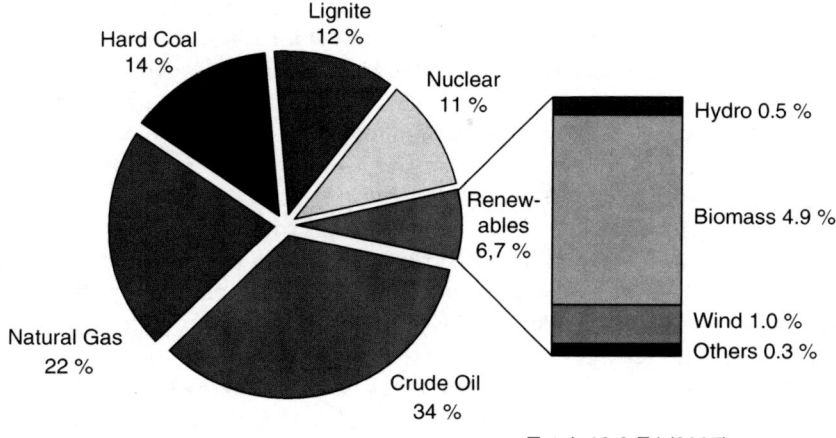

Figure 7.1 Primary energy consumption Germany (2007).[1]

since the mastery of fire a hundred thousand years ago. This is due to the available conversion technology, the relatively low costs and the easy access of biofuels, such as wood, dung and crop residues, in the poor rural areas of developing countries where about half of the world population still lives. However, soot from the burning of biomasses, largely wood and animal dung used for cooking by Asia's poor, has recently been identified by radiocarbon analysis as causing the brown cloud that hovers over the Indian Ocean and South Asia every winter. As well as being linked to global warming, the brown cloud is believed to lengthen droughts, exacerbate monsoons and further melt the Himalayan glaciers, which currently provide fresh water to billions of people.[2] In addition, however, more and more processed biofuels such as wood chips, wood pellets, biogas, biodiesel and bio-ethanol play an important role in many regions where they perform favourably compared to the given alternatives. The potential size of total biofuel resources is considerable throughout the world. Depending on how they are utilized, biofuels also have the advantage of being relatively environmentally sound compared to many other sources of energy, a characteristic that has led to a resurgence in interest throughout the world.

Cooking with biofuel is extremely rare today in rich countries, although there is significant use of biofuel for space heating in regions with easy access to forests. Specific space heat demand is generally decreasing because of improved insulation of old and new buildings due to regulations and other incentives in most industrialized countries. Average living area per citizen in most countries of the western world is still increasing, however, while population itself is not growing significantly or even decreasing. The net result is that space heat demand is likely to remain more or less stable in industrialized countries for the next decade. The household market in developing countries is still increasing due to increasing population and economic growth. The rapid urbanization common in many developing countries acts as a countervailing trend, however, because urban households tend to rely less on biofuels than do rural households. The net result is probably similar to that in developed countries, *i.e.* overall consumption of household biofuels is relatively stable, although in this case with large regional variations. China may well be an important exception because there seems to be an on-going trend to substitute coal for biomass in rural household applications.

The demand for electric energy is increasing in industrialized as well as in developing countries. Therefore, electricity production from biomass is often seen as one of the most important future markets for biomass worldwide. The same is also true, in principle, for the production of transportation fuels from biomass (see Chapter 8).

More broadly, expanded use of biomass in clean applications offers a way toward more sustainable energy systems in all countries, an issue of growing international concern. Because many renewable energy sources offer the potential to provide useful energy with reduced emissions of *e.g.* greenhouse gases and be additionally environmental advantageous compared to fossil fuel energy, their attractiveness has been increasing during the last decade. In particular, biomass is considered as environmentally and climatically sound because:

- if operated with high combustion efficiency and produced in a sustainable way, biomass fuel cycles are neutral to the climate, *i.e.* the carbon is completely recycled and there is – if no other greenhouse gases (like CH_4, N_2O) are released – no net increase of greenhouse gases in the atmosphere.;
- biomass combustion normally has low sulphur and nitrogen emissions with consequently small contributions to secondary particle formation and acid precipitation downwind;
- biomass has only a few other intrinsic contaminants that can pollute the environment such as the toxic elements mercury, lead, arsenic, fluorine, *etc.* found in some fossil fuels;
- in many circumstances, the residues produced during the processing and combustion of biomass (*i.e.* fermentation residues, ash) can be usefully recycled back onto the production area from which the biomass has been grown.

Among the various possibilities for exploiting solar energy, biomass shares with a hydropower reservoir the characteristic of being a form of "stored" solar energy. This offers significant advantages over wind or direct solar energy (like photovoltaic[3]), which need to be linked with expensive storage systems to be available reliably throughout the day, month and year. In addition, being a chemically based fuel, the energy density of biomass is fairly high, although not as high as that of most fossil fuels.

Biofuel is rarely directly affected by energy crises, being an indigenous energy carrier (*i.e.* biomass is, in most cases, produced close to the place where it is used) characterized in most cases by short supply chains with low

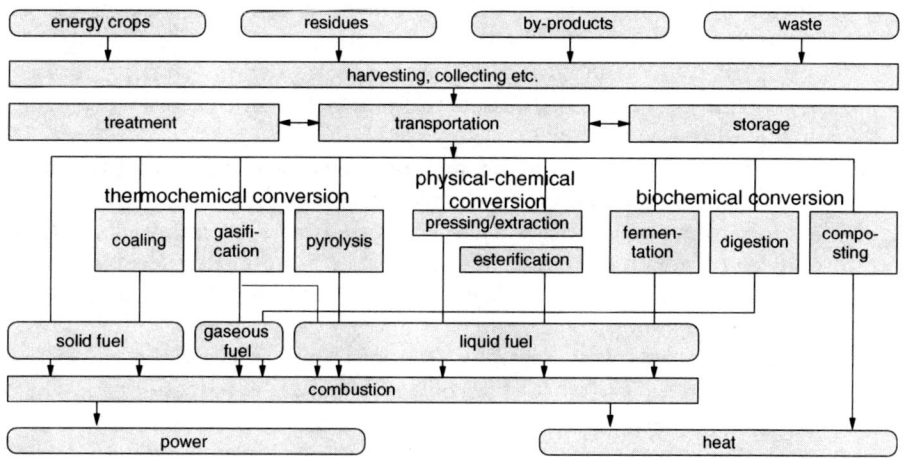

Figure 7.2 Possibilities to provide heat and/or power as well as fuels from biomass.[4]

risks of failure. This security of supply is increased by growing international markets for some types of biofuels with relatively high energy density (pellets, bio-ethanol, *etc.*). Also the production and utilization of biomass is widely accepted by the greater public (unlike nuclear and coal power in some countries) and offers benefits for rural areas related to employment, rural infrastructure, the conservation of cultivated areas and hence the attractiveness of rural regions.

Most biomass used for energetic purposes is directly combusted to produce heat and/or power, but a huge variety of additional possibilities are available to provide environmentally sound heat and/or electricity as well as transportation fuels from organic material. The most important conversion routes available now or in the near future will be discussed according to the framework shown in Figure 7.2.

It is useful to consider two major forms of biomass fuel: mechanically processed and chemically processed. For mechanically processed biofuel in which the material is used essentially in its natural form (as harvested) direct combustion usually supplies heat for cooking, space heating or electricity production, although there are also small- and large-scale industrial applications for steam raising and other processes requiring low-to-medium temperature process heat.

Chemically processed biofuels with clearly defined fuel characteristics allow for a technically easy and environmentally sound conversion into the desired useful energy. Such biofuels can then be traded easily and used with fewer problems to meet a supply task efficiently and comfortably. To ensure this, the following conversion routes are available (Figure 7.2).

- Thermo-chemical conversion summarizes all conversion processes of biomass into a solid, liquid or gaseous fuel based on heat. Therefore gasification, pyrolysis and charcoal production count as these processes. Of these possibilities only charcoal production is state of technology and widely used so far. But gasification in connection with electricity production seems to be a quite promising option which might become technically available in the near future.

- A physical-chemical conversion process provides a liquid fuel based on physical (such as pressing) and chemical (such as esterification) processes. The most important process used by the industry so far is vegetable oil production from oil seed and the transesterification of this vegetable oil to Fatty Acid Methyl Ester (FAME) as a substitute for petroleum-based diesel fuel.
- Bio-chemical conversion summarizes conversion processes based on biological processes. The most important possibilities are alcohol production from biomass containing sugar, starch and/or celluloses and biogas production from organic waste material. Both technologies are state of the art and widely used for energy provision.

These upgraded biomass fuels can be used in engines, turbines, boilers or ovens to provide thermal and/or mechanical energy (*i.e.* heat and power), which in turn can be converted into electrical energy. Additionally, liquid and gaseous fuels can be used directly or after treatment as transportation fuels.

Heat production is so far the most important use worldwide for biomass fuel. Direct combustion devices are widely distributed with thermal capacities ranging from a few kW in household stoves up to heating plants with several tens of MW. The conversion efficiencies vary from 8 to 18% for simple three-stone stoves up to approximately 95% and above for modern heating units with high-end condensing boiler technology. Electricity production – which is of minor importance on a world-wide scale – has been based mainly on the conventional steam cycle with efficiencies around 30% and capacities of a few MW and above.

7.2   Biomass Sources

7.2.1   Properties

Biomass consists primarily of the elements carbon (C), hydrogen (H) and oxygen (O) (Table 7.1). Additionally, trace elements can be found in some types of biomass. For example, straw can contain fairly high amounts of chlorine (Cl) and/or silicon (Si) and rape seed shows a relatively high amount of nitrogen (N). These trace elements can sometimes cause problems. For example, during combustion chlorine is responsible for high temperature corrosion in boilers, silica can lead to boiler slagging and nitrogen can cause high NO_x emissions.

In most practical applications the energy content of biomass is best described by the lower heating value. The lower heating value (LHV; also known as net

Table 7.1 Energy content and concentrations of some elements in untreated biomass compared with coal.[4]

Type of biomass	Heating value[1]	Ash content	Volatile compounds	C	H	O	N	K	Ca	Mg	P	S	Cl
	in MJ/kg	in %	in %	in % of dry substance									
Spruce wood (with bark)	18.8	0.6	82.9	49.8	6.3	43.2	0.13	0.13	0.70	0.08	0.03	0.015	0.005
Beech wood (with bark)	18.4	0.5	84.0	47.9	6.2	45.2	0.22	0.15	0.29	0.04	0.04	0.015	0.006
Poplar wood (short rotation)	18.5	1.8	81.2	47.5	6.2	44.1	0.42	0.35	0.51	0.05	0.10	0.031	0.004
Willow wood (short rotation)	18.4	2.0	80.3	47.1	6.1	44.3	0.54	0.26	0.68	0.05	0.09	0.045	0.004
Bark (softwood)	19.2	3.8	77.2	51.4	5.7	38.7	0.48	0.24	1.27	0.14	0.05	0.085	0.019
Rye straw	17.4	4.8	76.4	46.6	6.0	42.1	0.55	1.68	0.36	0.06	0.15	0.085	0.40
Wheat straw	17.2	5.7	77.0	45.6	5.8	42.4	0.48	1.01	0.31	0.10	0.10	0.082	0.19
Triticale straw	17.1	5.9	75.2	43.9	5.9	43.8	0.42	1.05	0.31	0.05	0.08	0.056	0.27
Barley straw	17.5	4.8	77.3	47.5	5.8	41.4	0.46	1.38	0.49	0.07	0.21	0.089	0.40
Rape straw	17.1	6.2	75.8	47.1	5.9	40.0	0.84	0.79	1.70	0.22	0.13	0.27	0.47
Corn straw	17.7	6.7	76.8	45.7	5.3	41.7	0.65					0.12	0.35
Sunflower straw	15.8	12.2	72.7	42.5	5.1	39.1	1.11	5.00	1.90	0.21	0.20	0.15	0.81
Hemp straw	17.0	4.8	81.4	46.1	5.9	42.5	0.74	1.54	1.34	0.20	0.25	0.10	0.20
Rice straw	12.0	4.4											
Husk	14.0	19.0											
Rye whole crop	17.7	4.2	79.1	48.0	5.8	40.9	1.14	1.11		0.07	0.28	0.11	0.34
Wheat whole crop	17.1	4.1	77.6	45.2	6.4	42.9	1.41	0.71	0.21	0.12	0.24	0.12	0.09
Triticale whole crop	17.0	4.4	78.2	44.0	6.0	44.6	1.08	0.90	0.19	0.09	0.22	0.18	0.14
Rye grain	17.1	2.0	80.9	45.7	6.4	44.0	1.91	0.66		0.17	0.49	0.11	0.16
Wheat grain	17.0	2.7	80.0	43.6	6.5	44.9	2.28	0.46	0.05	0.13	0.39	0.12	0.04
Triticale grain	16.9	2.1	81.0	43.5	6.4	46.4	1.68	0.62	0.06	0.10	0.35	0.11	0.07
Rape grain	26.5	4.6	85.2	60.5	7.2	23.8	3.94					0.10	
Miscanthus	17.6	3.9	77.6	47.5	6.2	41.7	0.73	0.72	0.16	0.06	0.07	0.15	0.22
Switch grass													
Sugar cane stalk (bagasse)	8.0	4.0	80	45	6.0	35	0.0					0.0	
Hay from various sources	17.4	5.7	75.4	45.5	6.1	41.5	1.14	1.49	0.50	0.16	0.19	0.16	0.31
Roadside green	14.1	23.1	61.7	37.1	5.1	33.2	1.49	1.30	2.38	0.63	0.19	0.19	0.88
Hard coal	29.7	8.3	34.7	72.5	5.6	11.1	1.3					0.94	<0.13
Lignite	20.6	5.1	52.1	65.9	4.9	23.0	0.7					0.39	<0.1

[1]LHV.

calorific value, net CV) of a fuel is defined as the amount of heat released by combusting a specified quantity at a reference state and returning the temperature of the combustion products to 150 °C. The LHV assumes that the latent heat of vaporization of water in the fuel and the reaction products is not recovered. It is useful in comparing fuels where condensation of the combustion products is impractical, or heat at a temperature below 150 °C cannot be put to use. By contrast, the higher heating value (HHV) (also known as gross calorific value or gross CV) includes the heat of condensation of water in the combustion products.

The LHV is greatly influenced by the water content. Figure 7.3 shows that the LHV of wood decreases from approximately 18.5 MJ kg^{-1} with increasing water content. Normally the water content of air-dried wood is between 12 and 20% yielding a heating value of 13 to 16 MJ kg^{-1}. Freshly harvested wood is characterized by a water content of about 50% or more. A low LHV is the result.

Besides the LHV a lot of additional quality aspects are relevant concerning biomass provision, storage, transportation and efficient conversion processes. Density and heat capacity, for example, are important parameters for unconditioned and for conditioned biomass. Particle density and particle-size distribution are of importance for chips and pellets, viscosity for liquid biomass.

For this reason during the last couple of years on a European level such parameters characterizing solid fuels have been standardized to allow for growing biomass-based fuel markets. Recently these activities have been also moved to the international level and an ISO-Standardization Committee has been founded to develop standards for biomass based fuels.

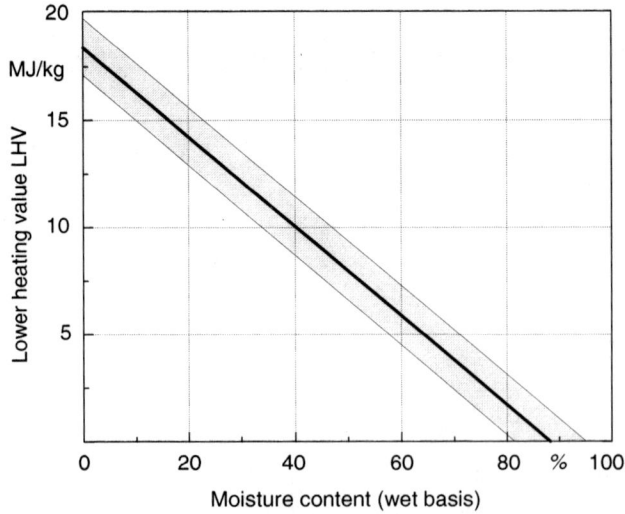

Figure 7.3   LHV of wood depending on the water content.[4]

7.2.2 Biomass Potential

For the energetic use of biomass a wide range of different resources is available in principle (*e.g.* forest wood, straw, cereal, liquid manure). They derive from the agricultural and silvicultural primary production respectively from the industries that follow the primary production as well as from waste management (*i.e.* when organic materials leave the production process).[5]

The amount of biomass available in total which is usable considering given technical restrictions is characterized by the technical potential. Besides technical restrictions for its estimation, structural and environmental restrictions (*e.g.* nature protection areas, biotope network areas) as well as the legal framework (*e.g.* legitimacy of hygienic precarious organic waste for use in biogas plants) are also taken into consideration because such restrictions may reduce the available biomass potential calculated under an exclusively technical point of view significantly.

Table 7.2 shows the present technical biomass potential based on the criteria mentioned above. Considering residues, by-products and waste they are differentiated in herbaceous biomass (*e.g.* straw, landscape conservation material), wood (*e.g.* forest residual wood, industrial residual wood) and other biomass (*e.g.* excrements, organic industrial waste). The forest potentials include the wood not used as a raw material (*i.e.* firewood, forest residual wood) and that share of the annual incremental growth which is not used at the moment.

Furthermore there are energy crops, which can be cultivated as annual or perennial cultures on agricultural land for exclusive energetic use.

Considering the current demand on food and fodder an available land area of about 2 Mio ha can be assumed for biomass production exemplarily within the Federal Republic of Germany as one country in Central Western Europe. On this area different types of plants are cultivated for different purposes.

- Thermo-chemical conversion: mixed cultivation of different lignocellulosic plants for the provision of solid biofuels;
- Physical-chemical conversion: rape-seed cultivation;
- Bio-chemical conversion: two-culture-system for the production of substrates for biogas production, cultivation of input products for the ethanol production.

Such a procedure leads to an estimation of an upper border, since in practice the selection of suitable cultivation cultures is significantly reduced due to the given local conditions.

However, the biomass declared (*e.g.* herbaceous biomass, energy crops) can be used only once (either thermo-chemical or bio-chemical or physical-chemical). Thus the entire fuel potential amounts for the German example to approximately 1,000 to 1,300 PJ a^{-1} (approximately 8% of the present German primary energy consumption).

Additional changes are to be expected with the potentials outlined here in the years to come. These variations might be relatively small for residues,

Table 7.2 Technical biomass potential exemplarily in the Federal Republic of Germany.[6]

	Energetically usable amount Mio mt$_{raw\ material}$ a^{-1}	Potential with thermo-chemical conversion PJ a^{-1}	Potential with bio-chemical conversion PJ a^{-1}	Potential with physical-chemical conversion PJ a^{-1}
Herbaceous residues, by-products and waste				
Straw	9.3	130	38–63	—
Grass from pastures etc.	2.6–4.0	37–56	15–23	—
Rural conservation material	0.9–1.8	11–22	8–16	—
Total	12.8–15.1	178–208	61–102	—
Wood residues, by-products and wastes				
Forest residual wood	13.7	169	—	—
Thinning	10	123	—	—
Additionally available wood	10.7	132	—	—
Demolition wood	6	78	—	—
Industrial residual wood	4	58	—	—
Landscape conservation wood	0.46	4	—	—
Total	45	563	—	—
Miscellaneous residues, by-products and waste				
Excrements and litter	162	—	96.5	—
Harvesting residues	7–14	—	9.1–18.3	—
Wastes from industries	3.1–4.7	—	6.4–12.2	—
Organic urban waste	7.6	—	12.5	—
Total	180–188	—	124–139	—
Gas from purification plants	—	—	19.5	—
Landfill gas	—	—	15–21	—
Total residues, by-products and wastes	—	741–770	219–282	—
Energy crops on 2 Mio ha		365	236^a–252^b	103^c
Total amount		1,106–1,135	455–533	103

[a]Biogas substrates.
[b]Ethanol from sugar beets (additionally biogas substrates (95 PJ a^{-1}) would be usable energetically).
[c]Vegetable oil respectively FAME from rape seed (additionally straw (125 PJ a^{-1}) and whole grain (65 PJ a^{-1}) would be useable energetically).

by-products and waste, whereas a clear increase is expected for energy crops – due to presumably declining land area requirements for the production of food and fodder caused by slightly fading population and ongoing yield increases. Thus an increasing biomass potential – and hence an increasing importance of cultivating energy crops – is likely in the future. According to forecasts the entire biomass potential might sum up to approximately $1500 \, \mathrm{PJ \, a^{-1}}$ in Germany in the year 2010 and increase to approximately $2000 \, \mathrm{PJ \, a^{-1}}$ in the year 2020. Improvements in the seeds for the energy crops can lead to even higher yields and thus higher potentials. Improved environmental and nature protection requirements (*e.g.* expansion of organic farming) might lead to a decrease of the agricultural area available and thus downsize possible potential increases. But even then an increase of the potential is expected. And this situation can be expected in most European countries.

7.3 Biomass Conversion

7.3.1 Thermo-chemical Conversion

7.3.1.1 Basics

The most widely spread use of solid biomass is the provision of heat released during combustion. During combustion, solid biofuels are oxidized primarily to carbon dioxide and water by releasing heat. Equation 7.1 shows the approximate process for wood, which can be described as $C_n H_m O_p$.

$$C_n H_m O_p + \left(n + \frac{m}{4} - \frac{p}{2} \right) O_2 \rightarrow n CO_2 + \frac{m}{2} H_2 O + \text{heat} \tag{7.1}$$

The biomass combustion consists of the following steps: heating and drying, pyrolytic decomposition, gasification and oxidation (Figure 7.4). These different steps are explained below.

- *Heating and drying.* Before any chemical reaction of the organic material can take place, water is evaporated at temperatures up to 200 °C. The water may leave the area where the chemical reaction takes place with the flue gas. Alternatively it may be converted to H_2. This process requires energy (*i.e.* it is endothermic).
- *Pyrolytic decomposition.* Within this step the biomass macromolecules are decomposed by heat in the absence of oxygen and the volatile compounds are driven out of the biomass material due to thermal effects. The process of the pyrolytic decomposition starts at about 200 °C with the decomposition of hemicellulose, which is a part of solid biomass like wood. Decomposition of cellulose, which is another component of wood biomass takes place at temperatures of 300 °C and higher. Lignin is decomposed at even higher temperatures. During this process the volatile components of the biofuel are vaporized at temperatures below 600 °C by a set of complex

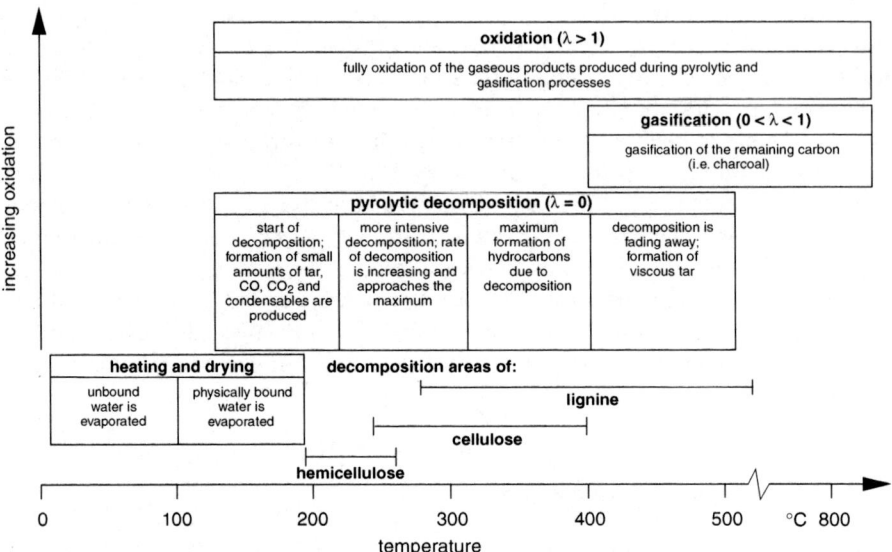

Figure 7.4   Example of thermo-chemical conversion for wood.[4]

decomposition reactions. As a result of such decomposition processes the following components are formed:

- volatile compounds, such as hydrogen (H_2), carbon monoxide (CO), methane (CH_4), carbon dioxide (CO_2), nitrogen (N_2) and steam (H_2O);
- carbon rich solid fraction (char);
- low molecular weight organic compounds and high molecular weight (condensable) compounds (*i.e.* liquid products).

Fixed carbon (char) and ash are the by-products that are not vaporized.

- *Gasification.* It is obvious that solid carbon necessarily remains as a product of the pyrolytic decomposition. In order to convert this solid carbon into a gas, an oxygen-containing agent such as air or pure oxygen is required. At a temperature range between 700 and 1,500 °C the solid carbon as well as gaseous products (CO, H_2, CH_4) are oxidized as follows:

Partial oxidation of solid carbon: $C + 1/2 O_2 \rightarrow CO$ $\Delta H = -111\,kJ/mol$

$$(7.2)$$

Complete oxidation of solid carbon: $C + O_2 \rightarrow CO_2$ $\Delta H = -197\,kJ/mol$

$$(7.3)$$

Oxidation of carbon monoxide: $CO + 1/2 O_2 \rightarrow CO_2$ $\Delta H = -283\,kJ/mol$

$$(7.4)$$

Oxidation of hydrogen: $H_2 + 1/2 O_2 \rightarrow H_2O$ $\Delta H = -242\,kJ/mol$ (7.5)

Oxidation of methane: $CH_4 + 2O_2 \rightarrow CO_2 + 2H_2O$ $\Delta H = -802\,kJ/mol$

$$(7.6)$$

Additionally, balance reactions take place. The most important are the reduction of CO_2 to CO (Boudouard reaction) and of H_2O to H_2 (heterogeneous water gas reaction). Simultaneously, carbon can be transformed to CH_4.

Boudouard reaction: $C + CO_2 \leftrightarrow 2CO$ $\Delta H = 173\,kJ/mol$ (7.7)

Heterogeneous water gas reaction: $C + H_2O \rightarrow CO + H_2$

$$\Delta H = 131\,kJ/mol \qquad (7.8)$$

Heterogeneous methane production: $C + 2H_2 \rightarrow CH_4$ $\Delta H = -87\,kJ/mol$

$$(7.9)$$

During the complicated mix of these and other reactions making up gasification, sometimes energy is needed and sometimes energy is released, depending on the given conditions. Additionally these reactions can also take place during the pyrolytic decomposition at higher temperatures because the biomass itself contains oxygen. Therefore the border line between these two conversion steps is not always clearly defined.

- *Oxidation.* Within this last step of the thermo-chemical conversion, the gaseous products produced during the steps already performed are fully oxidized to carbon dioxide and water releasing energy (*i.e.* they are exothermic).

If equation (7.1) occurs in one step, we speak of full oxidation. Under these conditions the excess air ratio is 1.0 or above (the excess air ratio is defined as the ratio between the amount of oxidizing agent fed to the conversion process and the amount of oxidizing agent needed to fully oxidize all reaction products; per definition the excess ratio is 1.0 if the conversion process is realized exactly as shown in equation (7.1)).

Oxidation can also be realized in two steps in which the excess air ratio of the first step is below 1.0. Under these conditions the reaction products can be further oxidized in a second step releasing the rest of the available energy. Carbon monoxide and/or hydrocarbons are typically produced at the first step and transported to another device for full oxidation. Within such processes the procedure described above is paused; for example after the gasification step (*e.g.* within a gasifier) is performed, the oxidizing step is realized at another time and at another place (*e.g.* within the engine).

- If in such a two-stage process the intermediate product is a liquid we call the process pyrolysis (the procedure described above is then suspended

Table 7.3   Excess air ratio and reaction product composition for thermo-chemical conversion processes.[4]

	Excess air ratio	Temperature in °C	Pressure in bar	Products Gas	Liquid	Solid
Oxidation	$\lambda \geq 1$	800–1300	1–30	$H_u = 0$		LHV $= 0$
Gasification	$0 < \lambda < 1^a$	700–900	1–30	$H_u > 0$		LHV ≥ 0
Pyrolysis, coalification	$\lambda = 0$	350–550	1–30	$H_u \geq 0$	$H_u > 0$	LHV > 0

^{a}In most cases $0.2 < \lambda < 0.5$.

after the pyrolytic decomposition). Under these conditions the excess air ratio of the pyrolysis process is zero. Solid, liquid and gaseous products are formed in varying amounts depending on the process conditions (*e.g.* temperature, heating rate, pressure, water content).

- If a gas is to be produced at the first step, the excess air ratio is between zero and one (Table 7.3). The gas, which often mainly contains carbon monoxide, is called producer gas.

7.3.1.2   Direct Combustion

The thermo-chemical conversion of solid biofuels produced from biomass into heat is called combustion. This heat released during the oxidation of organic material mainly into carbon dioxide and water can be used directly at the conversion plant (*e.g.* for cooking or space heating) or can be transported by means of a heat carrier (*i.e.* hot water, steam) to the place of consumption (*e.g.* district heating systems). The thermal energy can also be converted *via* a steam turbine or by co-generation (also known as combined heat and power [CHP] process) into electricity and low temperature heat. Because of economic limits, only relatively low conversion rates of biomass fuel energy into electrical energy (maximum of 30 to 35%) are possible at present. Therefore biomass co-firing in modern large-scale coal power plants with efficiencies up to 45% is regarded as a cost-effective option for the use of solid biofuels for power generation via direct combustion.[6]

The technology required to optimize combustion depends on the capacity, fuel consistency, water content, ash melting behaviour, trace element contaminants and other factors. Where these conditions vary, for example with changeable mixtures of biomass species in the feed, performance will inevitably suffer.

Due to the relatively high volatiles content and other characteristics of biomass, spatial separation is usually provided in modern small scale combustion devices between the fuel gasification process and the full oxidation of the produced gas into CO_2 and H_2O. The former occurs with primary air fed into the glowing fire and the latter through secondary air fed into the burning gas, preferably in an after-burning chamber. To achieve low airborne emissions, good mixing of air and combustible gas at high temperatures is necessary. Additional design features assist in reducing particle emissions, including those for keeping

Table 7.4 Typical combustion technologies and their characteristics.

	Combustion technology	Typical thermal capacity	Biofuels
Manually fed systems	open/closed chimney	2 to 15 kW	wood logs, briquette
	single stove	3 to 12 kW	wood logs, briquette
	tiled stove	2 to 15 kW	wood logs, briquette
	pellet stove	3 to 10 kW	pellets
	wood log stove	10 to 500 kW	wood logs, briquette
Automatic fed systems	gasification firing system	20 kW to 2 MW	wood chips
	under feed system	20 kW to 2 MW	wood chips, wood shavings and filings
	grate firing system for wood	150 kW to 15 MW	wood, bark
	fluidized bed system	from 10 MW	wood, bark, sewage sludge, black liqueur
	grate firing system for herbaceous biomass	50 kW to 20 MW	bales, chipped herbaceous biomass
	blow in firing system	200 kW to 50 MW	dust, shavings and filings

flue gas velocities low so that ash particles are not entrained. Table 7.4 gives an overview of combustion technologies primarily used in industrialized countries. Based on this a few major systems are discussed in more detail.

Pellet Combustor. Automatically fed combustion devices have been developed for wood pellets with standardized fuel characteristics. On the reverse side of such a device (Figure 7.5), a container is located to store fuel for automatic operation over a certain period of time. The fuel pellets are transported from this storage facility with a screw to a pipe from where the pellets fall into the combustion chamber. Primary air is fed by nozzles through the bottom of the combustion bowl. Secondary air is fed above the burning fuel *via* ring-shaped nozzles to ensure the complete conversion of the gaseous fuel components into CO_2 and H_2O. Additional air is fed into the system via the fall pipe to prevent fire flash back into the fuel container. This well-developed air-feeding system allows very low airborne emissions to be achieved. The ash produced during the combustion of the wood pellets is normally removed manually. During ordinary operation the combustion residues (*i.e.* ashes) contain less than 1% carbon and can be used as a fertilizer or taken to a landfill.

The thermal efficiency of such systems can attain 95% or more. Emissions are considerably lower compared to stoves fired with wood logs because

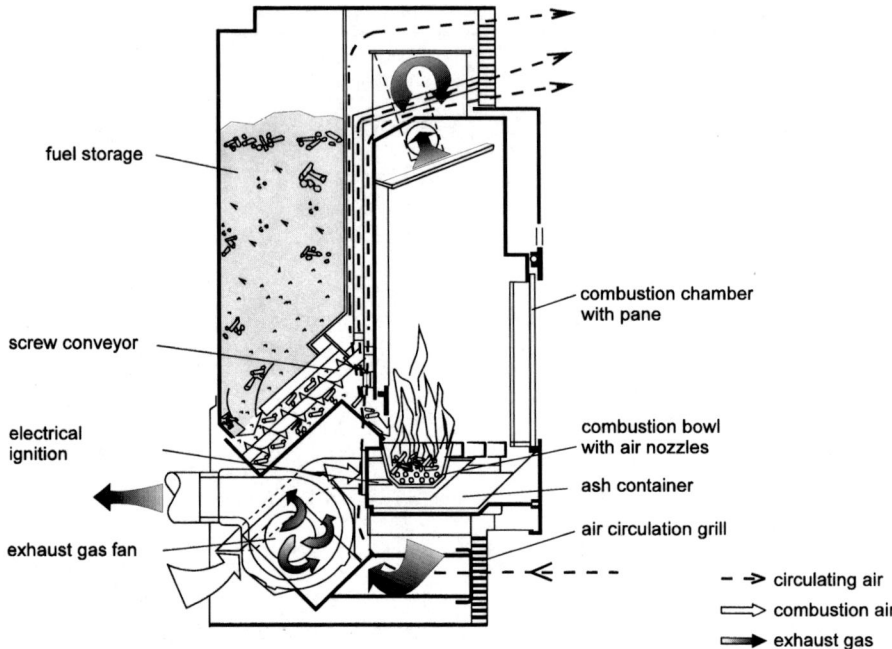

fuel storage

combustion chamber
with pane

screw conveyor

electrical
ignition

combustion bowl
with air nozzles

ash container

air circulation grill

exhaust gas fan

- -> circulating air
⇒ combustion air
⇒ exhaust gas

Figure 7.5   Combustion device fired with wood pellets.[4]

the combustion device can be optimally adjusted to the standardized and
well-defined pellets. Such heaters can be operated over a wide range of power
settings, *i.e.* between approximately 30 and 100% of the rated capacity.

Oven for Wood Chips.   A throw-charging system is used where wood chips
are transported from fuel storage with a stoker scroll and thrown with the
help of a centrifugal wheel into a combustion chamber equipped with a stiff
grate. Such a fully automatic feeding system allows the smaller fuel particles
to be combusted during the flight to the grate while more coarse fuel particles
are burned on the grate. The system also has the advantage that the fuel is
fed gently on the fire, thus reducing airborne emissions of particulate matter.

The primary air is blown with an automatic ventilator through holes in the
grate into the glowing fire. The secondary air is blown at the top of the burning
fuel. The air feeding system adjusts itself automatically according to information from a sensor fixed in the flue gas outlet, minimizing pollutant emissions.
No additional flue gas treatment is needed and ash can be used as a fertilizer or
put on a landfill site.

The heat is extracted from the flue gas *via* a heat exchanger located on top of
the combustion device. Within the heat exchanger, the flue gas is cooled down
and water is heated up – depending on the heat utilization system – either close
to boiling point (to be used in hot water systems) or to steam for *e.g.* electricity
generation.

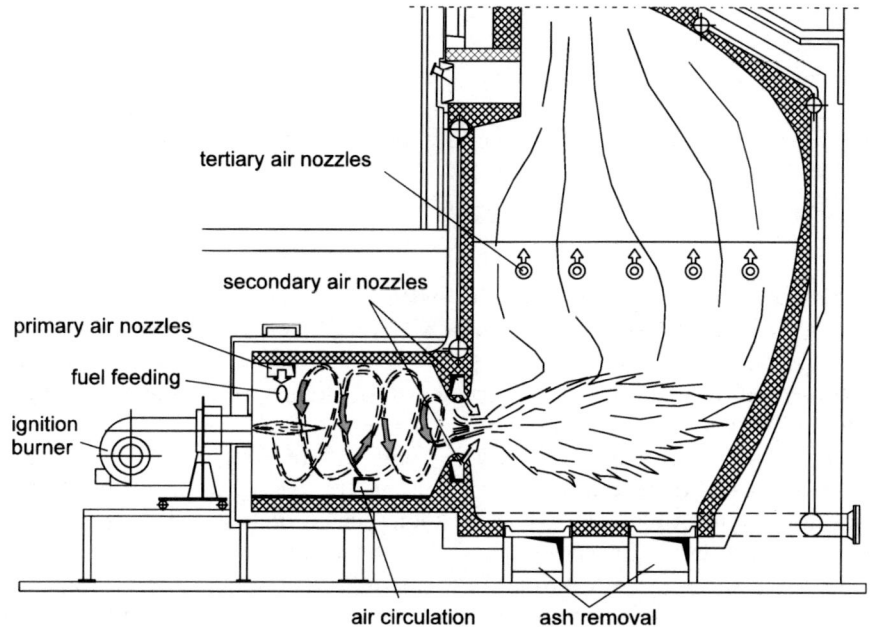

tertiary air nozzles

secondary air nozzles

primary air nozzles

fuel feeding

ignition burner

air circulation   ash removal

Figure 7.6   Blow-in combustion unit for wood dust.[4]

Such combustion systems are characterized by a fully automatic operation. They are available on the market in a range of thermal capacities from 300 up to approximately 3,500 kW and can be used either for heat provision for a single- or multi-family house or for industrial process heat.

Blow-in Combustion Units.   Blow-in combustion units have been developed for using dusty biomass residues (*e.g.* sawdust, shavings, filings) from the wood-processing industry. The fuel particles are transported pneumatically from the fuel silo directly into the combustion chamber or – as shown in Figure 7.6 – into a pre-combustion chamber. Additionally, primary air is blown into the pre-combustion or combustion chamber to ensure a full conversion of the fuel into gas. Secondary air is blown into the combustion chamber to ensure a complete combustion. To guarantee that the fuel particles ignite by themselves during their flight through the combustion chamber, fuel water content has to be below 15 to 20%. To first ignite such a unit, a burner based on light oil or natural gas is used to heat the pre-combustion or combustion chamber to 450 to 500 °C.

Such a combustion system is characterized by relatively low emissions of gaseous pollutants (such as NO_x or CO). Because fuel is fed in very small pieces, however, the particle content in the flue gas is relatively high. To meet particle emission regulations, cleaning of the flue-gas is necessary in most countries. Therefore such systems are often equipped at least with a cyclone or a multi-cyclone. In case cyclones do not sufficiently remove particle emissions a fabric or bag house filter or even an electrostatic dust removal system is additionally installed. Such combustion units are characterized by fully automatic operation and low emissions at thermal capacities of several MW up to some 10 and more MW and are mainly located at places where wood dust is produced as a residue, such as *e.g.* the furniture industry.

7.3.1.3   Coalification

In this process, woody biomass is heated up in a nearly oxygen-free environment. Up to about 200 °C, drying occurs, followed at higher temperatures with pyrolytic decomposition of the organic compounds. Substantial liquid and gaseous residues such as tar and carbon monoxide are released in the course of creating the final product, charcoal.

Such processes can be realized with quite different technologies. In developing countries, for example, charcoal kilns are usually made from earth or brick, although simple metal kilns are sometimes found. The efficiency of such devices is low (less than 25%) and the airborne emissions of volatile organic compounds are high. Additionally considerable amounts of toxic liquid residues are produced under certain frame conditions. In industrialized countries, charcoal is produced in large fully automatic devices. Here a differentiation is made between retort and flush gas processes. Within the former the charcoal is produced in a batch mode in a closed container (*i.e.* retort) where the thermochemical conversion from wood to charcoal takes place. The heat necessary to allow this process to take place is obtained from the combustion of the gaseous and liquid by-product. There are also continuous charcoal production systems where the wood is fed constantly through a reactor. Within such a reactor different zones with various settings of reaction conditions ensure that the charcoal is produced during the migration of the material through the reactor. Also here the gas and liquids produced as by-products are used as a source of energy to keep the coalification process going.

In general the importance of such processes is relatively low in industrialized countries because charcoal plays only a minor role within the energy systems of the countries of the western world. However, charcoal is used to a certain extent as a raw material (*e.g.* to produce activated charcoal for filtering) and as fuel for leisure. In developing countries charcoal is used as a clean fuel for heating and cooking.

7.3.1.4   Pyrolysis

The pyrolytic decomposing processes realized during pyrolysis are similar to those during charcoal production, but here the process conditions are set to ensure that the main product of the thermo-chemical conversion process is liquid rather than solid. There has been a broad variety of technologies developed during the last decades attempting to make pyrolysis oil for use in engines without any additional processing.

Most promising is flash pyrolysis in which fuel particles are heated very rapidly (more than $1,000\,°C\,s^{-1}$) and remain in the hot zone for a very short time (in general less than 1 s). After this very short time period, the liquid compounds produced from the solid biomass by decomposing the organic compounds (*i.e.* lignin, celluloses) have to be removed and cooled rapidly to avoid further decomposition into gases. To date, flash pyrolysis reactors have reached laboratory stage development level and the first pilot plants are available.

Reactors with ablative impact designs, for example, decompose the biological raw material primarily into liquid components on the surface of a hot rotation wheel (Figure 7.7). On the surface of this wheel the solid biomass is "melted". To avoid further decomposition, the produced components have to be removed quickly from the hot zone close to the rotating wheel. After that they are cooled down to produce the desired liquid energy carrier. Necessarily, gaseous and solid components are also produced, which are used to heat the wheel.

Pyrolysis oil is a mixture of different hydrocarbons (many of which are partly oxygenated) along with charcoal, ash and water. The actual composition is strongly dependent on the pyrolysis process as well as the specific process conditions and the type of biomass. This is especially true for the average composition and structure of the hydrocarbons (*e.g.* chain length, degree of

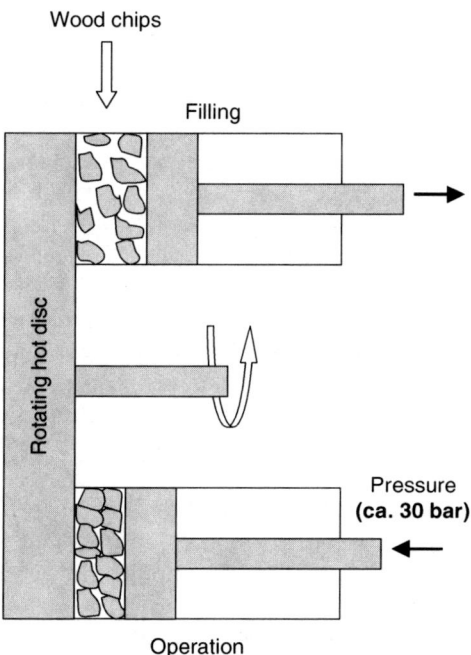

Figure 7.7 Pyrolysis system based on a hot rotating wheel.[7]

double bonding). This oily liquid can be toxic and is in most cases not stable in air. Additionally the conversion efficiency of the available fast pyrolysis processes is relatively low. The average heating value of pyrolysis oil is approximately 40% that of petroleum-based fuels.

The goal of pyrolysis is the provision of a liquid energy carrier that can be used directly in engines for *e.g.* the provision of heat and electricity. But so far the pyrolysis oil produced in the available plants cannot be used directly in existing engines. Therefore upgrading of the produced oil is necessary. For example, charcoal and ash particles need to be removed to ensure a sediment free fuel and, to increase stability, double bonds need to be reduced by adding hydrogen to yield saturated hydrocarbons which are then cracked or remain as paraffin waxes in the pyrolysis oil. Additionally other measures have to be taken to ensure the viscosity and combustion behaviour required by the respective engine. Like the fast pyrolysis process itself, these upgrading technologies are not fully developed yet. Because of these technical and economic constraints, there are no applications known for the provision of pyrolysis oil as a source of energy on a fully commercial basis.

Nevertheless few industrial applications are realized to produce liquids to be used as a raw material (*e.g.* as liquid smoke).

7.3.1.5 *Gasification*

Gasification describes the complete conversion of solid biomass at high temperatures to a gaseous fuel by adding a small amount of oxidizing agents to the gasification process. Unlike coalification and pyrolysis, gasification of biomass is realized in the presence of a limited amount of oxygen. The main objective of gasification is to transfer the maximum possible share of the chemical energy within the feedstock into a gas.

This so-called "producer gas" can be used on the one hand as a fuel for the provision of heat through direct combustion or used in engines, turbines or even fuel cells. On the other hand the producer gas can act as a feedstock for the production of liquid and/or gaseous fuels (like Fischer–Tropsch diesel, bio-methane, methanol and hydrogen).

Biomass gasification consists of the following, more or less spatially distributed, steps: heating and drying of the biomass, pyrolytic decomposition of the biomass (*i.e.* extracting the volatile components by heating) and gasification (*i.e.* partial oxidation of the biomass, partial reduction of the oxidation products [CO_2 and H_2O to CO and H_2] and simultaneous transformation of solid carbon to CO).

The physical and chemical processes of biomass gasification are carried out in a variety of different forms of equipment and technical concepts. Each of them offers certain advantages and disadvantages concerning feedstock possibilities, plant size and gas quality.

The gasification techniques can be distinguished related to different criteria such as reactor type (fixed bed or fluidized bed), gasifying agent (air, oxygen, or steam), heat supply into the reactor (directly or indirectly heated) and reactor

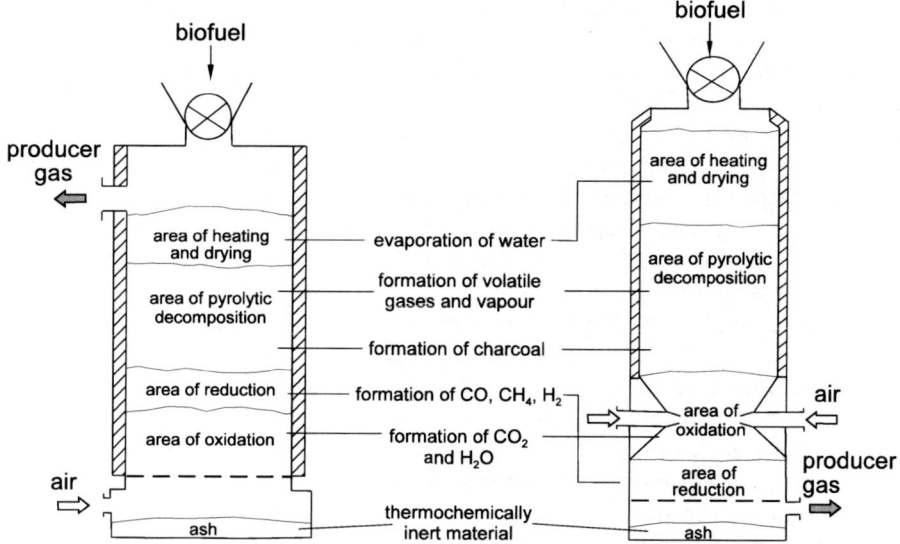

Figure 7.8 Principle of an updraft gasifier (left) and a downdraft gasifier (right).[4]

pressure (atmospheric or pressurized). Over the years, a considerable variety of biomass gasifiers has been developed.

- In fixed-bed reactors (Figure 7.8), the feedstock is exposed to the gasifying agent in a packed bed that slowly moves from the top of the gasifier to the bottom, where the ash is discharged. Fixed-bed gasifiers are dense-phase gasifiers characterized by a relatively large amount of fuel exposed to a limited amount of reactive gas. In fixed-bed reactors the feedstock occupies most of the reactor volume.
- Fluidized-bed reactors are lean-phase gasifiers having a low ratio of solid to reactor volume. Typically, the feedstock occupies only a small fraction of the total reactor volume. Fluidized-bed gasifiers are classified, depending on the intensity of fluidization, as bubbling fluidized-bed gasifiers, circulating fluidized-bed gasifiers or entrained flow gasifiers.
- Gasifiers have also been developed that cannot easily be classified as belonging to one of the former groups but have features of both.

Commonly, the goal of gasification is not to provide gaseous fuel itself but to provide an easy to handle and environmentally sound intermediate energy carrier with clearly defined characteristics that can be converted easily into another, more valuable, energy carrier, for example electricity or biomethane as one possible biofuel. To reach that goal, gas cleaning is usually necessary to ensure a long lifetime of the downstream conversion device because the gas produced within the gasifier does not usually match the fuel requirements in terms of condensable organic compounds and/or particles.

The gas leaving the gasification reactor is often called producer gas. The same name is also used if this gas is cleaned to match the requirements for combustion within *e.g.* an engine or a turbine for the provision of electricity (and heat). If the gas is conditioned to fulfil certain requirements concerning the composition and it is intended to use this gas for the synthesis of liquid or gaseous biofuels it is called synthesis gas (or in abbreviated form "syngas").

For the provision of a gas to be used as a fuel for CHP-systems (*e.g.* engines, turbines) a raw gas cleaning operation – either low temperature wet gas cleaning or, alternatively, hot gas cleaning – can be applied. The effectiveness of wet gas cleaning (*e.g.* cyclone and filter, scrubbing based on chemical or physical absorption and ZnO-bed) has been well proven for large-scale coal gasification systems. On the contrary, not all elements of hot gas cleaning (*e.g.* tar cracking, granular beds and filters, physical adsorption or chemical absorption, ZnO-bed, physical absorption) have reached technical maturity yet. Nevertheless, hot gas cleaning offers benefits for the overall energy balance and with regard to the avoidance of contaminated waste water.

Additionally the producer gas can be conditioned to fulfil the required gas characteristics especially for the conversion route to fuels (Figure 7.9). For that different system components can be applied: hydrocarbons within the product gas can be converted by means of an additional steam reforming step resulting in a higher H_2/CO ratio.

The cleaned producer gas can be used for the provision of electricity within already existing gasoline engines or gas turbines. The former is demonstrated

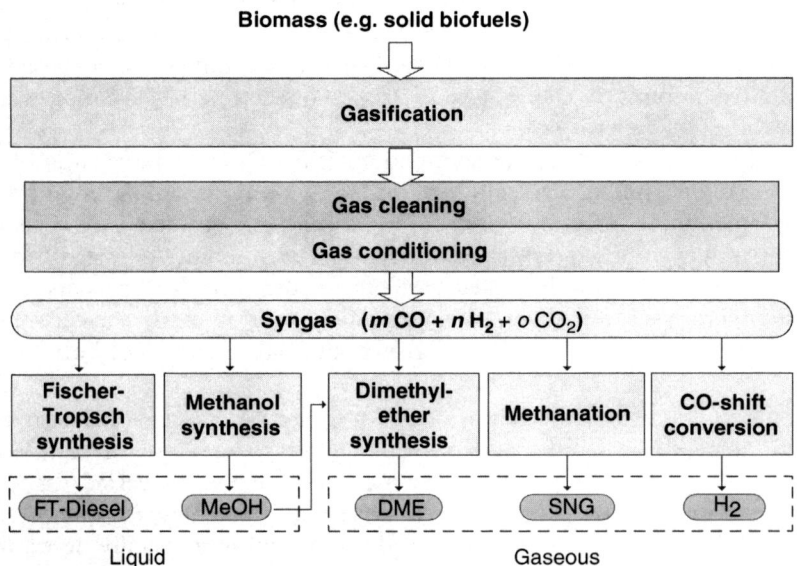

Figure 7.9 Possible routes to the production of liquid and/or gaseous fuels from producer gas derived from biomass.

within the gasification plant in Güssing, Austria, where wood chips are gasified and the cleaned producer gas is then used within an engine-based CHP-plant for the simultaneous provision of heat and electricity. Similar systems are in operation and also under planning in other European countries. The next development step will be biomass-based integrated gasification combined cycles (IGCC).[9]

Alternatively the cleaned and conditioned gas can be used for the provision of liquid and gaseous fuels. The necessary technology is currently under development (see Chapter 8).

7.3.2 Physico-chemical Conversion

Based on physico-chemical processes vegetable oil can be extracted from biomass to be used in the chemical and other industries and for energy generation. Compared to solid biofuels vegetable oil has some outstanding advantages. It is easy to store, has high energy content and can be used as an energy carrier *e.g.* in some engines or combustion units without major problems. However, just like crude oil is not generally useful in its raw or unprocessed form and thus needs processing in a refinery to yield kerosene, diesel fuel or fuel oils for energy generation, fats and oils often need a chemical treatment in order to meet the fuel characteristics of conventional diesel fuel to allow easy use in existing engines. The physico-chemical biomass conversion comprises the oleochemical unit operations:

- extraction (pressing, solvent extraction, rendering);
- chemical and/or physical refining;
- conversion into fatty acid methyl esters

described in more detail in Chapter 9.1.

The final step, *i.e.* conversion of refined vegetable oil into fatty acid methyl ester (FAME) is shown in Figure 7.10. According to this exemplarily shown process the mixture of cleaned vegetable oil together with a catalyst and methanol is pumped with a low velocity through a vertical pipe. The low velocity of the liquid ensures that the glycerine produced during the transesterification process can settle in the reactor and can be removed for use as a raw material. After the removal of the remaining methanol, the liquid is cleaned by a multi-stage washing process. The produced FAME is now ready to be used as a fuel either within power trains for cars and trucks or in engine-based CHP-systems.

Pure vegetable oil can be used in some existing engines either for mobile (*e.g.* in cars) or stationary application (*i.e.* in CHP-systems), but in most cases it lowers engine lifetime and increases maintenance requirements. Therefore only under very specific frame conditions could the use of crude vegetable oils be a promising option for the transportation sector or to provide power (or decentralized electrification and heat, if required) especially in rural areas. This is true for developing as well as for industrialized countries.

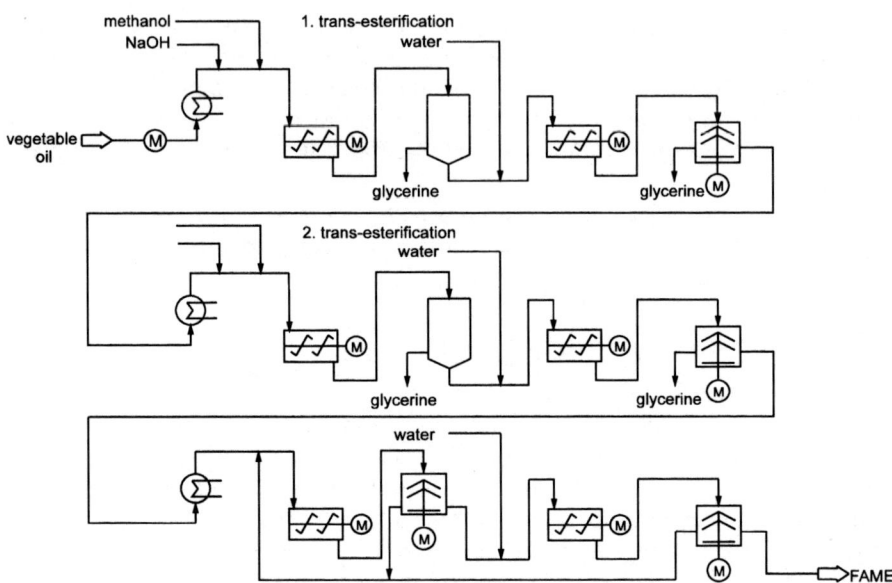

Figure 7.10 Continuously working transesterification process for FAME
manufacturing.[4]

Table 7.5 Selected properties of diesel fuel, FAME and rape oil (various
sources).

	Diesel fuel/light heating oil	*FAME*	*Rape oil*
Density (15 °C) in kg m⁻³	820–845	875–900	900–930
Viscosity (40 °C) in mm² s⁻¹	2.0–8.0	3.08	78.7
Flash point in °C	>55	130	min. 220
CFPP-value in °C (summer/winter)	max. 0 max. −20	max. 0 max. −20	
Sulphur content in mg kg⁻¹	max. 350	max. 100	max. 20
Cetan-value	min. 51	55	
Calorific value in MJ kg⁻¹	42.7	37.9	min. 35

FAME can be used directly as a substitute for diesel fuel in conventional compression ignition engines (see Chapter 8) or in CHP-systems with
installed capacities from some tens of kW up to several MW (Table 7.5).
The prerequisite for this is that the produced FAME fulfils the existing
standards. Normally this is easily possible with FAME based on rape oil
but it might become difficult using other types of vegetable oil (like crude
palm oil).

Besides the use of vegetable oil for the production of FAME also waste cooking oil (waste grease) can be used under certain conditions. Due to limited availability of waste grease this option is not playing a significant role in the market so far.

Available experiences with FAME based on rape oil (RME) have shown that there are no significant problems when using them in conventional diesel engines although the higher solvency of RME compared to conventional diesel fuel requires biodiesel compatible fuel lines.

7.3.3 Bio-chemical Conversion

Nearly all biomass is eventually decomposed naturally through biological processes. Some of these processes can be harnessed to produce fuels.

- Composting occurs if oxygen is available. During composting the biomass is degraded by bacteria mainly to carbon dioxide (CO_2) and water (H_2O) while releasing low-temperature heat that can, in principle, be used *via* a heat pump. But for the time being this option is only of theoretical importance.
- Under anaerobic (oxygen-free) conditions, a variety of degrading processes are employed by micro-organisms. Most relevant to energy are ethanol production *via* alcohol fermentation and biogas (methane) production *via* anaerobic digestion.

7.3.3.1 Alcoholic Fermentation

Sugar can be converted to ethanol based on micro-organisms. Because starch and even celluloses can be converted more or less easily into sugar, such biomass streams are also a potential resource for the production of bioethanol in addition to naturally sugar-containing crops like sugar cane and sugar beet (see Chapters 8 and 9.3).

In power trains for vehicles, ethanol can fully or partly substitute gasoline. But a larger volume is needed compared to gasoline because the LHV of ethanol is lower (Table 7.6). The combustion of ethanol, however, requires a lower air volume. Therefore the heating value of the mixture pressed into the cylinder is more or less the same for ethanol and gasoline. This is the reason why an engine powered by ethanol produces the same power as an engine driven by gasoline.

Internal combustion engines must be specially adapted to run on pure ethanol, because it shows different combustion behaviour compared to gasoline. Therefore in most countries ethanol is mixed with gasoline to a maximum of 10% in which form it can be used without any known problems in existing engines. Bioethanol is also being considered for powering fuel cells in future design and applications.

Table 7.6   Selected properties of fuels (various sources).

	Ethanol	*Gasoline*
Composition in %		
carbon	52	86
hydrogen	13	14
oxygen	35	0
LHV in MJ kg^{-1}	26.8	42.7
in MJ l^{-1}	21.3	*ca.* 32.0
Density (15 °C) in kg l^{-1}	0.794	0.72–0.78
Viscosity (20 °C) in $mm^2\,s^{-1}$	1.5	0.6
Boiling point in °C	78	25–215
Flame point in °C	12.8	–42.8
Ignition temperature in °C	420	*ca.* 300
Evaporation heat in kJ kg^{-1}	904	380–500
Minimum air volume in kg kg^{-1}	9	14.8
Octane-value	107	93

7.3.3.2   Anaerobic Digestion

During anaerobic digestion, organic material is decomposed in an oxygen-free atmosphere by bacteria that produce a gas, called biogas or, depending on where it is produced, landfill gas (LFG) or digester gas and usually containing methane (CH_4) and carbon dioxide (CO_2) plus some impurities of minor importance. Such decomposition occurs in nature, for example at the bottom of lakes within the sediments containing organic material, and takes place in landfills and dumps where the organic fraction of the waste material is converted into landfill gas.

In addition to degradable organic input and freedom from air, nutrients for the bacteria and absence of harmful, pathologic and inhibiting substances are needed for a successful biogas production. Also the organic material should be easily accessible to the bacteria. Well-suited for anaerobic digestion are agricultural residues (*e.g.* liquid manure, leaves from sugar beet), residues from the food processing industry (*e.g.* slurry of fruit and potato processing, some types of slaughterhouse residues) and sewage sludge. Only biomass containing lignin such as wood cannot be degraded by anaerobic bacteria.

During anaerobic digestion, organic matter is degraded by three different kinds of bacteria: fermentative, acetogenic and methanogenic. The first two bacteria families degrade the complex organic compounds of biomass into simpler intermediates (Figure 7.11). These intermediate products are then converted to methane and carbon dioxide by the methanogenic bacteria.

Anaerobic digestion relies on a dynamic equilibrium among the three bacterial groups, which is affected by *e.g.* temperature. Most digesters operate in the mesophilic temperature regime around 35 °C. Some operate in the thermophilic regime around 55 °C. Up to the temperature of peak microbial activity, higher operating temperatures produce greater metabolic activity within either regime. Additionally the pH-value and the composition of the

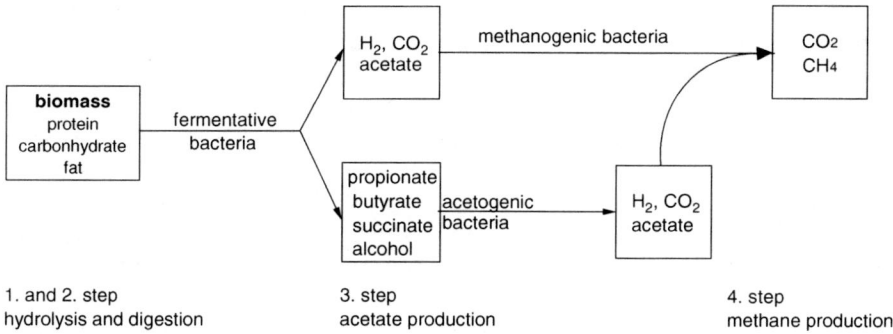

Figure 7.11 Anaerobic digestion process scheme.[10]

Table 7.7 Yield targets of biogas for different substrates (various sources).

Material	Yield of biogas in $m^3 \cdot mto^{-1}$ organic dry matter	*Material*	Yield of biogas in $m^3 \cdot mto^{-1}$ organic dry matter
Liquid manure from beef	250	Paunch content	420–520
Liquid manure from pork	480	Rey straw	300–350
Droppings from chicken	450	Potato herbs	560
Sewage sludge	400	Sugar beet leaves	550
Organic waste from households	170–220	Food residues	80–120
Waste fat	1040	Waste water from brewing industries	500
Roadside green	550	Waste water from sugar industries	650

biomass in relation to easily degradable biomass compounds affect the bacterial balance. For optimal fermentation, pH-values between 6.8 and 7.2 are often maintained by use of buffers.

The composition of the produced biogas is typically about two-thirds methane and one-third carbon dioxide. Additionally the biogas may contain trace substances such as H_2S, depending on the composition of the biomass, the process conditions and other parameters. The LHV of biogas, which depends mainly on methane content, ranges from 14 to 29 MJ m^{-3}.

Expected biogas yields for different substrates are given in Table 7.7. Already within the animal stomach digested substrates like animal manure and droppings show in general smaller yields than fresh material, *e.g.* maize silage or waste grease.

The bacteria use relatively little of the biomass energy for their own survival and only a very small amount of heat is produced (this is the reason why biogas plants have to be heated), making the conversion process of the organic material into biogas fairly efficient. Slurries of organic matter in more than 66% water ensure optimal process conditions with a good access of the bacteria to the organic material so that gas yield is maximized. To optimize production rates, the slurry should be kept at temperatures of 28 to 35 °C or 50 to 70 °C, which correspond to the ranges appropriate to different bacterial species. Production will also occur at lower temperatures, but essentially stops at 10 °C. Due to the low conversion rates such low temperature level has no importance for technical processes optimized to maximize the space time yield.

Typical feedstock for biogas production is organic waste material available with a high water content and/or – in some countries because of the feed-in laws existing in these countries – energy crops (like maize silage). Examples for the former are sewage sludge, animal manure and organic residues from some types of agricultural production and the food processing industry.

Within biogas plants, preparation of the feedstock is needed that might include short-time storage, sedimentation of mineral contaminants (like sand, stones), reducing the material into small pieces, mixing of different types of feedstock to maximize the gas yield and heating to the required temperature level. Then the feedstock is pumped into the biogas reactor where anaerobic fermentation takes place. For successful operation, the bacteria must always be well mixed with the organic material. It is also important to realize a good temperature distribution within the reactor. The biogas is removed from the top of the plant and, after removal of impurities like water, stored before it is used as a source of energy.

The digested material is removed from the reactor and stored in a tank where a small amount of additional biogas is produced. Therefore this storage facility should be covered gas tight to avoid the release of the produced gas into the atmosphere because biogas is harmful to global climate. This digested slurry is used as a fertilizer, because it contains the nitrogen that was originally part of the feedstock. Since biogas technology helps to close the nutrient cycle, this is of increasing importance in environmentally sound management of organic material (Figure 7.12).

Biogas can be used in a boiler, stove or engine for the provision of heat, mechanical power and/or electricity. A part of the heat and electricity produced is often applied to warming up the feedstock and running of the plant. This is a system layout most widely used in industrialized countries. Alternatively or additionally, biogas can also be fed into the natural gas system. To do so requires an upgrading of the biogas to fulfil the natural gas standards. This is done in Switzerland, Germany and Sweden, for example. Based on such a feed-in into the gas grid the use of the biogas – in the form of biomethane – as a transportation fuel is then easily possible. Nevertheless the biomethane can also be used for an environmentally sound and climate neutral high efficient heat provision in very densely populated areas (*i.e.* downtown areas in big cities) using *e.g.* existing condensing boilers.

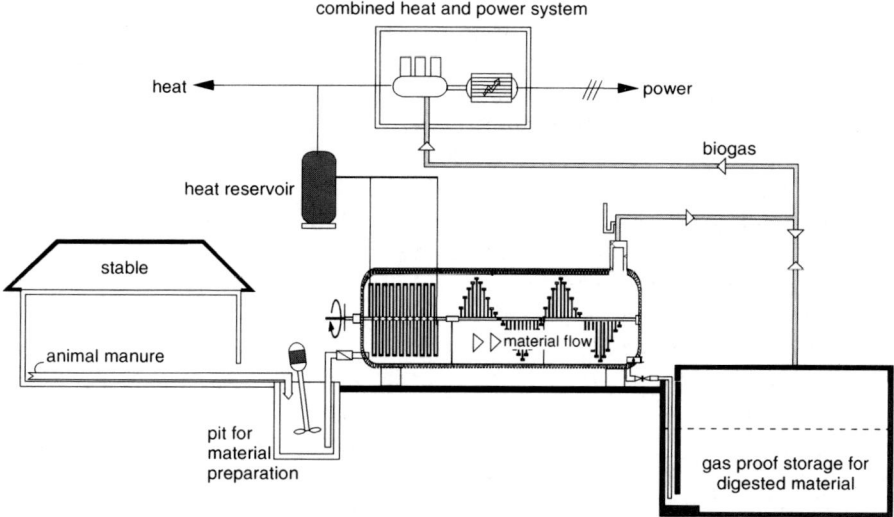

Figure 7.12 Biogas plant using animal manure as feedstock and commonly used in countries like Denmark, Germany and Austria.[11]

7.4 Biomass Use

The available biomass potentials discussed in Chapter 3 can be converted by the options outlined in Chapter 2 into thermal energy, electric energy and/or fuels. However, individual technologies for miscellaneous fields of application are in different states of technological development; some are industrially used or ready for the market and others are still at research or pilot plant level. This technical variety combined with the expected increase of the potentials thereby represents an ideal possibility for integrating biomass successfully into the energy system.

Considering the potentials mentioned exemplarily for Germany and the already realized energetic use of $675 \, \text{PJ} \, \text{a}^{-1}$ in 2007,[8] the available potentials are already used to a considerable extent. While residues, by-products and waste are the prior ranking biomass sources of energy at present, an increasing amount of energy crops are used as feed for the German energy network. Currently cultivated energy crops are mainly rape-seed and biogas substrates (*i.e.* energy maize), which are grown on a land area of roughly 1.5 Mio ha. This corresponds to approximately 10% of the German agricultural land area.

The current energetic use of biomass and the related requirements of area are described in Table 7.2 and can be summarized as follows:

- *Heat*. Heat is mainly supplied by small-scale combustion systems in households and (to a limited extent) by combined heat and power plants (CHP) (approximately 311 PJ of fuel energy in 2007). It has to be noted that the bigger share of this wood is not covered within the official felling statistics, but traded locally in an informal manner from various sources.

- **Power**. Electric energy is primarily produced *via* the direct combustion of solid biofuels within power plants equipped with a steam cycle as well as by making use of biogas in combined heat and power plants (*e.g.* gas engines). Furthermore biogas from sewage plants and landfills as well as liquid bioenergy carriers (mainly vegetable oil) is used within combined heat and power plants (CHP). In 2007 approximately 23.4 TWh have been generated by the use of biomass (including electricity generated from organic waste within waste incineration plants).

- **Fuels**. In 2007 mainly biodiesel (FAME) and to a minor extend bioethanol, which was processed to ETBE, were produced. Overall roughly 156 PJ has been sold on the German energy market.

The current use of biomass amounts to almost 783 PJ (primary energy equivalent) in 2007, which is about 5% of the primary energy consumption of Germany.[12] If the present frame conditions persist in the future, primarily the power production from biomass will increase in years to come. For the different possibilities of using biomass the following tendencies are recognizable.

- A market expansion is expected for first-generation biofuels, *e.g.* biodiesel and bioethanol, with the focus on bioethanol. Due to the given (land) restrictions in Germany it is likely that increased imports could result from this development. Assuming a consistent progress in technical development, BtL-plants might be available for commercial use around the year 2020 at the earliest.[13] And it is still not known to what extent gaseous bioenergy carriers, *e.g.* synthetic natural gas (Bio-SNG) or biogas, could gain importance within the traffic sector.

    In this respect biomethane seems to be a very promising energy carrier, because it can be fed into the existing natural gas grid and support the security of supply especially for Central and Western Europe. For the generation of biomethane two conversion pathways are available: the thermo-chemical biomass conversion with gasification and methanation (Bio-SNG) and the bio-chemical biomass conversion with anaerobic digestion (biogas).

    The conversion technologies for Bio-SNG and biogas differ significantly concerning *e.g.* the biomass feedstock, the conversion principle, the state of technology, the installed capacity per conversion unit, the available technical experiences and the R&D demand. But both conversion routes allow the provision of biomethane fulfilling the requirements for a feed-in into the existing natural gas grid. Therefore biogas as well as Bio-SNG has to be upgraded to guarantee a similar calorific value and chemical composition compared to natural gas. Only if this prerequisite is fulfilled can biomethane be used in a mixture with natural gas by using the same gas grid infrastructure without creating problems for the end user.

Both biomethane production paths complement one another in an ideal way. While the thermo-chemical route focuses on solid biofuels (*e.g.* wood, straw) the bio-chemical route uses wet biomass (*e.g.* animal manure, maize silage). The latter will be realized with plant capacities in the one-digit thermal MW-scale and the former in the two- to three-digit MW-scale. The provided product is basically similar and can be used together with natural gas in any mixture. The erection of the biogas and Bio-SNG conversion plants can be planned directly at the established gas grid.

- Based on the current support schemes for the generation of electricity, biogas from anaerobic digestion will increase further and might contribute significantly more to the generation of electricity from biomass; this will only happen if the support schemes are intensified and/or highly efficient techniques for generating electricity (*e.g.* by gasification) become market-ready and are available cost-efficiently. The use of liquid biofuels (*i.e.* vegetable oil) for the provision of electricity will be of minor importance in comparison with the transportation sector. Due to the expected dynamic price development within the global vegetable oil markets the market development is characterized by substantial uncertainties.

- The further development of the provision of heat from solid biofuels depends strongly on the energy price level of fossil fuels (*i.e.* crude oil price) as well as on the given frame conditions (*e.g.* a renewable heat law currently under discussion within the German government, the financial consequences of the carbon dioxide trading scheme, the carbon tax). Those developments are very difficult to predict.

- Additionally biomass has to be made available for the "new technologies" for the conversion of biomass into liquid or gaseous secondary energy carriers (synthetic biofuels (BtL) and synthetic natural gas (Bio-SNG)). These technologies could be available from 2015 to 2020 onwards at a (semi) commercial stage. In order to cover the upcoming biomass demand, increasing build-up of short rotation plantations is expected.

Based on these developments the overall energetic use of biomass might increase up to 850 to 950 PJ a^{-1} until the year 2010 (depending on heat utilization) in Germany, which is about 6 to 7 % of the expected primary energy demand.

7.5 Final Considerations

Against the background of the growing biomass markets – and for this reason an increasing use of the available potentials – competitions will necessarily arise.[14–18] These are expected on the following levels:

- Land area;
- Biomass;
- (End-)energy sources.

The development in Germany as well as in other European countries throughout recent years shows that all these possible competitions exist in reality. Thus competitions so far influence prices on all different levels. While land and biomass competition rather result in increasing prices due to establishing of additional demand, price increasing effects might be restrained in the field of (end-)energy sources with an increasing substitution of fossil fuel energy at a high energy price level because of reduced price volatility of fossil fuels. The different levels of competition are analyzed in the following exemplarily for Germany. But it is expected that the conclusions to be drawn are also true for most industrialized countries of the western world.

7.5.1 Competition Areas

Competition can arise on the following levels or areas:

Level "Land area". An increase of the use of biomass can lead to conflicts around the limited agricultural land area. Basically this kind of competition is only solvable by an expansion (such as in Australia or Argentina) or by a raise in the productivity of the already used agricultural land.

In Germany, for example, today there are already sporadic competitions about agricultural land of good quality for the cultivation of rape seed (for producing FAME) and biogas substrates (for electricity production); the consequences are increasing land lease prices. Even if the agricultural land available in the years to come for energy crops is known roughly, locally adjustment problems can occur and therefore lease prices could rise at least temporarily. Additionally interactions with the international agricultural markets (and also perhaps with the cost-attractive import options) can be expected.

Level "Biomass". The increased biomass use goes along with an additional demand: in Germany, for example, bioethanol, biodiesel and biogas plants demand an increasing amount of *e.g.* cereals and oil seeds from the regional, national and global agricultural markets, which so far have been used mainly for food and fodder as well as for chemistry. Additionally there is a demand from the energy markets for forest and industry (residual) wood to be used for the provision of solid bioenergy carriers; this stands in competition to the markets for wood as a raw material to be used *e.g.* in the pulp and paper industry as well as for the production of furniture. The results are – and this appeared in the past on the cereals and oil seed markets as well as on the wood markets – rising prices. Also, in the field of residues, by-products and waste, the price structure has partly been changed significantly due to the emerging new energy markets (*i.e.* for demolition wood or certain organic waste fractions, which can be used *e.g.* in biogas plants). So, for biomass resources national and even international markets were developed during the last years, which answer the resource competitions by quantity-price problems (which means the desired quantities are not available at the intended price at the market).

Competitions for the biomass are basically solvable by an expansion of the production, an import and a substitution with other (non-biogenic) products (*i.e.* substitution of woody materials by plastics or stones). However, they differ according to the organizational extent and the necessary time periods.

- *Expansion of production*. The (*a priori* limited) supply of products from the agricultural and forestry primary production can be expanded only by an expansion of production. This is possible by a more intensive production (*i.e.* higher yield per hectare) and/or a production on land not used on a commercial basis so far. For the latter option the needed land areas in Europe are available (see above). In contrast to that, emerging competitions about residues, by-products and waste can not be met by an expansion of the agricultural and forestry primary production or only to a very limited extent.
- *Import*. Since many products of the agricultural and forestry primary production are already today traded on international markets, increased imports to Germany are basically possible. This option can in principle be realized quickly and on a large scale. However, increased imports lead (from a global point of view) to an expansion of the land areas used for agriculture and/or forestry somewhere else. And these additional land areas are available on the long term only if they are managed in a sustainable way (therefore they can not be established immediately). Additionally, due to the strong influence of the climate on the agricultural and forestry yields, the international markets for wood and agricultural goods are characterized by partly significant price fluctuations. If a new high demander enters such a market (in this case the energy industry), the already unstable market equilibrium will be additionally disturbed; the expected result is a rise in the market prices. Due to such rising prices it is worthwhile for the farmer and the forester to expand the production with the consequence that the prices will probably fall again after a certain period of time. If thus the additional demand of the energy industry rises moderately, the prices could in the medium term move back to the original level. For residues, by-products and waste international markets are of clearly smaller importance.
- *Substitution*. Additionally substitution effects can appear (that means that one kind of biomass is replaced by another; *e.g.* replacement of woody biofuels by herbaceous biofuels). To what extent these options have an effect on the markets depends on the available or to be developed technical possibilities and the attainable respectively demanded prices on the market.

Level "(End-)energy Sources". On the background of an approximately stable energy consumption in Germany, (end-)energy sources provided from biomass necessarily compete with other energy sources providing a comparable energy service. These competitions result in changing market shares of the different available alternatives. This could result in a stabilization of the energy price level because of the declining demand for fossil energy sources.

The expansion of the use of biomass for heat provision in small scale units today already leads to noticeable substitution effects in using *e.g.* light fuel oil or even natural gas. This is intended by the government for environment and climate protection reasons and supported by numerous laws and regulations. Assuming a free energy market, the consumer will choose the most promising option. Thus supply and demand will be adjusted over the respective market price. Since the energy market is traditionally strongly influenced by political measures *e.g.* for environmental reasons, the definition of an appropriate framework ensures that the politically desired options have good development perspectives for the customer. Because the prices for energy for the final consumer will remain orientated on the fossil energy price level, the prices for end energy based on biofuels will also depend on the market. In Germany, for example, this is so far only true for the prices for the provision of heat and transportation fuels because within the field of electricity the prices are fixed by the feed-in law and therefore so far not noticeably affected by the price of the fossil energy sources. But even then the decoupling from the fossil energy price – if at all – is to be expected only in the medium to long term. Nevertheless it can come to effect in restraining increasing prices.

7.5.2   Effects on Competition

Altogether, there are wanted and unwanted competitions on very different levels, which can result in very different – intended and unintended – consequences. From this the following fundamental statements can be derived.

- Competitions do not lead *per se* to a lack of land areas or biomass. There are various compensation possibilities (*e.g.* expansion of the cultivated land areas, import, substitution, higher yields). Additionally there will be increasing price, which can be connected with appropriate feedback effects and therefore do not have to be assessed negatively *a priori*.
- Since the covering of the energy demand by biomass is supported by the legal framework, the increase of prices must be kept as small as possible. This is true because the markets for food, fodder and raw material (*e.g.* wood) coupled with the bioenergy markets should not be disturbed in a long lasting way.
- Presumably possible increase of prices can be kept small if the increase in the demand (concerning bioenergy) corresponds with the increase of supply (due to land areas no longer needed for food and fodder production). The speed of the expansion of the use of bioenergy within the energy system therefore should be oriented at the technical and organizational possibilities for the expansion of the biomass production; internationally it might be essential for the volatility of the markets.

- At the same time the market expansion must allow for an increased technical learning and must support as quickly as possible the development and market introduction of (energy) efficient systems. On the one hand this is necessary to allow the limited and valuable biomass to be used as efficiently as possible and on the other hand the cost reduction potentials can be exploited immediately. In parallel the development of a broad basis of the used biomass fractions is necessary to compensate possible price fluctuations of individual biomass fractions, for example as a result of climatically indicated yield variations.
- The rising prices, which can be expected in the case of overheated development of biomass and bioenergy use, could on the other hand support price stabilizing effects for the general price level of energy sources.

7.5.3 Configuration Approaches

The consequences of such competitions – and these are *e.g.* rising prices – can be reduced (but not completely avoided) if the political frame conditions for the further development of power production from biomass are set adequately. To identify the competitions least wished from a socio-economic and national point of view, appropriate goals for increased biomass use must be defined. Such goals could *e.g.* be:

- protection of climate and finite resources;
- economic development of rural regions;
- development of conversion technologies for the global energy markets to strengthen the national economy and to create (industrial) employment;
- security of supply (*i.e.* cost effective energy supply, sustainable energy system).

Against the background of the multiplicity of partly contrary possible goals – and therefore in the short term hardly possible definition of goals on which a political consensus might be reached – the identification of a comprehensive and widely acceptable strategy for the avoidance of unwanted competitions is almost impossible from the current point of view. Such a strategy can only be developed in the long term within a discourse throughout the overall society and/or a political process, which adequately considers the entitled (conflicting) interests of the groups concerned.

For the support of such a political process different indicators are presented exemplarily in Table 7.8 to be used for the evaluation of different targets. The indicators are limited to these material flows, for which from a present point of view a substantial "potential for competition" is to be expected. It has to be noted that the presented criteria are not final and not complete; certain aspects (*e.g.* aspects of import) are not considered. Nevertheless it becomes obvious that the different use paths show both strengths and weaknesses regarding defined goals.

Table 7.8 Analysis of different options for the use of biomass exemplarily for Germany.[14]

Usage path (Energy)	fuel**	Economy Specific actual costs (€ct kWh$_{output}$$^{-1}$)*	Economy Marginal costs on an oil basis^a (US$ barrel^{-1})	Climate and resources protection Level of fuel utilization - net^b (%)	Climate and resources protection Specific CO$_2$-equivalent-savings^c (g kWh$_{output}$$^{-1}$)*	Development of technology Development potential^d	Development of technology Specific investment costs (€ kW$_{output}$$^{-1}$)*	Development of rural regions Local processing possible	Security of supply Dependency from oil (calculations, orientated at "oil dependency" indicator of the World Bank)
("Pure") heat	solid biofuels	6-9	45-75	70-85	140-250	Low	300-1500	Energy provision and use	Medium
Power (CHP plants included)	Biogas	10-25	No direct correlation	40-55	480-550	Medium	2000-4500	Energy provision	Medium
	solid biofuels (combustion)	13-25		25-60	450-600	Medium	2400-7000	Energy provision	
	solid biofuels (gasification)	12-27		45-55	500-600	High	4000-6000		
	vegetable oil	15-37		20-60	170-320	Low	950-2350	Energy provision and use	
Fuel	vegetable oil	4-6(42-69)^f	40-75	35-65	170-180(85-95)^c	Low	40-380	Energy provision and use	High
	biodiesel	5-7(52-73)^f	55-80	50-60	185-250(95-125)^c	Low	120-210	Biomass provision	
	bioethanol from starch (corn)	8-12(68-104)^f	75-120	45-50	40-200(25-123)^c	Medium	2080-3740		
	bioethanol from lignocelluloses (straw)	~12(108)^f	~125	~45	210-360(100-190)^c	High	~2400		Very high
	synthetic biofuels (BtL)	8-10(86-93)^f	95-110	~45	200-205(100-150)^c	High	1900-2100		
	biogenous hydrogen	10-12(97-113)^f	110-125	~55	215-260(100-120)^c	High	6740-26,700		
Natural gas substitute (mainly fuel)	biogas	5-16(59-91)^f	60-100	45-85	160-245(100-150)^c	Medium	1470-5160	Energy provision	High
	synthetic natural gas Bio-SNG	~7-4(74)^f	~80	~65	160-180(100-110)^c	High	~990		

*for heat: kWh$_{th}$ respectively kW$_{th}$ (th: thermal); for power: kWh$_{el}$ respectively kW$_{el}$ (el: electrical); for fuels and natural gas substitute: kWh$_{fe}$ respectively kW$_{fe}$ (fe: fuel equivalent).

**If not indicated differently, the conversion of the fuel is based on combustion technology.

^{a}for liquid biofuels the costs for the distribution until the filling pump are included, but no profit margins for producers and traders.

^{b}rate of fuel utilization relating to the biomass (solid biofuel, corn, oil seeds) respectively the biogas potential of wet biomasses (manure, silage).

^{c}in relation to the appropriate references (heat: oil or gas burner/power: power mix in D; CHP is incorporated/fuel: petrol or diesel vehicles).

^{d}concerning an increase of the rate of utilization, CO$_2$ and cost reduction; development potentials for the mitigation of emissions (e.g. particulate matter) are not incorporated.

^{e}additional declaration in g km^{-1}

^{f}additional declaration in €ct l$_{fe}$$^{-1}$.

Regarding the political setting of the frame conditions within the (bio-)-energy system for minimizing unintended competitions the different, partly conflicting goals of an intensified biomass use for energy have to be evaluated and assessed in order to derive appropriate priorities.

It is rather unlikely – regarding the high complexity of the goals and interests – that there is a silver bullet, which can avoid all use competitions on all the different levels they might occur. This means that for the time being all promising pathways for the conversion of biomass into useful energy have to be developed further. Such an approach is also supported by *e.g.* the following considerations.

- Many biomass conversion pathways are still characterized by relatively new conversion technologies. By an appropriate and target-oriented research the costs and/or environmental effects, but also the used biomass fractions, might be optimized and the areas of application might broaden significantly. For example, the bioethanol production in Brazil, which has been developed continuously throughout the last three decades, today uses only a small fraction of the energy input and emissions compared to the time of its market introduction. Such improvements in technology, economics and environment are not fully foreseeable at the time, when decisions regarding market introduction or even intensified R&D are taken. Therefore the exclusion of certain technologies and/or conversion pathways is not recommended and thus a broad development of technologies for an efficient biomass use is promising.
- In the future larger energy crop potentials will be available. Therefore for the agricultural and – to a lesser extent – for the forestry primary production, advanced and improved crops as well as innovative cultivation systems are urgently needed. Presently discussed conversion pathways for biomasses containing special substances therefore need a certain biomass diversity and thus (compared with today) a larger range of energy crops used – with all advantages exerted by it.
- The international and EU-wide agricultural commodities markets showed in the past that – with certain restrictions – they can react flexibly to changing frame conditions. This does not lead to basic supply problems, but to increased prices for agricultural products. As a result the agricultural production might increase and prices decrease again. Therefore it should be possible in the future to serve growing markets by an expansion of the agricultural and forestry production – however, with initially and eventually durably higher prices.
- Finally the definition of acceptable political targets with high priority (*e.g.* security of supply versus rural development) might take a very long time. And the necessary discussion within society overall would slow down the further development of an increased use of biomass within the energy system. Furthermore the value of different political goals within society could change – depending on the political development and special events. Here the development of different goals in parallel, covering the majority

of the targets currently on the political agenda, offers certain stability against such changes and thus a better planning security for the players involved.

7.5.4 Conclusions and Recommendations

For environmental and climate protection reasons biomass should contribute significantly more to cover the demand for heat, electricity and fuels in the future energy system. Parallel to that biomass is to be used increasingly as a raw material (*e.g.* biorefinery) besides the raw material markets established for generations (such as *e.g.* the use as building timber or within the pulp and paper industry). And irrespective of that the globally increasing demand for food and fodder due to a still increasing population on Earth must be secured globally and locally. Therefore the question of whether biomass can fulfil all these partially very ambitious and competing goals adequately in time gains importance.

- By an increasing use of biomass – both as a source of energy and as a raw material – potential and real competitions of use can be caused.
- *A priori* this competition can not be avoided but conversely can stimulate innovative developments and hence help to exploit available efficiency potentials.
- To promote a socially acceptable and politically enforceable development path of an increased biomass use – under minimization of potential competitions – a clear political goal definition for the global human community would be helpful. Thereby the question has to be answered: which political goals in energy, environment, agriculture, economy and society can be realized with what priority under which frame conditions? But it is likely that considerable time will pass before the prevailing opinions converge in a certain direction.
- Therefore promising biomass conversion pathways should be developed and improved in order to be able to assess different options on a significantly improved database and, based on that, to propose political priorities based on facts and not on visions. These priorities would have to be updated regularly according to the latest technical developments and system analytic findings.
- Against the background of expanding biomass markets it can be expected that prices tend to rise on short notice. As a consequence due to the following national and international market adjustment processes more biomass could be made available. Additionally rising prices of the biomass help to exploit improvement potentials by the development of optimized plants and systems. This again has an influence on the biomass price.
- Finally price-stabilizing effects could be obtained on the use side by the substitution of fossil energy.

Altogether an expansion of the use of biomass as a source of energy leads – due to the huge variety of possibilities to use biomass within our society – inevitably to competitions on very different levels. This development thus necessarily disturbs the established markets first. Thus it comes to adjustment reactions and processes (*e.g.* prices rise on very different levels). This again supports an expansion of the primary agricultural and forestry production as well as the development of more efficient conversion systems and paths. In the medium term this development results in new market equilibriums – eventually even on, again, low price level. On the basis of such mechanisms, which can additionally be overlaid by other effects, competitions could indeed support and promote the further development of the use of biomass as a source of energy. Market expansion rates adjusted to the development of the potentials and political instruments supporting the market development aimed at planning dependability can help to reduce such increase of prices on the biomass markets. However, the possibilities to influence these markets by political measures are presumably significantly lower compared to oil price fluctuations and effects initiated by it.

References

1. German Federal Ministry for the Environment, ed., *Nature Conservation and Nuclear Safety*, Berlin, 2008, http://www.bmu.de/english/aktuell/4152.php.
2. Ö. Gustafsson, M. Kruså, Z. Zencak, R. J. Sheesley, L. Granat, E. Engström, P. S. Praveen, P. S. P. Rao, C. Leck and H. Rodhe, *Science*, 23 January 2009, **323**(5913), 495.
3. M. Kaltschmitt, W. Streicher and A. Wiese, ed., *Renewable Energy – Technology, Economics and Environment*, Springer, Berlin, Heidelberg, 2007.
4. M. Kaltschmitt, H. Hartmann and H. Hofbauer, ed., *Energie aus Biomasse*, 2. Aufl., Springer, Berlin, Heidelberg, 2009.
5. Fachagentur Nachwachsende Rohstoffe, ed., *Biokraftstoffe*, 3. Aufl., Gülzow, 2007.
6. D. Thrän, Personal communication, German Biomass Research Centre, Leipzig, 2008.
7. PYTEC Thermochemische Anlagen, Personal communication, Hamburg, 2008.
8. F. Müller-Langer, Personal communication, German Biomass Research Centre, Leipzig, 2008.
9. IEA Energy Technology Essentials, *Biomass for Power Generation and CHP*, ETE03, http://www.iea.org/Textbase/techno/essentials3.pdf (retrieved 23.01.2009).
10. M. Kaltschmitt, Lessons taught at Hamburg University of Technology, Winterterm, 2008/09.

11. J. Daniel, Personal communication, German Biomass Research Centre, Leipzig, 2008.
12. V. Lenz, M. Edel and M. Kaltschmitt, *Erneuerbare Energien – Stand 2007 in Deutschland*, BWK, 2008, **60**(4), 106–117.
13. A. Vogel, F. Müller-Langer and M. Kaltschmitt, *Chem. Eng. Technol.*, 2008, **31**(5), 755–764.
14. D. Thrän and M. Kaltschmitt, *Biotechnol. J.*, 2007, **2**, 1514–1524.
15. M. Thrän, M. Kaltschmitt, A. Kircherer and M. Piepenbrink, *Kriterienmatrix zur stofflichen und energetischen Nutzung nachwachsender Rohstoffe*, Erich Schmidt Verlag, Berlin, 2008.
16. M. Kaltschmitt and M. Weber, *Biomass & Bioenergy*, 2006, **30**, 897–907.
17. M. Kaltschmitt and D. Thrän, *Energiewirtschaftliche Tagesfragen*, 2008, **58**(7), 8–13.
18. M. Kaltschmitt and D. Thrän, *Zeitschrift für Energiewirtschaft*, 2008, **32**(2), 127–138.

CHAPTER 8

Green Fuels – Sustainable Solutions for Transportation

ECKHARD DINJUS,[a] ULRICH ARNOLD,[a] NICOLAUS DAHMEN,[a] RAINER HÖFER[b] AND WOLFGANG WACH[c]

[a] Forschungszentrum Karlsruhe, Institut für Technische Chemie, Postfach 3640, D-76021, Karlsruhe, Germany; [b] Cognis GmbH, Rheinpromenade 1, D-40789, Monheim, Germany; [c] Südzucker AG, Mannheim/Ochsenfurt, Wormser Str. 11, D-67283, Obrigheim/Pfalz, Germany

8.1   Introduction

The importance of biofuels is growing rapidly and this is reflected in steadily increasing research activities in both academia and industry as well as an increasing number of joint ventures comprising several institutions. Thus, not only is the extent of publications on this highly dynamic topic strongly increasing but so too, due to its obvious socioeconomic relevance, is public interest.

In the European Union (EU) the strategy with respect to the implementation of biofuels is outlined for the member states in several directives and communications such as the Renewable Energy Directive of the European Commission from 2003 (2003/30/EC), the Communications from the *Commission Biomass Action Plan* (COM(2005) 628 final), *An EU Strategy for Biofuels* (COM(2006) 34 final), *Towards a European Strategic Energy Technology Plan* (COM(2006) 847 final) and *An Energy Policy for Europe* (COM(2007) 1 final). Most recently, the European Commission has enacted a Renewable Energy

RSC Green Chemistry No. 4
Sustainable Solutions for Modern Economies
Edited by Rainer Höfer
© The Royal Society of Chemistry 2009
Published by the Royal Society of Chemistry, www.rsc.org

Directive due to boost EU renewable energy use to 20% by 2020. The latter also aims for a substitution of 10% of transport fuels by biofuels in 2020 as a mandatory target for the Member States. Furthermore, this directive also defines minimum sustainability standards, *e.g.* a 35% reduction of greenhouse gas emission and a focus on the type of land used for biomass cultivation.

A wide variety of biomass types based on vegetable organic materials can be used for the generation of biofuels. Food and feed crops, *e.g.* wheat, other cereals, rapeseed, sugar cane or sugar beet, and "energy crops" designed for non-nutritional consumption, *e.g.* willow, short rotation trees or grasses, can be used. To minimize competition with traditional applications, low-value materials such as agricultural residues and wastes, *e.g.* straw, bark, reclaimed wood, bagasse or waste paper, are largely regarded as the preferable resource. Value chain considerations, however, additionally credit value-added by-products like energy-rich proteins in the case of certain food and feed crops.

The biomass volume available for biofuel production in Europe was reported to be 95 Mio mto oil equivalent (Mt_{OE}) per year in 2005 with an estimated increase to a value between 112 and 172 Mt_{OE} in 2020[1] and 243 to 316 Mt_{OE} in 2030.[2] For comparison, predicted energy demands for the transport sector are 416.3 Mt_{OE} in 2020 and 437.2 Mt_{OE} in 2030.[2] Biomass yields under average conditions are assumed to be approximately 10 mto of dry biomass per hectare and year in the case of willow.

The term biofuels is used for all types of biomass-derived fuels employed in the transportation sector. Most biofuels can be applied either as a substitute for common fossil fuels such as gasoline and diesel or in blends with these fuels. Engines can need some modification as, for instance, in Flexible Fuel Vehicles (FFVs). Other biofuels, *e.g.* biomass-based natural gas, require major changes in both vehicle construction and fuel distribution infrastructure. Biofuels, mainly for historical reasons, are assigned to three main categories, the so-called *first-* and *second-generation biofuels* and the recently emerging *third-generation biofuels*.[3,4]

8.2   First-generation Biofuels

First-generation biofuels are obtained from cereal and oil crops using established technologies at commercial scales. Their CO_2 emission reduction is typically about half of the emission of fossil fuels. One reason for this is the energy-intensive production of fuels and crops. These biofuels comprise biodiesel, pure vegetable oil, ethanol, ethyl-tertiary-butylether and biogas. For historical reasons the use of wood gas as an alternative fuel might be mentioned. This was quite popular during and after the Second World War in several European and Asian countries because the war prevented easy and cost-effective access to oil. Wood gas (a mix of approximately 49 vol.% CO_2, 34 vol.% CO, 13 vol.% CH_4, 2 vol.% ethylene and 2 vol.% H_2) is produced by thermal gasification of wood in a wood gas generator.

8.2.1 Bioethanol[5]

Utilizing ethanol as a fuel is not a recent invention, but has been practised for almost a century and a half: Nikolaus August Otto developed his prototype of a spark-ignition engine in the 1860s using ethanol and was sponsored by the sugar factory of Eugen Langen envisaging an outlet for the mass production of ethanol.[6] Henry Ford expressed the vision "*to build a vehicle affordable to the working family and powered by a fuel that would boost the rural farm economy*". His 1908 Model T engine was originally designed to run on 100% ethanol. In the 1930s Rhenania-Ossag, Monheim (later Deutsche Shell AG), sold *Dynamin*, a blend of petroleum gas with 45% benzene and up to 10% ethanol from potatoes (called "Spiritus". or *Kartoffelsprit* in German; German *Reichskraftstoff* [RKS] developed shortly after the First World War even contained up to 25% Spiritus [95 vol.%] in a blend with 25% Tetralin® [tetrahydronaphthalin, manufactured by hydrogenation of naphtha][7] and 50% benzene[8]). Since the 1970s the Brazilian Proálcool initiative has become a kind of bonfire for sustainability in automotive transportation.[9,10] In Europe ethanol will also take a big share of the 10% minimum target set by the new EU Directive on the use of biofuels in transportation by 2020. The usual approach to meet requirements for biocomponents in the gasoline pool is to blend bioethanol directly (low blends and high blends up to 95% ethanol, called E95) with the effect of improved octane numbers of low blends. An alternative route is the use of ethanol as a raw material for ethers such as ethyl-tertiary-butyl-ether (ETBE), which can be blended with gasoline easily.[11] The use of biofuel in the transport sector in the form of bioethanol constitutes a cost-effective option to achieve reduction in greenhouse gas emissions.[12–14] In 2007 the worldwide ethanol production had reached 64,126 Mio litres with sales of 49,531 Mio litres for the fuel market.

8.2.1.1 Plant Raw Materials for Use in First-generation Bioethanol Production

Creation of ethanol in fact does not start in the factory but on farmers' fields *in planta* via plant photosynthesis. Photosynthesis, which mainly takes place in the green leaves of plants, leads to the final carbohydrates hexosephosphates. Hexosephosphates are converted to sucrose, the first free carbohydrate, which is the most important transport carbohydrate in plants.

The so called "storage organs" are very important parts of the plant which are strictly dependent on the supply with sucrose and other metabolites. A storage organ is generally any photosynthetic non-active part of the plant in which excess of energy (generally in the form of sugars, proteins or lipids) and nutrients are temporarily stored in order to release it again later – when it has become necessary – and use the stored material for future growth.

Storage organs with their high amounts of accumulated energy are those parts of the plants which are typically used as raw material in the bioethanol fermentation process. According to the nature of yeast (*Saccharomyces cerevisiae*), which assimilates glucose or sucrose (after cleavage by yeast invertase) as energy source, plant organs exclusively storing carbohydrates are most suitable for fermentation. As a result – and also because hexose-storing plants are scarce in nature – sucrose or starch accumulating plant organs are the best raw materials for bioethanol production. Consequently, one can distinguish between starch- and sucrose-based substrates for bioethanol production. Among the starch–based substrates the most important ones are the cereals, whose grains contain high amounts of starch (*e.g.* wheat, triticale, barley, maize, rice, sorghum). Indeed cereals are the most widely used substrates for bioethanol production in the USA, Canada and Europe; whereas maize (*Zea mays*) is predominant in the USA and eastern Canada, wheat (mainly *Triticum aestivum*) prevails in western Canada and Europe. Other starch-based substrates are tubers of potato (*Solanum tuberosum*) and taproots of cassava (*Manihot esculenta*); however, both play only a minor role as raw material for bioethanol production as compared to grains.

The most important sucrose-based plant raw material is sugar cane (*Saccharum officinarum*), which stores up to 15% of sucrose (w/w) in the stem internodes. Bioethanol production in South America, eastern Asia and Australia is almost exclusively based on the use of sugar cane molasses or sugar cane juice as substrate. Other sucrose-based substrates of importance are on one hand sweet sorghum (*Sorghum saccharatum*), which is closely related to sugar cane and, similar to cane, stores up to 15% of sucrose in the stem internodes. On the other hand there is sugar beet (*Beta vulgaris*), which store up to 20% sucrose in the taproot. Thick juice of sugar beet plays a significant role as a substrate in the European bioethanol production, since this crop has the highest bioethanol yield per hectare as compared to any other crop. However, difficulties in storage and the fact that sugar beet is not available over the whole year prevent a much wider use of this crop in bioethanol production.

8.2.1.2 *Production Processes for Bioethanol*

The overall production process for bioethanol has a clear advantage over some other industrial fermentation products. Due to the fact that all the input streams are converted to products such as bioethanol, gluten or a protein-rich feed stuff, so-called DDGS (distillers dried grains with solubles), fertilizer and a readily biodegradable wastewater the process can be seen as a zero-waste-concept.

8.2.1.2.1 Bioethanol Produced from Sugar Biomass. Production of bioethanol from sugar plants is a straightforward process: after conditioning of the feedstock an aqueous raw sugar juice (sucrose syrup) is extracted (see Figure 8.1 and Chapter 9.3) and directly submitted to fermentation.

8.2.1.2.2 Bioethanol Produced from Starch-containing Biomass. In the starch-producing process based on cereals the main components of the

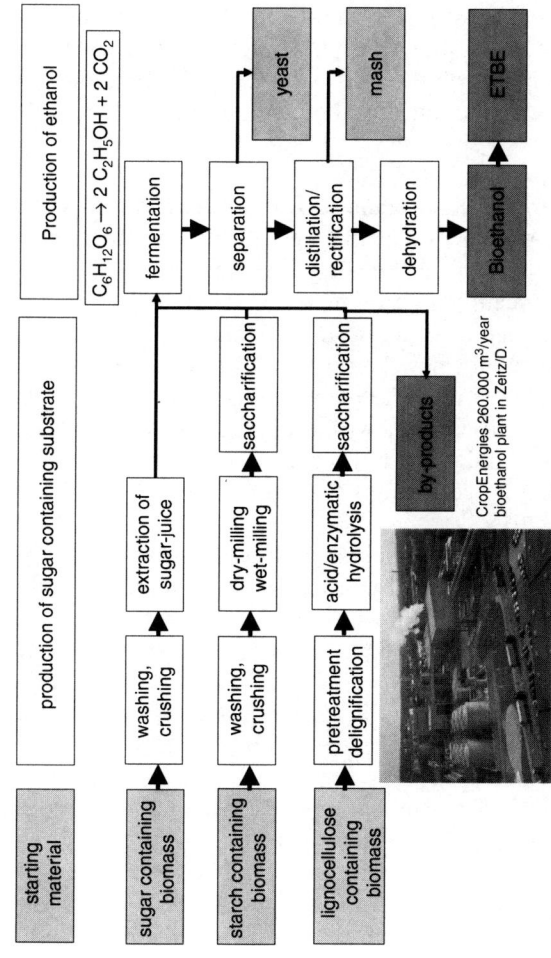

Figure 8.1 Bioethanol production processes (reproduced from J. Bigorra and R. Höfer, *Comunicaciones*, 38 Jornadas CED, Barcelona, 2008, pp. 179–200).

kernels, the endosperm containing starch granules, are separated from the bran and the germ by dry- or wet-milling (see Section 9.1.4.2.10). Wet-milling facilities are more flexible and have the ability to switch between the production of ethanol and the production of syrup and/or fructose. However, virtually all of the new starch bio-ethanol facilities being built are less costly dry-milling operations simply grinding the unprocessed heterogeneous seed into granules preferably using hammer mills.[15] Producing ethanol from starch-containing biomass instead of sugar adds an extra step to the process. This stage, called saccharification, consists in depolymerizing the amylose and amylopectin structures, which compose starch by enzymatic reactions into a glucose syrup or hydrolysate (see Chapter 9.2). In some concepts the saccharification and the fermentation take place simultaneously.[16,17]

8.2.1.2.3 Bioethanol from Lignocellulosic Raw Materials.

Ways to sustainably produce ethanol without diverting agricultural production away from food are seen in the utilization of lignocellulosic materials.[18,19]

Bioethanol from lignocellulosic raw materials is actually classified as bioethanol of the so-called second generation. The idea to produce ethanol from lignocellulose, however, is not new and a lot of work has been done since the 1920s. Most work and some industrial plants had stopped in the 1950s.[20] Today there are some demonstration plants with different new technologies especially driven by modern biotechnological approaches under construction. There is a strong agreement of all involved groups that technology will not be ready on an industrial scale before 2015.[21]

Lignocellulosic feedstock is biomass mainly consisting of a complex composite of the two structural carbohydrates, cellulose and hemicellulose, and phenolic lignin.[22] Cellulose is the most abundant organic compound on Earth (see Chapter 9.4) The orientation of the linkages and the additional hydrogen bonding make the cellulose polymer rigid. The crystalline structure of cellulose makes it insoluble and resistant to attack. Hemicellulose on the contrary has a branched, amorphous character and is relatively easy to hydrolyze. The name hemicellulose encompasses all polysaccharides based on polymeric hexosans generated from a variety of monomers (*e.g.* glucose, mannose, galactose) and all polymeric pentosans based on monomers like arabinose or xylose. The lignin, a non-carbohydrate polyphenolic substance encrusts the cell walls and cements the cells together (see Figure 8.2). This protective sheet of hemicellulose and lignin around the cellulose must be modified or removed by pre-treatment processes.

In comparison to the grain-ethanol process typical cellulosic-ethanol process steps require pre-treatment and enzymatic/chemical hydrolysis prior to fermentation. Pre-treatment basically refers to the mechanical and physical actions to clean and size the biomass, furthermore to destroy its cell structure by chemical, physical and/or biological methods to make it more accessible to further chemical or biological processes with the aim to increase the yield of hydrolysis (see Figure 8.3).

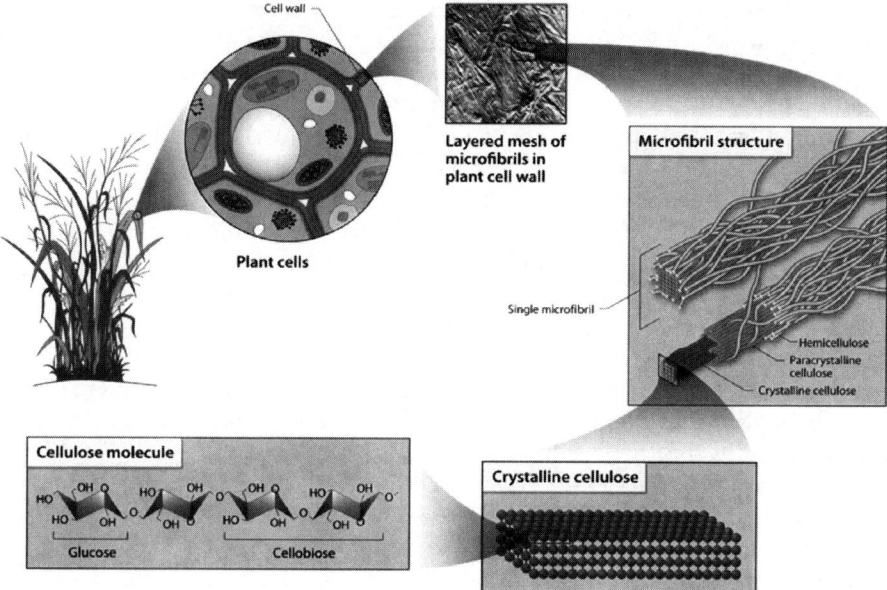

Figure 8.2 Cellulose structure and hydrolysis challenges (Genomics: GTL Roadmap, US Department of Energy Office of Science, August 2005, http://genomicsgtl.energy.gov/roadmap).

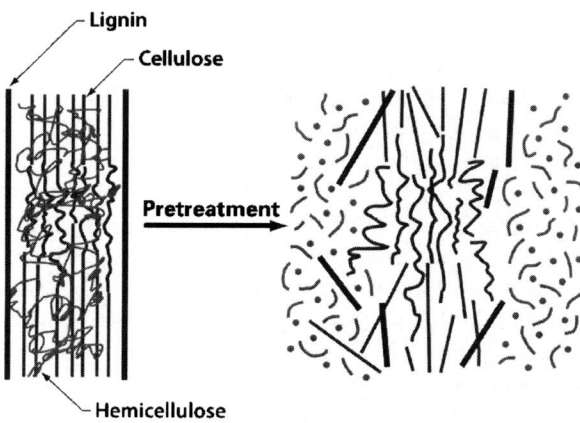

Figure 8.3 Pretreatment (Genomics: GTL Transforming Cellulosic Biomass, US Department of Energy, June 2006, http://genomicsgtl.energy.gov/biofuels/ and US DOE. 2006. Breaking the Biological Barriers to Cellulosic Ethanol: A Joint Research Agenda, DOE/SC/EE-0095).

The surrounding hemicellulose and/or lignin are removed, and the cellulose microfibre structure is modified. The hemicellulose polymer is hydrolyzed to a large extent in the pre-treatment step whereas cellulose hydrolysis needs an additional step (acid or preferably enzymatic).

Common chemical pre-treatment methods use dilute acid, alkaline, ammonia, organic solvent, sulphur dioxide or other chemicals.[23] Some pre-treatment methods can use only water such as steam explosion or liquid hot water (LHW).[24] Steam explosion is one of the most promising methods to make biomass more accessible to cellulose attack. The material is heated using high-pressure steam (20–50 bar, 210–290 °C) for a few minutes; these reactions are then stopped by sudden decompression to atmospheric pressure.

Several pre-treatment processes combine physical and chemical elements. Addition of dilute acid in steam explosion can effectively improve enzymatic hydrolysis, decrease the production of inhibitory compounds and lead to more complete removal of hemicellulose.

In hydrolysis, the cellulose is converted into glucose sugars. This reaction is catalyzed by dilute acid, concentrated acid or enzymes (cellulase). Two process concepts have been investigated, more than others, regarding ethanol production from lignocellulosic materials. On the one hand dilute sulphuric acid and on the other hand the use of cellulolytic enzymes. In the dilute sulfuric acid technology, which is mainly developed by NREL,[25] the raw material is treated with 0.1–3% (w/w) H_2SO_4 at temperatures normally ranging from 160 to 200 °C. It may be advantageous to perform dilute-acid hydrolysis in two steps since the hemicellulose fraction is more easily degraded than is the cellulose fraction. A disadvantage of the dilute acid process is the somewhat low ethanol yield and the necessity of using expensive construction materials that are resistant to corrosion by acid at high temperatures. The acid must also be neutralized, which leads to the formation of large amounts of gypsum, $CaSO_4$ or other compounds that have to be disposed of.

In the past the high costs of cellulolytic enzymes has been a major hurdle for a successful process development. A lot of effort has been made within the last five years from the world's leading enzyme companies Novozymes and Danisco Genencor during a funded US DOE project. Enzyme costs could be reduced significantly and are no longer prohibitive (Figure 8.4).

An important process modification made for the enzymatic hydrolysis of biomass was the introduction of simultaneous saccharification and fermentation (SSF). In this process, cellulose, enzymes and fermenting microbes are combined, with the intention of reducing equipment and improving efficiency.

The lignin will remain during processing and is in most concepts foreseen as solid fuel for power generation. Furthermore concepts to convert lignin into higher value products such as organic acids, phenols and vanillin are followed.

8.2.1.2.4 Ethanol Production by Yeast Fermentation. Many organisms have been exploited for ethanol production from saccharified starch or other sugars, however, the yeast *Saccharomyces cerevisiae* still remains the most important species.

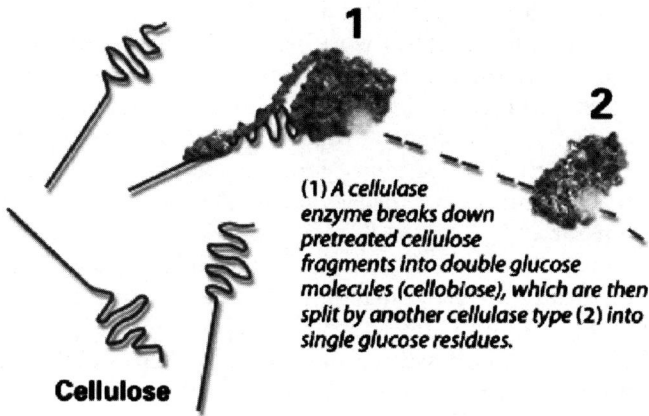

(1) A cellulase enzyme breaks down pretreated cellulose fragments into double glucose molecules (cellobiose), which are then split by another cellulase type (2) into single glucose residues.

Figure 8.4 Hydrolysis (Genomics: GTL Transforming Cellulosic Biomass, US Department of Energy, June 2006, http://genomicsgtl.energy.gov/biofuels/ and US DOE. 2006. Breaking the Biological Barriers to Cellulosic Ethanol: A Joint Research Agenda, DOE/SC/EE-0095).

The chemical reactions are shown below:

$$C_{12}H_{22}O_{11} \quad + \quad H_2O \quad \rightarrow \quad C_6H_{12}O_6 \quad + \quad C_6H_{12}O_6$$

Sucrose   Water   Fructose   Glucose

$$C_6H_{12}O_6 \quad \rightarrow \quad 2\ C_2H_5OH \quad + 2\ CO_2$$

Fructose/Glucose   Ethanol

Under aerobic conditions yeasts grow efficiently on sugar substrate making lots of biomass and carbon dioxide. However, when oxygen is absent, yeasts switch to the anaerobic metabolism. Under these conditions 1 mol of C-6 sugars are converted into 2 mol each of ethanol and carbon dioxide. Due to the fact that *Saccharomyces* produces some side products such as glycerol (osmoprotectant), acetic acid and other alcohols, the yield of ethanol is never 100%. Important for a high ethanol productivity of the yeast cells are the genetic composition as well as the amount and nature of fermentable sugars. Carbohydrates such as glucose, maltose and maltotriose derived from liquefied and saccharified cereal starch can be metabolized by *Saccharomyces cerevisiae*. Further, sucrose or galactose can easily be fermented by that yeast; however, large polysaccharides such as starch or oligo- and polysaccharides derived from cellulose can not be taken up in that form.[26] Additionally, *Saccharomyces* is not able to ferment either lactose or C-5 sugars such as arabinose or xylose, which makes it unsuitable for the production of ethanol from whey or saccharified biomass. Therefore a lot of R&D work on genetically modified micro-organisms has the task to valorize the C-5 fraction of the hydrolyzed biomass.[27]

Due to the fact that the extracellular proteolytic activity of *Saccharomyces* is negligible, high-molecular-weight nitrogenous materials can not be utilized by the cells. Therefore the application of inorganic ammonium ions (*e.g.* ammonium sulphate, di-ammonium hydrogenphosphate), urea, amino acids or small peptides is necessary.[28]

At higher ethanol concentrations the intracellular alcohol interferes with membrane organization, increasing its fluidity and permeability to ions and small metabolites and inhibiting transport of nutrients.[29] Especially Ca^{2+} and Mg^{2+} ions are able to increase the plasma membrane stability. It has been demonstrated that incorporation of unsaturated fatty acids and/or sterol(s) as well as proteolipids into cellular membrane of yeasts helps to alleviate ethanol tolerance.[30,31] For the synthesis of the unsaturated fatty acids the presence of traces of oxygen under fermentation conditions is required.[32,33] Further to Ca and Mg ions, other trace elements such as Co, Cu, Mn and Zn[34] and vitamins, *e.g.* pantothenate, thiamine, riboflavin, nicotinic acid, pyridoxine, biotin, folic acid and inositol, are essential for the growth and ethanol production by yeasts.[35]

During ethanol fermentations, yeasts suffer from various stresses. Ethanol as the major metabolic product of yeast fermentation accumulates in the cell and acts as a potent chemical stress towards the yeast cell. Further, temperatures higher than 35 °C, pH values below 3.5, acetic acid produced either by the yeast itself or by contaminants such as lactic acid bacteria or wild yeasts, lactic acid and osmotic pressure have a negative impact on the ethanol production by *Saccharomyces cerevisiae*.

8.2.1.2.5 Ethanol Fermentation with Bacteria. In the 1980s the use of thermophilic bacteria for the production of ethanol was the focus of study for a number of research groups. The idea behind it was the facilitated recovery of ethanol at elevated temperature, the lower costs for heating a thermophilic fermentation rather than cooling a mesophilic one, increased process stability, lower contamination risk and the higher substrate versatility of thermophilic bacteria compared to that of the mesophilic ones.[36] Thermophilic clostridia such as *C. thermocellum*, *C. thermohydrosulfuricum* and *C. thermosaccharolyticum* are able to use a wide range of substrate, from polymeric carbohydrates such as cellulose, pectin, xylan and starch to mono- and disaccharides such as glucose, cellobiose, xylose and xylobiose. A disadvantage of these micro-organisms is the product pattern. Besides ethanol the side products acetate, lactate, carbon dioxide and hydrogen are also formed in various amounts.

Another promising micro-organism for the production of fuel ethanol is *Zymomonas mobilis*.[37] This facultative anaerobic, Gram-negative bacterium degrades glucose by the so-called Entner-Doudoroff-pathway with the consequence that only 1 mol of ATP is produced during the breakdown of

1 mol glucose. This means that less biomass than by *S. cerevisiae* is generated and more carbon is converted to ethanol fermentation.[38] It was reported that the ethanol yield of *Z. mobilis* could be as high as 97% of the theoretical yield of ethanol to glucose.[39] However, despite these advantages *Zymomonas mobilis* is not applicable for the ethanol industry. First of all the species has a very specific substrate spectrum including just the three sugars glucose, fructose and sucrose. Maltose or maltotriose – generated by the liquefaction of starch by α-amylases – can not be converted by this bacterium. Another disadvantage is that its growth on sucrose is accompanied by the formation of extracellular fructooligosaccharides (levan) and sorbitol.[21] Finally, the use of *Zymomonas mobilis* is not commonly accepted as animal feed and this generates problems for its biomass disposal if being replaced by *Saccharomyces cerevisiae* in the industrial ethanol production. Therefore *Saccharomyces cerevisiae* is still the workhorse for the industrial synthesis of fuel ethanol.

8.2.1.2.6 Distillation/Rectification/Dehydration. After fermentation the mash should contain more than 12% of alcohol (V/V) and virtually neither sugars nor starches. The removal of the alcohol from the mash is performed by distillation.

The setup of the distillation unit can vary significantly with the type of process, raw material supplied and the desired product quality. Usually, the unit consists of at least two steps:

- removal of the ethanol from the mash (distillation);
- purification of the raw alcohol (rectification).

The distillation and rectification is conducted in columns with a variable number of trays, the construction and design of which vary significantly. The setup of the distillation unit is directly linked to the quality requirements for the ethanol produced. The distillation yields a raw alcohol with an alcohol content of approximately 85–87%. The rectification is necessary to increase the alcohol content and to remove so-called fusel oils, which are C_3-C_5 alcohols. In modern distillation units, these two basic steps are integrated to optimize the energy requirement for the process.

For the production of fuel-grade ethanol, the ethanol has to be "dried". Anhydrous ethanol cannot be produced by simple distillation because ethanol forms an azeotropic mixture with water. The maximum ethanol content achievable by distillation is approximately 97.2 vol.%, which is usually not sufficient for the application as fuel-ethanol. The residual water can be removed either by azeotropic distillation by the addition of, *e.g.*, cyclohexane or by the application of molecular sieves. Today, state-of-the-art plants operate with molecular sieves which provide considerable advantages in terms of investment and operating costs.

Whereas the aqueous alcohol is further rectified and dried, the stillage is separated in decanter centrifuges into a more liquid and a solid fraction.

The liquid is partly recycled to the flour to be added together with fresh water to get the slurry prior to liquefaction. The recycling degree must be as high as possible to avoid too much freshwater consumption but is on the other hand limited by the potential negative effect of components from back-stillage on the yeast.

This recycling of the thin stillage and the reduction of freshwater consumption are crucial to reduce energy consumption – besides other measures such as multistage distillation and rectification. In today's downstream processing an additional means is applied to save energy: the vapours from the consecutive stillage drying process are often used as an energy source for concentration of the thin stillage prior to drying. The dried stillage – named DDGS – is used as protein rich feed.

8.2.1.2.7 Genetically Modified Organisms. Traditionally, *Saccharomyces cerevisiae* is used in industrial ethanol fermentation; however, *S. cerevisiae* is not able to utilize pentoses, which present a significant fraction of the sugars present in lignocellulosic material.

Due to their prokaryotic nature the genetic manipulation of bacteria is much easier than that of yeasts. Nevertheless, much effort has been undertaken in the last years to manipulate the more ethanol tolerant yeasts towards the ability to metabolize C-5 sugars, especially xylose and arabinose.[31–35,40–44] Further, ethanol yields of about 20–30 g L^{-1} are still too low to make these organisms industrially profitable.

8.2.1.2.8 Ethanol Usage as Transportation Fuel. Energy content of ethanol is approximately two-thirds that of petrol. Nevertheless ethanol is worldwide the most used biofuel. Up to 20% of ethanol can be blended to fossil petrol without the need for engine manipulations. However, Flexible Fuel Vehicles (FFVs), tolerating ethanol, petrol and variable mixtures of both (more particularly the high blend E85 of 85% bioethanol and 15% petrol), have become widely accepted and are offered by several automobile manufacturers. In Brazil, as an example, 86% of cars and light commercial vehicles sold in 2006 were FFVs.

In Germany, however, bioethanol will mainly be used as low blend of bioethanol and petrol (E5, E10) and for the production of ETBE (ethyl-tertiary-butyl-ether). Similar to methyl-tertiary-butyl-ether (MTBE), ETBE is used to enhance the octane index and improve knock-resistance and combustion properties of gasoline.[45] It is less challenging to handle, does not induce evaporation of gasoline and does not absorb moisture like ethanol does. ETBE is produced by reacting ethanol and isobutylene *via* acid catalyst:

$$H_3C \underset{H_3C}{\overset{}{>}}\!\!=\!\!CH_2 \xrightarrow{\ HO-C_2H_5/\ H^{\oplus}\ } H_3C \underset{H_3C}{\overset{CH_3}{>}}\!\!<\!\!O\!-\!C_2H_5$$

8.2.1.2.9 Outlook for Bioethanol. Production of ethanol from starch-based crops such as wheat and corn is a well-known technology and is reaching a level of maturity. In contrast, processes using lignocellulosic raw materials are still under development and significant technology jumps with reductions in ethanol production cost can be expected in the mid-term. Feedstock infrastructure, pre-treatment and enzymatic processes have to be flexible regarding feedstock and feedstock compositions. High ethanol yields need to be reached. In the short term concepts with an integration of cellulose-to-ethanol processes into first generation plants are rather promising.

8.3 Lipid-based Biofuels

Lipids, in the form of adipose tissue triacylglycerol, intramuscular triglyceride and dietary-derived fatty acids from plasma triacylglycerol and very low-density lipoproteins, represent the largest store of nutrient energy in humans. As a stored source of energy, fat has an advantage over carbohydrate: the energy density is higher while the relative weight is lower. Fatty acids provide more adenosine triphosphate (ATP) per molecule than glucose. Ultimately, more energy can be derived per gram of fat ($9 \, \mathrm{kcal \, gm^{-1}}$) than per gram of carbohydrate ($4 \, \mathrm{kcal \, gm^{-1}}$) or protein ($4 \, \mathrm{kcal \, gm^{-1}}$). Although there is limited capacity for fat oxidation during exercise, endurance athletes and dieters are eager to burn more fat during exercise. Indeed, men and women with higher levels of energy expenditure due to sport are claimed to show a better lipid profile than their sedentary counterparts.[46]

The high energy density, the liquid character and the structural analogy to hexadecane (*cetane*, the high-quality standard of the ignition quality of a diesel fuel, *i.e.* cetane number = 100) make lipids likewise suitable as an "alternative" diesel fuel source.[47]

The dual capacity as energy source for human nutrition and as fuel for transportation emphasizes the important role of lipids on one side and reveals the conflict relative to sustainable availability for both food and fuel applications on the other.

8.3.1 Vegetable Oils as Transportation Fuels

"Oil from the Sun" has been envisaged as fuel since the very beginnings of automotive transportation. When Rudolf Diesel presented the prototype compression-ignition engine built by his French licensee Société Française des Moteurs à Combustion Interne at the 1900 World Trade Show in Paris he was awarded the "Grand Prix" because of the innovative concept, the efficiency, the performance and the economics of the engine. Diesel demonstrated that all kinds of (cheap) petroleum products like mineral oil, petroleum fuel, shale oil or coal tar dust could be used to fuel the engine.[48] He also reported a diesel engine shown at the 1900 World Trade Show, working with peanut oil, and demonstrated himself the suitability of peanut oil as an alternative to petroleum fuel in compression–ignition engines. When presenting the performance figures of the test, Rudolf Diesel actually made the visionary statement:

"The fact that fat oils from vegetable sources can be used may seem insignificant today. But such oils may become in the course of time as important as the petroleum and coal tar products of the present time" . . . "One cannot predict what part these oils will play . . . in the future. In any case, they make it certain that motor-power can be produced from the heat of the sun . . . even when all our natural stores of solid and liquid fuel are exhausted."[49]

Interest in energy self-sufficiency of tropical colonies or other remote areas and periodic petroleum shortages spurred research into vegetable oil as a petroleum diesel substitute during the 1930s and the Second World War, and again in the 1970s and early 1980s. However, the high viscosity of vegetable oils (about 35–60 cSt compared to 4 cSt for diesel fuel at 40 °C) in fact leads to poor atomization in the engine and incomplete combustion. As a result formation of carbon deposits, injector coking and piston ring sticking may occur. Moreover, high viscosity and low volatility risk give rise to poor cold starting, misfire and long ignition delay.[50] If *Straight Vegetable Oil* (*SVO*) also referred to as *Pure Plant Oil* (*PPO*) (the term used in the biofuels directive published by the European Commission is "*pure vegetable oil from oil plants*") is to be used in conjunction with diesel in a dual-fuel mode,[51–53] necessary modifications include an additional fuel tank, a system to allow switching between the two fuels and a heating system for the SVO tank and lines. If SVO is to be used exclusively, modifications would include a preheating system, an upgraded injection system and the addition of plugs in the combustion chamber.[9] Although the 1970s already saw the formation of the first commercial enterprise, Elsbett, to allow consumers to run straight vegetable oil in their automobiles,[54] R&D strategies started aiming at the adaptation of alternative fuels to existing compression–ignition technologies instead of developing new engines capable of being fuelled by vegetable oils instead of petroleum diesel.

8.3.2 Vegetable Oils as Biodiesel Feedstock

The most appropriate way to adjust the viscosity of vegetable oils or animal fats for use in unmodified diesel engines is by converting the triacylglycerides into lower molecular weight fatty acid mono-alkyl esters standardized as *biodiesel*, when meeting the requirements of ASTM D 6751 and Euro norm EN 14214, respectively. The term *biodiesel* refers to the pure biofuel (designed B100) before blending with diesel fuel. Biodiesel blends are denoted as "BXX" with "XX" representing the percentage of biodiesel contained in the blend (*e.g.* B20 is 20% biodiesel, 80% petroleum diesel).

The principles of biodiesel synthesis are relatively simple oleochemical reactions (see Chapter 9.1) and have been known and applied for many decades (Figure 8.5).[47,55] The basic technology consists in a catalyst induced transesterification of a vegetable oil in a batch, semi-batch or continuous process to create a fatty acid methyl ester (FAME). The catalyst used is a strong base,

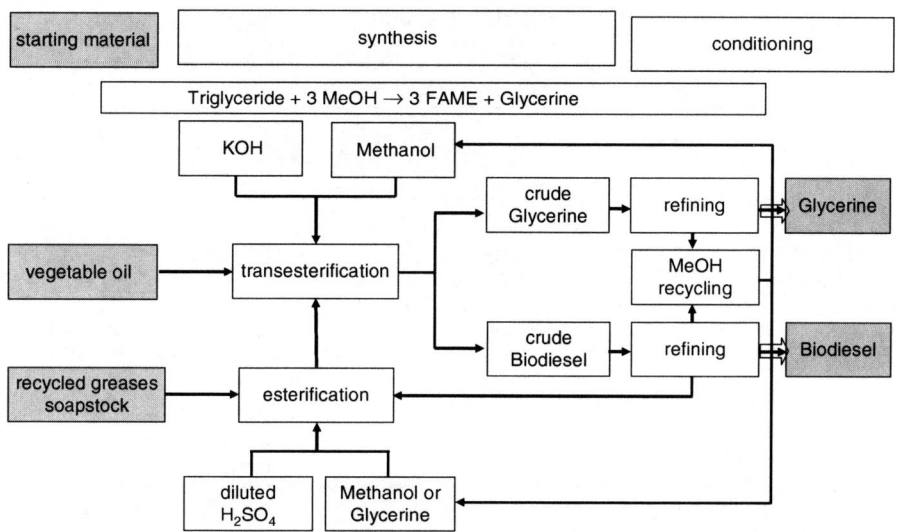

Figure 8.5 Production of biodiesel (reproduced from R. Höfer and J. Bigorra, *Green Chem. Lett. Rev.*, June 2008, **1**(2), 79–97).

such as sodium or potassium hydroxide. Once the ester chains are broken off, the leftover glycerine molecule is a valuable by-product of the reaction.[56] Different feedstock can be used, such as more or less any vegetable oil, including recycled oils and soapstock.[57] Process improvements such as higher reactivity, more complete conversion, less use of base initiator and faster separation time to increase throughput are claimed when performing the transesterification in the presence of a phase-transfer catalyst[58] or when using a stable ionic liquid as both solvent and catalyst.[59] Technical feasibility of bio-diesel synthesis by alcoholysis of vegetable oils catalyzed by commercial lipases has also been described.[60] Finally, a dramatic improvement in the rate of the transesterification reaction has been demonstrated in selected micro-reactor systems.[61]

Pure plant oils (PPO) can be used directly as a starting material for biodiesel synthesis *via* transesterification. To reduce free fatty acids, raw materials containing such are submitted to an esterification step prior to transesterification.

Composition of the oil feedstock, more particularly the fraction of saturated fatty acids, has a decisive influence on biodiesel specifications and properties; more particularly on viscosity, consistency and wax crystal settling, *i.e.* plugging at low temperature. Fuel properties (Table 8.1) and performance as engine fuel in comparison to petroleum diesel (PD) have continuously been studied following the development of new motor generations and machine applications including exhaust emissions and mutagenic potential of particulate matter as well as exhaust-gas limit value requiremens.[62] Compared with PD European biodiesel based on rapeseed methyl ester (RME) has similar viscosities and a remarkably higher cetane number. The caloric value of RME relative to the

Table 8.1 Influence of saturated fatty acids fraction on specifications of biodiesel.

	Oil and ester characteristics					
	Melting range (°C)		Iodine	Cetane	CFPP	Saturated fatty acids
Type of oil	Oil/fat	FAME	value	number	(FAME)	(%)
Rape seed	−5	−10	110–120	55	−20; −10; 0*	7–8
Soy bean	−12	−10	120–140	53	−4	12–15
Palm	30–38	14	44–54	65	10–14	45
Petroleum diesel fuel				min. 51	DIN EN 590: −20; −10; 0* *seasonal: Winter −20 transition −10 summer 0	

CFPP = cold filter plugging point, DIN EN 14 214, the temperature at which a fuel will cause a fuel filter to plug due to fuel components, which have begun to crystallize or gel.

mass is 13.4% below that of PD, relating to the volume 7.9%. As RME contains oxygen the minimal air consumption decreases by 13% regarding the mass, and by 7% regarding the volume. The caloric value of the air-fuel mixture, however, is almost identical in both cases.

RME is biodegradable and non-toxic. The flash point of 130 °C (compared to ≥ 55 °C for PD) ensures a safe storage. Compared to petroleum fuels RME has a more favourable combustion emission profile such as low emissions of carbon monoxide, sulphur and particulate matter. As a consequence of the so-called NO_x-particle trade-off phenomenon, emissions of NO_x are increased. Mutagenicity of RME emissions, however, is much lower compared to PD indicating a reduced health risk from cancer.[63]

8.3.3 Fats and Oils as BtL Raw Material

The hydrogenolysis of esters to alcohols is a reaction between esters and hydrogen which selectively splits a C–O bond adjacent to a carbonyl group. Hydrogenolysis of fatty acid methyl esters under methanol circulation and with glycerine and methanol recovery is actually the preferred technology for industrial manufacturing of natural fatty alcohols (see Chapter 9.1). Fatty alcohol synthesis starting directly from natural fats or vegetable oils yields 1,2-propanediol instead of glycerine as by-product. During fatty alcohol manufacturing reaction temperatures and choice of the catalyst are critical in order to avoid over-hydrogenation to hydrocarbons.

Whereas the biodiesel synthesis from natural fats or vegetable oils does not alter the ester structure of glycerolipids, the intentionally conducted over-hydrogenation has recently created considerable interest as a new synthesis

Table 8.2 Fuel property comparison (reproduced from http://www.green-carcongress.com/2006/03/neste_oil_and_o.html).

	NExBTL	*GTL*	*Biodiesel (RME)*	*Sulphur-free diesel*
		Fuel property comparison		
Density at +15 °C (kg m^{-3})	775 . . . 785	770 . . . 785	∼885	∼835
Viscosity at +40 °C (mm^2 s^{-1})	2.9 . . . 3.5	3.2 . . . 4.5	∼4.5	∼3.5
Cetane number	∼84 . . . 99	∼73 . . . 81	∼51	∼53
Cloud point (°C)	∼−5 . . . −30	∼0 . . . −25	∼−5	∼−5
Heating value (lower) (MJ kg^{-1})	∼44	∼43	∼38	∼43
Heating value (MJ litre^{-1})	∼34	∼34	∼34	∼36
Polyaromatic content (wt.%)	0	0	0	0
Oxygen content (wt.%)	0	0	∼11	0
Sulphur content (mg kg^{-1})	<10	<10	<10	<10

route to *Green Diesel*. Catalytic hydro-treatment comprising a Group VIII and Group VIB element catalyst (especially Pd, Pt, Ni, Ni-Mo or Co-Mo supported on Al_2O_3, SiO_2 or carbon) and conducted in a semi-batch or continuous process in a (depending on process conditions) broad temperature and pressure range of 300–360 °C and 6–80 bar results in a complete de-oxygenation, *i.e.* cracking of the triacylglyceride structure and converting the glycerol part into propane and the fatty acid moieties into linear hydrocarbon molecules similar or even identical to *cetane*.

$$CH_3(CH_2)_7CH=CH(CH_2)_7C(O) - OH + xH_2$$
$$\rightarrow nC_{17}H_{35} + xH_2O + yCO_2 + zCO$$

These are isomerized to branched paraffins in a second stage. The resulting middle distillate is suitable for use as diesel fuel without particular specifications. Some commercial examples of vegetable oil hydrotreatment are NExBTL (Neste), H-Bio (Petrobras), Ecofining (UOP) and the ConocoPhilips process.[64]

Neste Oil claims for its industrially very advanced NExBTL process that it generates less life-cycle CO_2 per kilogram oil equivalent than rapeseed-based biodiesel: 0.5–1.5 kg CO_2 kgoe^{-1} fuel for NExBTL, versus 1.6–2.3 kg CO_2 kgoe^{-1} fuel for biodiesel and 3.8 kg CO_2 kgoe^{-1} for petroleum diesel.

8.3.4 Lipid-based Jet Fuels

Recently Boeing 747 test flights of Virgin Atlantic Airways and Air New Zealand using "Biojet Fuel" blends have created public interest.[65] Actually, jet fuel is based on either an unleaded paraffin oil (*Jet A-1*) or a naphtha-kerosene blend (*Jet B*). It is similar to diesel fuel, and can be used in either compression ignition engines or turbine engines. Indeed, a specially adapted biodiesel fraction (*bio kerosene*) has been successfully tested already at the beginning of the 1980s in an EMBRAER turbo-prop powered aircraft flying from São José de los Campos to Brasilia.[66] While Virgin Atlantic tested a 20% blend of biojet fuel derived from babassu and coconut oil by transesterification, Air New Zealand used a 50% biojet fuel blend based on jatropha oil. Major technical challenges faced by biojet fuels up to now are propensity to freeze at normal operating cruise temperature and thermal stability characteristics in the engine.[67]

8.3.5 Conclusions for Lipid-based Biofuels

Fuels based on biomass may be the only solution when fossil fuels gradually become exhausted but a tremendous rift needs to be overcome before renewable resources can completely substitute fossil fuels. If the emerging energy gap and the fuel crisis are to be solved while at the same time providing sufficient feedstock for the chemical industry and foodstuffs for humans and animals then a well-organized, anticipatory cross-border cultivation policy including water management is required. In this scenario, natural fats and oils as biofuel sources offer well-established pre-treatment processes and energy density, physico-/chemical characteristics and handling properties similar to petro-fuel (Figure 8.6).

8.4 Methane *via* Anaerobic Digestion

Biomethane (biogas) is produced by anaerobic digestion or fermentation of biomass, manure or biodegradable wastes, *e.g.* sewage sludge. The resulting gas is predominantly composed of methane and carbon dioxide. Biomethane can be used for electricity production or heating (see Chapter 7) and, if cleaned and compressed, as fuel for combustion engines or fuel cells.[68,69]

Increasingly, it is aimed at feeding of biomethane to the natural gas grid as green gas. Thorough cleaning is necessary to meet the demanding specifications. In this context, biomethane is also used as a transportation fuel today; more than 5 Mio. methane-fuelled cars are in use worldwide today. Compared to all other biofuels, biomethane provides the highest yield of fuel equivalent per area. On the other hand long residence times of several days are required for biogas formation using mainly wet feed material, meaning the use of huge reactors and limited feedstock capacity by considering transport economics.

8.5 Second-generation Biofuels

Second-generation biofuels are produced from lignocellulosic materials, *e.g.* low-value crops and residues from agriculture or forestry[3,4] or purpose-grown

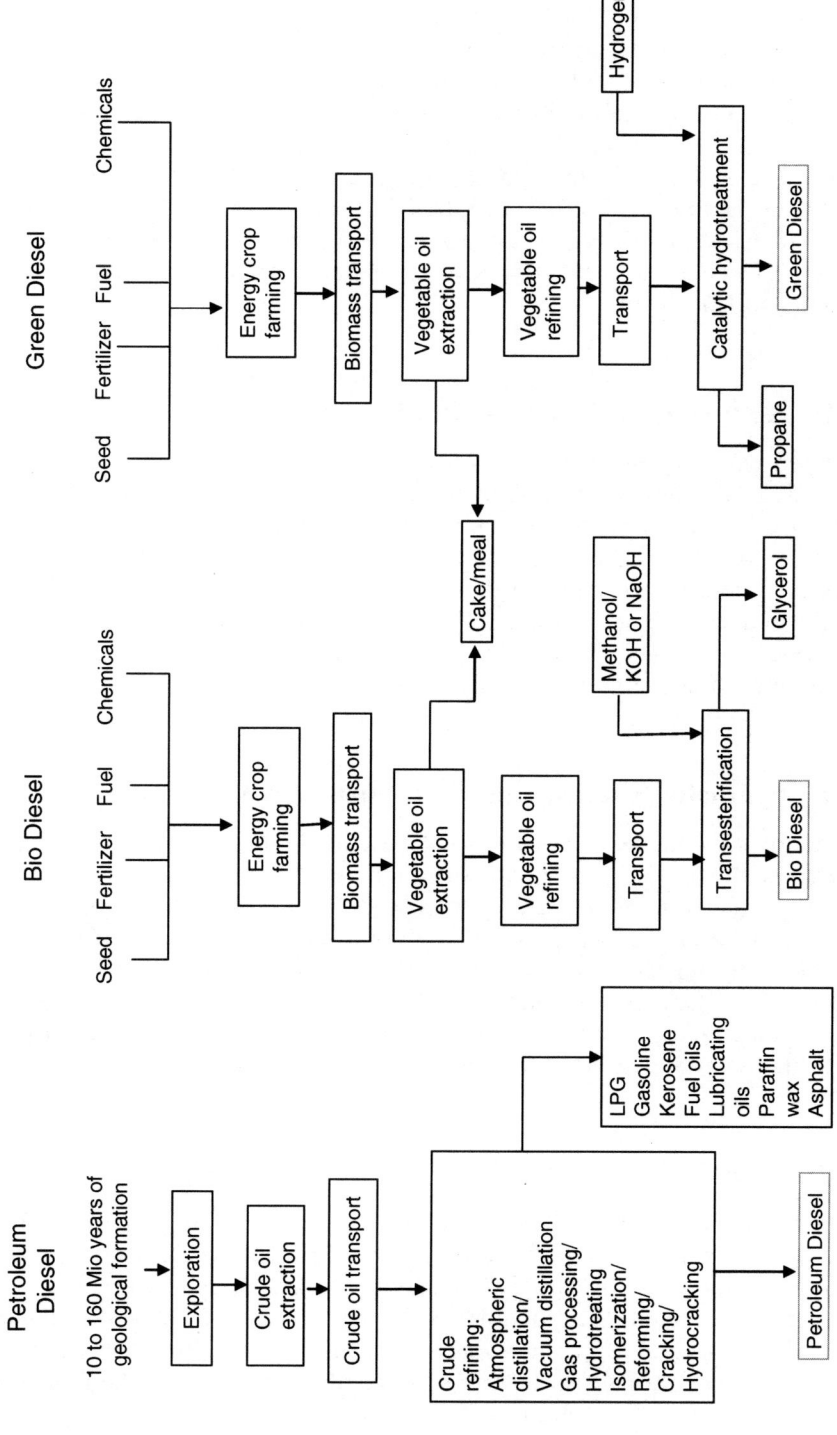

Figure 8.6 Flow diagram of lipid-based fuels compared to petroleum diesel.

plants from short rotation forestry. Compared to first-generation biofuels, extended raw material options offer a significantly improved production capacity and CO_2 reduction potential. Furthermore, such feedstock may be considered more sustainable and does not compete directly with food. Advanced production technologies are employed which have become available now, even on a commercial scale. The most important second-generation biofuels are hydrogen, synthetic natural gas (SNG), bioethanol from lignocellulosic raw materials (Section 8.2.1.2.3), butanol and HTU diesel as well as the extensive class of **B**iomass **t**o **L**iquid (BtL) fuels based on synthesis gas (syngas). The latter fuel type comprises methanol, ethanol, dimethylether, BtL gasoline and BtL diesel.

8.5.1   Hydrogen *via* Biomass Gasification

Hydrogen can be produced *via* gasification of biomass,[70] provided that the gasification procedure is optimized to yield a high hydrogen content in the resulting gas mixture. A downstream water gas shift reaction can be used to reduce carbon monoxide contents and thus increase hydrogen yields. Further options for hydrogen production from biomass are supercritical water gasification[71] and dark fermentation.[72–80] Crude biohydrogen has to be cleaned, compressed or liquefied, or stored in suitable storage media and can be used as fuel for appropriate combustion engines or fuel cells.

8.5.2   Synthetic Natural Gas *via* Biomass Gasification

In addition to methane production by anaerobic biomass digestion, methane can also be obtained by biomass gasification and the latter technology allows for the use of a wider variety of biomass types compared to the former. Thus-produced methane, so-called synthetic natural gas (bio-SNG), can be used, after cleaning and compression or liquefaction, as fuel for modified spark ignition engines and exhibits high octane numbers. Another option to produce SNG is supercritical water gasification of biomass.[78,79] Naturally, such a process is highly attractive, since wet biomass can be employed.

8.5.3   Biobutanol

Acetone, butanol and ethanol can be produced in a ratio of approximately 3 : 6 : 1 from sugars derived from food crops or sugars obtained from the degradation of cellulose using *Clostridium* bacteria (ABE fermentation). Butanol/diesel and also butanol/gasoline blends can be used in unmodified or slightly modified engines and represent another attractive option. However, as in the case of ethanol, cellulose conversion to fermentable sugars and the following fermentation have to be optimized.[80]

8.5.4 HTU Diesel

Conversion of biomass at a temperature of 300–350 °C and a pressure of 120–180 bar within the so-called **HydroThermal Upgrading (HTU)** process yields a mixture of hydrocarbons, carbon dioxide, water and dissolved organics, which can be further processed in a catalytic **HydroDeOxygenation (HDO)** step to yield diesel with characteristics similar to fossil diesel.[81] A major advantage is that wet biomass feedstocks can be employed without drying; in contrast, water at hydrothermal conditions acts as a solvent and reactant at the same time, leading to a product with less oxygen compared to biocrude prepared by pyrolysis.

8.5.5 Pyrolysis Oil

Pyrolysis of biomass produces gaseous, liquid and solid products by thermal cracking in an oxygen-free atmosphere. The burnable gas and the char may be used internally for heating the pyrolysis reactor. Char is also a suitable substitute for coke produced from fossil fuels. By applying fast or flash pyrolysis, high quantities of liquid condensates up to 60 wt.% on a water- and ash-free basis are formed, which principally may be used as a fuel. However, pyrolysis oil is not a suitable fuel being a mixture of *ca.* 400 different chemical compounds, including substantial amounts of water (15 wt.% water formed during fast pyrolysis reaction + the humidity of the biomass feedstock) and oxygenated compounds lowering further the heating value. Having tested as a fuel for diesel engines and turbines concerns arose in regard to fuel properties and in particular to corrosion, fouling and plugging of engine components.[82,83] Upgrading of bio-oils to fuel standards may be achievable, but requires expensive treatment *e.g.* by hydrotreating. However, fast pyrolysis may be a suitable method for biomass pre-treatment prior to efficient gasification (see Section 8.5.6.2).

8.5.6 Syngas-based Biofuels

Employing BtL technologies, almost every type of biomass, organic residue or waste can be used for the production of so-called (bio)synfuels. Depending on the respective process, these can be very similar or even superior to conventional petroleum-derived gasoline or diesel and can serve as direct substitute thus maintaining engine types and fuel distribution infrastructure. The technologies are similar to already established **Gas to Liquid (GtL)** or **Coal to Liquid (CtL)** processes and proceed *via* gasification of suitable biomass-derived materials to syngas. The gas is cleaned in the next step and the hydrogen/carbon monoxide ratio is adjusted to the respective requirements strongly depending on the desired product. The thus-obtained syngas is a valuable base for the generation of a whole range of products and fuels,[84–86] *e.g.* methanol, ethanol, dimethylether (DME), olefins, gasoline and diesel (Figure 8.7). Several catalytic processes are involved, *e.g.* the direct synthesis of methanol or ethanol from syngas, the methanol-based processes MTO (**M**ethanol **T**o **O**lefins), MTG

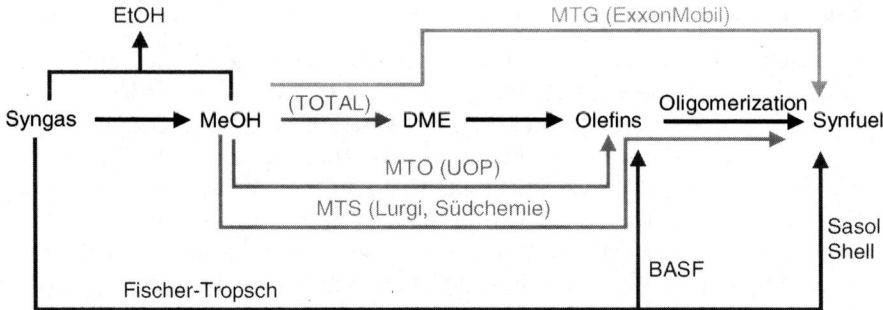

Figure 8.7 Production of biosynfuels from syngas: catalytic processes and some selected suppliers.

(**M**ethanol **T**o **G**asoline) and MTS (**M**ethanol **T**o **S**ynfuel), the production of fuels by oligomerization of olefins (**C**onversion of **O**lefins to **D**istillate, COD) or the synthesis of a wide range of hydrocarbons in Fischer–Tropsch processes. Some main features of these processes are outlined in the following sections.

Diesel or gasoline produced from syngas may be tailored towards new demands more easily than fuels produced from crude oil. This can be an advantage to meet new emission standards as well as for the development of advanced fuels for improved motor concepts *e.g.* for use in combined combustion system (CCS), *i.e.* engine concepts which combine the advantages of both diesel compression ignition and Otto spark ignition technologies.[87]

8.5.6.1 Syngas-based Products

Methanol. Methanol is produced from syngas at temperatures between 220 and 275 °C and pressures ranging from 50 to 100 bar using $Cu/ZnO/Al_2O_3$ catalyst systems. Today, the required syngas is obtained from natural gas in an upstream steam reforming process and the underlying chemistry is summarized, in a simplified way, in equations (8.1) and (8.2). Some major methanol technology suppliers are ICI, Lurgi and Mitsubishi.

$$CH_4 + H_2O \rightarrow CO + 3H_2 \qquad (8.1)$$

$$CO + 2H_2 \rightarrow CH_3OH \qquad (8.2)$$

As outlined above, biomass feedstock can also be employed for the production of syngas and, thus, methanol produced from biomass-derived syngas and fuels obtained through processing of biomass-based methanol can be considered as typical biofuels. Blends with petrol containing up to 20% of methanol can be used in combustion engines without elaborate modifications. However, the comparatively low energy density and safety concerns have limited so far broad applications of methanol as fuel. Another option is the use of methanol in fuel cells,[88–90] either directly in **D**irect **M**ethanol **F**uel **C**ells (DMFCs) or indirectly as hydrogen source.

Ethanol and Higher Alcohols from Syngas. Direct synthesis of ethanol from syngas is intensely investigated especially in North America.[91,92] Another approach in this context is the homologization of methanol, *i.e.* the reaction of methanol with syngas to yield ethanol. Higher alcohols can also be formed. The reactions are summarized in equations (8.3) and (8.4).

$$n\text{CO} + 2n\text{H}_2 \rightarrow \text{C}_n\text{H}_{2n+1}\text{OH} + (n-1)\text{H}_2\text{O} \tag{8.3}$$

$$\text{CH}_3\text{OH} + n\text{CO} + 2n\text{H}_2 \rightarrow \text{C}_{n+1}\text{H}_{2n+3}\text{OH} + n\text{H}_2\text{O} \tag{8.4}$$

However, ethanol selectivity is still moderate and there is lack of highly efficient catalysts for these reactions.

Dimethylether. Several strategies for the production of dimethyl ether (DME) are described, *e.g.* direct synthesis from syngas according to equation (8.5) or *via* dehydration of methanol according to equation (8.6). From a mechanistic point of view direct synthesis proceeds also *via* methanol formation and subsequent release of water but without procedural isolation of methanol. The process can also be designed to yield both methanol and DME. Established methanol catalysts are employed for methanol formation and typical dehydration catalysts are solid-acid catalysts, *e.g.* alumina, silica-, phosphorus- or boron-modified alumina, zeolite, (silico)aluminophosphates, tungsten–zirconia or sulfated-zirconia.[93,94]

$$3\text{CO} + 3\text{H}_2 \rightarrow \text{CH}_3\text{OCH}_3 + \text{CO}_2 \tag{8.5}$$

Since DME characteristics are similar to those of liquefied petrol gas (LPG),[95] it can be used in typical LPG applications, *e.g.* power generation, propellants, domestic cooking fuels or automotive fuels. If DME is employed as admixture, LPG properties are not significantly affected up to a DME content of around 20%. Compared to LPG, the cetane number is much higher (55–60 in contrast to 5 and 10 for propane and butane) and DME is, in principle, a suitable fuel for diesel engines. However, DME can not be blended with fossil diesel and its energy density is much lower, so that engines have to be adapted.

Gasoline via the MTG Process. Conversion of methanol to gasoline in the so-called MTG process is accomplished over zeolitic ZSM-5 catalyst systems either *via* fluidized- or fixed-bed technology. The former was demonstrated by Mobil, Union Rheinische Braunkohlen Kraftstoff AG and Uhde in a pilot plant at UK Wesseling and the latter was operated successfully by Methanex in New Zealand licensed by ExxonMobil. The Mobil process yields around 38 wt.% of gasoline, 4 wt.% of LPG, approximately 58 wt.% of water and a small amount of fuel gas.[96,97]

The underlying chemistry is complex and the multistage process is initiated by the formation of DME through dehydration of methanol equation (8.6). The following chain growth and cyclization reactions proceed *via* further release of water and can be described, in a very simplified manner, by equations (8.7) and (8.8). These reactions involve not only reactions of DME with itself equation (8.8) but also reactions of DME with methanol equation (8.7).

$$2CH_3OH \rightarrow CH_3OCH_3 + H_2O \qquad (8.6)$$

$$nCH_3OCH_3 + nCH_3OH \rightarrow \text{``}(CH_2)_{3n}\text{''} + 2nH_2O \qquad (8.7)$$

$$nCH_3OCH_3 \rightarrow \text{``}(CH_2)_{2n}\text{''} + nH_2O \qquad (8.8)$$

The resulting hydrocarbon product mixture is free of sulphur as well as nitrogen and exhibits low benzene content. Thus produced gasoline (Table 8.3) complies with European gasoline regulations. Further advantages of the process are minimal CO_2 emission and high energy efficiency. However, direct production of longer-chain alkanes for diesel and jet fuel using typical MTG catalysts is not possible.

The MTG process can be combined with MTO technology as realized in the so-called MOGD process (**M**obil **O**lefin to **G**asoline/**D**istillate). Employing this process, olefins are synthesized in the first step followed by olefin oligomerization to gasoline or diesel using a ZSM-5 catalyst. Hence, this route also affords, unlike the MTG process, the production of synthetic diesel or jet fuel.[96]

A similar process is the TIGAS process (**T**opsøe **I**ntegrated **Ga**soline **S**ynthesis) developed by Haldor Topsøe A/S and demonstrated at a pilot plant in Houston.[96] It is, in principal, an improved MTG process combining methanol, DME and gasoline production in a single synthesis loop thus circumventing intermediate production and storage of methanol. Different syngas compositions can be employed since methanol/DME synthesis is flexible in terms of syngas specifications.

Another approach is the MTS process (**M**ethanol **T**o **S**ynfuel) introduced by Lurgi GmbH, which is, in terms of process engineering, comparable to the MOGD process. However, primarily diesel is produced – besides a smaller

Table 8.3 Typical MTG products and gasoline composition.[97]

Hydrocarbon products	w/w%	Gasoline composition	w/w%
Light gas	1.4	Highly branched alkanes	53
Propane	5.5	Highly branched alkenes	12
Propene	0.2	Cycloalkanes	7
Isobutane	8.6	Aromatics	28
n-Butane	3.3		
Butenes	1.1		
C_{5+} Gasoline	79.9		

amount of gasoline – and therefore the process is, with respect to the product distribution, more like the Fischer–Tropsch process, which is described in the following section.

Fischer–Tropsch Diesel. Direct conversion of syngas to hydrocarbons succeeds in the Fischer–Tropsch (FT) process, discovered by Franz Fischer and Hans Tropsch in the 1920s.[98] The FT process is well established and currently operated *e.g.* by Sasol (Sasol Advanced Synthol, SAS process in Secunda, South Africa) and Shell (Middle Distillate Synthesis, MDS process in Bintulu, Malaysia). Coal and natural gas are employed in these processes as feedstock for the generation of syngas and a biomass-based process is already in an advanced stage (SunDiesel® *via* the Choren Carbo-V® Process).[99] FT syntheses can be conducted in slurry-phase, fixed- and fluidized-bed reactors, mainly depending on the catalysts and reaction temperatures. With respect to reaction temperatures one can distinguish between low- and high-temperature FT technologies (LT-FT and HT-FT). The former operate at temperatures around 220 °C and yield primarily long-chain hydrocarbons such as paraffins and waxes whereas the latter, operating at temperatures around 340 °C, produce mainly naphtha and olefins (Table 8.4). Long-chain products obtained *via* LT-FT processes are hydro-cracked in the next step to yield diesel in very high quality. Generally speaking, LT-FT syntheses combined with slurry-phase reactors are, at present, the preferred option.[100,101]

The Fischer–Tropsch reaction of carbon monoxide with hydrogen can be described, in a simplified way, according to equation (8.9). Chain growth proceeds at the surface of cobalt- and/or iron-based catalyst systems, *e.g.* Fe-, Fe/Co-, Fe/Co-Spinel-, Co/Mn-Spinel- or Cu-doped Co-catalysts. Iron catalysts, on the one hand, are cheaper than cobalt catalysts and more flexible as well as resistant with respect to syngas composition and quality. Cobalt catalysts, on the other hand, exhibit best performances at a $H_2:CO$ ratio of

Table 8.4 Comparison of product spectra obtained from different Fischer–Tropsch processes.[102]

Compound	LT-FT *Cobalt catalyst 220 °C*	HT-FT *Iron catalyst 340 °C*
Methane	5	8
Ethylene	0	4
Ethane	1	3
Propylene	2	11
Propane	1	2
Butylenes	2	9
Butane	1	1
C_{5+} Gasoline	19	36
Gasoil/Distillate	22	16
Heavy Oil/Wax	46	5
Oxygenates	1	5

2:1 and feature longer lifetimes as well as higher selectivities than iron catalysts.

$$nCO + (2n + 1)H_2 \rightarrow C_nH_{2n+2} + nH_2O \qquad (8.9)$$

As mentioned above, high-quality diesel can be obtained *via* FT processing. High cetane numbers of > 70 are reached and the fuel is almost free of sulphur, nitrogen or aromatic contaminants. Furthermore, engine tests revealed low NO_x, CO, hydrocarbons and particulate emissions. Compared to fossil diesel, FT diesel shows superior characteristics and can be employed instead of, or in blends with, fossil diesel without modification of engines and gas station systems.

Current research in the field of FT catalysis focuses on improved selectivities. Selectivities can be influenced not only by the catalyst but also by choice of the process and reactor type as well as favourable reaction conditions such as temperature, pressure, syngas composition, feeding of co-reactants into the reaction zone, *etc.* As an example, FT synthesis can be modified to yield particularly α-olefins[103] and these can be converted to diesel in a subsequent reaction.

8.5.6.2   Synthesis Gas Generation

Major uses of syngas, produced mainly from natural gas, coal or petroleum fractions today, is for hydrogen generation, methanol production, hydroformylation, and Fischer–Tropsch synthesis. However, a broad range of products can be derived from syngas opening access to many biofuels (see above) and platform chemicals.[104] As a result of the increasing interest in syngas-based products, many gasification methods, reactors and plant configurations are available or under development. While the chemical syntheses are mainly well established and commercially applied based on fossil feedstock, the front end processes to convert biomass into synthesis gas are still under development.

The main chemical equilibrium reactions for complete gasification *e.g.* at 1200 °C are compiled in Table 8.5.

By gasifying biomass or biomass-based intermediates producer or synthesis gas (or simply syngas) is generated with H_2, CO, CO_2 and CH_4 as the main components. The elemental composition of biomass feedstocks vary between 35 and 53 wt.% for carbon, 5 and 6 wt.% for hydrogen, 32 and 43 wt.% for oxygen, 0.1 and 1.4 wt.% for nitrogen, 0.01 and 0.08 wt.% for chlorine and 0.02 and 0.5 wt.% for sulphur. Therefore, an H_2/CO ratio in the order of 1:1 is obtained, which cannot directly be used for all syntheses. By the water gas shift reaction

$$CO + H_2O \rightarrow H_2 + CO_2 \qquad (8.10)$$

as a separate process step the ratio has to be adjusted to the requirements of the synthesis applied.

However, the gas produced by gasification contains impurities, depending on the type of feedstock and conversion technology applied. Typical are the

Table 8.5 Main chemical equilibrium reactions for gasification.

Heterogeneous key reactions for complete gasification including C, CO, CO₂, H₂, O₂, H₂O, CH₄ at T > 1200 °C			$\Delta_r H\,kJ\,mol^{-1}$
R1	Combustion	$C + O_2 \rightarrow CO_2$	−394
R2	Hydration	$C + 2H_2 \rightarrow CH_4$	−75
R3	H₂O-gasification	$C + 2\,H_2O \rightarrow CO_2 + 2H_2$	+ 119
R4	CO₂-gasification	$C + CO_2 \rightarrow 2CO$	+ 173
Other important gasification reactions as linear combinations of R1–R4:			
R5	Shift-reaction	$CO + H_2O \rightarrow CO_2 + H_2$	R4 − R3
R6	Partial oxidation	$2C + O_2 \rightarrow 2CO$	R1 + R4
R7	CO-oxidation	$2CO + O_2 \rightarrow 2CO_2$	R1 − R4
R8	H₂-combustion	$2H_2 + O_2 \rightarrow 2H_2O$	R1 + R4 − 2R3

Table 8.6 Maximum contamination levels tolerated by methanol synthesis.[105]

Contaminant	*Level*	
Tars	<0.1	mg Nm^{-3}
CH₄	<3	vol.%
NH₃	10	ppm
HCN	0.01	ppm
Total sulphur	0.5	ppm
Halides	0.001	ppm

organic impurities like tar and BTX (benzene, toluene and xylenes), the inorganic nitrogen-, sulphur- and chlorine-containing impurities NH_3, HCN, H_2S, COS and HCl, as well as volatile metals (regarding biomass in particular Na and K), dust and char or soot. Larger hydrocarbons, summarized as tars being produced at low temperature gasification processes, may cause fouling of downstream equipment, coat surfaces and plug pores in filters and sorbents. Other impurities are corrosive or act as catalyst poisons in the subsequent synthesis stages. Therefore, crucial toleration levels are set for contaminants; Table 8.6 shows *e.g.* those for methanol synthesis.[105]

For gas cleaning, conventional technology is available. Tars and BTX may be removed by either thermal or catalytic cracking, or by scrubbing with an oil-based medium. The other above-mentioned impurities are removed by standard wet gas cleaning technologies, *e.g.* the well-established Rectisol process. In that process, CO_2 and sulphur compounds are removed in separate fractions, resulting in a pure CO_2 product (*e.g.* for urea production) and an H_2S/COS enriched Claus gas fraction suitable for sulphur production. In a multi-stage process, the impurities are absorbed by cold methanol under elevated pressures between 30 and 60 bar. The contaminants are then separated or even fractioned by pressure release and temperature increase during solvent regeneration. In advanced dry, hot gas cleaning, the residual contaminants are removed by chemical adsorbents at elevated temperatures. However, for syntheses operated between 200 and 300 °C there is only little energy advantage, if syngas compression is required prior to synthesis (compression require cold inlet gas). In the case that gasification occurs already at pressures slightly above that of the subsequently following synthesis, dry pressurized gas cleaning predicts significant energetic benefits.

For biomass gasification, a variety of technologies exist.[106] In view of large scale production of high-quality synthesis gas as required for synfuel production mainly two types of gasifiers have to be considered here. Direct or indirect gasification using fluidized bed and entrained flow gasifiers is possible with potential plant capacities of up to several hundred MW fuel input capacity. Here, experience from commercially operated plants with coal and lignite as fuels is available and has already been transferred to the use of biomass as a feedstock for heat and electricity production *via* gasification.

For reasons of economy of scale, syngas production and further processing demand large-scale facilities; for comparison, a crude oil refinery has an input capacity in the order of $10 \, \text{Mio} \, \text{mto} \, \text{a}^{-1}$. On the other hand, biomass usually exhibits low volumetric energy densities, has to be harvested and collected from wide areas of agriculture or forestry and does not constitute a uniform feedstock. Chemical composition and the content of minerals and other inorganic material, usually denoted as the ash content, may vary significantly. There are four major sources of biomass: agricultural and forestry residues, municipal solid waste, industrial waste and purpose-grown bioenergy crops. The technologies applied should be able to be fed by different kinds of biomass (multi-feed) to achieve high throughput capacities improving economic viability of the plants.

In fixed bed reactors, the feedstock is exposed to the gasifying agent in a packed bed that slowly moves from the top of the gasifier to the bottom, where the ash is discharged. By moving through the reactor, the biomass passes distinct zones of drying, pyrolysis, oxidation and reduction. Usually the different types of fixed-bed reactors are characterized by the direction of the gas flow through the reactor and consequently are denoted as updraft, downdraft and horizontal (crossdraft) gasifiers. Depending on fuel and product gas application, a multitude of fixed bed gasifier designs exist. In general biomass fuelled fixed bed reactors are suited for district heat and power production up to fuel input capacities of 20 MW. For syngas production, substantial gas cleaning would be required, due to the relatively high amounts of tar, residual pyrolysis products, ash and dust particles, depending on the type of reactor applied (Figure 8.8).

In fluidized-bed gasifiers, biomass particles are rapidly mixed and heated by hot fluidized sand. Air, oxygen or steam is used as fluidization and gasification agent. The biomass is quickly decomposed into a combustible gas. In an autothermal process, a part of the incoming biomass is burned in the bed to maintain the reaction temperature at the desired value of typically 700–900 °C. Due to the intense mixing, the gasification reactions cannot be distinguished in local zones as in fixed-bed reactors, but occur throughout the whole bed leading to a uniform temperature distribution. The degree of fluidization can be small (bubbling fluidized bed, BFB) or high (circulating fluidized bed, CFB). BFB

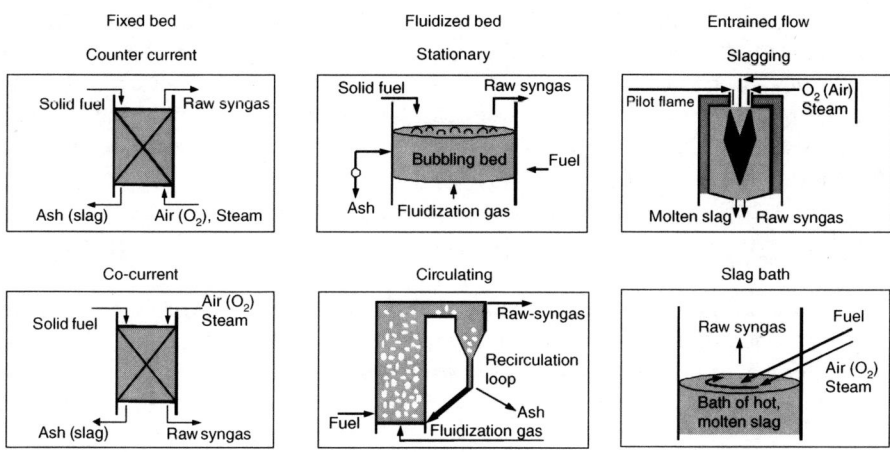

Figure 8.8 Commonly used types of gasifier.

gasifiers, having a well-defined interface between the reaction zone of the fluidized bed and the freeboard above the bed surface, are well known and commonly used because of their robust properties but show higher tar formation of the order of 1–2 wt.%. In a CFB gasifier, there is no distinct interface between the fluidized sand bed and the freeboard; the entrained media and char are recycled back to the gasifier *via* a cyclone. The carbon conversion is considerably better than in BFB gasifiers. In addition, a CFB gasifier can be operated at high pressures, which is advantageous and more economic in regard to hydrogen, liquid fuels and chemicals production.

In order not to poison the sensitive synthesis catalysts, synthesis gas must be of high purity and free from dust and tar. Unlike most fixed-bed and fluidized-bed gasifiers, entrained-flow gasifiers are able to generate a gas practically free from tar with only little methane at the high gasification temperatures above 1000 °C. When the operating pressure in the gasifier is above the synthesis pressure, there is no need for the technically expensive step of recompression of the synthesis gas, and gas purification is facilitated. An entrained flow gasifier can be slurry or particle fed using air, oxygen or steam as the gasification agent. Any feed material which can be pumped and sprayed pneumatically, and which has a heating value in excess of 10 MJ kg^{-1} is suitable in principle for entrained flow gasification. Also biomass powders can be transferred out of a pressurized entrained flow into the gasifier by a so-called high-density flow feed system. However, this requires periodic operation of a sophisticated system of locks, which becomes more complicated as the pressure rises. By virtue of the short residence time (about 1 s), high temperature (1200–1600 °C), high pressure (20–40 bar) and high capacity (> 100 MW), a tar-free synthesis gas with low methane content is produced.

To use also biomass and biogenic waste with higher ash and alkali contents, the gasifier type used may be operated at slagging conditions: depending on the alkali content ash softening or melting already occuring below 1000 °C. The liquid slag can be handled *e.g.* by a reactor equipped with a cooling screen, which originally had been developed for salt-bearing lignite at the DBI (Deutsches Brennstoff Institut, German Fuel Institute) of Freiberg, Saxony after 1976. This so-called GSP gasifier can tolerate high alkali chloride and ash contents and rapid fluctuations of quantities and softening points typical of straw and similar residues.

8.5.6.3   Current Developments

Depending on the plant size, the biomass feedstock used and the type of fuel to be produced different concepts for syngas-based fuels are under development today. However, other plant configurations and gasifiers up to 100 MW biomass input capacity exist but which are related to heat and power production today.[106]

SVZ, Sekundärrohstoff-Verwertungszentrum Schwarze Pumpe, Germany. At SVZ three different types of gasifiers with a total fuel capacity of 420 MW have been operated. The product gas was used for methanol (120,000 mto a^{-1}), heat and power (75 MW_e) production. The methanol plant consists of a three-stage fine cleaning of the syngas, a catalytic reactor operated at pressures up to 40 bar and a temperature about 250 °C with a recycling gas loop. Methanol cleaning is performed by a two-stage distillation. Being applied to town gas production from lignite, the gasification complex was turned to the use of secondary raw materials such as plastics, sewage sludge, paper, contaminated wood or municipal solid waste since 1996. The gasifiers in use were:

- SVZ fixed-bed gasifier for municipal solid waste and refuse derived fuels (RDF) in mixtures with coal (24 bar operation pressure);
- BGL fixed-bed gasifier for solid waste/coal mixtures;
- Entrained flow gasifier (GSP-type) for oil slurries, pastes, fuel mixtures from tar and sewage sludge *etc.* (25 bar operation pressure).

Coal was added to the secondary fuels to balance out the significant fluctuations in fuel properties, *e.g.* heating value, and to provide residual char as energy source for the endothermic gasification process (autothermal operation). In 2007 the Swiss Sustec Industries AG, then owner of SVZ, put the gasification plant out of service and discontinued the waste conversion operation because of insufficient profitability.

Värnamo demonstration plant, Sweden. Sycraft AB operated the pressurized biomass IGCC (Integrated Gasification Combined Cycle) for combined heat and power production (CHP). The circulating fluidized bed gasifier was operated between 1993 and 1999 for about 8000 hours, with a fuel capacity of

18 MW at 18 bar pressure. Wood as the main fuel aside from straw, RDF and bark was collected in a radius of 50 km. Using air as the gasification medium, about 50 vol.% of nitrogen is contained in the raw synthesis gas. After 2000, in a new EU project (CHRIGAS) the production of a hydrogen-rich synthesis gas was targeted. Therefore it is planned to refurbish the air-blown gasifier to make an oxygen and steam-blown gasification process. In addition, a hot gas filtration system will be used to improve the overall energy balance.

Güssing, Austria.   To demonstrate the production of bio-SNG from solid biofuels, the existing 8 MW fuel-input biomass gasifier in Güssing is equipped with a 1 MW (product capacity) methanation plant (construction start: January 2008) within the EU Bio-SNG project. On the way to the final product, bio-SNG, with the quality of natural gas that can be used for stationary and mobile applications, the synthesis gas has to be cleaned from dust, tars, sulphur compounds and other impurities. Subsequently hydrogen and carbon monoxide are converted into methane by the methanation reaction. The gasifier at Güssing is an FICFB (Fast Internal Circulating Fluidized Bed) steam gasifier so far being operated for combined heat and power production. Within its activities as an energy centre for the production of heat, power, substitute natural gas and liquid biofuels (coordinated by TU Wien) the gasification product gas has also been applied to fuel cells and to Fischer–Tropsch synthesis. A small-scale $1 \, \mathrm{h^{-1}}$ ($10 \, \mathrm{Nm^3 \, h^{-1}}$ syngas) Fischer–Tropsch plant together with the required gas cleaning section for water, chlorine and sulphur removal has been set up and operated in the frame of the EU RENEW project. The syngas, delivered at atmospheric pressure from the gasifier, had to be compressed before the synthesis by a two-stage compressor. Scale-up of the Fischer–Tropsch plant for demonstration is planned.

ECN, The Netherlands.   Another type of indirect gasifier, developed in regard to bio-SNG production by the Energy Research Centre of the Netherlands (ECN), is the MILENA gasifier. A new installation with a fuel input capacity of 1 MW has been commissioned in 2008. Indirect gasification means that the char is internally recycled to the combustor with the bed material. In the combustor (fluidized bubbling bed) the char is burned with air and heats the bed material. The bed material is recycled back to a riser, where the gasification takes place. Thus, a nitrogen-free product gas is obtained. Next, the whole bio-SNG production process, already tested successfully in lab scale ($5 \, \mathrm{kg \, h^{-1}}$ biomass input) will be scaled up to the 1-MW size.

Choren CarboV$^{\circledR}$, Germany.   In the BTL process of Choren Industries a three-stage gasification process is applied to convert wood chips, waste wood, *etc.* into a tar-free synthesis gas used for synfuel production. In the

1-MW alpha-plant the principal technical feasibility was demonstrated. The beta-plant, completed in 2008, is constructed for the production of 15,000 mto a^{-1} synthetic fuel. In the first stage of gasification, the air dried biomass (water content 15–20 wt.%) is carbonized through partial oxidation with air or oxygen at temperatures between 400 and 500 °C into a tar-containing gas and solid char (low temperature gasification). The tar-containing gas is partially oxidized for tar decomposition by oxygen and steam at temperatures of about 1400 °C in the second stage of the process. At that temperature, in the third gasification stage, the pulverized char is blown into the hot gasification medium to be converted to a tar-free synthesis gas. After appropriate gas cleaning and conditioning, the syngas is converted to BtL-synfuel (SunDiesel) by Fischer–Tropsch synthesis. Volkswagen, Daimler and Shell are involved in the project.

bioliq®-process of the Karlsruhe Institute of Technology (KIT). Biomass is distributed in many ways over large agricultural areas, pastures or forests. In order to take into account the widespread distribution, the low energy content and the very different properties of potential feed materials in a more effective way, the two-stage process bioliq® has been developed at the KIT.[107] According to the bioliq® concept, biomass is first liquefied in local facilities by fast pyrolysis into a pumpable biosyncrude by mixing the char and liquid condensates formed during the thermal degradation. Any kind of dry lignocellulose may be used for that process. The biosyncrude can be stored stably in tanks and is capable of safe and economic transport. The volumetric energy density of the biosyncrude in the case *e.g.* of straw is more than ten times higher than that of the initial biomass. Out of several such regionally distributed pre-treatment plants, a central large-scale facility is then supplied with biosyncrudes, *e.g.* by railway tank wagons. Biosyncrude gasification, raw syngas cleaning and conditioning are handled more efficiently, flexibly and with less environmental pollution and, above all, more economically in large facilities than in a large number of small-scale plants. For this type of chemical plant, the decremental cost exponent is on the order of 0.7, *i.e.* an increase in throughput by roughly one order of magnitude reduces by half the specific plant investment costs. For gasification, high pressure entrained flow gasification has been selected and tested in the 3–5 MW pilot plant of Future Energy in Freiberg (now Siemens Fuel Gasification Technology) in four gasification campaigns with different slurries and at different conditions up to pressures of 26 bar and temperatures above 1200 °C. The feasibility of fast pyrolysis for biomass pre-treatment by producing an intermediate suitable for efficient gasification was demonstrated in bench-scale plants (10 kg h^{-1}).

Based on the promising results a pilot plant covering the complete process chain is about to be constructed in cooperation with Lurgi on the KIT site.[107] The 2-MW (500-kg biomass input capacity) fast pyrolysis plant based on a transported bed (twin screw) reactor was commissioned in 2008. A 5-MW high-pressure entrained flow gasifier equipped with a cooling screen and suitable for operation up to 80 bar at temperatures above 1200 °C is presently under construction. The $1600\,Nm^3\,h^{-1}$ syngas obtained at pressures above synthesis pressure will supply a methanol production unit without expensive compression of the syngas. Subsequently, a methanol-to-synfuel plant will be added, which is currently under design.

8.5.6.4 *Conclusion for Second-generation Biofuels*

Except for heat and power production, no biomass gasification process is commercially established today. A variety of pilot and demonstration projects are on the way to show the technical and economic feasibility of biomass-based syngas processes. Benefit and experiences may be taken from already established coal and gas conversion technologies. However, specific differences from biomass conversion exist, demanding improved solutions. Issues to be addressed concern the broad range of potential feedstocks differing in chemical composition, consistency and availability. Many issues are related to the integration of gasifiers, gas cleaning and conditioning with the syngas conversion processes according to the requirements of the chemical syntheses applied. It is nearly impossible to separate the fuels and chemicals business; considering a BtL production complex, not only the most valuable mix of fuels and chemicals has to be produced but also co-production of energy, heat and power has to be included to achieve high thermal efficiencies. It is always more costly to produce only liquid fuels! In that respect, synthetic fuels production demands co-generation (polygeneration) of fuels, chemicals, heat and power within an integrated bio-refinery.

8.6   Third-generation Biofuels and Beyond

Getting photosynthetic organisms to produce fuel directly has long been a subject for researchers working with organisms that can absorb sunlight and CO_2 to produce plant oils or other fuels. Algae, such as *Botryococcus braunii* and *Chlorella vulgaris*, are relatively easy to grow and create interest as feedstock for energy and fuel production (Section 9.1.4.2.25).[108-110] Microalgae, produced in open ponds or more efficiently in photobioreactors, may contain up to 50 wt.% of lipids which can be converted to suitable fuels, meanwhile denoted as third-generation biofuels. Instead of growing their algae outside in the sunlight, researchers of Solazyme grow them inside dark stainless steel fermenters, in which the organisms convert sugars to oils.

Actually, the search for fuel-producing microbes is one of the hottest areas in synthetic biology today using synthetic biology to lend *E. coli*, yeast and other easily grown micro-organisms the ability to create mixtures of compounds that can be used to make various substances for use as fuels. A handful of start-up companies have leapt into the field, some even teaming up with major energy companies such as Chevron and BP. As an example, LS9, Inc. researchers are reengineering *E. coli* and other organisms to make what they refer to as renewable petroleum focusing on a pathway that converts sugars to fatty acids, which can then be converted to biodiesel. In another approach in San Francisco the University of California is working with biosynthetic bacteria like *E. coli* to make isobutanol.[111]

A recently discovered fungus, hidden within a stem from a scraggly tree in northern Patagonia, produces dozens of the same midlength hydrocarbons, which are also contained in gasoline, diesel fuel and jet fuel.[112]

Other approaches, already announced as fourth-generation biofuels, are the conversion of spent vegetable oil to gasoline[113,114] or the use of (modified) micro-organisms to produce suitable feedstocks from CO_2.[115]

References

1. F. Seyfried and K. Lenz, *Renewable fuels for advanced power train*, Final Report, European Commission 6th Framework Programme, SYNCOM, Ganderkesee, 2008, https://www.renew-fuel.com.
2. *Biofuels in the European Union – A vision for 2030 and beyond*, Final draft report of the Biofuels Research Advisory Council, 2006.
3. www.biofuelstp.eu.
4. www.refuel.eu/biofuels.
5. M. Roehr, ed., *The Biotechnology of Ethanol*, Wiley-VCH, Weinheim, 2001; J. Goettemoeller and A. Goettemoeller, *Sustainable Ethanol: Biofuels, Biorefineries, Cellulosic Biomass, Flex-fuel Vehicles, and Sustainable Farming for Energy Independence*, Prairie Oak Pub., Maryville, 2007; L. Olsson, *Biofuels*, Springer, Berlin, 2007; W. Soetaert, E.J. Vandamme, ed., *Biofuels*, John Wiley & Sons, Chichester, 2009.
6. D. Antoni, V. V. Zverlov and W. H. Schwarz, *Appl. Microbiol. Biotechnol.*, Nov. 2007, **77**, 23–25.
7. G. Felletschin, J. Knaut and M. Schöne, *Deutsche Hydrierwerke (DEHYDAG)*, Schriften des Werksarchivs, Bd. 12, Henkel KGaA, Düsseldorf, 1981.
8. Wa. Ostwald, *Zeitschr. für angew. Chemie*, 1922, **35**(46), 278–280.
9. World Watch Institute, *Biofuels for Transport. Global Potential for Sustainable Energy and Transport*, Earthscan, London, Sterling, 2007.
10. M. G. S. Andrietta, S. R. Andrietta, C. Steckelberg and E. N. A. Stupiello, *Int. Sugar J.*, 2007, **109**(1299), 195–200.
11. H. Honkanen, P. Uotila, M. Koskinen and M. Sourander, *Biofuels Technology*, 2008, **Q4**, 19–23.
12. I. Klenk and M. Kunz, *Sugar Industry*, 2008, **133**, 625–635.
13. I. Klenk and M. Kunz, *Sugar Industry*, 2008, **133**, 710–718.
14. International Energy Agency, *Biofuels for Transport – An International Perspective*, Paris, 2004.
15. D. R. Kelsall and T. P. Lyons, in *The Alcohol Textbook*, ed. K. A. Jacques, T. P. Lyons and D. R. Kelsall, Nottingham University Press, Nottingham, 4th edn, 2003, 9–21.

16. R. F. Power, in *The Alcohol Textbook*, ed. K. A. Jacques, T. P. Lyons and D. R. Kelsall, Nottingham University Press, Nottingham, 2003, 23–32.
17. V. Sharma, K. D. Rausch, M. E. Tumbleson and V. Singh, *Starch/Stärke*, 2007, **59**, 549–556.
18. M. J. Taherzadeh and K. Karimi, *BioResources*, 2007, **2**, 472–499.
19. B. D. Solomon, J. R. Barnes and K. E. Halvorsen, *Biomass and Bioenergy*, 2007, **31**, 416–425.
20. R. Katzen and D. J. Schell, in *Biorefineries – Industrial Processes and Products*, ed. B. Kamm, P. R. Gruber and M. Kamm, Wiley-VCH, Weinheim, 2006, vol. 1, pp. 129–138.
21. European Biofuels Technology Platform: *Strategic Research Agenda & Strategy Deployment Document*, CPL Press, Newbury, 2008.
22. B. Kamm, M. Kamm, M. Schmidt, T. Hirth and M. Schulze, in *Biorefineries – Industrial Processes and Products*, ed. B. Kamm, P. R. Gruber and M. Kamm, Wiley-VCH, Weinheim, 2006, vol. 2, pp. 97–149.
23. M. Galbe, P. Sassner, A. Wingreen and G. Zacchi, *Adv. Biochem. Engin./Biotechnol*, 2007, **108**, 303–327; C. N. Hamelinck, G. van Hooijdonk and A. P. C. Faaij, *Biomass and Bioenergy*, 2005, **28**, 384–410.
24. J. Larsen, M. O. Petersen, L. Thirup, H. W. Li and F. K. Iversen, *Chem. Eng. Technol.*, 2008, **31**, 765–772.
25. M. P. Tucker, K. H. Kim, M. M. Newman and Q. A. Nguyen, *Appl. Biochem. Biotechn.*, 2003, **105–108**, 165–177.
26. G. M. Walker, *Yeast Physiology and Biotechnology*, Wiley & Sons, Chichester, 1998.
27. C. A. Abbas, in *The Alcohol Textbook*, ed. K. A. Jacques, T. P. Lyons and D. R. Kelsall, Nottingham University Press, Nottingham, 4th edn, 2003, pp. 41–57.
28. I. Russell, in *The Alcohol Textbook*, ed. K. A. Jacques, T. P. Lyons and D. R. Kelsall, Nottingham University Press, Nottingham, 2003, pp. 85–120.
29. Z. Ciesarova, *et al., Folia Microbiol*, 1996, **41**, 485–488.
30. S. C. Sharma, D. Raj, M. Forouzandeh and M. P. Bansal, *Appl. Biochem. Biotechnol.*, 1996, **56**, 189–195.
31. T. D'Amore, C. J. Panchal, I. Russell and G. G. Stewart, *Crit. Rev. Biotechnol.*, 1990, **9**, 287–304.
32. M. Knauf and K. Kraus, *Sugar Industry/Zuckerindustrie*, 2006, **131**, 753–758.
33. H. Alexandre, I. Rousseaux and C. Charpentier, *FEMS Microbiol. Lett.*, 1994, **124**, 17–22.
34. N. Kosaric and F. S. Vardar, in *The Biotechnology of Ethanol*, ed. M. Roehr, Wiley-VCH, Weinheim, 2001, pp. 90–98.
35. S. Alfenore, C. Molina-Jouve, S. E. Guillouet, J. L. Uribelarrea, G. Goma and L. Benbadis, *Appl. Microbiol. Biotechnol.*, 2002, **60**, 67–72.
36. S. E. Lowe, M. K. Jain and J. G. Zeikus, *Microbiol. Mol. Biol. Rev.*, 1993, **57**, 451–509.

37. P. S. Panesar, S. S. Marwaha and J. F. Kennedy, *J. Chem. Technol. Biotechnol.*, 2006, **81**, 623–635.
38. F. W. Bai, W. A. Anderson and M. Moo-Young, *Biotechnol. Advances*, 2008, **26**, 89–105.
39. G. A. Sprenger, *FEMS Microbiol. Lett.*, 1996, **145**, 301–307.
40. J. Becker and E. Boles, *Appl. Environ. Microbiol.*, 2003, **69**, 4144–4150.
41. K. Otterstedt, C. Larsson, R. M. Bill, A. Ståhlberg, E. Boles, S. Hohmann and L. Gustafsson, *EMBO Rep.*, 2004, **5**, 532–537.
42. M. Sonderegger, M. Jeppsson, C. Larsson, M. -F. Gorwa-Grauslund, E. Boles, L. Olsson, I. Spencer-Martins, B. Hahn-Hägerdal and U. Sauer, *Biotechnol. Bioeng.*, 2004, **87**, 90–98.
43. K. Karhumaa, B. Wiedemann, B. Hahn-Hägerdal, E. Boles and M.-F. Gorwa-Grauslund, *Microbial Cell Factories*, 2006, **5**, 18.
44. B. Hahn-Hägerdal, K. Karhumaa, M. Jeppson and M. -F. Gorwa-Grauslund, *Adv. Biochem. Eng. Biotechnol.*, 2007, **108**, 147–177.
45. www.ethanol-gec.org.
46. T. Scheers, R. Philippaerts, L. Van Langendonck, W. Duquet, N. Duvigneaud, L. Matton, M. Thomis, K. Wijndaele and J. Lefevre, *Publ. Health Nutr.*, 2008, **11**, 1098–1106; J. A. Hawley, *Med. Sci. Sports Exerc.*, 2002, **34**(9),1475–1476; J. A. Hawley, *The Physician and Sports Medicine*, 1998, **9**, 26.
47. J. Van Gerpen, B. Shanks, R. Pruszko, D. Clements and G. Knothe, *Biodiesel Production Technology*, Subcontractor Report NREL/SR-510-36244, Golden, 2004.
48. Exposition universelle internationale de 1900 à Paris, Rapports du jury international, Groupe IV – *Matériel et procédés généraux de la mécanique.* Deuxième partie, Classe 20, Imprimerie nationale, Paris, 1904, http://cnum.cnam.fr/PDF/cnum_8XAE568.2.pdf.
49. R. Diesel, *Proc. Inst. Mech. Eng.*, 1912, pp. 179–280; R. Diesel, *Die Entstehung des Dieselmotors*, Springer, Berlin, 1913. Faksimile mit einer technikhistor. Einführung und einem Lebensbild von Rudolf Diesel von H.-J. Braun, Steiger, Moers, 1984; W. R. Nitske and C. M. Wilson, *Rudolf Diesel, Pioneer of the Age of Power*, University of Oklahoma Press, Norman, 1965; G. Knothe, *Inform*, 2001, **12**(11), 1103–1107; G. Knothe, J. Van Gerpen, J. Krahl, ed., *The Biodiesel Handbook*, AOCS, Champaign, 2005; Stadt Augsburg, Hrsg., *Augsburg feiert 150 Jahre Rudolf Diesel*, Jubiläumsschrift, WEKA info Verl., Mering, 2008.
50. J. Walton, *Gas Oil Power*, 1938, **33**, 167–168; I. Mormino, *Vegetable oils and fats for use in compression ignition engines: chemical and physical properties, engine performance and emissions*, Thesis presentation, FloHeaCom, Universiteit Gent, 2008.
51. T. Breuer and A. Becker, *Berichte über Landwirtschaft*, 2008, **86**, 226–241.
52. A. Demirbas, *Fuel*, 2008, **87**, 1743–1748.
53. A. W. Schwab, M. O. Bagby and B. Freedman, *Fuel*, 1987, **66**, 1372–1378.
54. R. Olbert, *Kultur und Technik*, 1990, **2**, 62–71.

55. B. Gutsche, *Fett/Lipid*, 1997, **99**(12), 418; Ch. Breucker, V. Jordan, M. Nitsche and B. Gutsche, *Chem.-Ing.-Techn.*, 1995, **67**, 430; A. Zellner, *Katalytische Herstellung von Rapsölmethylester*, Dissertation Uni GH Duisburg, 1989.
56. M. Pagliaro and M. Rossi, *The Future of Glycerol: New Usages for a Versatile Raw Material*, RSC Publ., Cambridge, 2008.
57. S. Sato, W. Bueno de Almeida and A. Shigueu Araújo, WO 2006/050589, 2006 (Cognis).
58. H. Hüsken, B. Gutsche and J. Richter, EP 1 870 446, 2006 (Cognis); M. Halpern, WO/2007/111604, 2007 (PTC Organics).
59. J. M. Earle, K. R. Seddon and N. V. Plechkova, WO/2006/095134, 2006 (The Queen's University of Belfast).
60. M. Nielsen, J. Brask and L. Fjerbaek, *Eur. J. Lipid Sci. Tech.*, 2008, **110**(8), 692–700; R. Rodrigues, G. Volpato, K. Wada and M. Ayub, *JAOCS*, 12/2008, 925–930.
61. G. N. Jovanovic, B. K. Paul, J. Parker, A. Al-Dubabian, WO/2007/142983, 2007 (State of Oregon); S. R. Buddoo, N. Siyakatshana and B. Pongoma, CSIR Biosciences, Modderfontein, http://researchspace. csir.co.za/dspace/bitstream/10204/2680/1/Buddoo_P_2008.pdf (retrieved 11.01.2009).
62. G. Vellguth, *Grundl. Landtechnik*, 1982, **32**, 17–186; M. Mittelbach, M. Wörgetter, J. Pernkopf and H. Junek, *Energ. Agr.*, 1983, **2**, 369–384; M. Wörgetter und Mitarbeiter, *Pilotprojekt Biodiesel*, LT 2/87, Teil 1 + 2, Bundesanstalt Landtechnik, BM Land Forst Wasser, Heft Nr. 25 + 26, Wieselburg, Dezember 1991; A. Munack and J. Krahl, ed., *Beiträge Fachtagung Biodiesel – Potenziale, Umweltwirkungen, Praxiserfahrungen*, Braunschweig, 2002, Sonderheft 239, Landbauforschung Falkenrode, http://literatur.fal.de/fallitdok_extern/zi027595.pdf; A. S. Ramadhas, S. Jayaraj and C. Muraleedharan, *Renew. Energ.*, 2004, **29**, 727–742; J. Krahl, A. Munack, Y. Ruschel, O. Schröder and J. Bünger, *SAE Transactions*, 2008, **116**, 931–937.
63. J. Krahl, A. Munack, O. Schröder, H. Stein, M. Dutz and J. Bünger, in *Fuels 2003: 4th International Colloquium TAE*, ed. W. J. Bartz, Ostfildern, 2003, pp. 115–123, http://www.ufop.de/downloads/Biodiesel_comparison. pdf; Bundesforschungsanstalt für Landwirtschaft (FAL), *Jahresbericht 2003*, Braunschweig, 2004, http://www.fal.de/nn_789864/SharedDocs/07__TB/ DE/Downloads/tb__e3__jb03__de,templateId=raw, property= publication File.pdf/tb_e3_jb03_de.pdf.
64. B. Smith, H. C. Greenwell and A. Whiting, *Energy Environ. Sci.*, 2009, published on the web 25th November 2008, DOI: 10.1039/b814123a; S. Lestari, I. Simakova, A. Tokarev, P. Mäki-Arvela, K. Eränen and D. Y. Murzin, *Catal. Lett.*, 2008, **122**, 247–251; M. (R. W.) Snåre, *Development of Next Generation Biodiesel Technology*, Faculty of Technology – Department of Chemical Engineering, Åbo/Turku, 2006; J. Myllyoja, P. Aalto and E. Harlin, WO 2007003708, 2006 (Neste); D. Y. Murzin, I. Kubickova, M. Snåre, P. Mäki-Arvela and J. Myllyoja, WO 2006075057,

2006 (Neste); A. Kocal, *New Mexico Biodiesel Policy Summit*, 'Albuquerque, 2008, http://www.biodieselnewmexico.com/pdfs/UOP_NMSummit2008.pdf.

65. J. St. John, http://www.greentechmedia.com/articles/biofuel-powers-air-new-zealand-test-flight-5436.html (retrieved 13.01.2009).

66. E. Parente, *Lipofuels: Biodiesel & Biokerosene*, www.nist.gov/oiaa/TECHBIO1.pdf (retrieved 13.01.2009).

67. D. L. Daggett, R. C. Hendricks, R. Walther and E. Corporan, *Alternative Fuels for Use in Commercial Aircraft*, The Boeing Company, 2007, http://www.boeing.com/commercial/environment/pdf/alt_fuels.pdf (retrieved 13.01.2009).

68. www.biogas.org.

69. D. Deublein and A. Steinhauser, *Biogas from Waste and Renewable Resources*, Wiley-VCH, Weinheim, 2008.

70. www.hfpeurope.org.

71. A. Kruse, *Biofuels Bioprod. Bioref.*, 2008, **2**, 415–437.

72. Y. -C. Lo, S. -D. Chen, C. -Y. Chen, T. -I. Huang, C. -Y. Lin and J. -S. Chang, *Int. J. Hydrogen Energy*, 2008, **33**, 5224–5233.

73. H. Argun, F. Kargi, I. K. Kapdan and R. Oztekin, *Int. J. Hydrogen Energy*, 2008, **33**, 1813–1819.

74. J. K. Kim, L. Nhat, Y. N. Chun and S. W. Kim, *Biotechnol. Bioprocess Eng.*, 2008, **13**, 499–504.

75. M. Krupp, *Biohydrogen production from organic waste and wastewater by dark fermentation – a promising module for renewable energy production*, Shaker Verlag, Aachen, 2007.

76. K. Nath and D. Das, *Industrial Biotechnology*, 2006, **2**, 44–47.

77. D. Das and T. N. Veziroğlu, *Int. J. Hydrogen Energy*, 2001, **26**, 13–28.

78. J. Yanik, S. Ebale, A. Kruse, M. Saglam and M. Yüksel, *Int. J. Hydrogen Energy*, 2008, **33**, 4520–4526.

79. J. Yanik, A. Ebale, A. Kruse, M. Saglam and M. Yüksel, *Fuel*, 2007, **86**, 2410–2415.

80. T. C. Ezeji, N. Qureshi and H. P. Blaschek, *Curr. Opin Biotechnol.*, 2007, **18**, 220–227.

81. P. Lako, D. J. Gielen, L. Dinkelbach and R. van Ree, *Biomass for Energy and Industry*, 1998, 533–536.

82. Z. Qi, J. Chang, T. Wang and Y. Xu, *Energ. Convers. Manag.*, 2007, **48**, 87–92.

83. A. V. Bridgwater, *Therm. Sci*, 2004, **8**, 21–49.

84. D. C. Elliott, D. Beckman, A. V. Bridgwater, J. P. Diebold, S. B. Gevert and Y. Solantausta, *Energy & Fuels*, 1991, **5**, 399–410.

85. G. A. Mills, *Fuel*, 1994, **73**, 1243–1279.

86. A. V. Bridgwater, *Chem. Eng. J.*, 2003, **91**, 87–102.

87. H. Heinrich, *Sustainable Fuels for Road Traffic*, Solvsafe Prelude Convention, Düsseldorf, 2007, www.solvsafe.org; H. Heinrich, *Die Kraftstoffstrategie von Volkswagen für eine nachhaltige Mobilität*, Wiener

Energiegespräche, Wien, 2006, http://www.eeg.tuwien.ac.at/events/egs/pdf/egs061128_heinrich.pdf (retrieved 26.02.2009).

88. H. Liu, C. Song, L. Zhang, J. Zhang, H. Wang and D. P. Wilkinson, *J. Power Sources*, 2006, **155**, 95–110.
89. L. Carrette, K. A. Friedrich and U. Stimming, *ChemPhysChem*, 2000, **1**, 162–193.
90. S. Wasmus and A. Küver, *J. Electroanal. Chem.*, 1999, **461**, 14–31.
91. V. Subramani and S. K. Gangwal, *Energy & Fuels*, 2008, **22**, 814–839.
92. J. Hu, Y. Wang, C. Cao, D. C. Elliott, D. J. Stevens and J. F. White, *Catal. Today*, 2007, **120**, 90–95.
93. F. S. Ramos, A. M. Duarte de Farias, L. E. P. Borges, J. L. Monteiro, M. A. Fraga, E. F. Sousa-Aguiar and L. G. Appel, *Catal. Today*, 2005, **101**, 39–44.
94. F. Yaripour, F. Baghaei, I. Schmidt and J. Perregaard, *Catal. Commun.*, 2005, **6**, 147–152.
95. www.total.com.
96. F. J. Keil, *Micropor. Mesopor. Mater.*, 1999, **29**, 49–66.
97. http://nzic.org.nz/ChemProcesses/energy.
98. F. Fischer and H. Tropsch, US 1746464, 1930.
99. www.choren.com.
100. K. R. Radtke, M. Heinritz-Adrian and C. Marsico, *VGB PowerTech.*, 2006, **86**, 78–84.
101. P. L. Spath and D. C. Dayton, *NREL Technical Report NREL/TP-510-34929*, 2003.
102. A. Steynberg and M. Dry, ed., *Fischer–Tropsch Technology*, Elsevier, Amsterdam, 2004.
103. B. W. Hoffer, E. Schwab, G. Kaibel, D. Neumann, J. Bürkle and T. Butz, DE 102005056784, 2005 (BASF).
104. I. Wender, *Fuel Process. Tech.*, 1996, **45**, 189–297.
105. W. Torres, S. S. Pansare and J. G. Goodwin Jr, *Cat. Rev.*, 2007, **49**, 407–456.
106. H. A. M. Knoef, ed., *Handbook of Biomass Gasification*, BTG, Enschede, 2005.
107. N. Dahmen, E. Dinjus and E. Henrich, *Biofuels, Bioproducts & Biorefining*, 2009, **3**, 28–41.
108. J. N. Rosenberg, G. A. Oyler, L. Wilkinson and M. J. Betenbaugh, *Curr. Opin Biotechnol.*, 2008, **19**, 430–436.
109. H. Xu, X. Miao and Q. Wu, *J. Biotechnol.*, 2006, **126**, 499–507.
110. A. Melis and T. Happe, *Plant Physiol.*, 2001, **127**, 740–748.
111. S. Atsumi, T. Hanai and J. C. Liao, *Nature*, 2008, **451**, 86–89.
112. G. Strobel, B. Knighton, K. Kluck, Y. Ren, T. Livinghouse, M. Griffen, D. Spakowicz and J. Sears, *Microbiology*, 2008, **154**, 3319–3328.
113. W. Charusiri and T. Vitidsant, *J. Energy*, 2003, **5**, 58–68.
114. F. A. Twaiq, N. A. M. Zabidi and S. Bhatia, *Ind. Eng. Chem. Res.*, 1999, **38**, 3230–3237.
115. K. K. Treseder and M. F. Allen, *New Phytol.*, 2000, **147**, 189–200.

CHAPTER 9
Biomass for Green Chemistry

KARLHEINZ HILL AND RAINER HÖFER

Cognis GmbH, Rheinpromenade 1, D-40789, Monheim, Germany

Coal tar dyes mark the beginning of the chemical industry in the UK, in France and in the Rhine Valley with chemists like August Wilhelm Hofmann and William Henry Perkin and company names like Ciba, BASF, Hoechst and Bayer. Since the last century crude oil and natural gas have become the key raw material sources for the modern chemical industry yielding all base chemicals like olefins, acetylene and more particularly aromatics. In the future, as in the past, a reliable and economically reasonable supply of carbon and hydrogen as raw material source for basic chemicals and their derivatives will be the basic requirement for the chemical industry.[1]

Fossil raw materials such as crude oil, gas and coal are limited, though. In contrast, nature is producing 170 billion mto of biomass every year simply by photosynthesis.[2] Thereby nature provides a remarkably wide range of renewable raw materials from agricultural, marine and forestry production of varying material properties and differing chemical compositions, which constitute a substantial potential for industrial production and energetic use besides the traditional areas of food, feed, construction and garments.[3] The term *biomass* means all organic materials of biogenic, non-fossil character and comprises also matter living and growing in nature and waste materials resulting from both living matter and organic matter that is already dead (Table 9.1).

While petrochemical feedstock in a modern national economy like Germany accounts for 17 Mio mto of the chemical industry's raw materials mix, the use of renewable resources represents approximately 2.7 Mio mto, *i.e.* 14%.

RSC Green Chemistry No. 4
Sustainable Solutions for Modern Economies
Edited by Rainer Höfer
© The Royal Society of Chemistry 2009
Published by the Royal Society of Chemistry, www.rsc.org

Table 9.1 Agrarian raw materials for use in the chemical industry, Germany, 2006. For comparison: Petrochemical raw materials: 17 Mio. mto. With the permission of Fachagentur Nachwachsende Rohstoffe, Gülzow; source: FNR, VCI, meó Consulting Team, Mantau/ University Hamburg, BFH; http://www.fnr.de/cms35/Industrielle-Nutzung.1709 + M54721168da5.0.html#content3341 (retrieved 18.08.2008).

Raw material	Consumption (mto)
Vegetable oils	800,000
Animal fats	350,000
Starch	630,000
Cellulose/dissolving pulp	320,000
Sugar	295,000
Natural fibres	176,000
Rosin and natural waxes	31,000
Other (incl. proteins)	86,000 (55,000)
Total	2,688,000

Actually, biomass is mainly used to complement petrochemicals in applications, where renewable products can display advantages due to their chemical structure or properties. While fossil raw materials coal and crude oil are biomass, which has been reduced over millions of years to carbon or hydrocarbons, renewable raw materials are already composed of a universe of lower and higher molecular weight structures. It is a formidable challenge for chemical research to use these pre-manufactured renewable moieties as starters for alternative manufacturing processes eventually employing new green catalysts, green solvents or white biotechnology; to develop degradable polymers for packaging, bonding and de-bonding on command or agricultural purposes, for example; to synthesize environmentally benign safer chemicals, and biodegradable or recyclable products.[4] Alternatively, white biotechnology can transfer complex biomass structures yielding platform chemicals like ethanol, glycerol, succinic or lactic acid. Gasification technologies can convert biomass into synthesis gas – short-cutting the millions of years of geological coal and crude oil formation and subsequent gasification and cracking of these fossil base-stocks. Last but not least, anaerobic digestion or fermentation of biomass yields biogas (also called landfill gas or digester gas), which primarily comprises methane and carbon dioxide.

With high oil prices the interest in alternative sources for hydrocarbon feedstock has increased and technologies to convert bio-ethanol into ethene (see Figure 9.1) are moving to commercial scale (see Chapter 9.3). In combination with new processes, renewable feedstock represents alternative routes to petrochemical production which are becoming competitive at an oil price above 40 US$ per barrel.[5]

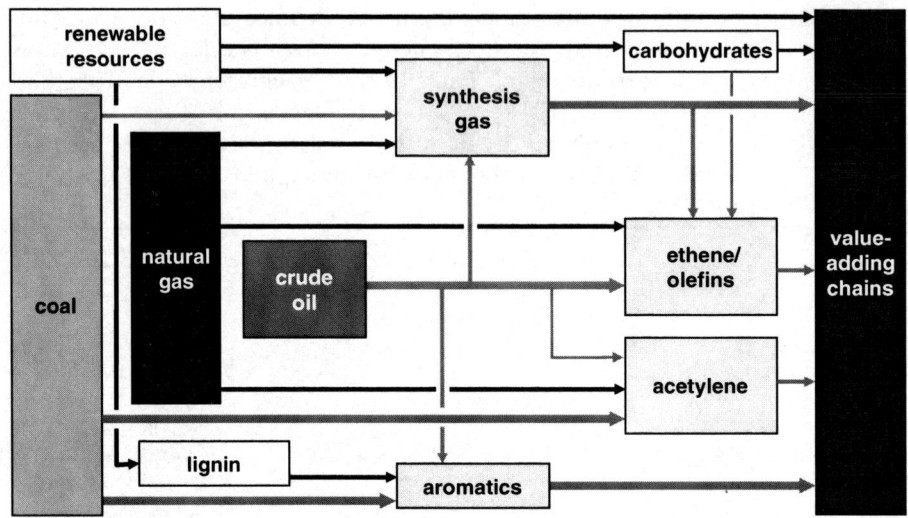

Figure 9.1   Basic raw materials yielding key intermediates for the chemical industry.

References

1. J. -D. Arndt, S. Freyer, R. Geier, O. Machhammer, J. Schwartze, M. Volland and R. Diercks, *Chem. Ing. Techn.*, 2007, **79**(5), 521.
2. W. Umbach, in *Perspektiven nachwachsender Rohstoffe in der Chemie*, ed., H. Eierdanz, VCH, Weinheim, 1996, p. XXIX.
3. meó Consulting Team, Institut für Energetik und Umwelt, Faserinstitut Bremen, *Marktanalyse Nachwachsende Rohstoffe* (Teil I + II), ed., Fachagentur Nachwachsende Rohstoffe, Gülzow, 2006.
4. J. H. Clark, *Green Chem.*, 2006, **8**, 17.
5. S. Andre, C. Günther, S. Sanghvi and A. Vogel, *McKinsey on Chemicals*, 2008, **1**, 31.

CHAPTER 9.1

Natural Fats and Oils[1]

KARLHEINZ HILL AND RAINER HÖFER

Cognis GmbH, Rheinpromenade 1, D-40789 Monheim, Germany

9.1.1 Introduction

Depending on geography, climate and soil conditions, fish and venison from wildlife, milk, butter, meat from domestic animals and vegetable oils are all part of the human diet and are essential daily items for food and non-food applications. Vegetable oils are frequently classified into two main groups, according to their source: pulp oils (palm, olive, avocado) and seed oils (soybean, cottonseed, peanut, sunflower, rapeseed, sesame, palm-kernel, coconut, linseed, castor and other minor). The major edible animal fats, also termed as meat fats, are lard and tallow, butter and marine oils. Lard and rendered pork fat are produced from the fat of pigs. Edible tallow is produced mainly from the fat of cattle (beef tallow) or sheep (mutton tallow). The importance of fats for humans, animals and plants lies in their high energy contents. Their calorific value is more than double that of carbohydrates and proteins. In addition, fats allow humans and animals to consume fat-soluble vitamins and provide them with essential fatty acids. Securing the consistent availability of fats and oils in sufficient quantity with equitable geographic distribution is required to fight hunger and under-nourishment on Earth.

Besides utilization for food and feed, fats and oils are one backbone for green industrial chemistry and key components of bio-refinery concepts.[2] Vegetable oils and their by-products, cysts, shells, hulls and parings as well as waste and

RSC Green Chemistry No. 4
Sustainable Solutions for Modern Economies
Edited by Rainer Höfer
© The Royal Society of Chemistry 2009
Published by the Royal Society of Chemistry, www.rsc.org

recycling products can serve as energy sources for the generation of thermal or electrical power and as raw material base for alternative fuels. Efforts to manage production of fats and oils involving the entire crop and its biomass as a zero waste operation throughout the value chain from sowing, harvesting and processing, to shipping, trade and disposal are important elements of modern industrial ecology[3] and of the Chemical Industry Responsible Care® – Global Charter.

9.1.2 Paradigm Changes in Global Fats and Oils Production, Use and Trade

Vegetable and animal oils and fats actually represent the main portion of around 43% of all renewable raw materials used in the chemical industry of a country like Germany (see Chapter 9, Table 9.1).

Their production at a global level has undergone fundamental changes over time. During the 1930s oils and fats were already traded globally; worldwide oilseeds reached an average annual production of 49 Mio mto, and fat and oil trade reached an average of 20 Mio mto including up to 540,000 mto (1938) sperm whale oil.[4,5] At that time, sperm oil was highly thought of as the finest lubricant, and the cleanest burning oil. It was also the first source for saturated and unsaturated fatty alcohols until catalytic high pressure hydrogenation of fatty acid esters was established. After the Second World War efforts to protect endangered whales evolved into one of the first wildlife campaigns and have perhaps made greater progress than any other international effort to protect endangered species. In 1946 the International Whaling Commission (IWC) was formed to protect the future of whales. Commercial whaling has been banned *via* an "indefinite moratorium" set down in 1986. The controversy over whaling, however, is also an example of conflicting ideas about human-environment relations, resource management and the very concept of sustainability. This applies more particularly to Greenland, where whales have been a part of the marine-based economy of Inuit people and their ancestors since their expansion from the Canadian Arctic into Greenland over four thousand years ago.[6,7]

By 1960, world production of oils and fats had reached 30 Mio mto. Driven mainly by population growth, rising *per capita* requirements for human nutrition, changing consumer behaviour regarding the fatty acid composition of the diet, and also by increasing industrial demand major changes in fats and oils production have occurred in the following decades. While most of the fossil crude oil products serve for energy or transportation purposes and < 10% for chemical applications, natural fats and oils are mainly used for human food and animal feed. In recent years, however, biofuels are creating additional demand and have become a new segment of vegetable oil use creating a noticeable impact on the applications for food and chemistry purposes (Figure 9.1.2).

Table 9.1.1 World production of fats and oils.[8]

Oilseed Production 2006/2007 (Mio mto)	USA	Canada	Mexico	Brazil	Argentina	South America[3] Total	EU – 27	Other EU	Russia	Ukraine	Uzbekistan	Total Commonwealth of Independend States (CIS)[4]	Turkey	China	India	Pakistan	Philippines	Malaysia	Indonesia	Thailand	Africa	Oilseed Total World (Mio mto)	Oil, Total World (mto)	Harvest Area, World (Mio. ha)
Oil seeds & seed oils																								
Soybean	86.8	3.5	–	58.7	48.3	115.3	1.3	–	0.8	0.9	<0.1	1.8	–	16	8	–	–	–	–	0.7	1	237.2	36.6	94.3
Cottonseed	6.7	–	0.3	2.3	0.5	2.8	0.6	–	–	–	2	3	1.3	12.2	9.3	4.4	–	–	–	<0.1	2.7	44	5	34
Groundnut[1]	1.2	–	–	0.1	0.5	0.7	–	–	–	–	–	–	–	10	3.7	<0.1	–	–	–	<0.1	5	23	4.1	22
Sunflower	1	0.2	–	0.1	3.5	3.8	6.3	0.5	6.4	5.5	–	12.5	0.8	2	1.4	0.3	–	–	–	–	0.7	30	11.2	24
Rapeseed[2]	0.6	9.0	<0.1	<0.1	<0.1	0.3	16.1	0.1	0.5	0.6	–	1.3	<0.1	12.7	6.2	0.3	–	–	–	<0.1	<0.1	47.8	18.5	27.2
Sesameseed	–	–	–	–	–	–	–	–	–	–	–	–	–	0.5	0.7	–	–	–	–	–	0.9	3.4	0.86	7.5
Palmkernel	–	–	–	–	–	–	–	–	–	–	–	–	–	–	–	–	–	4	4	–	0.8	10	4.4	10.5
Copra, coconut oil	–	–	0.2	–	–	–	–	–	–	–	–	–	–	–	–	–	1.7	<0.1	1.4	0.2	0.1	4.7	3	9
Linseed	0.3	1	–	<0.1	<0.1	<0.1	0.178	<0.1	<0.1	–	–	0.1	–	0.24	0.2	–	–	–	–	–	<0.1	3	0.7	2.9
Castorseed	–	–	–	0.1	–	ca. 0.1	–	–	–	–	–	–	–	0.2	0.8	–	–	–	–	–	<0.1	1.2	0.52	1.3
Corn oil (maize oil)	1.2	–	<0.1	<0.1	<0.1	<0.15	0.25	0.1	<0.1	<0.1	–	<0.1	<0.1	0.2	–	–	–	–	–	–	<0.1	–	2.32	–
Total seed & seed oils	**97.8**	**13.7**	**0.6**	**61.5**	**52.5**	**123.6**	**24.73**	**1.2**	**7.6**	**7**	**2**	**18.7**	**2.1**	**54.54**	**30.9**	**5**	**1.7**	**4.2**	**7**	**0.7**	**10.2**	**405**	**87.3**	**233[5]**
Pulp oils																								
Olive oil	–	–	–	–	<0.1	<0.1	2.1	–	–	–	–	–	0.1	–	–	–	–	–	–	–	0.3	–	3	–
Palm oil	–	–	–	0.2	–	1.2	–	–	–	–	–	–	–	–	–	–	–	16	17.2	1	1.6	–	38	–
Total pulp oils	**–**	**–**	**–**	**0.2**	**–**	**1.2**	**2.1**	**–**	**–**	**–**	**–**	**–**	**0.1**	**–**	**–**	**–**	**–**	**16**	**17.2**	**1**	**1.9**	**–**	**41**	**–**
Grand Total oil seeds & vegetable oils	**97.8**	**13.7**	**0.6**	**61.7**	**52.6**	**124.8**	**26.83**	**1.2**	**7.6**	**7**	**2**	**18.7**	**2.2**	**54.54**	**30.9**	**5**	**1.7**	**20.2**	**24.2**	**1.7**	**12.1**	**405**	**128.3**	**–**
Animal Fats & Oils																								
Fish	<0.1	–	–	–	–	0.5	0.14	0.1	–	–	–	0.45	–	0.1	–	–	–	–	–	–	–	–	1	–
Butter	0.5	<0.1	–	–	–	0.2	1.8	<0.1	0.2	–	–	0.4	–	–	2	0.5	–	–	–	–	–	–	6.9	–
Lard	0.5	0.1	–	0.4	–	–	2	<0.1	0.2	–	–	0.2	–	3.6	–	–	–	–	–	–	–	–	7.6	–
Tallow & Greases	3.8	0.4	0.1	0.6	0.2	0.8	1.1	<0.1	0.1	–	–	0.2	–	0.9	0.2	0.1	–	–	–	–	–	–	8.5	–
Total animal fats & oils	**4.9**	**0.5**	**0.1**	**1**	**0.2**	**1.5**	**5.4**	**0.1**	**0.5**	**–**	**–**	**1.1**	**–**	**4.6**	**2.2**	**0.6**	**–**	**–**	**–**	**–**	**–**	**–**	**24.0**	**–**
Grand total oils & fats																						**405**	**152.3**	**233**

Sources: ISTA Mielke, *Oil World Annual 2007*, private communication Thomas Mielke. mto = metric tons; Mio = Million; ha = hectare (2.471 acres) [1]shelled basis, 70% of unshelled; [2]including canola; [3]Argentina, Bolivia, Brazil, Colombia, Ecuador, Paraguay, Uruguay, Venezuela; [4]Russia, Ukraine, Kazakhstan, Belarus, Moldova, Armenia, Aserbaijan, Georgia, Kyrgystan, Tajikistan, Turkmenistan, Uzbekistan; [5](for comparison 1983/84: 150 Mio. ha)

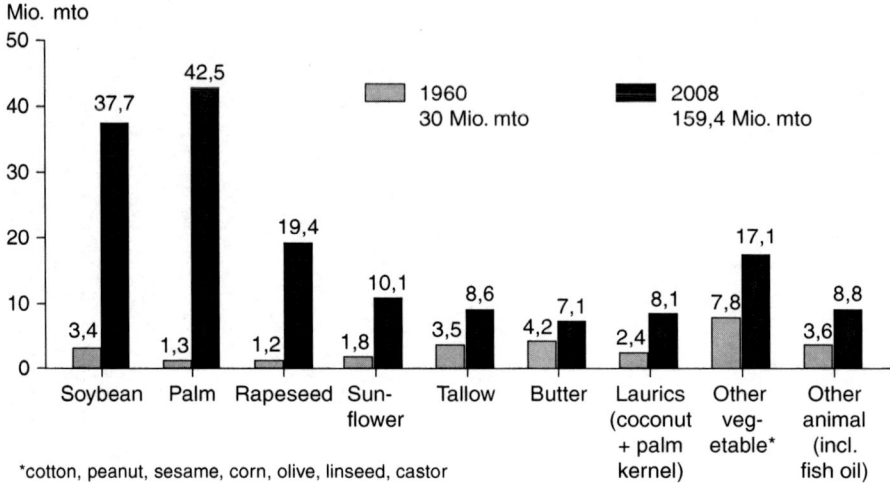

Figure 9.1.1   Development of world oils and fats production 1960/2008.[8]

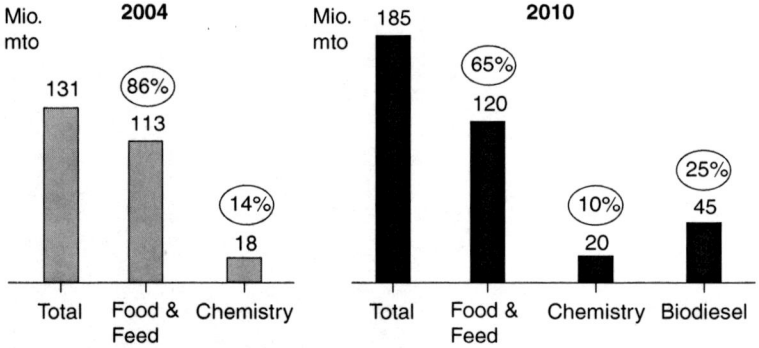

Figure 9.1.2   World consumption of oleochemical raw materials 2004/2010. Forecast for biodiesel in 2010: 45 Mio mto;[9] a more conservative forecast envisages 32 Mio mto in 2012;[10] for comparison: world consumption of mineral oils in 2004: approximately 4 bn mto.

On a global basis annual production of oil seeds reached 405 Mio mto in 2006/2007[8] (Figure 9.1.3, Table 9.1.1). In the 2007/2008 season, though, world output of oilseed dropped back to 391 Mio mto. After more than 20 years of more or less steady growth the global area devoted to ten oilseeds declined by 1 Mio ha to 232.6 Mio ha reflecting the high interdependence and growing volatility of grain and oilseed farming and their respective outlets to the food, dairy, livestock and biofuels markets. Production of fats and oils reached 152.3 Mio mto in Oct./Sept. 2006/2007, *i.e.* more than five-fold higher than 1960, and production is estimated to be 159.4 Mio mto in 2007/2008. World production of oil meals was 255.8 Mio mto in Oct./Sept. 2006/2007 and reached an estimated 261.5 Mio mto in 2007/2008.

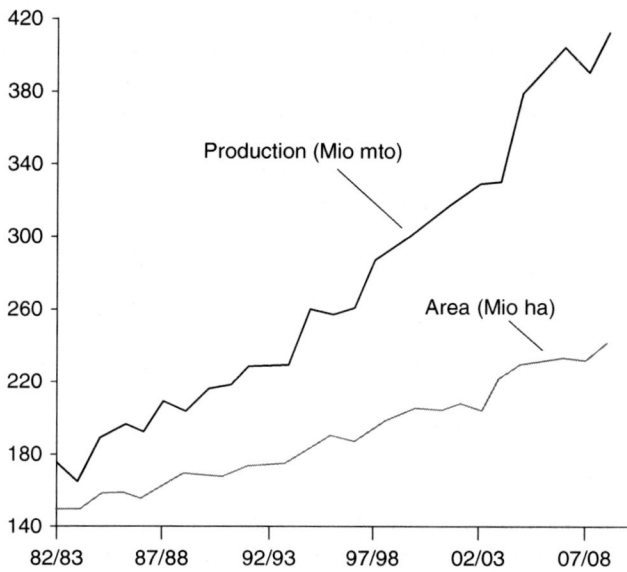

Figure 9.1.3 Oilseeds: world area and production[8] (reproduced with permission of ISTA Mielke Oil World, Hamburg).

Possibly even more dramatic than the sky-rocketing of overall production are regional changes in cultivation and alterations in the types of fats and oils which have occurred. In 1960 butter and tallow ranked number 1 and 2 in world production, respectively. Since then they have been largely surpassed by vegetable oils, *i.e.* soybean, palm, rape seed and sunflower oils (Figure 9.1.1).

Europe is only one example for these changes. Towards the end of last century large areas of European landscape changed from the green of grassland to the brown-yellow of rapeseed and sunflower crops. This alteration in postcard colours reflects a fundamental change in agro-production in Central and Northern Europe, which traditionally had been focusing on farmyard products like milk, butter and beef tallow. Permanent grassland decreased significantly, accentuated by the fall in numbers of livestock after the EU imposed milk quotas in 1984 in order to stop subsidized overproduction of milk (which had resulted in so-called milk-lakes and butter-mountains). This released land that could be farmed for the production of crops. Areas sown leapt from just a few thousand hectares in the early 1980s to 4.8 Mio ha by 1990 and 10.2 Mio ha by 2007. Additionally, the reformed Common Agricultural Policy, CAP, and a considerable body of legislation and energy-related policies have provided incentives to encourage the use of energy crops for the production of renewable energy, in particular biofuels. Meanwhile the E-27 became the world's largest producer of rapeseed oil and number 2 (behind the CIS) in sunflower seed and oil, which is dominant in the southern European countries.

Regional changes of similar magnitude have happened in South America. Enlarged production of soy beans and livestock farming of cattle for meat production helped the Brazilian national economy to significantly improve its international financial position and to participate in the dynamics of global growth. Brazil succeeded in reaching record highs in exports, trade balance and, since 2003, a consecutive yearly current account surplus resulting in a significant improvement of Brazilian external sustainability indicators to the best levels of the historical series.[11]

At the same time soya has overtaken illegal logging and ranching as the main engine of deforestation of virgin rainforest to make room for the crop. It needs, however, to be noted that the Brazilian Agro Energy Plan 2006–2011 envisages optimizing the use of areas affected by human action on natural vegetation, maximizing the sustainability of the production systems, discouraging unjustifiable expansions of the agricultural frontier and encroachment upon sensitive or protected systems.[12]

In Southeast Asia the ambitious expansions of oil palm plantations in the tropical areas have made this region by far the main palm oil producing region in the world outdistancing natural rubber cultivation as the key agro resource in Malaysia (Figure 9.1.4).

The accelerated extension of oil palm plantations has also initiated discussions about the social and environmental impacts, more particularly violation of land rights of indigenous people, destruction of community-based economies, deforestation of rainforests, oil palm cultivation as false pretence for illegal logging, biodiversity loss including risk of orang-utan, Asian elephant, clouded leopard, Sumatran and Javan rhino extinction,[14] fragmentation and loss of Sumatran tiger habitat and climbing pollution levels due to forest fires for land-clearance.[15]

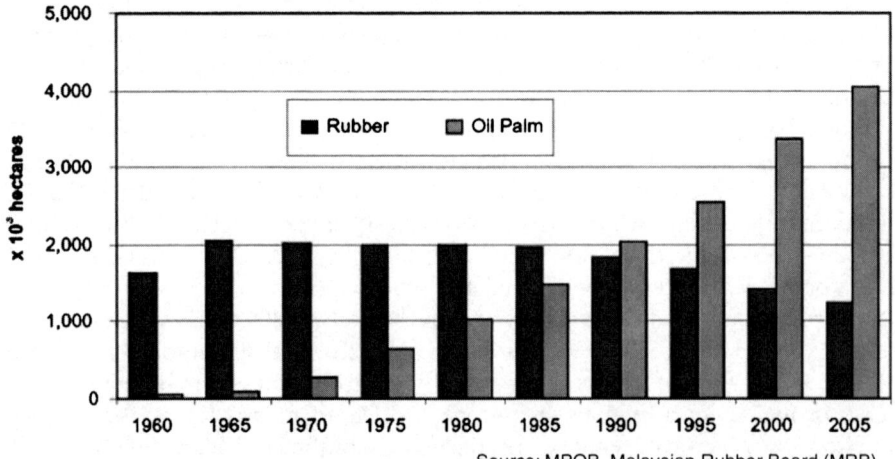

Source: MPOB, Malaysian Rubber Board (MRB)

Figure 9.1.4 Planted area for oil palm and natural rubber in Malaysia[13] (reproduced with permission of Wiley-VCH Verlag GmbH & Co. KGaA, Weinheim).

Petroleum oil reserves are concentrated in a relatively small number of countries. World consumption has surged steadily upward. The rapidly industrializing economies of China and India are projected to increase their demand dramatically. The consumption of Asia including Japan is already at the level of North America. The crude oil main streams flow from a few countries in the Persian Gulf, West Africa, the Gulf of Mexico and the Caribbean plus Russia to the highly industrialized regions in North America, Europe and Asia[16] (Figure 9.1.5).

Compared with crude oil, biofuels are expected to reduce many of the vulnerabilities associated with today's highly concentrated energy economy.[17] Developing countries will have a competitive advantage in biofuel production due to lower land and labour costs, warm tropical climates and longer growing seasons. International trade channels will follow those already established between grain and oilseed exporting countries like Brazil, Argentina, USA and palm oil exporters like Malaysia and Indonesia and consumers in Europe and Asia (Figure 9.1.6).

Africa has enormous potential to become a grain and vegetable oil exporting region if ever political and administrative frameworks become sustainable and the concept of *good governance*[18] is established within each country. However, it needs to be noted that with growing importance of biofuels the markets for food, fuel and energy, when based on biomass, will increasingly become inter-twined.

Downstream processing will eventually move away from historical centres of technology to the sites of origin, a phenomenon which is happening already for fatty acid, fatty alcohol and fatty acid ester production.[20] Indeed, there is no guarantee for low freight charges supporting global trade and global

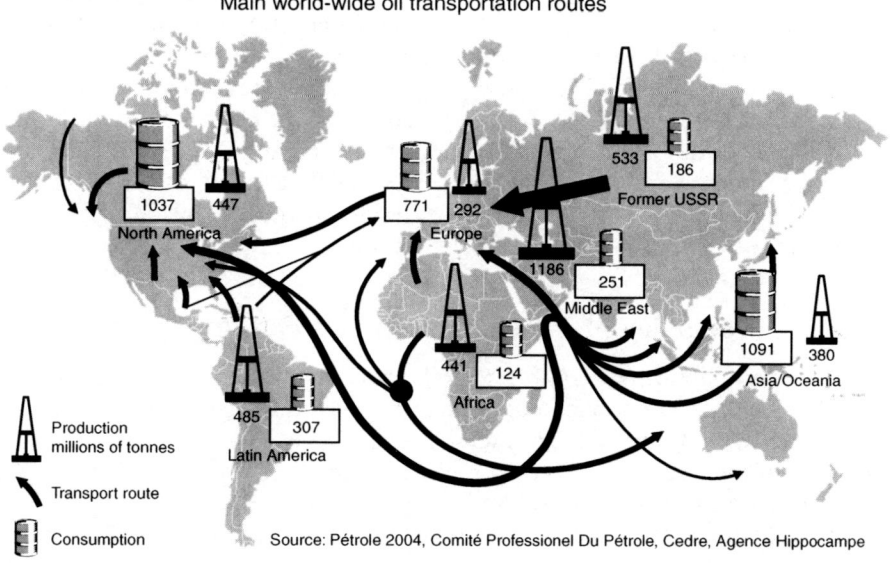

Figure 9.1.5 Global trade fossil oil.[16]

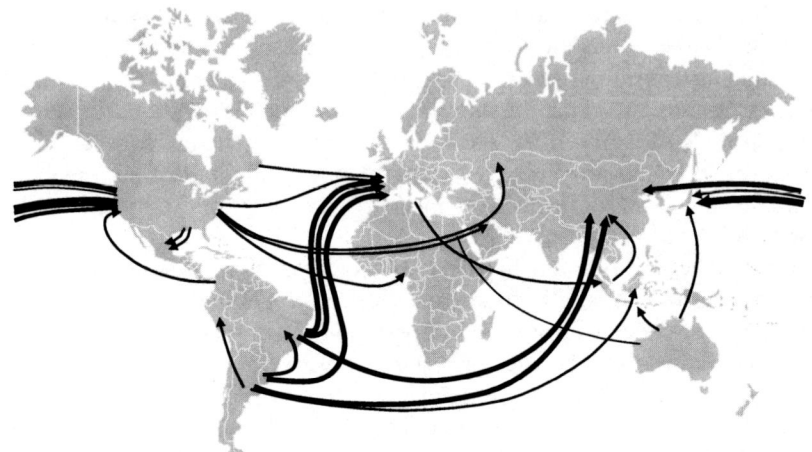

Figure 9.1.6 Global trade grains and oilseeds (Reproduced with the permission of Rabobank).[19]

transportation forever. On the contrary, increasing crude oil prices will sooner or later impact on transportation cost. This could compensate the lower energy density of biomass compared to fossil crude and further push towards decentralized bio-refinery concepts.

9.1.3 Production of Oils and Fats

Sustainable production of oils and fats starts from sustainable crop plantations and livestock husbandry. Based on Good International Industry Practice (GIIP) as technical reference, Environmental, Health and Safety (EHS) Guidelines for crop plantation and crop production regulate environmental issues such as stress on water resources, soil erosion, pesticide use, eutrophication of aquatic environments, biodiversity impacts, crop residues and atmospheric emissions. Additionally, EHS Guidelines for Vegetable Oil Processing are applicable to facilities that extract and process oils and fats from vegetable sources.[21]

9.1.3.1 Production of Vegetable Oils and Fats

Modern manufacturing of vegetable oils can be considered as a set of unit operations, which are applied depending on the composition of the crop, local or regional conditions and the intended end use. Because of the quick deterioration of the fruit, pulp oils are processed immediately after harvesting close to or on the oil fruit plantation.

Oilseeds, unlike pulp oil fruits, can easily be transported and stored over long periods in ventilated silos. Industrial extraction usually takes place close to the location where the crude oil is further processed. Preparatory steps prior to oil extraction comprise preparation and conditioning, i.e. conveying the seed from the silos to the factory processing equipment; cleaning; drying; removal of husks, hulls, shell or seed coat from the oilseed kernel; crushing, sometimes followed by rolling to improve oil extraction by increasing the surface area of the seed.

After pre-treatment of the seeds, the oil is extracted and separated from the meal. Oil extraction can be performed mechanically (cold pressing or expelling) or chemically by solvent extraction.

During solvent extraction, in the majority of cases hexane is used to wash the conditioned raw materials, typically in a counter current extractor. Hexane is removed from the oil through distillation, and from the cake through steam stripping in a multi-stage counter current toaster and recovered for reuse. EHS issues during solvent extraction are odour and volatile organic compounds, VOC. Extraction plants and processes should in each case comply with the solvent emission directives in force.[22] Efforts to replace hexane as the solvent of choice by aqueous enzymatic methods or supercritical fluids have not yet found industrial acceptance.

Mechanical extraction does not use solvent extracts. It is made the traditional way, *i.e.* separation of the oil from the solid particles by squeezing the oil out of the crushed mass of seeds using several different types of mechanical extraction often in small-scale operations.[23] This method is typically used to produce the more traditional oils (*e.g.* olive oil) and it is preferred by most health-food customers.

Crude oils and fats, which are obtained by expelling, extraction or rendering (in the case of animal raw materials) still contain components that are undesirable for taste, smell, stability and appearance. The removal of these undesirable non-triglyceride compounds is called refining. Physical and chemical unit operations exist for refining (Figure 9.1.7), degumming being the first one. Degumming removes hydratable (HP) and part of non-hydratable phosphatides (NHP). Crude oil containing a considerable amount of gums is subject to **water degumming** immediately following extraction. Water or steam is added to hot oil as a result of which a phosphatides-containing gum layer is formed. The layer is separated from the oil by decantation (settling) or continuously by centrifugation and – in the case of soybean refining – processed into commercial lecithin. **Dry acid degumming** applies for low gum contents such as in palm oil, palm kernel oil, coconut oil or animal fats. Phosphoric or citric acid is added to pre-heated crude oil followed by vigorous agitation. The conditioned gums are absorbed into bleaching earth and separated by filtration.

Enzymatic degumming was first introduced by Lurgi as the EnzyMax® technology.[24] Reduced gum volume due to formation of water-soluble lysophosphatide, avoidance of soap stock formation, hence increased oil yield, and reduced environmental impact have made enzymatic degumming based on phospholipase enzyme an alternative to conventional degumming with increasing commercial interest (Figure 9.1.8).[25]

Chemical refining (reduction of free fatty acids and oxidation products of free fatty acids, which would make the oil or fat rancid; removal of the residual

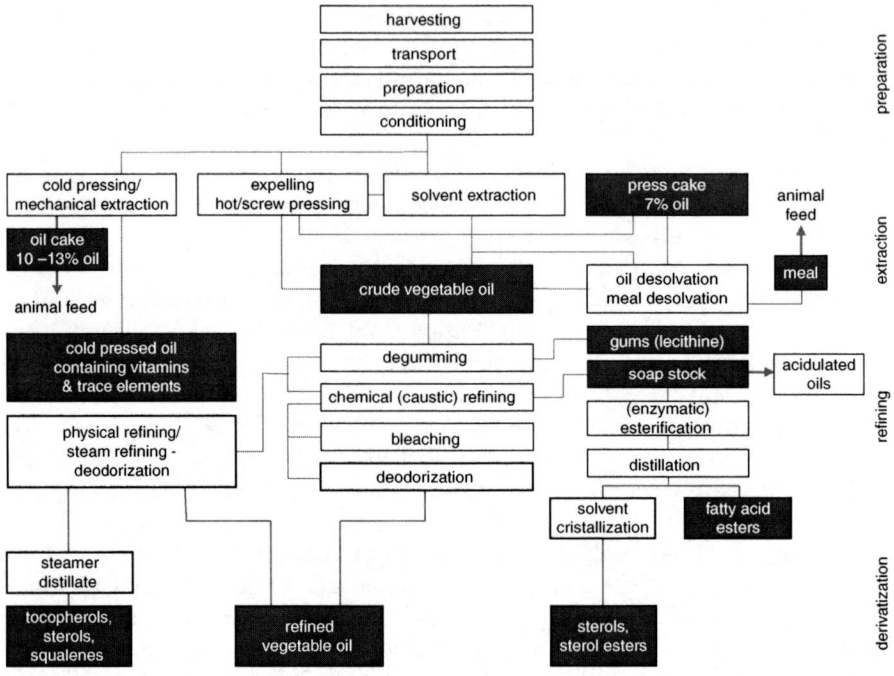

Figure 9.1.7 Vegetable oil processing unit operations.

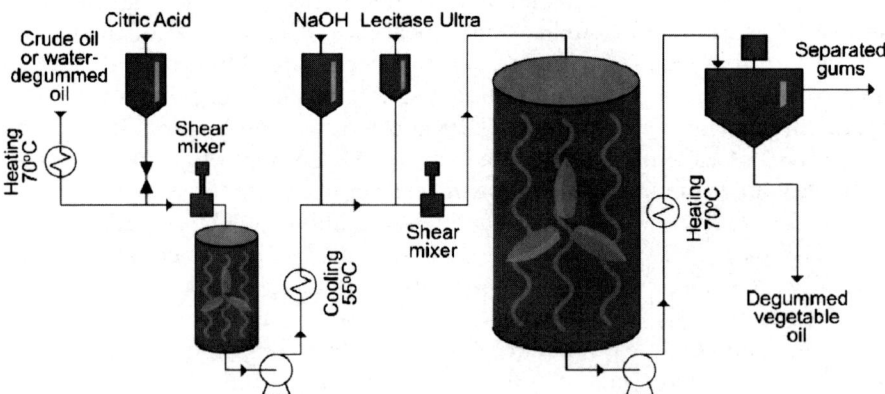

Figure 9.1.8 Layout of enzymatic degumming (reproduced with the permission of Novozymes).

phosphatides) comprises the dispersion of an acid, *e.g.* phosphoric acid in water de-gummed oil or crude oil and the addition of slight excess of caustic soda liquor. By this treatment soap stock is formed as a second phase. The soap stock contains the free fatty acids originally present in the untreated oil, some triglyceride oil and other compounds such as sucrolipids and lipoproteins. The alkali refined, so-called neutral oil is then bleached by heating under reduced pressure with bleaching earth which is subsequently removed by filtration. The alkaline neutralization process has major drawbacks: the yield is comparatively low and oil losses occur due to emulsification and saponification of neutral oil.

In physical refining, (Figure 9.1.9) the fatty acids are removed by a steam distillation (stripping) process similar to deodorization. In practice, a temperature of 230–250 °C, a pressure of 2 mbar and sparge steam of 1 to 2% are applied for 60 to 90 min to reduce the free fatty acid content of the refined oil to a level of about 0.03–0.05%. As a prerequisite, special degumming processes may be required for physical refining that phosphatides be removed to a level below 5 mg phosphorus/kg oil. In physical refining the gas exiting the steam distillation is made up of a mixture formed by stripping gas, malodorous compounds and valuable by-products such as free fatty acids (FFA), triglycerides, sterols, tocopherols and squalenes.[26]

After refining **hydrogenation** of the unsaturated fatty acids in the presence of a nickel fixed bed or a (supported) nickel slurry phase catalyst transforms the liquid oil into a (semi-)solid fat. When Wilhelm Normann invented the liquid phase hydrogenation of oils,[27] which he called fat-hardening (*Fetthärtung* in German), he initiated a paradigm shift in the diet of industrialized nations. Hardened fats became a storage stable and economic alternative to butter but

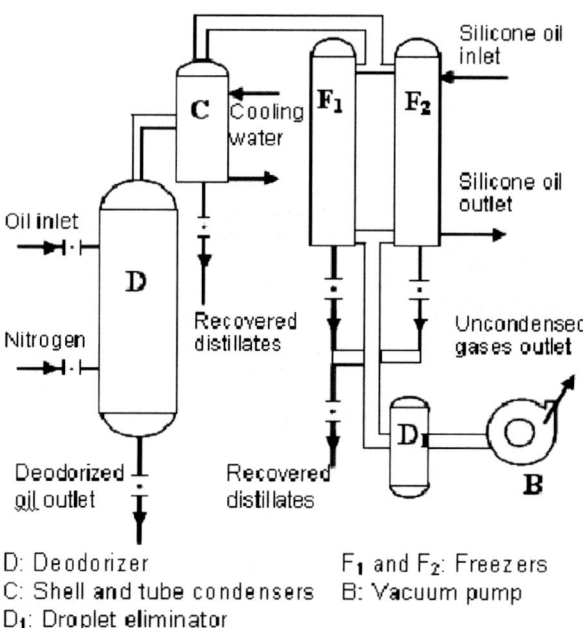

Figure 9.1.9 Physical refining, schematic diagram (reproduced with the permission of Mᵃ Manuela Prieto Gonzalez, Universidad de Oviedo, Departamento de Energía; manuelap@uniovi.es).

also, going beyond food applications, in the manufacturing of candles, industrial lubricants, fabric softeners, thixotropes, mould release agents, cosmetics and toiletries, and household and automotive polishes.

9.1.3.2 Production of Animal Oils and Fats

As in the processing of vegetable oils the extraction of fats and oils from animal sources can be regarded as a sequence of unit operations (Figure 9.1.10).[28] In modern economies milk and butter are predominantly used as dairy products[29] (besides e.g. casein paints and glues) whereas tallow and marine oils are used for food as well as for industrial purposes.

Milk can be regarded as a dilute O/W-emulsion of butter fat stabilized by membranes of phospholipids and casein proteins. Changing whole milk to butter is a process of transforming the diluted fat-in-water emulsion to a water-in-fat emulsion (butter). Churning physically agitates the cream until it ruptures the fragile membranes surrounding the milk fat. Once broken, the fat droplets can join with each other and form clumps of fat, or butter grains, and coalesce. In the end, there are two phases left: a semi-solid mass of butter and the liquid left over, which is the buttermilk. The buttermilk is drained off, and the remaining butter is kneaded to form a network of fat crystals that becomes the continuous phase with water dispersed in it. Butter oil is the fat concentrate obtained primarily from butter or cream by the removal of practically all the water.

Animal fats are produced batch-wise with the use of manual techniques or at industrial scale using batch, semi-continuous or continuous processes called rendering. Rendering involves the splitting of the fat bearing raw material into

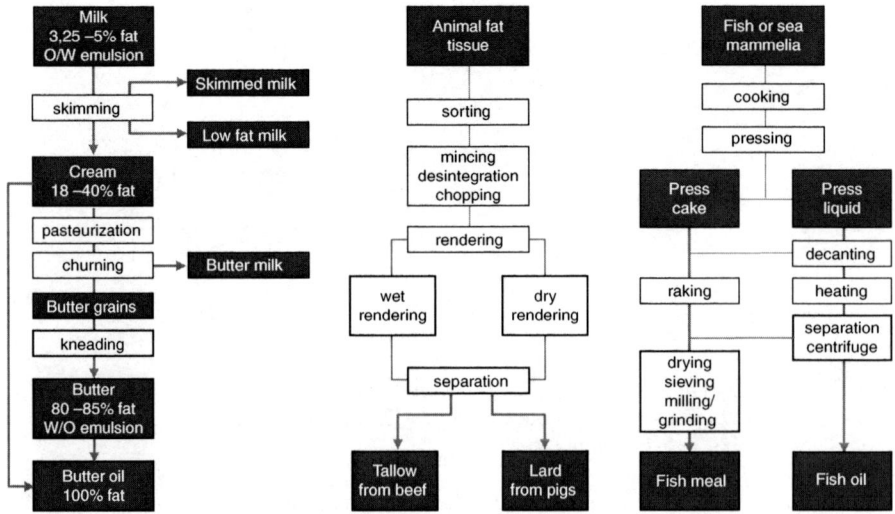

Figure 9.1.10 Animal fat processing unit operations.

its three major components comprising water, fat and solid proteinaceous meal in a rendering cooker. Edible rendering consists of heating ground fat trimmings in a melting tank to about 43 °C, pumping the melted tissue in a disintegrator, which ruptures the fat cells, followed by separation of the proteinaceous solids from the melted fat and water by centrifugation. The melted fat and water are heated to about 93 °C and in a second-stage centrifuge the edible fat is separated from water.

Fish oil and fish meal are produced from fish that are caught specifically for this market, by-catch from fishing activities and solid waste from filleting and canning. Raw materials are transported by screw conveyors to a cooking process which acts to coagulate the protein. The cooked mixture is screened, and then pressed to remove most of the water from the mixture. The press cake is shredded and dried. The pressed liquid generated from the previous processes passes through a decanter to remove most of the sludge, which is fed back to the meal dryer. Fish oil is separated from the liquid by centrifuges, polished and refined.

Over-fishing of the oceans is a deep-rooted concern of ecologists and the public. Although there is wide agreement between experts that fisheries must be managed with the least impact on stocks, food-webs and habitats, under current fisheries management most fish are taken as soon as they can be caught regardless of size or age for use as both marine food and fish oil. Thus a paradigm shift is requested and promoted by the marine stewardship council (MSC) for world fisheries towards an ecosystem-based fisheries management, EBFM.[30] A recent study of the Leibniz Institute of Marine Sciences in Kiel, Germany, the Rhodes University, Grahamstown, South Africa, and the Aquatic and Fisheries Sciences Centre in Rennes, France, has shown that fishing has much less impact on stocks if fish are caught after they have reached the size where growth rate and cohort biomass are maximum. Such management would allow juvenile and adult fish to better fulfil their ecological roles and prevent an increase in over-fished or collapsed stocks.[31]

9.1.4   Chemical Composition of Fats and Oils

Chemically, natural fats and oils are triglycerides, also called acylglycerols; that is fatty acid triesters of glycerine with different even-numbered, linear fatty acids. The fatty acids may be saturated or unsaturated. There are very few exceptions to this (Figures 9.1.11 and 13). The most well-known exceptions are castor oil, glyceroltriester of 12-hydroxyoleic (ricinoleic) acid, vernonia oil, sperm oil, jojoba oil and tall oil.

Unsaturated fatty acids containing double bonds may have the *cis*- or *trans*-configuration (Figure 9.1.12). In *cis*-fatty acids, all the hydrogen atoms adjacent to the double bonds are on the same side of the longitudinal carbon axis. In *trans*-fatty acids (TFA), the hydrogen atoms adjacent to the double bonds occur on alternate sides of the main axis. The *trans*-configuration is chemically more stable. With non-treated vegetable oils the *cis*-configuration prevails.

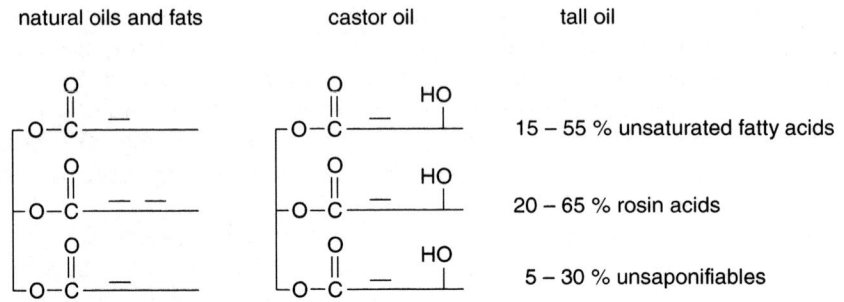

Figure 9.1.11   Chemical composition of oleochemical feedstock.

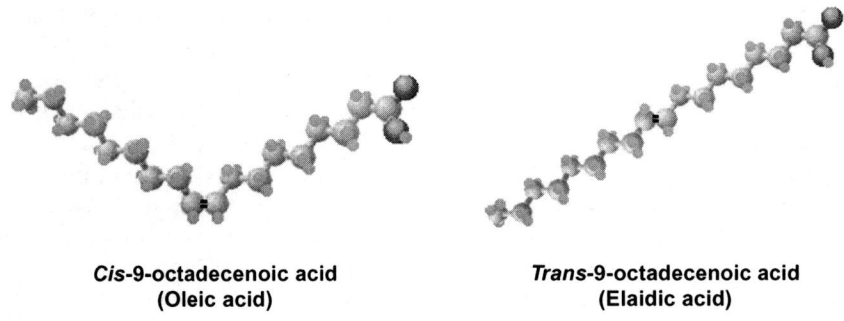

Cis-9-octadecenoic acid   *Trans*-9-octadecenoic acid
    (Oleic acid)   (Elaidic acid)

Figure 9.1.12   *Cis*- and *trans*-configuration of unsaturated fatty acids.

Due to interaction with a hydrogenation catalyst the *cis*-configuration can be transformed into the *trans*-configuration. This typically happens during partial hydrogenation of polyunsaturated vegetable oils and (as a function of time and temperature) during physical refining. In response to published human studies that show that intake of TFA increases low-density lipoprotein-cholesterol (LDL-C) in the blood and risk of coronary heart disease, health authorities worldwide recommend that consumption of *trans*-fat be reduced to trace amounts.

If natural fats and oils are solid or semisolid at 20 °C they are designated as fats ("edible fats" in food technology); if they are liquid at 20 °C they are described as oils (correspondingly "edible oils" in food technology). Naturally occurring oils and fats always contain minor constituents such as free fatty acids, phospholipids, sterols, hydrocarbons, pigments, waxes and vitamins. Careful extraction of by-products like tocopherols or sterols and addition to foodstuff as industrially produced dietary supplements is the basis of modern neutraceutical or functional food manufacturing. Such by-products are, for example, extracted from soybean oil steamer distillate,[26,32] from lecithin after degumming[33] or from crude tall oil, CTO.[34]

Vernonia oil, is a triglyceride of vernolic acid, an epoxy oleic acid. Vernonia oil is extracted from the seeds of *Vernonia galamensis* (or ironweed), a plant native to Eastern Africa. Vernonia seed contains about 40 to 42% oil of which 73 to 80% is *vernolic acid*.

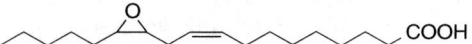

Vernonia oil has been discussed as natural alternative to epoxidized soya bean oil or epoxidized linseed oil for applications such as reactive diluent for low VOC alkyd paints.

Tall oil is a by-product of sulphate pulping; Crude tall oil (CTO) mainly consists of a mixture of unsaturated fatty acids (TOFA), rosin acids and insaponifiables.

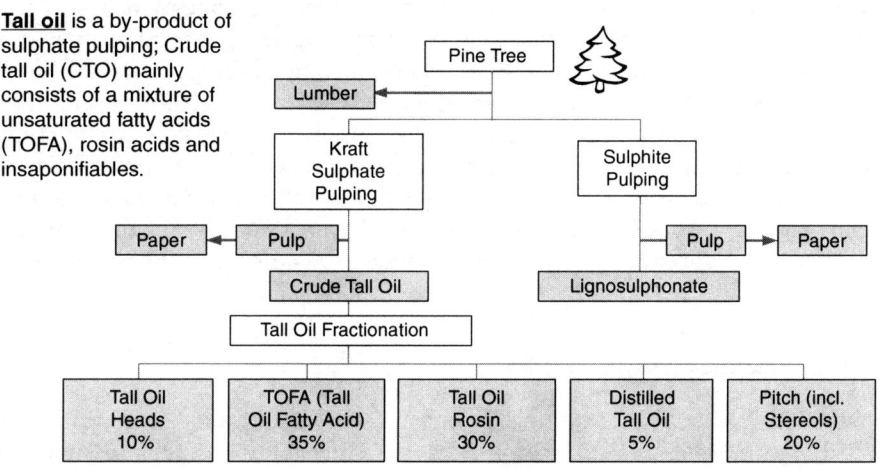

Jojoba oil is a vegetable oil obtained from the crushed bean of the jojoba shrub. The jojoba shrub is native to the Sonoran Desert of Northwestern Mexico and neighbouring regions in Arizona and Southern California. Jojoba has flat gray-green leathery leaves and a deep root system that make it well adapted to desert heat and drought. It has a life span of 100 – 200 years. Like sperm whale oil, jojoba oil is composed almost entirely of liquid wax esters of straight chain mono-unsaturated cis $C_{20} - C_{24}$ fatty acids and unsaturated higher fatty alcohols. Main component is Docosenyl-eicosenoate (C42) and it's chemical structure is

$$CH_3(CH_2)_7CH=CH(CH_2)_9C(O)O(CH_2)_{12}CH=CH(CH_2)_7CH_3(37\%).$$

Jojoba oil is similar to, and miscible with, sebum, which is secreted by human sebaceous glands to lubricate and protect skin and hair. Obtained by cold pressing, its peculiar chemical properties make it the only vegetable oil in nature having the same characteristics as sperm oil.

Sperm oil occurs in the blubber and in the head cavities of the sperm whale (this cavity helps the whale keep part of his head above water to breathe). The oil contains a solid fraction (spermaceti, cetin, cetylpalmitate) which can be cristallised. The liquid fraction mainly contains oleyloleate and higher mono-unsaturated fatty acid monesters of unsaturated fatty alcohol.

Figure 9.1.13 Natural oils with unusual chemical composition.

Essential oils are usually derived from the non-seed parts of the plants with chemical composition other than the triglyceride structure of natural fats and oils. They refer to the subtle, aromatic liquids extracted from the flowers, seeds, leaves, stems, bark and roots of herbs, bushes, shrubs and trees through distillation. Essential oils are concentrated liquids containing volatile aromatic compounds. They are used in perfumery, aromatherapy, cosmetics, incense, medicine, household cleaning products and for flavouring food and drink. Their use in aromatherapy and other health care areas is growing.

Natural waxes may be secretions of plants or animals with properties between those of a fat and a resin, namely plastic (malleable, pliable) behaviour at ambient temperatures, polishable under slight pressure, insoluble in water, hydrophobic. They consist mainly of esters of long-chain alcohols and acids ($> C_{22}$), free primary straight chain aliphatic C_{24} to C_{36} alcohols, free fatty acids and other components like paraffins, sterols or neutral resins. Natural waxes include those of animal origin like beeswax, woolwax (lanolin), shellac and spermaceti as well as vegetable waxes like carnauba wax, candelilla wax, sugar cane wax, orange peel wax and floral waxes, *i.e.* by-products of the extraction of fragrance oils from flowers such as jasmine, rose and lavender.[35]

9.1.4.1 Animal Fats and Oils

Animal derived fats can be categorized as milk, rendered fats and marine oils.

Milk fats have an unusually broad fatty acid distribution ranging between C_4 and $C_{20/22}$, mainly C_{16} and C_{18}. Land-animal fats mainly have a fatty acid chain length of C_{16} or C_{18} and ruminant fats contain 5–10% *trans*-fatty acids, which are produced from linoleic and linolenic acid in the rumen. Marine oils of industrial importance mainly come from three different species: herring, pilchard and menhaden. Fish oils are distinguished from land-animal fats by their high content of highly unsaturated (4–6 double bonds) C_{20}, C_{22} and C_{24} fatty acids including ω-3 fatty acids. However, fish do not actually produce omega-3 fatty acids, but instead accumulate them by either consuming micro-algae (phytoplankton) that produce these fatty acids, as is the case with prey fish like herring and sardines or, as is the case with fatty predatory fish, by eating prey fish that have accumulated ω-3 fatty acid from micro-algae (Table 9.1.2).

9.1.4.2 Vegetable Fats and Oils

Vegetable oils according to their major fatty acid composition can be sub-classified into laurics and oleics. Laurics comprise coconut oil and palm-kernel oil. Laurics mainly contain saturated C_{12} and C_{14} fatty acid triglycerides, the raw material basis for surfactants. Oleics mainly contain unsaturated C_{16} and C_{18} fatty acid triglycerides. Apart from food, which is the predominant use, they deliver lubricants and base oils, which can be used for energy generation, fuel, nutrition and specialty chemical derivatives.

Table 9.1.2 Fatty acid compositions of natural oils and fats.*

Column groups: **vegetable** = castor oil … Jatropha; **animal** = lard, tallow; **sec animal** = sperm whale oil, herring oil. Bold figures are average values; natural figures indicate the range.

systematic name	trivial name	double bonds	formula	castor oil	coconut oil	palm kernel oil	palm oil	peanut oil	cotton seed oil	soy bean oil	sun-flower oil	HO sun-flower oil	linseed oil	rape seed high erucic	rape seed low erucic	Jatropha	lard	tallow	sperm whale oil	herring oil
hexanoic acid	caproic acid	0	C6H12O2		**0.5** 0.2-0.8	**0.3** 0-1														
octanoic acid	caprylic acid	0	C8H16O2		**8** 6-9	**3.5** 3-5														
decanoic acid	caprinic acid	0	C10H20O2		**6** 4-8	**3.5** 3-5														
dodecanoic acid	lauric acid	0	C12H24O2		**48** 46-52	**48** 46-50	**0.2** 0-0.4										**0.2** 0-0.5			
tetradecanoic acid	myristic acid	0	C14H28O2		**18** 15-20	**16** 14-17	**1.5** 0.5-3		**1** 0.5-2								**1.5** 0.3	**2.5** 1-4	**16** 14-18	**8** 6-10
hexadecanoic acid	palmitic acid	0	C16H32O2	**2** 1-2	**9** 8-10	**8** 6-9	**44** 40-47	**10** 7-12	**21** 20-27	**10** 9-11	**6** 5-7	**4** 3.5-4.5	**7** 6-7	**3** 2-4	**4.5** 4-5	**14.7** 10-17	**26** 21-29	**28** 21-29	**13** 12-14	**12** 10-14
octadecanoic acid	stearic acid	0	C18H36O2	**1** 1-2	**2.5** 1-3	**2** 1-3	**4** 2-7	**3** 2-6	**2** 1-3	**4** 3-5	**4** 3-6	**4** 3.5-5	**3.5** 3-5	**1** 0.5-2	**1.5** 1-2.5	**6.1** 5-10	**16** 13-18	**20** 15-24	**9** 8-10	**1** 0.5-1.5
eicosanoic acid	arachidic acid	0	C20H40O2		**0.1** 0-0.4	**0.1** 0-0.4	**0.3** 0-0.5	**3** 2-4	**0.5** 0.2-1				**0.5** 0.3-1	**0.7** 0.5-1	**0.5** 0.1-1	**0.1**	**0.3** 0-1	**0.3** 0-0.5	**2** 1-3	
docosanoic acid	behenic acid	0	C22H44O2					**2** 0.1-3						**1** 0.4-1.5	**1** 0.4-1.5	<0.1				
tetracosanoic acid	lignoceric acid	0	C24H48O2					**2** 1-3					<0.1							
dodecenoic acid	lauroleic acid	1	C12H22O2																	
tetradecenoic acid	myristoleic acid	1	C14H26O2														**0.3** 0-1	**0.5** 0-1	**4** 3-5	
hexadecenoic acid	palmitoleic acid	1	C16H30O2				**0.2** 0-0.4	Sp.	**0.5** 0-2	<0.5				Sp.	Sp.	<1	**3** 2-4	**3** 2-4	**12** 11-13	**9** 7-11
octadecenoic acid	oleic acid	1	C18H34O2	**3** 2.5-4	**6.5** 5-8	**16** 13-19	**38** 36-42	**50** 35-70	**29** 20-35	**28** 20-35	**24** 18-28	**83.5** 82-90	**22** 12-34	**15** 11-24	**60** 50-65	**40.7** 36-64	**41** 39-45	**38** 33-46	**14** 12-16	**10** 9-11
eicosenoic acid	gadoleic acid	1	C20H38O2					Sp.						**9** 8-11	**1.5** 1-2.5	<0.1	**1** 0-2	**0.5** 0-1	**17** 15-18	**13** 11-15
docosenoic acid	erucic acid	1	C22H42O2											**47** 41-52					**7** 5-8	**18** 15-21
12-hydroxy-octadecenoic acid	ricinoleic acid	1	C18H34O3	**87** 86-92																
9,12 octadecadienoic acid	linoleic acid	2	C18H32O2	**3** 3-8	**1.6** 0-2.5	**2.5** 1-3	**11** 7-12	**30** 20-35	**45** 42-54	**53.5** 50-56	**64** 60-68	**5** 2-10	**17** 14-26	**13** 10-15	**21** 18-25	**36** 18-45	**7.5** 4-9	**4** 2-7		**1**
9,12,15 octadecatrienoic acid	linolenic acid	3	C18H30O2				**0.3** 0-0.5			**8** 7-10			**52** 35-65	**8** 7-13	**10** 8-11.5	**0.2**	**0.7** 0.2	**0.7** 0-1.5		**0.5** 0-1
5,8,11,14,17 eicosapentaenoic acid		5	C20H32O2														0-1			**10** 8-12
4,7,10,13,16,19 docosahexaenoic acid		6	C22H34O2																	**7** 6-8
iodine value (after Wijs)				81-91	7.5-10.5	14-23	44-54	84-100	99-113	120-140	126-136	80-100	155-205	97-108	110-120	96	53-57	40-48	70	123-142
saponification value				177-197	250-264	245-255	195-205	188-195	189-198	189-195	186-194	186-194	188-196	170-180	180-190		190-202	190-200	140-144	179-194
melting point °C				-10…-15	23-26	24-26	30-40	-2	-2 to +2	-10	-15	-10	-20	-9	-5		33-46	40-46		

Analytic numbers of oils.

* **Source:** Henkel KGaA, Hrsg., Fettalkohole, Düsseldorf, 1983; R. Höfer, D. Feustel, M. Fies, Welt der Farben 11 (5/1997), updated, based on proprietary analysis and C. Thywissen, Neuss/D, K. Becker, G. Francis, Univ. Hohenheim/D for Jatropha, bold figures are average values, natural figures indicate the range of fatty acid composition in %. Analytical methods: DGF C6 10A + 11D

9.1.4.2.1   Soya Bean Oil, Soybean Oil

Historical Development and Economic Importance.   Since 1965 global soybean production has increased from 30 Mio mto to 237 Mio mto and soya oil production from 5 Mio mto to 37 Mio mto in 2007. This change in the pattern of soya seed production is reflecting both the growing importance of soybean oil and also the growth in importance of oilcake and meal as a protein source for human nutrition and animal feed.

Soya beans originate from China (Figure 9.1.14). Since the 1930s they have been grown in the Midwest, the Corn Belt of the USA. In 1996 the United States produced about 49% of the world's soybeans. To feed growing demand, new agricultural frontiers have been opened up in Brazil and Argentina and meanwhile MERCOSUR countries have become by far the largest regional producers. Although Chinese soybean acreage ranks fourth in the world, its productivity is still lower than in the other soya-growing regions. This makes China actually a net importer for soybean oil – imports in 2007 actually doubled national production.

Genetic Modification.   The soya bean success story would not have been possible without weed control pesticides on one side and modern genetic engineering technologies on the other. In 1995 Monsanto introduced Roundup Ready® (RR) soybeans that are genetically modified.[36] RR soybeans have greatly improved agricultural efficiency allowing a farmer to spray widely the herbicide glyphosate in soya plantations in order to kill unwanted plants like annual grasses, perennial weeds and other pests, to reduce tillage or even to sow the seed directly into an unploughed field. The transgenic RR plant is resistant against glyphosate, N-(phosphonomethyl)glycine. Glyphosate is a non-selective, post-emergence, systemic broad-spectrum herbicide, absorbed through the leaves, and was first sold as the isopropylammonium salt (IPA salt) under the trade name Roundup®.

$$
\begin{array}{ccc}
 & O & O \\
 & \parallel & \parallel \\
HO - C - CH_2 - N - CH_2 - P - OH \\
 & | & | \\
 & H & OH
\end{array}
$$

glyphosate, N-(phosphonomethyl)glycine

Glyphosate is only effective on actively growing plants and is not effective as a pre-emergence herbicide. The herbicidal mode of action is to inhibit the shikimate pathway enzyme 5-enolpyruvylshikimate 3-phosphate (EPSP) synthase, the functionality of which is absolutely required for the biosynthesis of aromatic amino acids and the survival of plants.[37] Like weeds and grasses conventional soya would thus not withstand treatment with glyphosate. RR plants carry the gene coding for a glyphosate-insensitive form of this enzyme, obtained from *Agrobacterium* sp. strain CP4. Once incorporated into the plant genome, the gene product, CP4 EPSP synthase, confers glyphosate-tolerance to the crop. Adoption of genetically modified (GM) soya has been extremely rapid

Soybean botanical   Soybeans

Figure 9.1.14 Soybeans.

and now makes up approximately 90% of all soya crops in the USA.[38] Also Brazil and Argentina are going towards 90% adoption of GM soya crops. Soybean tolerant to glyphosate has also been approved for use, import and processing in Europe since 1996 under Directive 90/220/EEC.[39]

Crop Protection. Already by the early 1900s, a serious fungus disease, caused by the pathogen *Phakopsora pachyrhizi*, had been taking hold in various soybean producing countries.[40] This disease was first discovered in 1902 in Asia (China, Japan, *etc.*), which gave rise to the name "Asian Rust". As a result of the windblown transport of spores by means of powerful currents of air, the disease spread from Asia to various African countries. From Africa it spread to South America, and lastly to Middle and North America. *Phakopsora pachyrhizi* is a bound pathogen (it does not survive outside a vegetable host) capable of infecting, in addition to soybean, about a hundred species of plants belonging to the same family as the soybean. This allows it to survive in the period when crops are not present, thus becoming a source of infection to be feared all year round. Although genetic research is attempting to develop new soybeans that are resistant to the disease, currently the only instrument of defence is the use of fungicides belonging to different chemical categories, more particularly the triazole category (*e.g.* cyproconazole, tebuconazol, tetraconazole), the strobilurine category, chloronitriles (like chlorothalonil) and/or carboxamides.

Composition and Uses. Soybeans are distinguished from other oil crops in that they only yield approximately 21% oil but 40–50% protein. Because of the high content in proteins soybeans historically have been a protein rather than an oil seed crop. The market for soybean products is largely driven by soy meal applications, more particularly by the food and the livestock feed industries (Figures 9.1.15 and 9.1.16). The majority of soy protein is a

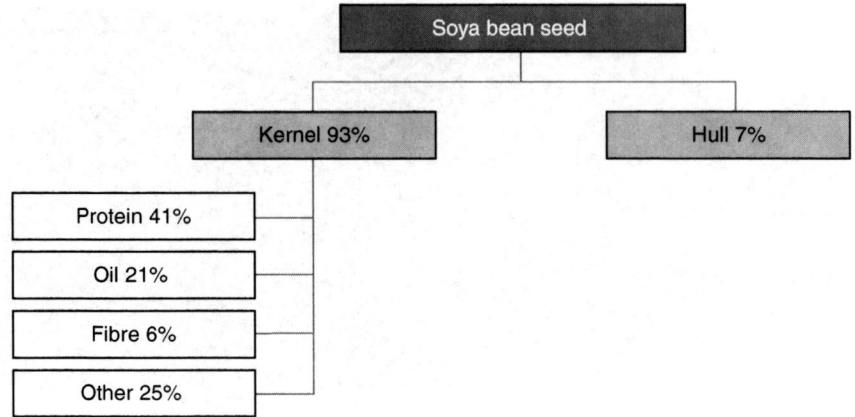

Figure 9.1.15 Composition of soya bean seed.

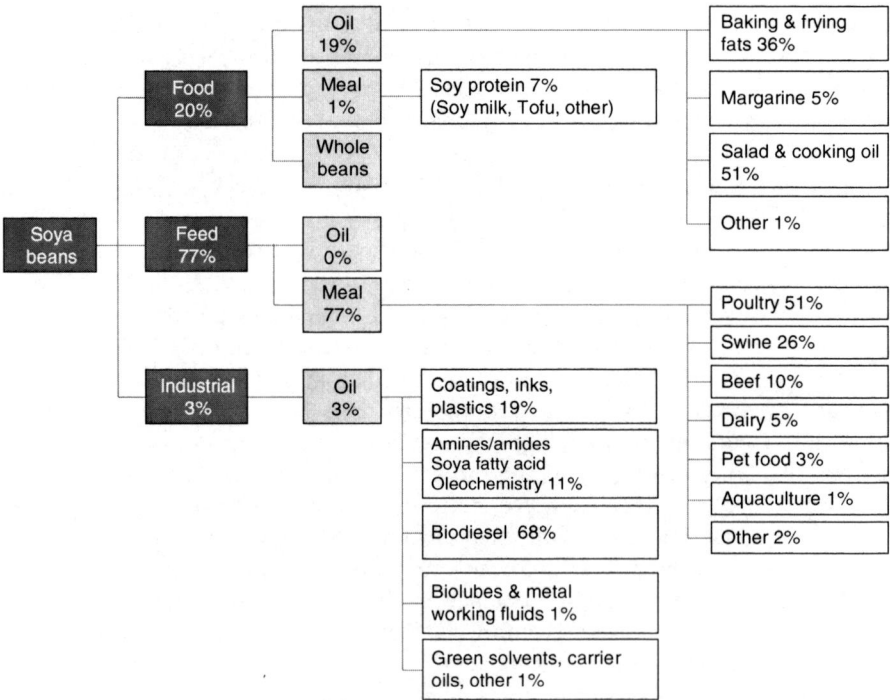

Figure 9.1.16 Soybean applications, regional US market.[42]

relatively heat-stable storage protein held in discrete particles called protein bodies. This heat stability enables soy food products requiring high temperature cooking, such as tofu, to be made. Soya increases the protein content of processed meat products. It replaces them altogether in vegetarian foods.

In the USA 77% of soybeans are used in agricultural feeds for open land animal husbandry. Indeed, soy protein is what has made the global factory farming of livestock such as poultry, swine and cattle a possibility. More recently soy protein is increasingly used as a supplement to fish meal in aquacultures.[41]

20% of soybeans are used as oil for human consumption and 3% for industrial applications.[42] Soya oil has moved into markets once served by cottonseed oil, peanut oil and animal fats. Among the food uses, baking and frying fats plus salad and cooking oils are the main applications. Among the 11 Mio mto soya oil applied to non-food purposes about 19% are employed in coatings, inks and plastics applications in the form of:

- soya alkyd resins, urethane oils, oil-modified phenolic or urea resins;
- epoxidized soybean oil (ESO) applied as plasticizer for PVC and nitro-cellulose or as sticker in agrochemical formulations;
- soya polyols (Sovermol®) as building blocks for polyurethanes and as radiation curing monomers (Photomer®).

About 11% are processed to yield soya fatty acid, amines, amides, esters, soaps and other oleochemical derivatives.

The production of methyl soyate for environmentally friendly solvents and for biodiesel fuel (in the USA) is becoming a significant outlet for soy oil in non-food applications. Further progress in the demand for soy biodiesel will result in additional soybean meal supplies, which will considerably increase soy meal competitiveness as feedstock for livestock farming and aquaculture.

9.1.4.2.2 Palm Oil and Palm-kernel Oil

Historical Development and Economic Importance. Even more impressive compared to soya is the growth development of palm oil. The oil palm[43] is a native of West Africa. From the beginning and throughout the twentieth century oil palm cultivation and development efforts in Zaire (formerly Belgian Congo, now Democratic Republic of the Congo) had achieved significant increases in yield and oil quality. Congo oil palm plantations did expand to a peak of 250,000 ha in 1958. Later, political unrest and mismanagement of the country's economy following the army coup in 1960 led to stagnation and decline in the industry. This is in marked contrast to developments occurring at the same time in Southeast Asia. The optimal soil, rainfall and sunshine conditions of Southeast Asia resulting in high yields and uniform quality and the strong-willed support of local governments especially in Malaysia helped the industry to grow quickly, more particularly during the last 30 years. Since 2004 palm oil surpassed soybean oil as the leading vegetable oil and reached annual production of 41 Mio mto in 2008 spearheaded by Malaysia and

Table 9.1.3   Comparison of oil yield per hectare of five major oilseeds (plus Jatropha).

Oilseeds	Oil Yield $(mto\,ha^{-1}\,annum^{-1})$
Palm oil (Malaysian)	3.93[a]
Rapeseed (EU)	1.33[b]
Soybean (USA)	0.46[b]
Sunflower (Argentina)	0.66[b]
Jatropha	3.00[c]
Coconut (Philippines)	0.66[d]

Source: [a]MPOB (2006); [b]Khoo (2001); [c]H. A. Khoo (2007), Steffan Preusser (2006); [d]Oil World (2006).

Indonesia. The leading role of palm oil compared to other oil crops arises from the high yield of oil per hectare.[44,45]

Further increase of palm oil production by yield increase, *e.g.* by better seed material, as an alternative to land expansion, is one of the challenges faced by the palm oil growing industry (Table 9.1.3).

Sustainable Palm Oil Cultivation.   For a number of reasons palm oil has become the focus for concerns dealing with the basic means of sustainable development:

- The oil palm flourishes in the humid tropics between 7 or eventually 10 degrees north and south latitude, in other words in the central belt of tropical rainforests. Rainforest destruction with its attendant elimination of wildlife and biodiversity is a concern not only of environmentalists but of large parts of the world community.
- Palm oil production has been a basic source of income of the rural poor and their indigenous community-based economies; large-scale palm oil plantations operated by multinational companies and financed by globally operating financial institutions are regarded as a threat for the survival of smallholder operations and are suspected of the alienation of indigenous people who often do not have the formal government-recognized documentation to prove their ownership of the land.[46]
- Large oil palm plantations and intense palm oil production are accused of polluting the environment, of using unsafe fertilizers and pesticides in excess and of paying poor attention to workers' health and safety.
- Palm oil, when traded and used as a transportation fuel and in electrical and thermal power generation risks provoking a conflict with the traditional role as foodstuff.

In order to address these concerns without creating difficulties for the developing nations to continue their economic and social progress, to promote the sustainable production and use of palm oil and to meet the challenges from NGOs and corresponding demands from consumers the Roundtable on

Sustainable Palm Oil (RSPO) was formally established in 2004. The RSPO is a global initiative having organizations carrying out their activities in and around the entire supply chain for palm oil as stakeholders. Palm oil producing countries, companies and organizations like the Malaysian Palm Oil Council (MPOC) are well aware that "sustainable palm oil is now a business requirement, which no company can overlook" (Dato' Seri Lee Oi Hian) and is turning into a key trade issue.[47] As a consequence, pursuing certification of sustainably produced palm oil to a globally accepted standard with full traceability has been recommended.[44,48] Accordingly, the RSPO Verification Working Group has elaborated and launched an Audit Programme and a roadmap.[49]

The RSPO Principles and Criteria for Sustainable Palm Oil Production[50] stipulate:

- public availability of health and safety plans, land titles and user rights;
- compliance with local, national and international laws and regulations including the UN declaration of indigenous people;
- business plans to achieve long-term economic and financial viability;
- water management programs to minimize peat soil subsidence;
- appropriate integrated pest management (IPM) techniques;
- development, implementation and monitoring of plans to reduce pollution and emissions, including greenhouse gases;
- a minimum age of workers, which should be not less than 15 years;
- involvement of governments with the problem of stateless persons.

The RSPO multi-stakeholder process has gained recognition as a valid mechanism to develop and implement social and environmentally responsible management practices towards sustainable development and could eventually serve as a blueprint for multi-stakeholder production and supply-chain processes for other critical regionally or globally traded commodities.[51]

Distinctive Features of Palm Oil and Palm-kernel Oil Production. The oil palm is a perennial crop. It bears fruit 30 months after planting and is economically viable for up to 30 years. The oil palm bunch is harvested manually using a knife attached to a pole which dissevers the stalk of the fruit bunch from the palm (Figure 9.1.17). The harvested bunches are termed fresh fruit bunches (FFBs). The oil palm is distinguished from other oil crops as it delivers two vegetable oils with different chemical composition: palm oil and palm-kernel oil (Figure 9.1.18).

Palm oil is the pulp oil extracted promptly after harvesting from the fleshy mesocarp of the oil palm fruit using continuous screw presses. The liquid coming from the press is a mixture of oil, water and non-oily solids (NOS). After separation from sludge in a settling tank, clarification, purification and vacuum drying, crude palm oil (CPO) is yielded from the press liquid.

Palm-kernel oil is the seed oil contained in the kernel of the palm fruit. The kernel is the endosperm of white cellular mass coated with a tough black

Palm trees   Palm fruit botanical

Figure 9.1.17   Oil palm and palm fruit.

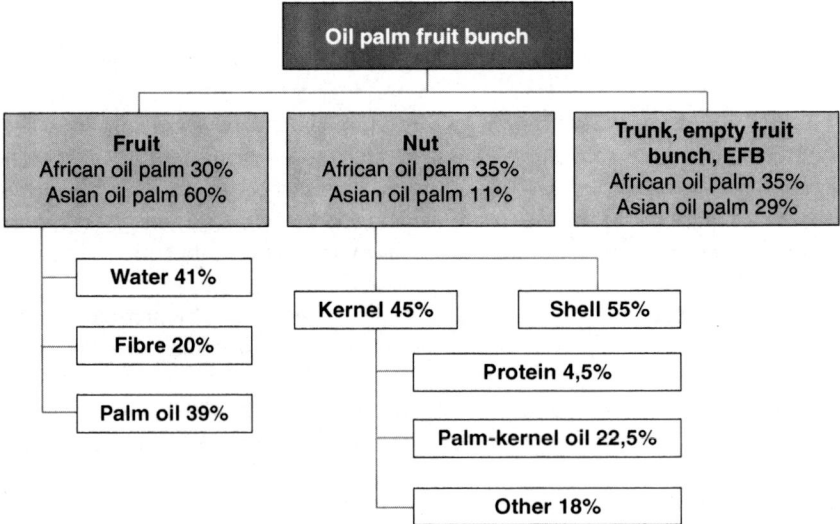

Figure 9.1.18   Composition of oil palm fruit.

membrane or testa, which is encased in a thick shell or endocarp of the oil palm
seed nuts (Figures 9.1.17 and 9.1.18). When extracting the palm oil, nuts and
fibres remain in the press cake. Nuts are separated from the fibres and cracked.
After separating kernel and shell the palm-kernel oil is extracted from the
kernels by unit operations employed for vegetable seed oil extraction.

Palm oil is traded as CPO and, increasingly, after physical/steam refining or chemical/alkali refining as processed palm oil (PPO).

PPO = processed palm oil, stands for different categories of treatment, *i.e.* neutralized, bleached, deodorized and/or fractionated; according to the degree of processing different export duty tariffs may apply: N/RPO = neutralized/refined palm oil, BPO = bleached palm oil, CPL = crude palm olein, N/RBPO = neutralized/refined bleached palm oil, N/RPL = neutralized refined palm olein, BPL = bleached palm olein, N/RBDPO = neutralized refined bleached deodorized palm oil, N/RBPL = neutralized refined bleached palm olein, N/RBDPL = neutralized/refined bleached deodorized palm olein, Dp = duty on PPO and Dc = duty on CPO; source: A. Mohd Nasir.[52]

Because of its fatty acid composition, which includes 50% saturated and 50% unsaturated fatty acids, palm oil is a semi-solid oil, which sediments at room temperature even in tropical countries. Fractionation of CPO or RBDPO by separating palm oil into a low melting fraction (palm olein) and into a higher melting fraction (palm stearin) significantly extends the application. One of the main objectives of fractionating palm oil is to obtain palm olein of low cloud point for cooking oil or further processing, whereas palm stearin is used as a component of harder frying fats or for the production of shortening, margarine and vanaspati.

Three different technologies can be applied for palm oil fractionation:[53]

- winterization or dry fractionation;
- hydrophilization or rewetting, also called wet fractionation or detergent process;
- solvent fractionation.

Winterization is claimed to be the cheapest technology in terms of cost per unit of oil fractionated followed by the rewetting process. Due to the continuous development of the dry fractionation process, a whole variety of products formerly produced by solvent fractionation with hexane can now be obtained with a high degree of selectivity with winterization. As the crystallization operates in the bulk, viscosity problems limit the degree of crystallization in one single step, and multi-step operations are currently used. Additionally, combining proper crystal development with highly efficient separation by using membrane press filters allows compression of the stearin cake to obtain as much occluded liquid olein as possible.

Solvent fractionation is the most expensive process, and has been almost abandoned, except for the production of high-premium mid-fractions for cocoa butter replacers using an acetone-based double-stage process.

In a well-run palm oil mill, it is expected that each 100 mto of FFB processed yields 20 to 24 mto of crude palm oil and about 3 mto of palm-kernel oil. Thus between 72 and 76% of the FFB comes out at various stages of the process as waste. The solid wastes that result from the milling operations are empty fruit bunches, palm fibre and palm kernel shell. With the aim of achieving zero-waste operations the waste products are all put to economically useful purpose:[54] wet, empty bunches are dried and used as fuel. Another economic use for the empty bunches is to return them to the plantation as a mulch to enhance moisture retention and organic matter in the soil. Residual shell is disposed of as gravel for plantation roads maintenance. The fibre recovered from the nut/fibre separation stage constitutes the bulk of material used to fire the large boilers that deliver superheated steam for electrical power generation. Boiler ash is recycled as fertilizer and factory floor cleaning agent. Large- and medium-scale mills produce copious volumes of liquid waste from the sterilizer, clarifying centrifuges and hydro-cyclones. This effluent must be treated before discharge to avoid serious environmental pollution. Liquid waste treatment involves anaerobic fermentation followed by aerobic fermentation in large ponds until the effluent quality is suitable for discharge. In some of the mills the treated effluent is used on the farm as manure and as a source of water for irrigation. The sludge accumulating in the fermentation ponds is periodically removed and fed to the land.

Composition and Uses.   **Palm oil** is an oleic oil, which is distinguished by its high percentage of the saturated palmitic acid. Palm-kernel oil is a lauric oil and to some degree exchangeable with coconut oil due to a similar fatty acid composition.

Traditionally, about 80% of palm oil has been for edible use and 20% for non-edible applications. Products for food are used as a medium for frying; as a major component of non-dairy shortenings, margarines and vanaspati; in tailored applications for bakery products and confectionery fats; and in reduced fat spread, mayonnaise and salad dressings.[55] Non-food applications comprise the direct use of the triglyceride yielding palm oil polyols for polyurethane foams.[56] CPO is processed to yield glycerine, palm oil or hydrogenated palm oil fatty acids, soaps, amides, esters and other oleochemical derivatives for industrial and human care chemicals.[57] The global trend associating "natural" with plant-derived products and belief that these are milder and healthier has created a special Asian lifestyle favourable for palm oil derived cosmetics. Religious considerations also mean consumers prefer plant-derived products since they are regarded as kosher or halal.

This applies, of course, also for **palm kernel oil** (PKO), which is used either on its own or in combination with palm oil products in the manufacture of specialty fats. Incorporation of palm kernel oil or its derivatives, particularly in table margarine and spreads, improves the melting characteristic of the products. Non-food applications include soaps, which in Southeast Asia are primarily made from palm oil and PKO or coconut oil. A typical blend of soap noodles would be 80% palm oil and 20% PKO, giving about the right balance of lather, cleansing ability and hardness. PKO may also be used in a multitude of other cosmetic and body-care products for its moisturizing properties. Palm-kernel oil and coconut oil are distinguished from other vegetable oils because of their high content in lauric and myristic acid. Both oils have a relatively sharp melting point at 24–26 °C. It may be noted, however, that coconut oil has more of the short-chain fatty acids while PKO has more of the longer-chain fatty acids. Hydrogenation and fractionation of PKO fatty acid methylester yields lauryl alcohol and lauryl alcohol-based plant-derived surfactants for cosmetic, detergent and industrial applications.

Although actually (2007) only 1% of world biodiesel is based on palm oil, because of the high yield per ha palm oil could become a highly competitive raw material for biodiesel manufacturing.[45,58] Palm biodiesel obtained by transesterification of CPO has a pour point of 15 °C. The relatively high pour point and poor cold stability hamper the use as fuel and as a green solvent or carrier oil. Using winterization as well as rewetting of palm biodiesel a palm olein methyl ester with properties comparable to methyl soyate or methyl canolate is yielded, besides palm stearin methyl ester. Economies of palm biodiesel are improved, when sulphonating the palm stearin methyl ester by-product with sulphur trioxide to yield α-sulphonic fatty acid methyl ester surfactants (MESs). MESs based on palm stearin methylester are light-coloured washing detergents, which provide good cleaning ability, are less sensitive to water hardness and have better biodegradability characteristics compared to other detergents like linear alkyl benzene sulphonates.[59]

Non-refined cold pressed crude palm oil, also called "Red Palm Oil", has been consumed in Central and West Africa since time immemorial, as it has also been in Central America and Brazil. Red palm oil has a wonderful rich flavour; it is the richest known natural product in tocotrienols and carotenes, which give rise to the red colour. Development of techniques like integrated extraction technologies to recover these functional components from palm oil has contributed to the emergence of the nutraceutical industry in Malaysia that produces tocopherol and tocotrienol capsules, carotene-rich palm oil and palm carotene concentrate.[60]

9.1.4.2.3 Coconut Oil

Perhaps more than anything else, the coconut palm has contributed in creating the image of the islands in the Pacific Ocean and their beauty.

An adult coconut palm reaches 30 m or more in height. In some areas of the tropics, the palm grows with a minimum of attention through a lifespan of more than 50 years. About 96% of coconuts are grown by smallholders tending small farms. Every part of its fruit can be utilized, including the fibre for coir products, the shell for charcoal, the milk as a beverage and the meat as food. However, its most important product is copra, the meat after drying. Copra is a source of food, edible oil and protein meal (Figures 9.1.19 and 9.1.20).

Together with palm kernel oil and babassu oil, coconut oil forms the group of important lauric oils. They contain >40% of lauric acid and *ca.* 15% of

Coconut palm   Coconut botanical

Figure 9.1.19   Coconut palm, coconut botanical.

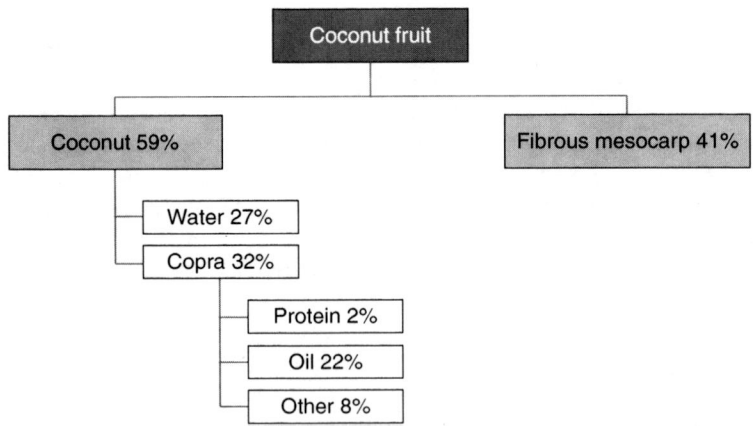

Figure 9.1.20   Composition of coconut fruit.

myristic acid besides a significant fraction of shorter-chain fatty acids. Fatty acids with 8 to 12 carbon atoms are classified as medium-chain fatty acids (MCFA, in German *Vorlaufffettsäuren*; to distinguish from short-chain fatty acids, SCFA, with carbon atom range from 2 to 6). With >15% C6, C8 and C10 fatty acids, coconut oil is the richest source of MCFA.

The intrinsic qualities of coconut oil make it very suitable for household culinary preparations, as frying oil, in non-dairy products, coffee whiteners and as spraying oil for biscuits and breakfast cereals. The cosmetic applications of coconut oil include hair and skin oil (when mixed with herbal oils and different scents), natural shampoos and skin and massage oils. The industrial use as oleochemical base-stock absorbs up to 60% of coconut oil. The oleochemicals sector comprises a myriad of acid, ester, alkanolamide and surfactant applications like isopropyl laurate (oil with solvent properties and wetting capacity for aerosols, decorative cosmetics and hair-care preparations), coconut fatty acid monoethanolamide (booster for detergents, foam stabilizer and thickening agent for shampoos and bubble baths), sodium lauryl sulphate (foaming agent for toothpaste, emulsifier for emulsion polymerization), sodium lauryl ether sulphate (base for manufacture of liquid shampoos, bubble bath, high-quality dishwashing agents) to name but a few.

9.1.4.2.4   Babassu Oil[61]

The 20-m high babassu palm covers extensive areas of Brazil's pre-Amazon region and is an inherent part of the regional economy of north-central Brazil. The nuts consist of a stone-hard protective shell that can be cracked only with immense efforts. As a result babassu nuts are gathered in the traditional way without any mechanization, and for the production of one mto of babassu oil nearly ten tons of stone-hard nutshells have to be removed and disposed of. Thousands of households derive a simple livelihood from collecting and processing babassu nuts. Babassu oil is very similar to coconut oil and had long been regarded as a promising alternative until palm oil took off even in the remote north of Brazil. As a consequence palm-kernel oil took over as the leading lauric oil.

9.1.4.2.5   Cuphea Oil

Like coconut, palm kernel and babassu, many species from the genus *Cuphea* have potential as sources of medium-chain triglycerides. These plants are native to the New World, from southern USA to northern South America. Most are herbaceous annuals that will grow in many locations. Table 9.1.4 illustrates the diversity in fatty acid composition available in *Cuphea* germplasm.

9.1.4.2.6   Rapeseed Oil, Canola Oil

Historical Development and Economic Importance.   Oilseed rape (also called rape or rapeseed) due to its ability to germinate and grow at low temperatures is one of the few edible oil crops that can be cultivated at northern latitudes. In agriculture the crop is also grown as a winter-cover crop. It provides good coverage of the soil in winter, and limits nitrogen run-off. The plant is ploughed back in the soil or used as bedding. Rape is likely to have been among the earliest domesticated food and fodder crops known since prehistoric times and cultivated in Europe since the thirteenth century (Figures 9.1.21 and 9.1.22).

Table 9.1.4 Fatty acid composition of some *Cuphea* seed oils in comparison to coconut oil.[62]

	Distribution (% of total fatty acids)				
Species	*8:0 caprylic*	*10:0 capric*	*12:0 lauric*	*14:0 myristic*	*Others*
C. painteri	73.0	20.4	0.2	0.3	6.1
C. hookeriana	65.1	23.7	0.1	0.2	10.9
C. koehneana	0.2	95.3	1.0	0.3	3.2
C. lanceolata		87.5	2.1	1.4	9.0
C. viscosissima	9.1	75.5	3.0	1.3	11.1
C. carthagenensis		5.3	81.4	4.7	8.6
C. laminuligera		17.1	62.6	9.5	10.8
C. wrightii		29.4	53.9	5.1	11.6
C. lutea	0.4	29.4	37.7	11.1	21.4
C. epilobiifolia		0.3	19.6	67.9	12.2
C. stigulosa	0.9	18.3	13.8	45.2	21.8
Coconut	8	7	48	18	19

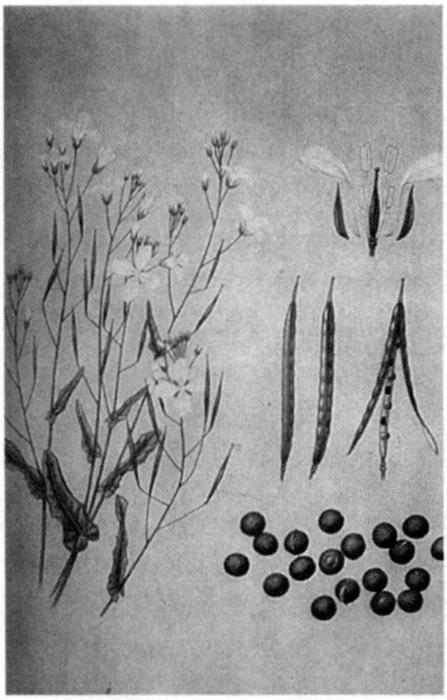

Rapeseed botanical

Rapeseed field

Figure 9.1.21 Rapeseed.

The oil pressed from the seed historically has been used in oil lamps and (because of the bitter taste) as poor people's cooking oil, later as lubricating oil in steam-powered ships and trains.[63] For two reasons, the nutritional aspects of rapeseed have been questioned: for the high fraction of eicosenoic and erucic acids in the oil on one hand, and the high glucosinolate content in the meal on the other. Glucosinolates (Figure 9.1.23) are thioethers, which contribute a bitter, "hot" taste to condiments (mustard, horseradish) and give flavour and

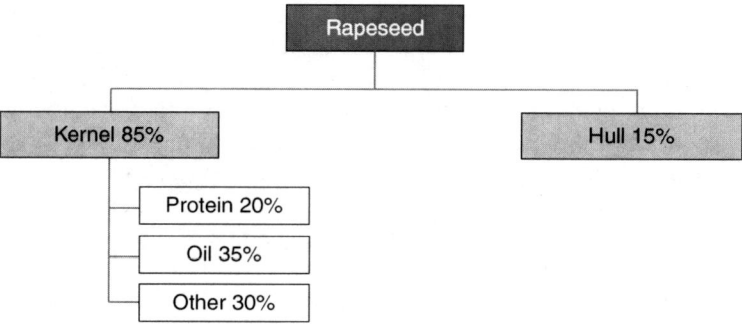

Figure 9.1.22   Composition of rapeseed.

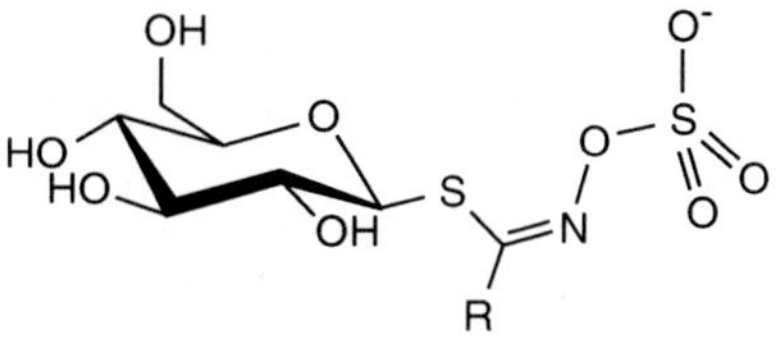

Figure 9.1.23   Glucosinolate, chemical structure.

odour to cabbage, broccoli, cauliflower and turnips. They act as defence compounds against herbivores. The high levels of erucic acid in traditional rapeseed oil were blamed, *inter alia*, for producing health concerns like fatty deposits in the heart.

Production of rapeseed took off in North America, when Canadian plant breeders at the University of Manitoba by 1974 succeeded in developing a new rapeseed cultivar, 0-rape or CANOLA, from the original high erucic acid rape seed, HEAR (Canola is a syncopated form of the abbreviation "Can.O., L-A.", **Can**adian **O**ilseed, **L**ow-**A**cid, now a trade name of the Canola Council of Canada; "canola" is mainly used in the American continent and in Australia, whereas Europe and other countries use the term "rapeseed"). The improvements later resulted in "double low" ("double zero", "00-rape", LEAR) varieties, *i.e.* reduction in erucic acid level to <2% of the measured fatty acid composition and a glucosinolate level less than 30 μmol g^{-1} of air dried, oil-free meal. By 1980 Canada had largely converted to the double-low varieties. It was more difficult and took nearly a further ten years to introduce the low erucic acid trait into European rapeseed lines of the winter type and to overcome the original problems of lower yield. Nonetheless, since 1991, virtually all rapeseed production in the European Union has shifted to 00-rape. Genetically modified (GM) herbicide resistant rapeseed, which enables a more efficient and effective approach to weed control, has been grown in Canada since 1996 and made up three-quarters of Canada's rapeseed crop in 2005. Although many field trials have been conducted, GM rapeseed is not yet being grown commercially in Europe.

Composition and Uses. The traditional brassica variety (HEAR) is distinguished by the high fraction of >60% unsaturated gadoleic (C20:1) and erucic (C22:1) acids, besides >35% oleic, linoleic and linolenic acids. The long-chain fatty acids limit the usefulness as food but have preserved an important niche market for HEAR in technical applications. At present, world consumption is estimated to be 57,000 mto per year. The plastics industry consumes a major part in the form of erucamide, a high melting wax, applied as lubricant for PVC processing and as slip and anti-blocking agent for polyolefin film. Hardening of erucic acid or splitting of hydrogenated HEAR oil with subsequent fractionation yields behenic acid. Hydrogenation of methyl behenate yields behenyl alcohol.

In the modern 00-rape the ≥ C:20 fatty acids have been reduced to trace fractions in favour of oleic, linoleic and linolenic acids, resulting in a fatty acids spectrum similar to olive oil. Characteristics of 00-rape are the most appropriate to fulfil the European biodiesel standards. As a result, the EU-27 are now the by far largest rapeseed oil consumers, and the astounding growth in European demand has been driven almost exclusively by the expansion of the biofuels sector over recent years. Fifty percent of European rapeseed production in 2005 has been turned into biodiesel, 37% has been used for human consumption, 7% for oleochemical products and lubricants; 6% has been exported. In 2007 the rape oil demand for biofuels has accounted for 60% of rape oil consumption. Rape-based biofuels include biodiesel and direct use as refined oil.

9.1.4.2.7 *Sunflower Oil*

Historical Development and Economic Importance. Sunflowers grow best in fertile, moist, well-drained soil in a moderate climate with temperatures mainly between 20 and 25 °C and full sun. The plant originates from the Americas. The seed was introduced into Europe by the Spanish explorers returning to the continent at the beginning of the AD 1500s. In Russia, it was introduced by Czar Peter I the Great, who, having seen sunflowers in the Netherlands, took seeds to Russia. Originally mainly regarded as a decorative flower, it was in Russia where the most important development took place in the use of sunflower as both a food and oil source (Figure 9.1.24). Starting from these origins Russia has built a leading position in sunflower growing. Including Ukraine and Russia the CIS States produce 40% of sunflower seed worldwide, followed by the EU-27 (20%) and Argentina (11%).

Composition and Uses. The sunflower kernel represents 75% of the seed, with an oil content of approximately 55%, amounting to 41% with respect to the whole seed (Figure 9.1.25). Sunflower oil contains >85% unsaturated fatty acids with linoleic acid accounting for 60%. The linoleic acid content in the original "linoleic sunflower oil" is very much dependent on climatic conditions and can deviate as much as ±20% from the average. Selected parcels

Sunflower plant   Sunflower botanical

Figure 9.1.24 Sunflower plant, sunflower botanical.

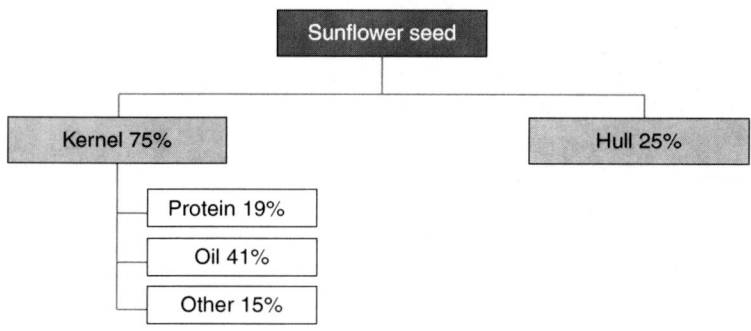

Figure 9.1.25 Composition of sunflower seed.

of oil, also called high-PUFA, contain >69% linoleic acids. Sunflower is one of the most important edible oils and is generally considered to be premium oil compared to most other natural oils because of its light colour, flavour, high level of linoleic acid and absence of linolenic acid.

High oleic sunflower oil (HOSO) is a cultivar, which has been developed through conventional breeding. Introduced in the mid 1980s, it is usually defined as having a minimum 80% monounsaturated oleic acid. The oil has a very neutral taste and provides excellent stability towards oxidation without hydrogenation. Demand for HO sunflower oil has been increasing as consumers look for healthier vegetable oils in response to concerns over saturated fats and *trans* fats in traditional edible oils and in part because of requirements for labelling of *trans* fats. A diet rich in high-oleic-acid sunflower oil favourably alters low-density lipoprotein cholesterol.[64]

NuSun is the name chosen by the US National Sunflower Association (NSA) to represent an entirely new class of sunflower seed and oil to meet the needs of the food industry.[65] Needs were identified as "oil that had a pleasing taste, stability without needing partial hydrogenation and low saturated fat levels". Initial testing of the new sunflower type (also developed through conventional plant breeding methods) began in 1996. In 2007, it is estimated that 85 to 90% of the sunflower acreage was NuSun with the rest equally divided between high oleic and traditional linoleic sunflowers. NuSun, also called mid-oleic sunflower oil, is lower in saturated fat (9%) than linoleic sunflower oil and has higher oleic levels (55–75%) with the remainder being linoleic.

Besides food uses the original linoleic sunflower oil as well as HOSO are used in industrial applications because of their specific properties.[66] For example, in view of the higher oxidative stability high-oleic sunflower oil is used as diesel and gasoline engine lubricant. Containing around 70% linoleic acid, sunflower oil is a semi-drying oil. Insofar as is economically feasible, sunflower oil may replace soybean oil in the manufacture of resins or carrier oils for paint and ink formulations.[67]

9.1.4.2.8 Cottonseed Oil

Cotton is a crop which delivers food (cottonseed oil), feed (meal) and fibre (cotton lint). In analogy to the importance of cotton for the development of the textile industry (see Chapter 9.6), cottonseed oil dominated the United States vegetable oil market for almost 100 years until the Second World War. On a global scale cottonseed oil production has been growing from 3.5 Mio mto in 1985 to 5 Mio mto in 2007. Cottonseed oil is mainly used for cooking, frying and in salads and as an alternative to lard in shortenings. Cottonseed oil has also been selected by Procter & Gamble to create Olestra, a sucrose-based artificial fat that isn't broken down by the body, and, as a result, adds no calories to foods.

9.1.4.2.9 Peanut Oil

Peanut oil (also known as groundnut oil, arachis oil or earthnut oil) is obtained from the seed kernels of the peanut plant, native to South America (Figure 9.1.26). Crude peanut oil has a nutlike flavour, which is removed by refining. It is mainly used for edible purposes in the preparation of shortening, margarines and mayonnaise, and as a salad-, cooking- and frying oil. Peanut butter is made by grinding specially roasted peanuts and homogenizing the mash with addition of liquid and possibly also hydrogenated peanut oil. Because of the high smoke point (229 °C) refined peanut oil is often used in deep-fat frying. Besides food, peanut oil is used as carrier oil in crop protection and in mineral oil-free pulp and paint defoamers.[68] Admixture of grain or seed with edible oil such as peanut oil is recommended to protect maize, sorghum, wheat or rice from insect damage.[69]

9.1.4.2.10 Corn Oil, Maize Oil

Maize[70] is, after wheat and rice, the most important cereal grain in the world (Figure 9.1.27). As food, the whole grain may be used or the maize may be

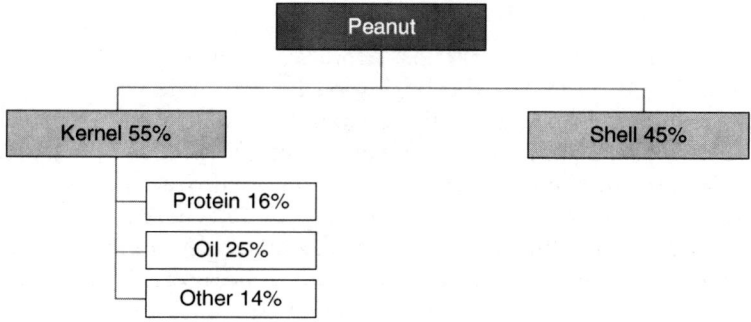

Figure 9.1.26 Composition of peanuts.

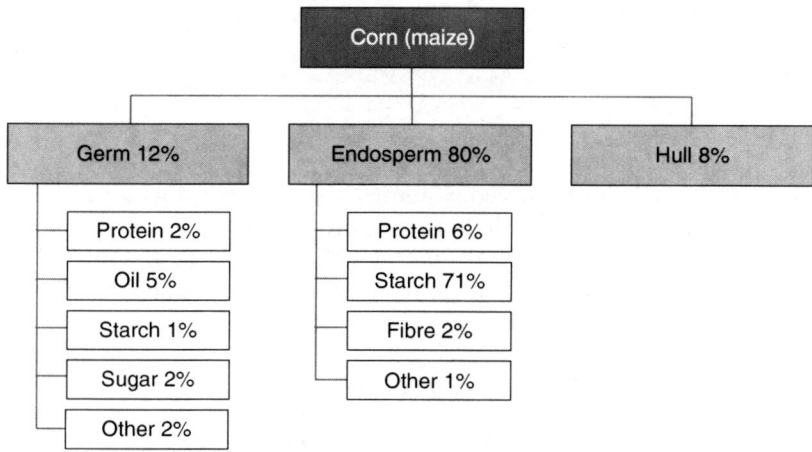

Figure 9.1.27 Composition of corn.

processed by two basic methods known as "dry milling" (to produce bio-ethanol, see Chapter 8) and "wet milling". In the wet milling process the cleaned corn is conveyed to steep tanks where it is soaked for 30–50 h at elevated temperature in an aqueous sulphurous acid solution resulting in the softening of the corn kernels. The corn germs can now be removed from the water-soaked kernels by a series of mills and cyclone germ separators. While the corn kernel is further processed yielding starch (see Chapter 9.2), syrup, corn gluten feed and sweeteners, the germ is dried and cold pressed through screw presses or expellers before solvent extraction with hexane.

The golden yellow corn oil has a distinctive musty flavour and odour and a composition similar to sunflower oil. However, it has a different tocopherol composition with mainly γ-tocopherol, which makes it more heat stable. Corn oil is used especially as a salad oil and in salad dressings as well as in cooking, where its high smoke point (unrefined 160 °C, refined 232 °C) makes it valuable as frying oil.

9.1.4.2.11   Rice Bran Oil

Rice oil, also called rice bran oil, has been used extensively in Asian countries. It is the preferred oil in Japan as frying oil for its subtle flavour and odour. During rice milling bran and germ are separated from the starchy endosperm (Figure 9.1.28).[71] The bran has an oil content of 12–18%. This means that the potential volume of rice bran oil is 0.8% of the total rice produced.

The fatty acid spectrum of rice bran oil is 22–25% palmitic acid, 37–41% oleic acid and 37–41% linoleic acid. More recently, interest in rice oil escalated with its identification as a "healthy oil" that reduces serum cholesterol. Rice bran is a good source of antioxidants including vitamin E and oryzanol (ferulic esters of sterols and triterpene alcohols), cholesterol-lowering waxes and anti-tumor compounds like rice bran saccharide.[72] Besides applications in nutrition and in phyto-chemicals, rice bran oil has traditionally been used for industrial applications, such as dimer acid manufacturing, depending on pricing for alternative vegetable oils.

9.1.4.2.12   Sesame Oil

Sesamum indicum is an herbaceous annual that reaches about 1 m in height (Figure 9.1.29). Sesame oil, with an annual production of 860,000 mto in 2007, is higher in quantity than linseed oil and safflower oil. Traditionally, China, India, Sudan, Burma and Uganda are the world's major sesame seed producing

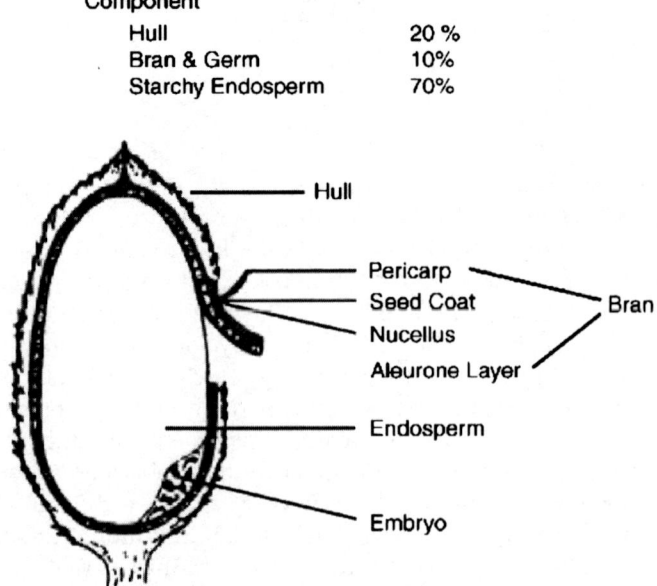

Component	
Hull	20 %
Bran & Germ	10%
Starchy Endosperm	70%

Figure 9.1.28   Rice kernel structure adapted from Orthoefer.[71]

Sesame plant   Sesame flowers

Figure 9.1.29 *Sesame indicum.* Sources: Sesame plant, Koeh-129.jpg from Wikipedia. Sesame flower (reproduction with permission of Karlheinz Knoch Botanik-Fotos, Bruchsal).

countries. Sesame oil is one of the most stable edible oils despite its high degree of unsaturation: oleic acid and linoleic acid constitute more than 80% of the total fatty acids in sesame oil with the percentages of oleic acid (36–42%) and linoleic acid (42–50%) close to each other. The presence of lignan type natural antioxidants (sesamin, sesamolin and their transformation products sesamol and sesaminol, respectively[73]) accounts for both the superior stability of sesame oil and the beneficial physiological effects of sesame.

More than any other vegetable oil crop, sesame is embedded since ancient times in a mythological background that lasts until today. Accounts from ancient history document early recognition of sesame seed as a source of high-quality food. Sesame was the symbol of immortality in early Hindu legend. Women of ancient Babylon would eat halva, a mixture of honey and sesame seeds to prolong youth and beauty. "*Open sesame*", the magic words used by Ali Baba to open the treasure cave in "*The Thousand and One Nights*", reflects the distinguishing feature of the sesame seed pod, which bursts open when it reaches maturity. As part of modern lifestyle, sesame is applied in food and cuisine, cosmetics and wellness treatment and in alternative medicine.[74] Sesame oil plays a prominent role in Ayurvedic medicine. The oil absorbs quickly, penetrates through the tissues and keeps the skin supple and soft. This makes sesame oil an outstanding massage oil and carrier oil for cosmetic and therapeutic activities and explains its use during abhyanga, a form of Indian massage. Abhyanga is believed to improve energy flow and help free the body of impurities. Used on baby skin, sesame seed oil will protect the tender skin, and more generally will cause young facial skin to have and display natural good health.

9.1.4.2.13   Safflower Oil

Safflower oil is the seed oil of the thistle-like safflower plant, thriving in the west of the USA, Mexico, North Africa and India. The plant can be grown under fairly arid conditions. The seeds resemble small sunflower seed kernels and can be harvested mechanically. Like corn oil the fatty acid composition is similar to that of sunflower oil. Safflower oil has a high oxidative stability and is being used increasingly in salad oils and dietetic margarines because of its high content of linoleic acid.

9.1.4.2.14   Cocoa Butter

Cocoa[75] is the dried and partially fermented fatty seed of the cacao tree from which chocolate is made. Cocoa production has increased from 1.5 Mio mto in 1983/1984 to 3.5 Mio mto in 2003/2004. Over 90% of the world's cacao is grown by smallholder farmers. Main producing countries are Côte d'Ivoire, Ghana, Indonesia, Cameroon, Nigeria, Brazil, Ecuador and República Dominicana. Cocoa butter, as the fat is called, becomes available during further processing of cocoa beans by fermentation (to develop the cocoa flavour), roasting, dehulling, grinding and expelling. The fat content of dehulled kernel is about 55–60% in dry matter (Table 9.1.5).

Cocoa butter is used extensively in the production of chocolate and other confectionery products, and to a lesser extent in the pharmaceutical and cosmetics industries.

9.1.4.2.15   Shea Butter

The shea or karité nut tree is found in the savannahs of West and Central Africa known as the Sahel region and produces its first fruit when it is about 20 years old, reaches its full production when the tree is about 45 years old and produces nuts for up to 200 years after reaching maturity. The fruit resembles large plums with a mucous pericarp generally containing only 1 seed, the shea

Table 9.1.5   Fatty acid composition of cocoa butter in %.

	Typical	Range
C16:0	26	22–30
C18:0	36	30–37
C18:1	33	30–36
C18:2	3	1.5–4.0
Iodine value	37	34–40

Table 9.1.6 Fatty acid composition of shea butter in %.

	Typical	*Range*
C16:0	4	3–6
C18:0	42	36–44
C18:1	44	42–52
C18:2	7	4–8
Iodine value	57	55–65

nut, source of shea butter. Shea butter is similar to cocoa but somewhat higher in oleic acid content. Shea butter has up to 11% unsaponifiable matter containing 4–10% kariten [$(C_3H_8)_n$, mp 63 °C], mainly a *trans*-1,4-polyisoprene like gutta-percha (Table 9.1.6).

Shea butter is one of the most affordable and widely used vegetable fats in the Sahel and an important ingredient for indigenous medicine and body care. The refined fat is sold as baking fat, margarine and other fatty spreads. Shea butter is traditionally used in medicines, particularly for the preparation of skin ointments, as a rub for rheumatism and to treat inflammation, rashes in children, dermatitis and sunburn. The kariten latex in shea is responsible for the sealing property that shea butter has and that may aid in protecting skin and preventing sun allergies. In cosmetics shea butter serves as a moisturizer and emollient and as a base for cosmetics that prevent skin drying; good-quality lipsticks also use shea butter. As a result, cosmetic industries in the Sahel and elsewhere market this ingredient in many soap, shampoo and skin-cream preparations.

9.1.4.2.16   Evening Primrose Oil

Evening primrose is a North American native biennial plant. With an estimated 90% of global supply, China has actually become the major producer of evening primrose seed worldwide. Evening primrose oil has a long history of use as an alternative medicine and is offered for healthy, active-lifestyle diets. It contains up to 10% of γ-linolenic acid (GLA). Formulations of GLA apply for a range of nutritional, cosmetic and pharmaceutical products, including some used in the treatment of rheumatoid arthritis (Table 9.1.7).

9.1.4.2.17   Olive Oil

Olive oil is obtained from the fruit of the olive tree. It is an important component in the diet of Mediterranean people and in Mediterranean cuisine. About 80% of total olive oil production comes from the European Community, with the Near East contributing *ca.* 7% and North Africa supplying about 11%. The cultivation of the olive tree is known since biblical times. It is one of the oldest signs of civilization in the world. The Mediterranean world has regarded the olive as sacred for thousands of years. In many religions and

Table 9.1.7 Fatty acid composition of evening primrose oil in %.

	Typical
C16:0	8
C18:0	2
C18:1	11
C18:2	69
C18:3n–6	9
Iodine value	150

Adapted from N. A. M. Eskin, *Eur. J. Lipid Sci. Technol.*, 2008, **110**(7), 651–654 and P. Clough, *Lipid Technol.*, 2001, **13**, 9–12.

cultures, the olive tree or its branches have been a symbol of peace, life and salvation. Besides food, olive oil has been used for religious rituals and medicine, as fuel in oil lamps as well as in skin care applications. In Ancient Greece, athletes ritually used olive oil to rub over their bodies. The Bible, the Quran and Hadith contain manifold references to olive oil. Still today the King of Saudi Arabia gives it as a traditional gift to pilgrims to Mecca.

The olive fruit contains oil of very similar composition in the pulp (approximately 20%) and in the kernel (approximately 12%). Genuine olive oil contains up to 83% oleic acid, between 4 and 21% linoleic acid and 8 to 25% saturated palmitic and stearic acids. The proportion of polyunsaturated fatty acids is low. Olive pomace oil is the oil obtained by solvent extraction of olive pomace. Pomace is the ground flesh and pits after pressing.

Internationally recognized quality definitions were promulgated by the International Olive Oil Council (IOOC) and have been adopted in the Joint FAO/WHO Food Standard Programme. The best virgin oils (those called *extra virgen* in Spanish) are cold-pressed, which produces a natural level of low acidity (under 1%) and a minimum organoleptic rating of 6.5 or more as determined by the IOOC method. Extra virgin olive oil can be used in endless ways in the kitchen, and has been a traditional ingredient in everything from antipasti to desserts. It is best used raw in salads, in order to enjoy its real flavour.

9.1.4.2.18   Linseed Oil, Flax Oil

Flax is a versatile, blue-flowered crop. It is most likely one of the oldest cultivated plants, grown either for the oil extracted from the seed or for fibre from the stem (Figure 9.1.30). There is well-documented evidence of the use of flax fibres for textiles going back to prehistoric times (see Chapter 9.6). However, in contrast to olive oil, no precise indication of the traditional use of linseed and linseed oil for nutrition or other applications may be found. Indeed, the first reliable source for utilization of linseed oil, namely for paintings, dates back to the eleventh century, when the German monk Theophilus in his "*Schedula diversarum artium – De diversibus artibus*" described the use of linseed oil for oil paints and amber varnishes.[76]

Flax botanical   Flax plant

Figure 9.1.30   Linseed, flax. Sources: flax botanical, Koeh-088.jpg; *Linum usitatissimum*, released under the GNU Free Documentation Licence, both from Wikipedia.

Flax seed has a warm, earthy and subtly nutty, butter flavour. The seed can be eaten entirely, and has traditionally been used as an add-on to cereal or bread. Today flaxseed is experiencing a renaissance among nutritionists, the health conscious public, food and feed processors (see Chapter 6) and chefs alike. It is one of the richest sources of lignan and α-linolenic acid (> 50%), one type of fatty acid in the polyunsaturated ω-3 fatty acid (PUFA) family, considered essential fatty acids. Lignans, like isoflavones, are one of the major classes of phytoestrogens, which are estrogen-like chemicals and act as antioxidants.

Linseed oil, or flax seed oil, is cold-pressed from the dried ripe seeds of flax. For medicinal purposes recently expressed linseed oil, *oleum lini*, is preferable.

In order to enlarge the breadth of the linseed application profile crop breeding efforts have led to new flax species, called Linola™ and NuLin™. Linola is high in oleic and linoleic acid (> 85%) and low in linolenic acid (approximatley 2%). Linola™ received GRAS status from the US FDA. It is a substitute to sunflower oil, to be used in margarines and in salad and cooking oils. NuLin™ on the contrary is a high ω-3 fatty acids flax (> 65% linolenic acid, IV > 205) and targets industrial applications and health food such as bakery products.[77]

9.1.4.2.19 Camelina Oil

Camelina sativa, usually known in English as gold-of-pleasure or false flax, also occasionally linseed dodder and Siberian oilseed, is a flowering plant in the family *brassicaceae*, which also includes rapeseed. The crop is now being researched due to its exceptionally high level of ω-3 fatty acids. Over 50% of the fatty acids in cold-pressed camelina oil are polyunsaturated. The major components are α-linolenic acid – C18:3 (ω-3 fatty acid, approximately 35–45%) and linoleic acid – C18:2 (ω-6 fatty acid, approximately 15–20%). The oil is also very rich in natural antioxidants making this highly stable oil very resistant to oxidation and rancidity. It has an almond-like flavour and aroma and is well suited for use as cooking oil. Because of apparent health benefits and its technical stability camelina oil is being added to the growing list of foods considered as functional foods.

9.1.4.2.20 Calendula Oil

Calendula, also known as Pot Marigold, has historically been grown as an ornamental and medicinal plant. Its seed oil contains up to 60% calendic acid an (8t,10t,12c-18:3) ω-6 trienoic acid synthesized in the plant from linoleate by a Δ12-oleate desaturase:

linoleic acid   calendic acid

Historically, calendula flowers were considered beneficial for reducing inflammation, wound healing and as an antiseptic. Calendula oil is used as a soothing therapeutic oil. It is a popular element of aromatherapy massage and as an ingredient in homemade skin care products. Calendula is a drying oil and considered as a replacement for tung oil in the coating industry.

9.1.4.2.21 Tung Oil

Tung oil, also called China wood oil, comes from the pressed seed from the nuts of the Tung tree, a deciduous shade tree native to China. Tung oil is a drying oil. When applied as wood surface coating, it provides a tough, highly water-resistant finish which does not darken noticeably with age, as does linseed oil. In fact, the "teak oil" sold for fine furniture is usually refined tung oil. Some woodworkers consider tung oil to be one of the best natural finishes for wood. Tung oil is composed primarily of (9c,11t,13t-18:3) α-elaeostearic acid, with smaller amounts of oleic, linoleic and palmitic glycerides.

9-*cis*,11-*trans*,13-*trans*-octadecatrienoic acid
(α-eleostearic acid)

Table 9.1.8 Approximate fatty acid composition of selected nut oils compared to olive oil.

Oil	Saturated fat %	Mono-unsaturated fatty acid % (MUFA)	Poly-unsaturated fatty acid % (PUFA)	Linolenic fatty acid %	Linoleic fatty acid %
Almond	8.2%	69.9%	17.4%	0.0%	17.4%
Brazilnut	24.4%	34.8%	36.4%	0.0%	36.0%
Cashew	19.8%	58.9%	16.9%	0.0%	16.5%
Hazelnut	7.4%	78.0%	10.2%	0.0%	10.1%
Macadamia	15.0%	78.9%	1.7%	0.0%	1.7%
Olive	13.5%	73.7%	8.4%	0.6%	7.9%
Peanut	16.9%	46.2%	32.0%	0.0%	32.0%
Pecan	8.0%	62.3%	24.8%	1.0%	23.6%
Walnut	9.1%	22.8%	63.3%	10.4%	52.9%

Source: USDA nutrient database; http://www.netrition.com/nuts.html.

9.1.4.2.22 Nut Oils

The latest of the different oils to be the darlings of the cooking world are cold or expeller pressed tree nut oils, more particularly hazelnut oil, walnut oil and almond oil. Most nut oils taste like the nut from which they were extracted. They make a delicious, tasty, unique flavour statement, even when blended in a small amount to mild-flavoured food oil such as canola. Not only do the nut oils add great flavour to food, they also carry health benefits because of a similar ω-fatty acid balance as olive oil. Nut oils in general are excellent used in salad dressing, over pasta with some cheese, in baked goods or for dipping French bread into (Table 9.1.8).

Refined nut oils, when used as massage oils, maintain a faintly nutty aroma. They are highly penetrative, have great moisturizing qualities and are heralded as helping to tone and tighten the skin while strengthening capillaries and assisting in cell regeneration.

Walnut oil has early been favoured by medieval oil painters like Jan van Eyck and Peter Paul Rubens for its drying properties, since it dries quickly and evenly throughout, has a light colour, has excellent endurance and yellows less with age than other drying oils such as linseed oil.

9.1.4.2.23 Castor Oil

Castor oil is obtained from the seed of the castor plant, *Ricinus communis*. The seeds of castor plants are the size of kidney beans and have beautiful and intricate designs (Figure 9.1.31). Harvesting castor beans is not without risk. Allergenic compounds found on the plant surface can cause permanent nerve damage. Castor beans also contain two toxins that are poisonous to people, animals and insects. One is ricin, a water-soluble protein. Ricin is a potent cytotoxin. The other toxic protein in the castor bean, RCA (*Ricinus communis*

| Castor oil botanical | Castor oil seed | Castor oil plant |

Figure 9.1.31   Castor oil plant, castor seed. Sources: *Ricinus communis*, botanical Koeh-257.jpg, from Wikipedia.

agglutinin), is a powerful hemagglutinin, *i.e.* it agglutinates red blood cells. Both RCA and ricin fully remain in the meal or cake after castor oil extraction.

The castor plant is also known as "Palma Christi" for its palm-shaped leaves and the magic healing power castor oil has been credited with in folk medicine.

A large part of world production and development of castor oil is by members of the International Castor Oil Association (ICOA). India (besides China, Brazil, Russia, Thailand, Ethiopia and the Philippines) is the world's largest producer of castor seed and meets most of the global demand for castor oil, contributing over 60% of the entire global production.

The most difficult and labour-intensive operation in the growing of castor was for many years and still is that of harvesting. Although newer dwarf-sized cultivars with dehiscent fruits enable mechanized harvesting, fruit collection and de-hulling is mainly done manually from wild or naturalized plants by women and children.

Uses.   Medicinal castor oil is prepared from the yield of the first pressing. The oil is classified as an oleaginous vehicle. It is present in some topical pharmaceuticals and is used as a purgative and laxative. The castor bean contains 50–55% oil. Among vegetable oils, castor oil is distinguished by its high content (>85%) of ricinoleic acid. As a result, in contrast to other vegetable oils castor oil is highly reactive for derivatization in the hydrocarbon moiety of the acylglycerol molecule.[78] It reacts with isocyanates to form polyurethanes, especially in the field of surface coatings, potting and encapsulation compounds. Ethoxylates of castor oil are well known as emulsifiers for cosmetic and technical emulsions. Rather than alkali catalyzed alkoxylation, castor oil alkoxylates suitable for PUR slab stock foams for mattresses require special "double metal cyanide" (DMC) catalysis.[79]

Hardening of castor oil yields hydrogenated castor oil (HCO, glycerol 12-hydroxy tristearate also known as castor wax). HCO has a broad application profile covering cosmetic applications like hair wax,[80] topical ointment compositions perceived as pharmaceutically elegant,[81] internal lubricants for PVC and other plastics;[82,83] saponification of HCO yields Li-, Ca- or Al-soaps which have been known for many years as thickeners or gelling agents for the preparation of lubricating greases and biodegradable oleogels;[84] when micronized, castor wax acts as thixotropic agent for organic surface coatings and sealants.[85]

Non-drying castor oil is converted into a drying oil when the hydroxyl group and adjacent hydrogen are removed from each ricinoleic acid chain by action of an acid catalyst at 230 to 260 °C. An additional double bond is created, resulting in two double bonds on each chain, of which 25% are in the conjugated position. The result is a light-coloured, odourless drying oil known as dehydrated castor oil (DCO). Varnishes, alkyds and resin systems based on DCO are noted for fast-drying, flexibility, excellent chemical resistance, adhesion to metals, high gloss and water-resistance.

Dehydrated castor oil, DCO

Sulphation of castor oil with sulphuric acid or oleum yields an easily water-soluble or dispersible sulph(on)ated castor oil also called Turkey red oil, one of the first synthetic detergents, which has been used for textile dyeing, softeners and metal working.[86]

Castor oil is also a raw material for the "green polyamide" PA 11, Rilsan®.

9.1.4.2.24   Jatropha Oil

Jatropha is a very diverse genus which includes herbaceous perennials like *Jatropha integerrima*, a widespread ornamental garden plant, and *Jatropha curcas*, a multipurpose plant with many attributes and considerable potential. It is a genus of crops like the castor plant belonging to the family *euphorbiaceae* (Figure 9.1.32). *Jatropha curcas* generates seeds containing from 36% up to 40% oil and may produce for over 35 years. Its frugality and durability even with poor soil and unfavourable climatic conditions has created interest as a crop that could combat desertification, provide employment in remote areas, improve the environment and enhance the quality of rural life.[87] The jatropha seed provides an oil similar to cotton seed and soybean oil. This has made

Jatropha integerrima   Jatropha curcas

Jatropha botanical   Jatropha botanical   Jatropha seeds

Figure 9.1.32   Jatropha.

jatropha interesting as an energy crop, for example when grown for biofuel production thus satisfying local needs for farm machinery and even for export.

9.1.4.2.25   Algal Oil

Algae range from small, single-celled organisms to multi-cellular organisms, some with fairly complex form. Algae are common in terrestrial as well as aquatic environments. Their cultivation does not impact freshwater resources. Crop growing can use ocean and wastewater and can be done without displacing arable land currently used for food production.[88]

Macro-algae, commonly known as seaweed, can reach sizes of up to 60 m in length. Commercial farming of seaweed has a long history, particularly in Asian countries. Macro-algae have been used for some time as food and for the extraction of hydrocolloids, *i.e.* as an excellent source of thickening, emulsifying and gel-forming agents (agars, alginate, carageenan).

However, the majority of algae that are intentionally cultivated fall into the category of micro-algae. **Micro-algae** are microscopic photosynthetic organisms, 1 to 5 μm in size and also referred to as phytoplankton. Micro-algae contain anywhere between 2% and 40% of lipids. They have much faster growth-rates than terrestrial crops, which makes them interesting as an energy feedstock. For that reason algae have been exhaustively studied during the US Department of Energy's Aquatic Species Program (ASP) initiated by the Carter administration in 1978. At that time algae species, which looked promising in the laboratory for biodiesel production, were not robust under conditions encountered in the field. Nonetheless, the potential of algal oil appeared highly attractive, and in the 2008 US Department of Energy Biomass and Biofuels Update to Congress[89] and other R&D programs algae reappear as a potential non-food feedstock (see Chapter 8).

Marine micro-algae provide the food base which supports the entire animal population of the open sea. They are sourced for nutritional and pharmaceutical use because of their nutrient-dense components including: carotenoids (β-carotene), sterols (4-α-methyl sterols, 5-α-stanols), lipids, polyunsaturated fatty acids (PUFAs) and polyphenols. Micro-algae contain a wide range of fatty acids in their lipids.[90] Of particular importance is the presence of significant quantities of the essential polyunsaturated fatty acids, ω-6 linoleic acid (C18:2) and ω-3 linolenic acid (C18:3), and the highly polyunsaturated ω-3 fatty acids, octadecatetraenoic acid (C18:4), eicosapentaenoic acid (EPA, C20:5) and docosahexaenoic acid (DHA, C22:6), which are found in fish oils, and which have been shown to lower blood pressure and ultimately lead to the reduction of cardiovascular disease risk. Concern about depletion of fish stocks could be addressed by alga-culture of micro-algae and extraction of algal oil as a sustainable alternative to fish oil production.

9.1.5 The Value Chain of Fats and Oils – Industrial Non-food Uses

The main biological functions of lipids include energy storage and acting as structural components of cell membranes. More particularly glycerolipids serve as a reserve or long-term source of energy. Fats also serve as an insulation material to allow body heat to be conserved. In addition, fats are essential to the proper maintenance of cell membranes in the body. Since an unsaturated fat contains fewer carbon–hydrogen bonds than a saturated fat with the same number of carbon atoms, saturated fats will yield more energy during metabolism than unsaturated fats with the same number of carbon atoms. Saturated fats can stack themselves in a closely packed arrangement, so they can freeze easily and are typically solid at room temperature. The combination of saturated and unsaturated moieties in the hydrocarbon chain of naturally occurring fats imparting a ductile, malleable solid texture such as in lard or tallow eventually represent the best compromise to fulfil the different biological tasks. Additionally, certain glycerolipids like polyunsaturated conjugated fats exercise useful physiological activities.

Based on this role in the metabolism of living organisms, glycerolipids and their derivatives offer unique structure/performance profiles for industrial purposes as well, namely:

- the low volatility combined to ready biodegradability and ability to compost;
- the polar molecular structure resulting in plasticizing and lubricating properties;

- their ability to protect and lubricate the skin, leaving a pleasant after-feel;
- the long, hydrophobic, chemically nearly inert and water insoluble hydrocarbon chain of saturated fatty acids;
- the facility for functionalizing or splitting the hydrocarbon chain *via* reaction at the double bond of unsaturated fatty acids;
- the hydrophilic/lipophilic structure hydrolyzing to yield fatty acid molecules as precursors for soaps and surfactants;
- the glycerol group as a C3-building block.

9.1.5.1   Fats and Oils as Precursors for Biopolymers

The capability of highly unsaturated oils to dry and form a polymer film by simple action of atmospheric oxygen explains the historic importance and traditional usage of drying oils in lacquers and varnishes. The actual chemical reactions involved in the drying process are still incompletely understood; they do, however, involve oxygen attack on the fatty acid chains at or near the double bonds, catalyzed by certain organic salts of multivalent metals, called dryers or siccatifs. Similar chemical reactions are involved with operations like manufacturing of stand oils, blown oils and conjugated oils, as well as with formation of linoleum from linseed oil.

Oxidation of unsaturated triacylglycerides by action of performic or peracetic acid means, to a certain degree, incorporating reactive oxygen in the molecular structure of the oil. By this, the polarity and the reactivity of the molecule increase significantly. Whereas the natural oils themselves have mainly solvent properties, epoxidized natural oils are well known as phthalate-free, non-volatile, extraction and migration resistant plasticizers.[83,91] Due to the reactivity of the oxirane group and due to the high degree of oxirane functionality, epoxidation of oils or oleic acid esters and (partial) ring opening of the oxirane group reacting with water, mono- and polyfunctional alcohols or acids yield so-called *oleochemical polyols* used as polymer building blocks for plastic and coating applications (Figure 9.1.33).[92]

Oleochemical polyols have been commercial since the 1980s. They are used for polyurethane foams,[93] for solvent-free heavy duty multilayer coatings in concrete protection (Figure 9.1.34) and for high-performance anticorrosion coatings. Alternatively, hydroformylation of biodiesel followed by hydrogenation yields a hydroxymethylated fatty acid methylester (FAME) monomer, which can be transesterified with polyfunctional alcohols to yield natural oil-based polyesters useful for the manufacturing of flexible moulded PUR foams.[94] Reaction of epoxidized oils with acrylic acid yields radiation curable oleochemical polyols, *e.g.* Photomer® 3005. Acrylated vegetable oils serve also as matrix resins for natural fibre reinforced biopolymer composites.[95]

Unsaturated and saturated short-chain dicarboxylic acids such as maleic, succinic and adipic acid are well-known industrial chemicals manufactured on a petrochemical basis.[96] Besides hydrogenation of maleic anhydride or oxidation of tetrahydrofuran, white biotechnology additionally opens a new route to

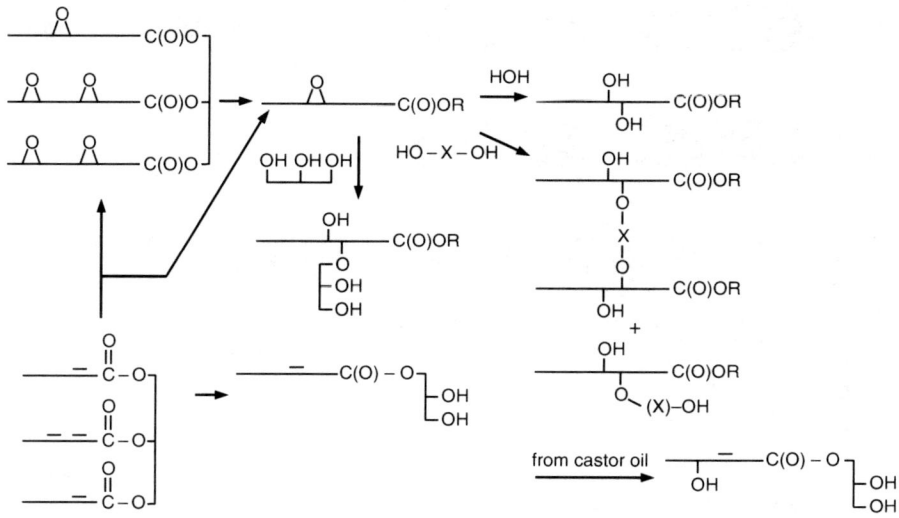

Figure 9.1.33 Schematic pathways to oleochemical polyols. (Source: T. Roloff, U. Erkens and R. Höfer, *Polyols based on renewable feedstocks: A significant alternative*, Urethanes Technology, August/September 2005, 29.)

Figure 9.1.34 Self levelling multilayer PUR-flooring.

succinic acid from fermentation of agricultural carbohydrates.[97] Natural oils supplement these to provide longer-chain diacids or ω-functional long-chain monoacids using the double bonds of unsaturated acids as reaction centre or *Sollbruchstelle* (Figure 9.1.35). For example, pyrolysis of castor oil or ricinoleic acid methyl ester under reduced pressure yields heptaldehyde and undecylenic acid.[78] Alternatively, ω-unsaturated monocarboxylic acids can be synthesized by metathesis of sunflower, HOSO or rapeseed fatty acid methyl ester and ethen yielding 9-decylenic acid methylester.[98] Bromination of the ω-double

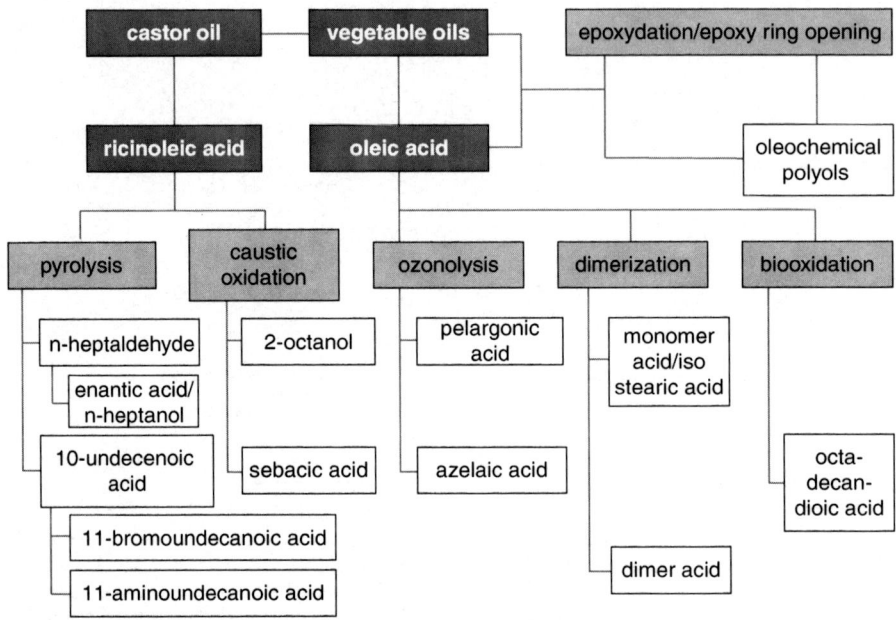

Figure 9.1.35 Polymer building blocks based on vegetable oils.

bond in undecylenic acid yields 11-bromoundecanoic acid, which reacts with NH_3 yielding ω-aminoundecanoic acid. Sebacic acid and 2-octanol are produced from ricinoleic acid or castor oil by oxidative cleavage in an alkaline medium. Oleic acid reacts with ozone forming azelaic acid and pelargonic acid.[99] However, oleic acid can also be oligomerized catalytically under heat. This industrial process leads to a mixture of branched C18 monomer fatty acids (the raw material for isostearic acid[100]), C36 dicarboxylic acid, C54 trimer fatty acids and higher oligomers from which the dimer fatty acids can be separated off by distillation.[99,101] Azelaic acid, sebacic acid, ω-aminoundecanoic acid and dimer fatty acids serve as polymer building blocks for polyamide-based high-performance engineering plastics, high-speed assembling, shoe hotmelt adhesives, printing inks and coating resins. They are marketed as green alternatives to the butadiene-derived dodecandioic acid and laurolactam. Biotechnological processes to manufacture higher molecular weight dicarboxylic acids as an alternative to these, in part, highly energy-intensive syntheses are under development. ω-Octadecandioic acid is to date only available *via* biooxidation technology,[102] whereas dimethyl 9-octadecylendioleate can also be synthesized by cross-metathesis of methyloleate with methylacrylate.[103]

9.1.5.2 Fatty Acids – Keystones of Oleochemistry

Historically, industrial oleochemistry began when Ernst Twitchell at Emery (now Cognis and Emery Oleochemicals) in Cincinnati successfully performed the first catalytic acid cleavage of fat.[104] Likewise, enzymatic fat splitting has been known since the beginning of last century. By catalytic fat splitting the chemically reactive fatty acids were readily available, in contrast to alkaline saponification which only provided lard soaps of low chemical reactivity (Figure 9.1.36).

Fat splitting can be done batch-wise or continuously in counter-current splitting columns. Crude fatty acids from splitting are still contaminated with colour bodies, partial glycerides, oxi- or partially polymerized fatty acids, unsaponifiable materials, free glycerine, *etc.* They may be purified by batch-wise or continuous distillation or separated into their component fatty acids by high-vacuum fractional distillation. Tall oil fatty acids are obtained by fractional distillation of crude tall oil (CTO). Industrially, fatty acids derived from palm oil and tallow represent the lion's share with coconut and palm-kernel coming next, followed by tall oil, soya and sunflower oil. These seven oils represent >97% of fatty acid feedstock. Global fatty acid production capacity grew from 2.5 Mio mto in 1985 to 7 Mio mto in 2006 with 60% now in Asia.[20,105]

Fatty acids are predominantly used as intermediates. Main applications are water soluble soaps for household cleaning, personal care, industrial and institutional (I&I) cleaning and synthetic rubber manufacturing by emulsion polymerization. Soaps are made by reaction of fatty acids with caustic alkalis, alkali carbonate or ammonia or (>90%) by direct saponification of the triglyceride oil.[106] Another important group of fatty acid soaps are dry, water-insoluble metal soaps used as lubricants or stabilizers for PVC and other plastics and aqueous calcium stearate dispersions applied as paper coating

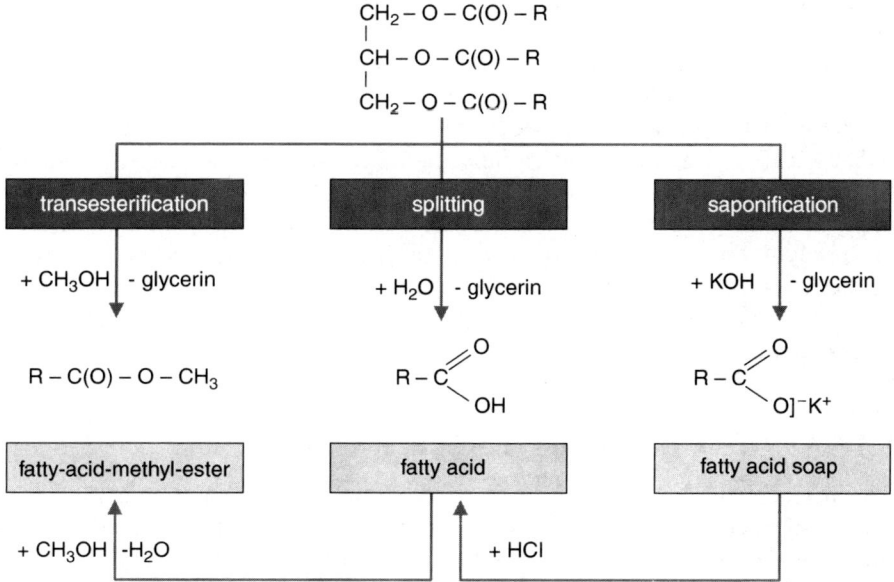

Figure 9.1.36 Processing of fats and oils.

lubricants and as friction agents to avoid dusting and picking during calendering and printing processes.

Fatty acids are raw materials for fatty amines, ethoxylated fatty amines, quaternary ammonium compounds (Quats), fatty acid-polyamine condensates, fabric softeners based on esterquats, fatty acid amides, alkyl keten dimers for paper sizing and amphoteric surfactants. Binding of oleic acid anhydride to cellulose fibre by chemical grafting is a recently introduced sustainable process to protect construction timber for outdoor use, such as pinewood shutters, from moisture and outside attack. Impregnation of the wood involves two stages: firstly in an autoclave, where vacuum and pressure ensure its penetration, and then in a tank where the wood is soaked in a bath of anhydride, to encourage grafting of the anhydride on the wood.[107]

9.1.5.3   Fatty Acid Esters

Fatty acid esters of mono- and polyfunctional alcohols are the workhorses of oleochemistry. In many fields of application fatty acid methyl esters replace fatty acids because they are less corrosive. Chemical reactions can often be carried out under milder conditions. They have lower boiling points and require less energy to distil and to fractionate than the corresponding fatty acids. The elimination of methanol from the reaction products can be more easily achieved than that of water. Therefore fatty acid methyl esters are primarily used for the production of saturated and unsaturated fatty alcohols. Methyl esters are manufactured by acid catalyzed esterification of fatty acids in countercurrent reaction columns or by alkaline transesterification starting directly from the triglyceride oils in a batch, semi-batch or continuous process (Figure 9.1.37).[108,109]

Together with fatty acids and glycerine, they form the raw material base for the oleochemistry value chain (Figure 9.1.38).

Methyl esters of soya, rapeseed and palm oil fatty acids used as biodiesel (Chapters 7 and 8) are preferentially manufactured by transesterification.[110] The acid catalyzed esterification and the alcoholysis of triacylglycerides apply also for the manufacture of higher alcohol monoesters as well as for fatty acid polyol esters. Due to their oily consistency and to their solubilizing properties fatty acid esters of low Mw alcohols are components in odorous substances like perfumes, and cosmetic and drug formulations such as ointments, creams or lotions.

9.1.5.4   Green Lubricants and Carrier Oils

The increasing prices of mineral oils mean that the substitution of mineral oils by fatty acid esters is becoming economically viable. This applies especially to the lubricant sector, where rising economic competitiveness on the one hand and ecological safety on the other have helped biogenic lubricants achieve the

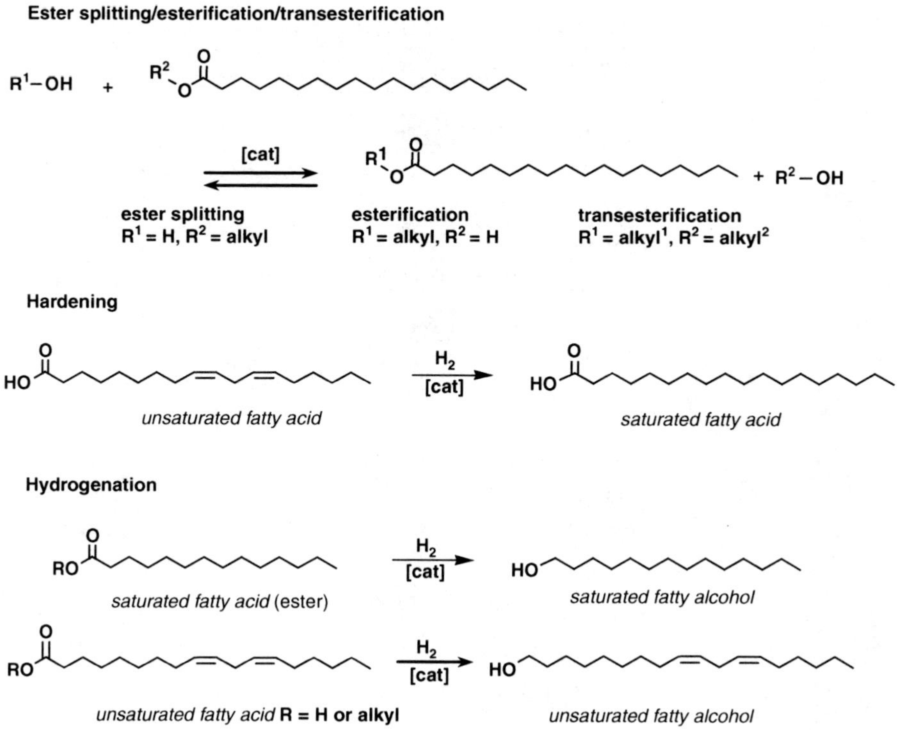

Ester splitting/esterification/transesterification

ester splitting	esterification	transesterification
R^1 = H, R^2 = alkyl	R^1 = alkyl, R^2 = H	R^1 = alkyl^1, R^2 = alkyl^2

Hardening

| unsaturated fatty acid | | saturated fatty acid |

Hydrogenation

| saturated fatty acid (ester) | | saturated fatty alcohol |

| unsaturated fatty acid **R = H or alkyl** | | unsaturated fatty alcohol |

Figure 9.1.37 Value adding basic reactions of oleochemistry.

breakthrough.[111] Lubricants form a film on metal surfaces to reduce or prevent friction between machinery parts in motion and minimize wear between interacting surfaces. Synthetic esters based on renewable sources have outstanding tribological properties which can, among others, be seen from the very good viscosity-temperature behaviour. The evaporation loss of such biolubes is much lower than the evaporation loss of mineral oils. Lubricants and functional fluids are omnipresent. Because of their widespread use in industrial machinery, in mining, metal working, fibre and textile manufacturing, agriculture, forestry, in construction, road construction and automotion, they pollute the environment in small, widely-spread quantities and rarely in large, locally fixed amounts.[112] Thus the selection of rapidly biodegradable lubricants[20,113] *inter alia* will reduce the expenses of oil spillage or disposal.

Esters of saturated, long-chain fatty acids with long-chain fatty alcohols, such as stearyl stearate, have a waxy characteristic. These waxes develop lubricating effects in plastics processing;[82,83] which means they prevent friction between plastic molecules and thus permit faster transport of the molten plastic mass through the processing machines (internal lubrication). At the same time they reduce the friction of the molten plastic with the surrounding machine parts (external lubrication).

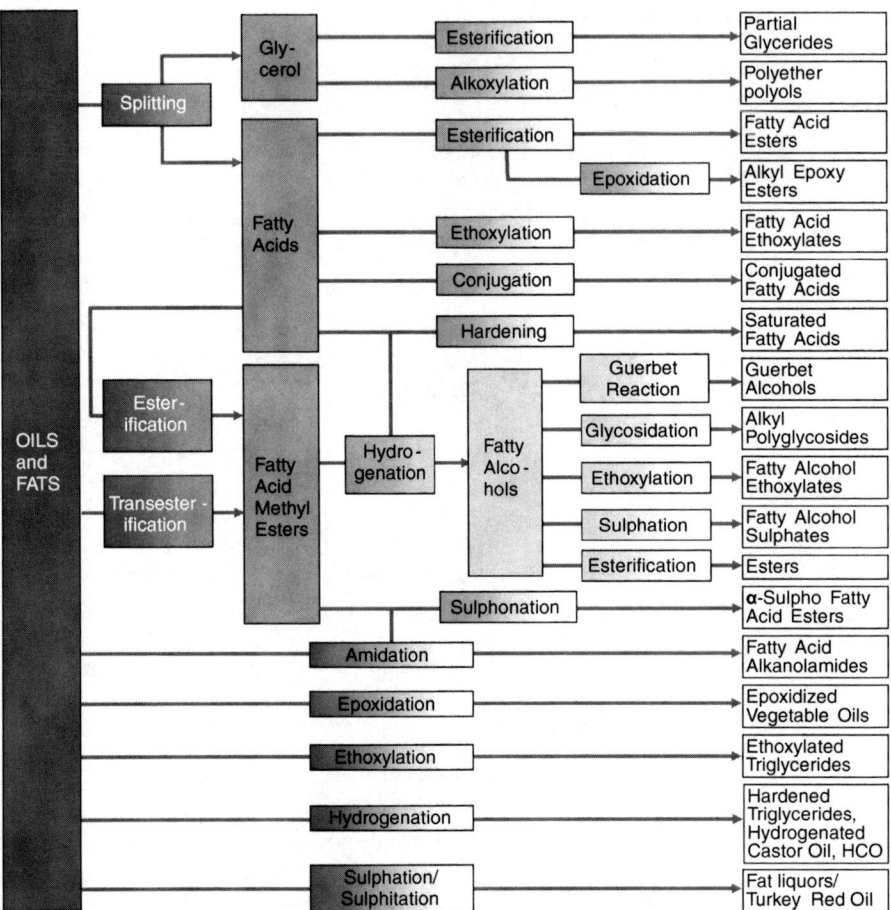

Figure 9.1.38 Oleochemistry value chain.

In industrial applications liquid fatty acid esters are well known as plasticizers for pyroxyline varnishes and as secondary plasticizers for PVC.[91] They are used as thinners and carrier oils and have become established as *green solvents* (see Chapter 10) for printing inks (Table 9.1.9) and for industrial surface cleaning due to their low volatility, environmental compatibility and contribution to work hygiene and industrial safety.[20,67]

As green carrier oils or solvents – together with green surfactants – they also play a key role as "inerts", *i.e.* non-active, inert formulation aids in modern crop protection, such as pesticide microemulsions or emulsifiable concentrates.[114] Besides fatty acid esters, certain liquid fatty acid amides, more particularly capryl dimethylamide, are also increasingly creating interest as green solvents and carrier oils.

Table 9.1.9 Specifications of oleo-based oils in comparison to mineral oil based printing ink distillates.

	Titer (pour point) (°C)	Flash point (°C)	Boiling point (°C)	Visc. At 25°C (Mpas)	Kauri butanol value	Anilin point	Iodine value (Wijs)
Short-chain fatty acid methylester	−28	75	204–244	2.7	107	<−20	<1
Rapeseed fatty acid methylester	−3	155	335–365	6	66	−1,7	90–100
Soybean fatty acid methylester	−6	170	335–365	10	67	3.69	115–135
n-Butylstearate	22	190	344–385	8.7	42	10	<1
Isobutylstearate	19	>170	341–381	8.4	41	10	2
Soybean fatty acid-2-ethylhexylester	−13	180	369–389	9.9	46	−7	85–95
Epoxy stearic acid methylester	–	–	348–386	13.5	>150	−19	–
Di-n-octylether	7	139	298	3.5	41	27	<1
Capryldimethylamide	–	–	294	6.6	>150	<−20	76
Propylene glycol monooleate	–	210	284	30	–	84	–
Printing ink distillate PKWF 4/7 af new[1]	−12	>100	n.d.		25	84	–
Printing ink distillate PKWF 6/9 af new[1]	−9	>110	n.d.		23	90	–

[1]Fa. Haltermann, Hamburg

9.1.5.5 Glycerine as C3-Building Block

The manufacturing of fatty acids by fat splitting and of fatty acid methylesters by alcoholysis of fats and oils yield approximately 10% to 13% of glycerine (glycerol) as by-product. Established glycerine markets are applications as hydrotrope, humectant, solvent, solubilizer, antifreeze and plasticizer for aqueous systems in cosmetics, pharma, food and industrial markets; as humidity stabilizer and preservative for tobacco; as raw material for glycerol esters, for polyetherpolyols, for synthetic resins, and nitroglycerine.[115]

The emergence of biodiesel had a major impact on the glycerol market. In the European 300 to 400,000 mto a^{-1} glycerol market appeared an additional offer of 200 to 400,000 mto a^{-1} crude glycerol. Glycerol price was cut to half between 2003 and 2006 reaching the level of ethylene glycol or below and one-third approximately of propylene glycol. As a result glycerine became of interest as a commodity solvent and C3-building block. More particularly, synthetic glycerine made in a three-step process by oxychlorination of propene and hydrolysis of the epichlorohydrin intermediate nearly disappeared from the market. Glycerine was cheap enough to make epichlorohydrin in an alternative way, *i.e.* starting from glycerine instead of propene (Epicerol® process, Solvay; GTE process, DOW). Similarly, hydrogenolysis of glycerine to manufacture 1,2-propylene glycol is being explored by several companies. Glycerol formal, glycerol butyral or glycerol carbonate as green solvents, glycerol tert.-butylether (GTBE) as a gasoline additive, triacetine as a plasticizer, dihydroxyacetone as a sunless tanner and 1,3-propanediol (1,3-propylenglycol) by bioconversion (as alternative to the acrolein-3-hydroxypropionaldehyde route[116]) are other existing or potential applications for a low-cost glycerol feedstock. Meanwhile a recovery in glycerine quotations can be noticed. These fluctuations of glycerol pricing in fact are indicating the upcoming dependency of glycerine valuation on tax regulations and state policies for the transportation and energy sectors.

9.1.5.6 Fatty Alcohols

While fatty acids are cornerstones for soaps, oleochemical esters, oils and lubricants, medium-chain fatty alcohols (C_{6-10}) are the basis for oleochemical solvents and plasticizers and C_{12-18} fatty alcohols since their discovery have laid the ground for green surfactants and detergents (Figure 9.1.39). Natural fatty alcohols are monobasic, straight-chain, primary, saturated or unsaturated, which explains the favourable ecotox-profile of fatty alcohols and fatty alcohol derived surfactants.[117] Approximately 75% of all consumed C_{12-18} fatty alcohols are used for the manufacturing of surfactants with household detergents covering the lion's share. Alkylpolyglycosides and polyglycolethers of natural fatty alcohols and their sulphates, phosphates and sulphosuccinates follow a rapid and complete degradation mechanism under aerobic and anaerobic conditions leading to the conclusion that they form a sustainable base set of green surfactants.

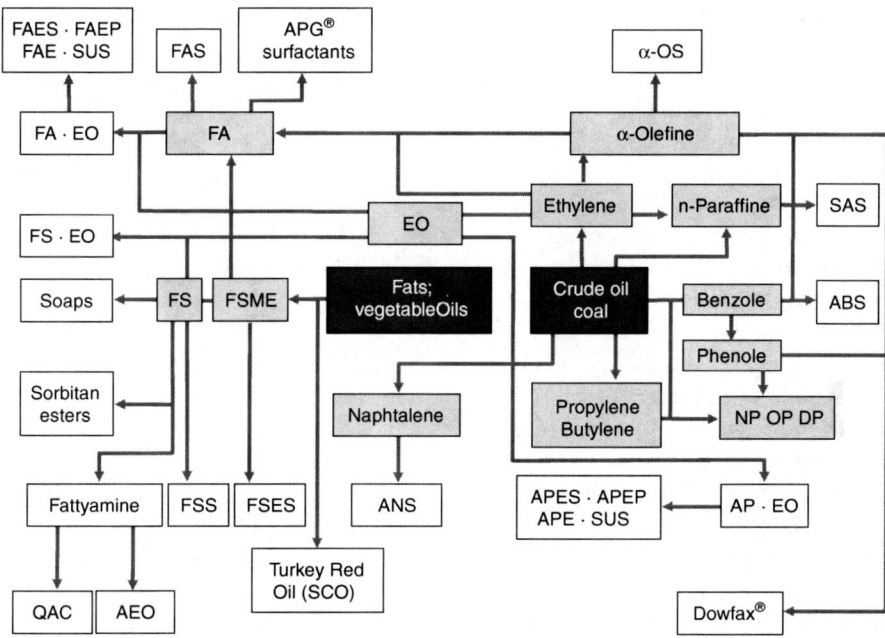

Figure 9.1.39 Fossil *vs.* renewable base stocks for surfactants.
Legend: αOS alpha olefin sulphonates, SAS sec. alkane sulphonates, ABS alkyl benzene sulphonates, NP nonylphenol, OP octyl phenol, DP dodecyl phenol, AP.EO alkylphenol polyglycolethers, APES alkylphenol ethersulphates, APEP alkylphenol etherphosphates, APE.SUS alkylphenolpolyglycolether sulphosccinic acid monoesters, ANS alkyl naphthalene sulphonates, SCO sulphated castor oil (Turkey Red Oil), FSME fatty acid methylester, FS fatty acid, FSS fatty acid sulphonates, FSES fatty acid methyl ester sulphonates, QAC quaternary ammonium compounds, AEO fatty amine ethoxylates, FS.EO fatty acid polyglycol esters, FA fatty alcohols, FA.EO fatty alcohol polyglycolethers, FAS fatty alcohol sulphates (f. ex. sodium lauryl sulphates), APG$^{®}$ surfactants alkyl polyglycosides, FAES fatty alcohol ethersulphates, FAEP fatty alcohol etherphosphates, FAE.SUS fattyalcoholpolyglycolether sulphosuccinic acid monoesters, Dowfax$^{®}$ alkyldiphenyloxide disulphonate.

Since 1931 Deutsche Hydrierwerke AG (DEHYDAG, now part of Cognis' company history) has been the first to synthesize natural fatty alcohols at an industrial level. Manufacturing today is realized by catalytic high-pressure hydrogenation of fatty acid methyl esters yielding saturated or unsaturated fatty alcohols depending on process conditions. Starting from wax esters, long-chain fatty alcohols ($C_{>22}$ fatty alcohols) such as lanolyl alcohol (also called hydrogenated wool fat) can be made by hydrogenolysis in a continuous flow process using *n*-decanol as reaction medium.[109] Alternatively, synthetic fatty alcohols have been produced since the 1950s according to:

- Ziegler-technology based on organic aluminium compounds in a hydro-carbon solvent (K. Ziegler, 1953; Alfol-Process, Conoco; Epal-process, Ethyl Corp);
- Hydroformylation (Oxo-process, O. Roelen, Ruhrchemie, 1938; world-wide there are three variants of the Oxo-process: 1. the classical synthesis using $HCo(CO)_4$ as catalyst; 2. Shell's SHOP-process, based on a cobalt carbonyl – phosphine complex; 3. UCC's process, using a rhodium catalyst).[118]

The Ziegler technology produces a broad Poisson-distribution of more or less even-numbered fatty alcohols which are fractionated into blends and single cuts. The Oxo-process results in more or less highly branched, even- and odd-numbered alcohols.

Based on easily available cheap ethylene or olefins, the Oxo- and Ziegler technologies have shaped the global fatty alcohol market until the 1990s. They are still the predominant technology for short- and medium-chain fatty alcohols. The fact that Asian nations have started to process their indigenous vegetable oils into fatty alcohols, though, has originated a significant increase in global fatty alcohol capacities and has clearly shifted the balance between synthetic and natural fatty alcohols to the renewable resources side.[20]

9.1.5.7 Green Surfactants

Bathing in ass's milk and honey like Cleopatra did in Roman times was the precursor to modern skin-care baths. Today's modern-trend cosmetics include natural oils such as olive, evening primrose and almond oil as well as natural waxes such as beeswax and carnauba wax. Emulsifiers in bath-oil concentrates distribute fatty substances or oils in water and prevent them being deposited on the bathtub. Bath oils are enriched with essential oils and aromatic substances for aromatherapy. Depending on the type of essential oil or corresponding mixtures, relaxing, invigorating or antispasmodic effects can be achieved. Essential oils must be well distributed in water by the surfactant components since they can irritate the skin in concentrated form. Natural surfactants based on renewable raw materials[119] combine a high emulsifying power with good electrolyte stability and extremely good skin compatibility. Examples for such natural surfactants are acylated proteins and amino acids (protein–fatty acid condensation products, Figure 9.1.40), which can be obtained from both animal source (*e.g.* leather waste) and from plants, such as wheat, rice and soybean, with the protein part representing the hydrophilic moiety in the surfactant structure whereas the hydrophobic part is formed by a fatty acid group. They favourably interact with the collagen in the skin, feel especially mild even in the smallest amounts and reduce any irritant effects of other

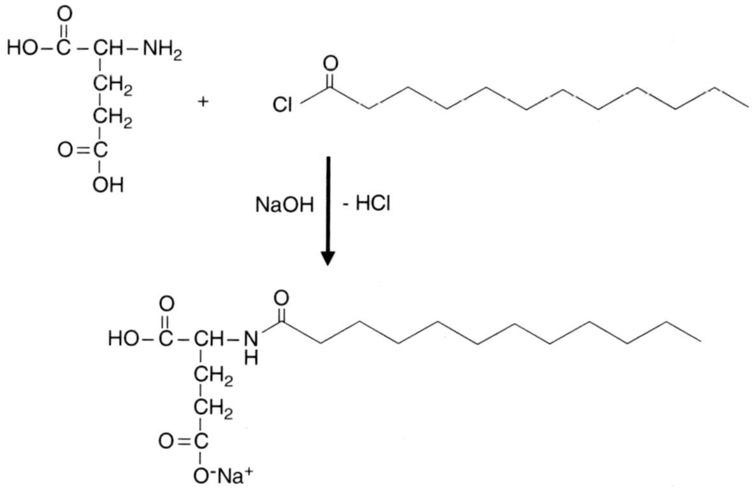

Figure 9.1.40 Protein-fatty acid condensates.

constituents of the formulation. The products are obtained industrially by aminolysis according to "Schotten-Baumann".[120]

Other than proteins, carbohydrates are also able to build up the hydrophilic part of surfactants,[121] for example glucose is derivatized selectively by acetalization with fatty alcohols to produce alkyl glucosides; N-methylglucamides are prepared by reductive amination with methylamine and subsequent acylation. Both products have proved to be highly effective surfactants in washing and cleansing agents, but only the alkyl glucosides are produced on an industrial scale at present[122] (Figure 9.1.41).

Special combinations of tensioactives are aimed at producing a feeling of mildness, relaxation, care, health and the modern term *wellness* in skin-care concepts.[123] The development of new surfactants and new formulations that conform to these demands require application technology testing but also toxicological and sensory testing to ensure skin compatibility and to rule out skin and mucous membrane irritation. In order to protect living beings, testing methods on alternative membranes including those of red blood cells and chicken egg yolk membranes or skin models are being developed to partly or completely replace the conventional animal tests of the past.[124]

Besides personal care and fabric care, surfactants are also key ingredients in industrial and institutional cleaning processes on glass, ceramic, plastic and metal surfaces. The chemical stability of surfactants against acids, alkalis and oxidizing agents is a determining factor for particular cases like bottle-cleaning installations. Here specially designed surfactants, so-called end-capped nonionic tensides, manufactured by Williamson-etherification of fatty alcohol polyglycol ethers with linear short-chain alcohols, combine key performance properties like alkali stability, low foaming and biodegradability.[125] In addition to the specific cleaning function in metal working operations surfactants are

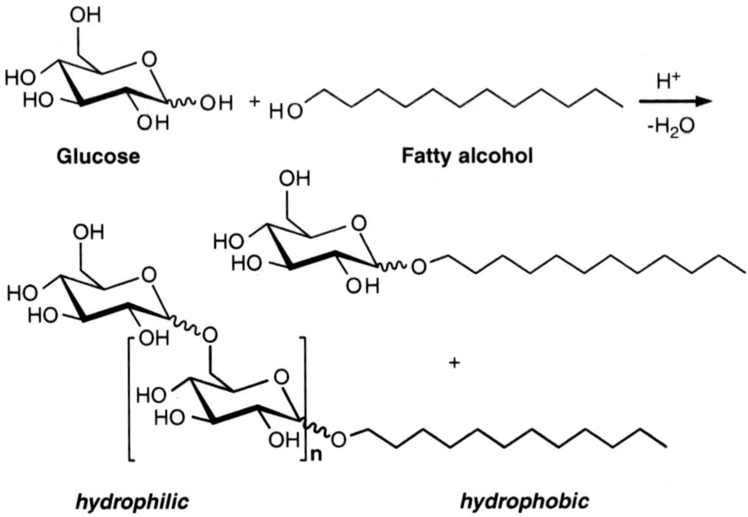

Figure 9.1.41 Synthesis of APG® surfactants.

applied in cooling lubricants, tempering oils, hydraulic emulsions, polishing pastes and mould release formulations.

Cleaning operations under highly complex conditions are required in crude oil extraction. In the lines leading from the wells to the central processing facility over a period of time, deposits of heavy hydrocarbonaceous materials and finely divided inorganic solids, referred to as "schmoo", deposit on the inner surfaces of the lines. The schmoo is a slimy, oily substance which adheres to almost any surface with which it comes in contact, and is difficult to remove. Removal of such deposit can be achieved by the use of a surfactant composition consisting essentially of a caustic aqueous solution of alkyl polyglycoside, ethoxylated fatty alcohols and an alkyl alcohol as solubilizer.[126] Their high pH tolerance and their surfactant activity at high temperatures under the specific conditions of crude oil extraction have made APG® surfactants a very appropriate component in tenside blends for well operations in general. Their excellent wetting action at pH 4 or less makes alkyl polyglycosides applicable in acidic cleaning solutions capable of dissolving carbonate and calcite scales. On the other hand, in combination with caustic material they are particularly effective in removing oil-based drilling fluids and other oil-based materials found in well bores. Additionally their stability at high pH makes APG® surfactants effective wetting, dispersing and emulsifying agents for cement slurries used in well-bore operations enhancing cement hydration and adhesion to the well-bore walls.[127]

Since the 1970s polymer dispersions manufactured by emulsion poly-merization have increasingly substituted solvent-borne paints, coatings, adhesives and inks. Water-based emulsion systems have made a significant contribution to VOC-reduction in corresponding industries. The emulsifier in emulsion polymerization has three key functions, namely: stabilizing the monomer droplets during the first stage of the emulsion polymerization, supplying surfactant micelles as the site of the polymerization reaction (literally the micelles can be regarded as some kind of micro-reactors) and stabilizing the latex particles at the end of the emulsion polymerization process pending transportation, storage and handling until the latex is intentionally destabilized by evaporation of water and formation of a polymer film (Figure 9.1.42).

Whereas the world of surfactants for emulsion polymerization 25 years ago mainly consisted of petrochemical species, the growth of renewable raw materials in general, and the growing importance of vegetable oils in particular, hand in hand with a growing sensitivity for sustainability initiated a move away from petrochemical emulsifiers towards surfactants derived from natural fats and vegetable oils. In the meantime green surfactants clearly dominate the emulsion polymerization technology. The breakthrough of green surfactants for emulsion polymerization happened when equally performing alternatives to alkylphenol ethoxylates (APE) and alkylphenol ether sulphates – until then the workhorses in emulsion polymerization technology – became available and since the FDA approved Disponil® FES-type surfactants according to FDA § 178.3400. Modern surfactant systems for polymer dispersion manufacturing consist of anionic fatty alcohol sulphates or fatty alcohol ethersulphates and green non-ionic surfactants like Disponil® A or Disponil® AFX.[128] To a lesser extent fatty alcohol ether sulphosuccinates and fatty alcohol ether phosphates are used.[129] Monomer emulsion stability is improved when Dowfax 2A1 is added.

Surfactants become a problem when they are discharged into the environment if they are not readily biodegradable. There are practically no

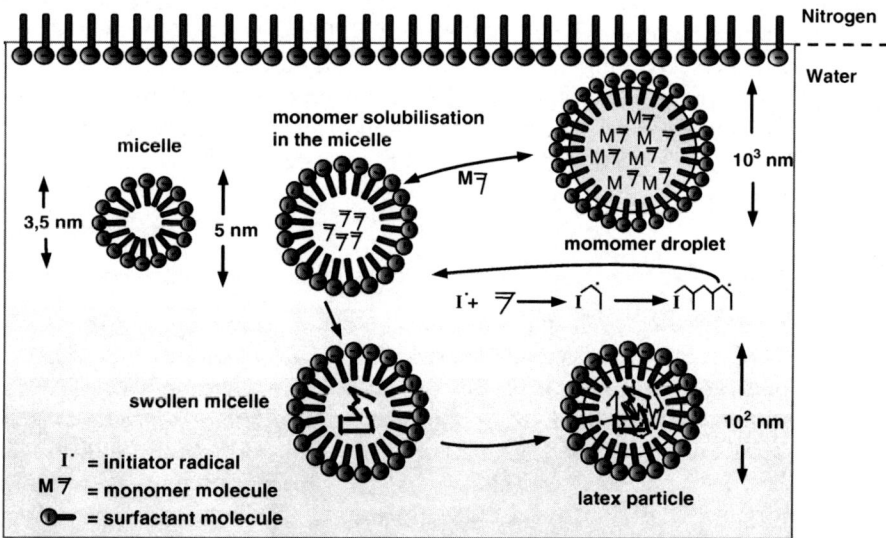

Figure 9.1.42 Emulsion polymerization process.

applications for surfactants in which they are recovered and recycled. Besides the use in detergents, direct release into the environment happens when surfactants are used as inerts in agrochemical applications. Modern crop protection aims at releasing the active pesticide straight at the point of action, improving its activity and reducing the dose rate and wastage of the active. Adjuvants in a pesticide formulation should be readily biodegradable, and should allow safe handling without causing damage to the environment or creating a health risk for the workforce or for untrained people, who risk coming into contact with crop protection chemicals when applied. Green surfactants such as (alkoxylated) alkyl polyglycosides are particularly qualified to meet this demand.[130]

9.1.6 Perspectives

Natural fats and oils are extremely well positioned to produce environmentally compatible and powerful products in the sense of a sustainable development. Oil crops are a unique combination of useful base stock materials. They supply essential protein for human food and animal feed. They need neither refining (in the sense of crude oil refining) nor cracking, before being processed for food or feed or being transferred into useful oleochemical specialties. The high biocompatibility of oleochemical surfactants, solvents and biolubes will even further increase their importance for all applications, which come into direct contact with the environment be it as fabric or body care detergents, additives for paints, inks or adhesives, lubricating oils, agrochemicals and even oil-well drilling adjuvants. Hulls, shells and other by-products are biomass for energy generation. They eventually carry extractives, which can be used as specialty chemicals, and their potential to be converted in new chemical products *via* biochemical technologies is by far not exploited. It can clearly be assumed that in future further possibilities for using renewable resources will be intensely investigated. Here the combination of different types of vegetable raw materials, such as vegetable oils, carbohydrates and proteins, to form new products and intelligent product concepts in order to meet market and consumer needs will be a challenge for research and development.

References

1. H. Schönfeld, ed., *Chemie und Technologie der Fette und Fettprodukte*, Verl. Julius Springer, Wien, 1936; M. Bockisch, *Fats and Oils Handbook*, updated and revised translation of the original German work, AOCS Press, Champaign, 1998; F. Shahidi, ed., *Bailey's Industrial Oil and Fat Products*, John Wiley & Sons, Hoboken, 6th edn, 2005; M. P. Malveda, R. Gubler, K. Yagi, *Fats and Oils Industry Overview*, Chemical Economics Handbook – SRI Consulting, 2005; J.-O. Lidefelt, ed., *Handbook of Vegetable Oils and Fats*, AAK AB, Karlshamn, 2nd edn, 2007; A. Thomas, in *Ullmann's Encyclopedia of Industrial Chemistry*, Wiley-VCH, Weinheim, 7th edn, 2007.

2. J. H. Clark, V. Budarin, F. E. I. Deswarte, J. E. Hardy, F. M. Kerton, A. J. Hunt, R. Luque, D. J. Macquarrie, K. Milkowski, A. Rodriguez, O. Samuel, S. J. Tavener, R. J. White and A. J. Wilson, *Green Chem.*, 2006, **8**, 853; B. Kamm, P. R. Gruber and M. Kamm, ed., *Biorefineries – Industrial Processes and Products*, Wiley-VCH, Weinheim, 2006.
3. S. Erkman, *vers une écologie industrielle*, ed. Charles Léopold Mayer – La librairie FPH, Paris, 1998; D. T. Allen and D. R. Shonnard, *Green Engineering*, Prentice Hall PTR, Upper Saddle River, 2002; M. M. Srivasatava and R. Sanghi, ed., *Chemistry for Green Environment*, Narosa Publ., New Delhi, 2005.
4. G. Hatje, in *Oil Crops of the World*, ed. G. Röbbelen, R. K. Downey and A. Ashri, McGraw-Hill, New York, 1989, p. 1.
5. F. Bohmert, *Vom Fang der Wale zum Schutz der Wale*, Schriften des Werksarchivs der Henkel KGaA, Bd. 14, Düsseldorf, 1982.
6. R. A. Caulfield, *Greenlander, Whales, and Whaling: Sustainability and Self-determination in the Arctic*, Dartmouth College, University Press of New England, Hanover, 1997.
7. J. Diamond, *Collapse*, Penguin Books, London, 2005.
8. B. S. Mielke, *Fett Wiss. Techn. Fat Sci Techn.*, 1987, **89**, 99; ISTA Mielke GmbH, ed., *Oil World Annual 2007*, vol. 1, Hamburg, 2007; ISTA Mielke GmbH, ed., *Oil World Annual 2008*, vol. 1, Hamburg, 2008.
9. R. Gubler, *Biodiesel*, ed. SRI Consulting, CEH Marketing Research Report, Nov. 2006.
10. S. Hansen, *6th European Motor BioFuels Forum*, Rotterdam, 2008.
11. J.-P. Langellier, *Le Brésil devient un pays créditeur après deux siècles d'endettement*, Le Monde (27. Février 2008); Banco Central do Brazil, *Focus-BC Report: Brazilian External Sustainability Indicators – Recent Evolution* (2/29/2008), http://www.bcb.gov.br (retrieved 05.03.2008).
12. Ministry of Agriculture, Livestock and Food Supply, *Plano Nacional de Agroenergia/Brazilian agroenergy plan 2006–2011, Embrapa Publ.*, Brasília, 2006, http://www.wilsoncenter.org/topics/docs/brazil.govt.agroenergyplan2006-2011.eng.pdf.
13. Y. Basiron, *Eur. J. Lipid Sci. Technol.*, 2007, **109**, 289.
14. S. Trent, ed., *The Politics of Extinction*, Environmental Information Agency/Telapak, London, 1998; http://www.eia-international.org/old-reports/Forests/Indonesia/PolExtinction/intro.html.
15. R. Carrere, Gen. Coordination, *The Bitter Fruit of Oil Palm: Dispossession and Deforestation*, World Rainforest Movement, Montevideo, 2001, www.wrm.org.uy/plantations/material/OilPalm.pdf (retrieved 06.03.2008).
16. Main worldwide oil transportation routes, http://www.black-tides.com/uk/oil/transport-oil/main-oil-transport-routes.php.
17. Worldwatch Institute, *Biofuels for Transport*, German Federal Ministry of Food, Agriculture and Consumer Protection, ed., Earthscan, London, Sterling, 2007.
18. http://www.unhchr.ch/development/governance.html.

19. S. Hansen and K. Ahti, Rabobank, private communication.
20. R. Höfer and J. Bigorra, *Green Chem.*, 2007, **9**, 203.
21. European Commission, *Integrated Pollution Prevention and Control Reference Document on Best Available Techniques in the Food, Drink and Milk Industries*, August 2006; IFC, *Environmental, Health, and Safety Guidelines Plantation Crop Production*, April 30, 2007; IFC, *Environmental, Health, and Safety Guidelines for Vegetable Oil Processing*, April 30, 2007.
22. Council Directive 1999/13/EC, March 11, 1999.
23. E. Ferchau, *Equipment for decentralized cold pressing of oil seeds*, http://www.folkecenter.dk/plant-oil (retrieved 21.04.2008).
24. K. Dahlke, *Oléagineux, Corps Gras, Lipides*, 1997, **4**(1), 55.
25. H. C. Holm, Novozymes, private communication; http://www.novozymes.com/NR/rdonlyres/66481D20-BE7E-4C12-ADD2-DC2C3E4D53CD/0/degumming.swf (retrieved 02.05.2008); C. L. G. Dayton, K. P. Staller and T. L. Berkshire, WO 2005063950, 2004 (Bunge Oils).
26. M. M. Prieto, J. C. Bada, M. Leon-Camacho and E. Graciani Constante, *Eur. J. Lipid Sci. Tech.*, 2008, **110**(2), 91; W. de Greyt, *Deodorization and physical refining*, IUPAC-AOCS Workshop on fats, oils & oilseeds analyses & production, Tunis (2004).
27. W. Normann, DRP 141 029, 1902 (Herforder Maschinenfett und Ölfabrik Leprince & Siveke); W. Normann, *Chemiker Zeitung*, 1937, **61**, 20.
28. R. L. Earle and M. D. Earle, *Unit Operations in Food Processing – the Web Edition*, http://www.nzifst.org.nz/unitoperations.
29. R. Jost, in *Ullmann's Encyclopedia of Industrial Chemistry*, Wiley-VCH, published online: 15 July, 2007.
30. J. C. Castilla and O. Defeo, *Science*, 26 August 2005, **309**(5739), 1324c; H. Mathies, *kontinente*, 4-2009, **14**, 06; www.msc.org.
31. R. Froese, A. Stern-Pirlot, H. Winkler and D. Gascuel, *Fish. Res.*, 2008, **92**, 231.
32. L. Jeromin, W. Johannisbauer, B. Gutsche, V. Jordan and H. Wogatzki, EP 0656894, 1993 (Henkel).
33. S. Both, T. Alexandre, B. Gutsche, J. Kray, C. Beverungen and R. Eickenberg, US 7 091 012, 2006 (Cognis).
34. S. Sato, H. J. Sousa Sales, H. Peloggia, P. Kempers, S. Both and U. Schörken, WO 2004/080 942, 2004 (Cognis).
35. S. Puleo and T. P. Rit, *Lipid Technol.*, July/Aug. 1992, 82–90; U. Wolfmeier, H. Schmidt, F.-L. Heinrichs, G. Michalczyk, W. Payer, W. Dietsche, K. Boehlke, G. Hohner and J. Wildgruber, in *Ullmann's Encyclopedia of Industrial Chemistry*, Wiley-VCH, Weinheim, 2002.
36. S. R. Padgette, K. H. Kolacz, X. Delannay, D. B. Re, B. J. LaVallee, C. N. Tinius, W. K. Rhodes, Y. I. Otero, G. F. Barry, D. A. Eichholtz, V. M. Peschke, D. L. Nida, N. B. Taylor and G. M. Kishore, *Crop Science*, 1995, **35**(5), 1451; J. Carpenter and L. Gianessi, *AgBioForum*, 1999, **2**(2), 65.

37. T. Funke, H. Han, M. L. Healy-Fried, M. Fischer and E. Schönbrunn, 2006, *PNAS*, **103**, 13010; L. Moran, *The Molecular Basis of Roundup*® *Resistance*, http://sandwalk.blogspot.com/2007/03/molecular-basis-of-roundup-resistance.html (retrieved 09.07.2008).
38. J. E. Carpenter and L. P. Gianessi, *Agricultural Biotechnology: Updated Benefit Estimates*, NCFAP Reports, January 2001, www.ncfap.org/pubs.htm; S. Bonny, *Agron. Sustain. Dev.*, 2007, 28.
39. *GVO-Zulassungen nach EU-Recht: Stand der Dinge*, Press Release 04/17 (28/01/2004), http://europa.eu/rapid/pressReleasesAction.do?reference= MEMO/04/17&format=HTML&aged=0&language=DE&guiLanguage =en (retrieved 07.05.2008).
40. Anonymous, *Tetraconazole: a Success Story*, ISAGRO Focus, February 2005.
41. Y. Dersjant-Li, in *Memorias del VI Simposium Internacional de Nutrición Acuícola*, ed. L. E. Cruz-Suárez, D. Ricque-Marie, M. Tapia-Salazar, M. G. Gaxiola-Cortés and N. Simoes, Cancún, 2002; A. Allodi, *Sustainability – European Aquaculture Perspectives: Will feed be a bottleneck?* FAO/ FEAP Joint Workshop, Rome, 2007.
42. M. Fitzpatrick, *Soybean Use and Trends 2001–2006*, U.S. Meat Export Federation, October 30, 2007, www.usmef.org/TradeLibrary/files/ 07_1030_Soybeans.pps (retrieved 07.05.2008).
43. R. H. V. Corley and P. B. Tinker, *The Oil Palm*, Blackwell, Oxford, 2003.
44. C. F. Rober and S. Menon, *Trade and Environment Dimensions in the Food and Food Processing Industries in Asia and the Pacific*, http:// www.un-trade-environment.org/documents/asia/study%20paper/ Case%20study%20Palm%20oil%20Malaysia.pdf (retrieved 15.05.2008).
45. Y. M. Choo, C. W. Puah and M. Basri Wahid, *Outlook of Palm Biodiesel in Malaysia*, Malaysian Palm Oil Board, 2007; Y. M. Choo, H. L. N. Lau and C. L. Yung, *Biodiesel Standard Development in Malaysia & Impacts of Palm Biodiesel on Engines and Emissions*, www.tistr.or.th/APEC_website/ APEC_pdfs/A-9_Dr.%20Choo_Bangkok_APEC%20-%2025%20Oct% 202007.pdf (retrieved 14.05.2008).
46. O. Ong'wen and S. Wright, *Small Farmers and the Future of Sustainable Agriculture*, Ecofair Trade Dialogue Discussion Papers, No. 7, March 2007.
47. R. Butler, *Malaysian palm oil industry puts sustainability in the spotlight*, mongabay.com, April 17, 2008.
48. Working Group on Developing Sustainability Criteria and Standards for the Cultivation of Biomass used for Biofuels, *Compilations of existing certification schemes, policy measures, ongoing initiatives and crops used for bioenergy*, 2007, http://www.uneptie.org/energy/act/bio/doc/Working% 20Paper_Developing%20Standards%20and%20Criteria%20for%20 Biomass%20Production_June%202007.pdf.
49. RSPO *Certification Systems – Final document, June 2007*, http://www.rspo. org/RSPO_Certification_Scheme_for_Sustainable_Palm_Oil_Launched_at_ RT5.aspx; M. Salleh, *Roundtable on Sustainable Palm Oil (RSPO) and*

the Supply Chain, PORAM FORUM 2007, http://www.poram.org.my/
database/Dato%20Mamat%20Salleh.pdf.

50. RSPO, *Principles and Criteria for Sustainable Palm Oil Production*,
October 2007, http://www.rspo.org/resource_centre/RSPO%20Principles%
20&%20Criteria%20Document.pdf.

51. M. W. Vis, J. Vos and D. van den Berg, *Sustainability Criteria and Certification Systems for Biomass Production – Final Report*, Report Nr. 1386,
btg, Enschede (2008).

52. A. Mohd Nasir, *Oil Palm Industry Economic Journal*, 2003, **3**(2), 21.

53. W. Stein, *JAOCS*, 1968, **45**(6), 471; L. Jeromin, N. Bremus, G. Friederici
and P. Pfeiffer, US 3950371, 1974 (Henkel); E. Deffense, *JAOCS*, 1985,
62(2), 376; O. Zaliha, C. L. Chong, C. S. Cheow, A. R. Norizzah and M.
J. Kellens, *Food Chem.*, June 2004, **86**(2), 245; M. Kellens, V. Gibon, M.
Hendrix and W. De Greyt, *Eur. J. Lipid Sci. Technol.*, 2007, **109**, 336.

54. K. Poku, *Small-Scale Palm Oil Processing in Africa*, FAO Agricultural
Services Bulletin 148, Rome, 2002; N. Thomma, *Migros World Tour*,
2002, 118.

55. Y. Basiron and K. W. Chan, *J. Oil Palm Research*, June 2004, **16**(1), 1.

56. D. Reed, *Urethanes Technology*, Oct./Nov. 2003, **20**(5), 41; K. S. Norin
Zamiah, T. L. Ooi, A. Salmiah, *J. Oil Palm Research*, June 2006, **18**, 198.

57. 66. Y. Akaike, *JAOCS*, **62**, February 1985, (2), 335.

58. Y. M. Choo, A. N. Ma, K. W. Chan and Y. Basiron, *J. Oil Palm
Research*, June 2005, **17**, 47.

59. K. Schmid, H. Baumann, W. Stein and H. Dolhaine, in *Proceedings of the
World Surfactants Congress*, München, vol. II, Kürle, Gelnhausen, 1984,
p. 105; B. Fabry, U. Kratzel, W. Schmidt and G. Kreienfeld, EP 0524996,
1991 (Henkel); A. E. Sherry, B. E. Chapman, M. T. Creedon, J. M.
Jordan and R. L. Moese, *JAOCS*, 1995, **72**(7), 835; S. Ahrnad and Y.
Basiron, *All Business*, March 1, 2003; W. B. Sheats, N. C. Foster, *Con-
centrated Products from Methyl Ester Sulfonates*, Published Papers, 2007
(Chemithon).

60. B. Tan and M. H. Saleh, US 5 157 132, 1990 (Carotech); Y. A. Tan,
R. Sambanthamurthi, K. Sundram and M. B. Wahid, *Eur. J. Lipid Sci.
Technol.*, 2007, **109**, 380.

61. F. L. Jackson and H. E. Longenecker, *Oil & Soap*, March 1944, 74;
K. S. Markley, *Economic Botany*, 1971, **25**(3), 267; Babassu Breakers,
http://www.tve.org/ho/doc.cfm?aid=912 (retrieved 03.06.2008).

62. R. Kleiman, in *Advances in New Crops*, ed. J. Janick and J. E. Simon,
Timber Press, Portland, 1990, p. 196.

63. R. K. Downey, in *Agricultural Biotechnology: Economic Growth Through
New Products, Partnerships and Workforce Development*, ed. A. Eagle-
sham and R. W. F. Hardy, NABC Report 18, 2006, http://nabc.cals.
cornell.edu/pubs/nabc_18/NABC18_Downey.pdf (retrieved 11.02.2008).

64. E. L. Ashton, J. D. Best and M. J. Ball, *J. Am. Coll. Nutr.*, 2001, **20**(4),
320; A. Allman-Farinelli, K. Gomes, E. J. Favaloro and R. A. Petocz,
JADA, July 2005, **105**(7), 1071.

65. L. W. Kleingartner, in *Trends in New Crops and New Uses*, ed. J. Janick and A. Whipkey, ASHS Press, Alexandria, 2002.
66. A. Westfechtel, in *Nachwachsende Rohstoffe für die Chemie*, Schriftenreihe Nachwachsende Rohstoffe, Bd. 18, Landwirtschaftsverlag, Münster, 2001, p. 297; G. P. Vannozzi, *HELIA*, 2006, **29**(44), 1.
67. S. Z. Erhan, M. O. Bagby and H. W. Cunningham, *JAOCS*, 1992, **69**(3), 195; R. Höfer and M. Fies, *Paintindia*, 2001, **51**(11), 65; J. Baro and P. Bene, *Farbe Und Lack*, 10/2007, **113**, 42.
68. R. Höfer, F. Jost, M. J. Schwuger, R. Scharf, J. Geke, J. Kresse, H. Lingmann, R. Veitenhansl and W. Erwied, in *Ullmann's Encyclopedia of Industrial Chemistry*, **A11**, VCH Weinheim, 1988, p. 456; M. J. Owen, in *Kirk-Othmer Encyclopedia of Chemical Technology*, John Wiley & Sons, Article Online Posting, 2001, p. 236.
69. E. Shaaya, M. Kostjukovski, J. Eilberg and C. Sukprakarn, *J. Stored Prod. Res.*, 1997, **33**(1), 7; M. B. Isman, *Crop Protection*, 2000, **19**(8–10), 603.
70. Anonymous, *Maize in Human Nutrition*, FAO, Rome, 1992, http://www.fao.org/docrep/T0395E/T0395E00.htm#Contents (retrieved 04.06.2008).
71. F. T. Orthoefer, *Food Technology*, 1996, **50**(12), 62; J. S. Godber and B. O. Juliano, in *Rice Chemistry and Technology*, ed. E. T. Champagne, American Association of Cereal Chemists, St. Paul, 3rd edn, 2004, p. 163.
72. R. Schramm, A. Abadie, N. Hua, Z. Xu and M. Lima, *J. Biol. Eng.*, 2007, **1**, 9.
73. J. S. Kim, *Einfluss der Temperatur beim Rösten von Sesam auf Aroma und antioxidative Eigenschaften des Öls*, Dissertation, Berlin, 2001.
74. C. K. Lyon, *JAOCS*, 1972, **49**, 245; L. A. Johnson, T. M. Suleiman and E. W. Lucas, *JAOCS*, 1979, **56**, 436; W. Collinge, *The American Holistic Health Association Complete Guide to Alternative Medicine*, Warner Books, New York, 1996; G. Mateljan, *The World's Healthiest Foods – Essential Guide for the Healthiest Way of Eating*, gmf pub, Seattle, Washington, 2007.
75. T. Beckett, *The Science of Chocolate*, RSC Publ., Cambridge, 2008.
76. R. Höfer, *6000 Jahre Sinn fürs Schöne – 6000 Jahre Lack*, Schriftenreihe VILF-Vorträge, Bd. 7, 2005, p. 23.
77. P. Dribnenki, *Flax for Industrial Uses*, Canadian Plant Bio-Industrial Oils Workshop, Saskatoon, 2006.
78. F. C. Naughton, *JAOCS*, 1974, **51**(3), 65.
79. J. Müller, S. Quaiser, P. Saling, V. Manea, J. E. Maloney and C. Bradley, *PU Magazin*, Sept./Okt. 2008, **8**, 253.
80. P. Gross and E. Flemming, EP 0301197, 1988 (Wella).
81. D. P. Jones, D. W. Hobson and P. P. Pilar, US 6479060, 2001 (Healthpoint).
82. K. Worschech, in *Becker/Braun – Kunststoff Handbuch*, ed. K. Felger, Hanser, München, Wien, 2nd edn, vol. 2/1, 1985. p. 570.
83. R. Höfer, in *Perspektiven nachwachsender Rohstoffe in der Chemie*, ed. H. Eierdanz, VCH, Weinheim, 1996, p. 91.

84. J. A. Waynick, US 5133888, 1990 (Amoco); R. A. Fletcher, WO/1997/003152, 1996 (Shell); Yu. L. Ishchuk, *Lubricating Grease Manufacturing Technology*, New Age Intern., New Delhi, 2005; R. Sánchez, J. M. Franco, M. A. Delgado, C. Valencia and C. Gallegos, *Green Chem.*, 2009, **11**, 686–693.

85. R. Höfer, H.-G. Schulte and J. Schmitz, in *H. Kittel – Lehrbuch der Lacke und Beschichtungen*, 2. völlig neu bearb. Aufl., vol. 4, ed. M. Ortelt and S. Hirzel Verl. Stuttgart, 2007, p. 277.

86. K. Lindner, *Tenside-Textilhilfsmittel-Waschrohstoffe*, Wiss. Verlagsges., Stuttgart, 1964.

87. R. K. Henning, *Jatropha curcas L in Africa*, http://www.under utilizedspecies.org/Documents/PUBLICATIONS/jatropha_curcas_ africa.pdf; *The Jatropha System*, http://www.jatropha.de (retrieved 03.02.2009).

88. D. Bowles, ed., *Micro- and Macro-Algae: Utility for Industrial Applications*, York, 2007, http://www.epobio.net/pdfs/0709AquaticReport.pdf (retrieved 09.11.2008).

89. Anonymous, *A look back the US Department of Energy's Aquatic Species Program: Biodiesel from algae*, http://www.biodieselamerica.org/files/articles/Biodiesel%20from%20Algae.pdf (retrieved 27.06.2008); Office of Biomass Program, *U.S. Department of Energy Biomass and Biofuels Update to the United States Congress*, 2008, http://www.eere.energy.gov/news/pdfs/may_2008_hill_briefing.pdf (retrieved 27.06.2008).

90. M. Alvarez Cobelas and J. Zarco Lechado, *Grasas y Aceites*, 1989, **40**, 118–145.

91. K. Brudermanns, M. Graß, R. Höfer, J. Koch and N. Scholz, in *H. Kittel – Lehrbuch der Lacke und Beschichtungen*, 2. völlig neu bearb. Aufl., vol. 4, ed. M. Ortelt, S. Hirzel Verl., Stuttgart, 2007, p. 203.

92. R. Höfer, P. Daute, R. Grützmacher and A. Westfechtel, *JCT*, June 1997, **69**(869), 65; R. Höfer, *ECJ*, 3/2000, 26; T. Roloff, U. Erkens and R. Höfer, *Urethanes Technology*, August/September 2005, 29.

93. B. Gruber, R. Höfer, H. Kluth and A. Meffert, *Fett Wiss. Technol., Fat Sci. Technol.*, 1987, **89**(4), 147; R. Miel, *Urethane Technologies International*, Oct./Nov. 2007, 40.

94. F. M. Casati, B. Dawe, S. Fregni and Y. Miyazaki, *PUMagazin*, Sept./Okt. 2008, 8, 234; Z. Lysenko, A. K. Schrock, D. A. Babb and A. Sanders, US 2006/0276609, 2004 (DOW).

95. C. Priebe, M. Skwiercz and K. Böge, WO 00/06632, 1998 (Cognis); M. Skwiercz, in *Nachwachsende Rohstoffe für die Chemie*, Schriftenreihe Nachwachsende Rohstoffe, Bd. 18, Landwirtschaftsverlag, Münster, 2001, p. 441; U. Riedel and J. Nickel, in *Biopolymers, Vol. 10 – General Aspects and Special Applications*, ed. A. Steinbüchel, Wiley-VCH, Weinheim, 2003, p. 1.

96. K. Weissermel and H.-J. Arpe, *Industrielle Organische Chemie*, 5, vollst. überarb. Aufl., Wiley-VCH, Weinheim, 1998.

97. J. G. Zeikus, M. K. Jain and P. Elankov, *Appl. Microbiol. Biotechnol.*, 1999, **51**(5), 545.

98. M. Biermann, in *Schriftenreihe des Fonds der Chem. Industrie, Heft 26, Fette und Öle*, 1986, p. 26; S. Warwel, P. Bavaj, M. Rüsch gen. Klaas and B. Wolff, in *Perspektiven nachwachsender Rohstoffe in der Chemie*, ed. H. Eierdanz, VCH, Weinheim, 1996, p. 119.

99. R. G. Fayter, in *Perspektiven nachwachsender Rohstoffe in der Chemie*, ed. H. Eierdanz, VCH, Weinheim, 1996, p. 107.

100. D. V. Kinsman, *JAOCS*, Nov. 1979, **56**, 823A.

101. A. G. Hinze, in *Schriftenreihe des Fonds der Chem. Industrie, Heft 26, Fette und Öle*, 1986, p. 19.

102. L. Dudley Eirich, D. L. Craft, R. W. Frayer and C. R. Wilson, *Biooxidation – A Platform to Produce Specialty Chemicals*, Soc. for Industrial Microbiology – Annual Meeting, Minneapolis, 2003; U. Schörken, Cognis, private communication; T. Roloff, U. Nagorny and U. Erkens, *Kunststoffe*, 5/2004, **94**, 104; R. Höfer, S. Winkels and H.-J. Pütz, *Pitture e Vernici, European Coatings*, 16/2004, 28.

103. M. Meier and W. J. Feder, in *Nachhaltigkeit in der Chemie – 13. Internat. Sommerakademie St. Marienthal, Initiativen zum Umweltschutz*, Bd. 70, ed. F. Brickwedde, R. Erb, M. Hempel, M Schwake and Erich Schmidt, Verlag, Berlin, 2008, p. 172.

104. E. Twitchell, US 601 603, 1897 (Emery).

105. K. H. Nottinger, *6th World Surfactant Congress*, CESIO, Berlin, 2004.

106. K. Schumann and K. Siekmann, in *Ullmann's Encyclopedia of Industrial Chemistry*, Wiley-VCH, published online 15 June 2000.

107. M. Magne, S. El Kasmi, M. Dupire, M. Morard, C. Vaca Garcia, S. Thiebaud-Roux, S. J. Peydecastaing, E. Borredon and A. Gaset, US 20050163935, 2003 (Lapeyre); T. Lucas, *L'Usine Nouvelle*, 4 Mai 2006 [3008], 42.

108. Ch. Breucker, V. Jordan, M. Nitsche and B. Gutsche, *Chem.-Ing.-Techn.*, 1995, **67**, 430.

109. B. Gutsche, H. Rößler and S. Würkert, in *Handbook of Heterogeneous Catalysis*, ed. G. Ertl, H. Knoezinger, F. Schueth and J. Weitkamp, Wiley-VCH, Weinheim, 2nd edn, 2008, p. 3329.

110. B. Gutsche, *Fett/Lipid*, 1997, **99**(12), 418.

111. F. Bongardt, *Henkel-Referate*, 1993, **29**, 112; Bundesministerium für Verbraucherschutz, Ernährung und Landwirtschaft, ed, *Bericht über biologisch schnell abbaubare Schmierstoffe und Hydraulikflüssigkeiten*, Bonn, 2002; M. Scherer and D. Rettemeyer, *Fluid Spezial*, 2007, 32.

112. R. Luther, *in Ullmann's Encyclopedia of Industrial Chemistry*, Wiley-VCH, Weinheim, 2002.

113. S. Boyde, *Green Chem.*, 2002, **4**, 293.

114. B. Abribat and R. Höfer, *Info Chimie Magazine*, Mai 2007, **479**, 44; Juin/Juillet 2007, 480, 38.

115. F. Bohmert, 75 Jahre Henkel Glycerin, Schriften des Werksarchivs, Bd. 18, Henkel KGaA, Düsseldorf, 1985; R. Christoph, B. Schmidt, U.

Steinberner and W. Dilla, in *Ullmann's Encyclopedia of Industrial Chemistry*, Wiley-VCH, Weinheim, 6th edn, 2000; A. Westfechtel, *Glycerol Derivatives – An Overview*, 26th ISF World Congress, Prag, 2005; J. N. Chheda, G. W. Huber and J. A. Dumesic, *Angew. Chemie, Int. Ed.*, 2007, **119**(38), 7298; M. Pagliaro and M. Rossi, *The Future of Glycerol: New Usages for a Versatile Raw Material*, RSC Publ., Cambridge, 2008.

116. T. Haas and D. Arntz, EP 0577972, 1993 (Degussa).

117. H. Sanderson, S. E. Belanger, P. R. Fisk, Ch. Schäfers, G. Veenstra, A. M. Nielsen, Y. Kasai, A. Willing, S. D. Dyer, K. Stanton and R. Sedlak, *Ecotoxicol. Environ. Saf.*, 2008, doi:10.1016/j.ecoenv.2008.10.006; H. Sanderson, S. E. Belanger, P. R. Fisk, Ch. Schäfers, G. Veenstra, A. M. Nielsen, Y. Kasai, A. Willing, S. D. Dyer, K. Stanton and R. Sedlak, *Ecotoxicol. Environ. Saf.*, 2008, doi:10.1016/j.ecoenv.2008.07.013.

118. http://www.zenitech.com/documents/new%20pdfs/articles/All%20about%20fatty%20alcohols%20Condea.pdf (retrieved 11.11.2008); B. Brackmann and C.-D. Hager, *6th World Surfactant Congress*, CESIO, Berlin, 2004.

119. K. Hill, in *Catalysis for Renewables*, Wiley-VCH, Weinheim, 2007, p. 75; K. Hill *Pure Appl. Chem.*, 2007, **79**, 1999.

120. A. Sander, E. Eilers, A. Heilemann and E. von Kries, *Fett/Lipid* 1997, **99**, 115; *Henkel Referate*, 1998, 34, 14.

121. K. Hill and O. Rhode, *Fett/Lipid*, 1999, **101**, 25; K. Hill and C. LeHen-Ferrenbach in C. Carnero Ruiz, ed., *Sugar-based Surfactants*, Surfactant Science Series, CRC Press, Boca Raton, 2009, **143**, 1–20.

122. K. Hill, W. von Rybinski, G. Stoll, *Alkyl Polyglycodies* VCH, Weinheim, New York, Basel, Cambridge, Tokyo (1996); K. Hill, *Household and Personal Care Today*, 2008(2), XVII.

123. N. Kurth and C. Weichold, *SÖFW-Journal*, 2007, 133, 51; A. Mehling, G. Pellón and H. Hensen, *Cosmetics & Toiletries*, June 2008, **123**(6), 53.

124. M. Schäfer-Korting, U. Bock, A. Gamer, A. Haberland, E. Haltner-Ukomadu, M. Kaca, H. Kamp, M. Kietzmann, H.C. Korting, H-U. Krächter, C-M. Lehr, M. Liebsch, A. Mehling, F. Netzlaff, F. Niedorf, M. K. Rübbelke, U. Schäfer, E. Schmidt, S. Schreiber, K.-R. Schröder, H. Spielmann and A. Vuia, *ATLA: Alternatives to Laboratory Animals*, 2006, 34, 283; A. Mehling, M. Kleber and H. Hensen, *Food Chem. Toxicol.*, 2007, **45**, 747.

125. R. Piorr, R. Höfer, H. J. Schlüßler and K. H. Schmid, *Fett Wissenschaft Techn. – Fat Sci. Techn.*, 1987, **89**(3), 106; S. Both, F. Bauer and M. Weuthen, *SÖFW-Journal*, 2008, **134**(10), 10.

126. A. F-C. Chan, W. M. Bohon, D. J, Blumer and K. T. Ly, WO/1999/041342, 1999 (Atlantic Richfield/Arco).

127. A. F. Chan and K. T. Ly, US 5830831, 1996 (Atlantic Richfield).

128. C. Baumann, D. Feustel, U. Held, R. Höfer, *Henkel-Referate*, 1997, **33**, 121; J. Bigorra, W. H. Breuer, S. Heldt, Ll. Llaurado, Comunicaciones,

Jorn. Com. Esp. Deterg., 2005, **35**, 155–162; A. M. Fernandez, U. Held, A. Willing and W. H. Breuer, *Progr. Organic Coatings*, 2005, **53**, 246; BDI-Umweltpreis 2005/2006, www.bdi-online.de.

129. W. H. Breuer and R. Höfer, *Tenside Surf. Det.*, 4/2003, **40**, 208–214; Z. Dou, J. P. Ruiz, Y. Li, Z. Zong and F. Trezzi, *PPCJ*, March 2008, 22–26.

130. M. Aven, EP1023832, 2000 (BASF); M. J. Hopkinson, C. E. Moore and J. D. Fowler, WO/2003/022049, 2002 (Syngenta); S. B. Cush and M. J. Hopkinson, EP1603388, 2004 (Syngenta); G. Frisch, G. Schnabel and J. Rude PCT/EP 2005/005525, 2005 (Bayer Cropscience); B. Abribat, *Specialty Chem. Magazine*, April 2008, 22.

CHAPTER 9.2

Starch: A Versatile Product from Renewable Resources for Industrial Applications

ANDREA GOZZO[a] AND DETLEV GLITTENBERG[b]

[a] Cargill R & D Centre Europe, Havenstraat 84, B-1800 Vilvoorde, Belgium;
[b] Cargill Deutschland GmbH, Düsseldorfer Strasse 191, D-47809 Krefeld, Germany

9.2.1 Markets

Starch and its derivatives can be produced from several raw materials including corn, wheat, pea, potato, and tapioca and, in general, its production follows local raw material availability. In 2006 European (EU-25) production of starches, modified starches and refinery products was approximately 8.6 Mio tons.[1] The main consumer, accounting for 4.5 Mio tons, was the food industry, followed by the paper and board sector with 2.4 Mio tons, the pharmaceutical and chemical sector with 1.2 Mio tons and the industrial binders[i] sector with 0.6 Mio tons (Figure 9.2.1).

The botanical source of the primary starch products, from which also refinery products are derived, has been roughly as follows: corn starch 4.0 Mio tons, wheat starch 2.9 Mio tons, potato starch 1.8 Mio tons.[2]

---

[i] Industrial Binders includes the adhesives, gypsum boards and textile industries.

RSC Green Chemistry No. 4
Sustainable Solutions for Modern Economies
Edited by Rainer Höfer
© The Royal Society of Chemistry 2009
Published by the Royal Society of Chemistry, www.rsc.org

The total raw material processed to produce these starches and derivatives has been estimated at 12.5 Mio tons of potatoes and 10 Mio tons of cereals (corn and wheat).[3]

The perishable nature of potatoes limits the market of this raw material to the region of production: what is produced in Europe is predominantly consumed in Europe. On the other side, cereals are commodities traded globally and therefore subject to the dynamics of a global market. A recent demonstration of these global dynamics was the temporary shortage of cereal on the market caused, among others, by an increased demand for food coming from China and India, bad weather conditions and the use of cereals as feedstock for "bioethanol" production.

It is worthwhile noting that there are regional differences in the use of crops for bioenergy. In the USA, bioenergy is mostly bioethanol, which reflects on the high share (~25%) of corn crop used for energy production.[4] In Europe, bioenergy mainly comes from biodiesel, which is derived from oilseeds and has no impact on corn availability. According to the forecast of the European Commission the use of corn for bioethanol production in Europe will be very limited (see Figure 9.2.2 and Table 9.2.1).

9.2.2 Starch and Derivatives

Native Starch: Structure and Properties

Starch is the material used by Nature to store and provide energy for plants in the form of a glucose polymer. It is synthesized and stored by all plants, but it is industrially extracted mainly from cereals and tubers such as corn, wheat, potato and tapioca.

Starch occurs as discrete granules whose size and shape depends on its botanical origin. However, all starch granules have a semi-crystalline structure, hydrated only to a very small extent, insoluble in cold water, and very dense.

The term "starch" identifies a mixture of two different polysaccharides: amylopectin and amylose (Figure 9.2.3). Both are homopolymeric α-1,4-glucans based on D-glucose as monomeric unit, but they differ in molecular weight and in the ramification of the polymer.

Amylose has a molecular weight in the range 100,000–1,000,000 Daltons, is only lightly branched and in aqueous solutions assumes the shape of a random coil consisting of helical segments joined by segments with no specific conformation.

Amylopectin has a molecular weight in the range 1,000,000–10,000,000 Daltons, and is highly branched. According to Manners,[6] amylopectin has a 1,6 branch point every 5 to 9 glucose units.

While amylopectin constitutes the backbone of the granule and contains sections in which neighbouring chains within clusters intertwine into parallel double helices to form crystals, amylose has a largely amorphous structure.[7] These two polymers are held tightly together *via* a dense network of internal hydrogen bonds that optimizes the spatial arrangement and guarantees efficient storage and availability of energy reserves, as glucose, to the plant.

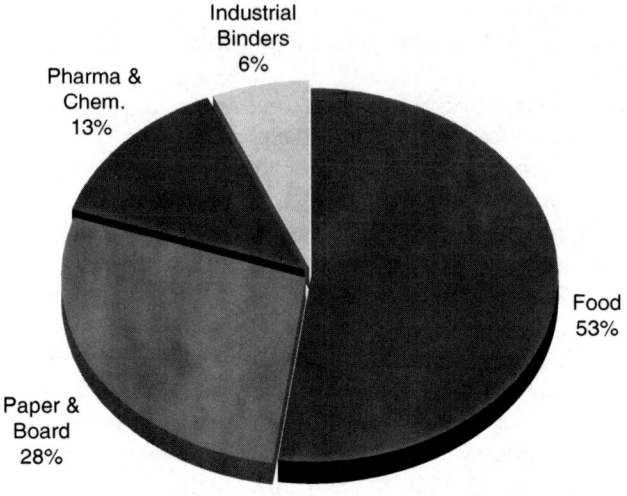

Starch consumption EU 2006	Tons
Food	4,500,000
Paper & Board	2,400,000
Pharma & Chemicals	1,150,000
Industrial Binders	550,000

Figure 9.2.1 Starch products consumption by sector. Europe 2006.

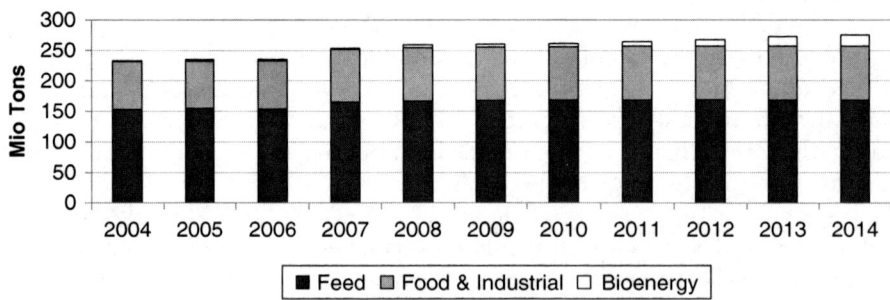

Figure 9.2.2 Total cereal market projections –EU-27.[5]

The internal hydrogen bond network also provides a barrier to water penetration inside the granules that are therefore insoluble in cold water. However, when heat is applied, the surrounding water molecules acquire enough energy to massively penetrate the network of hydrogen bonding and start competing with it, resulting in the hydration and dissolution of the polymer chains. This process is called *gelatinization* because it eventually leads to a starch gel (or paste) with

Table 9.2.1   Total cereal market projections – EU-27.

	Total cereal market projections for EU-27 (Mio MT)										
	2004	*2005*	*2006*	*2007*	*2008*	*2009*	*2010*	*2011*	*2012*	*2013*	*2014*
Feed	153.2	154.9	153.7	165.2	166.5	167.8	167.9	168.6	168.8	168.6	168.5
Food & Industrial	78.4	76.9	78.6	86.6	87.9	88	88.1	88.1	88.2	88.3	88.3
Bioenergy	0.7	2.7	2.5	1.9	4.8	4.5	5.5	7.5	10.3	15.7	18.4

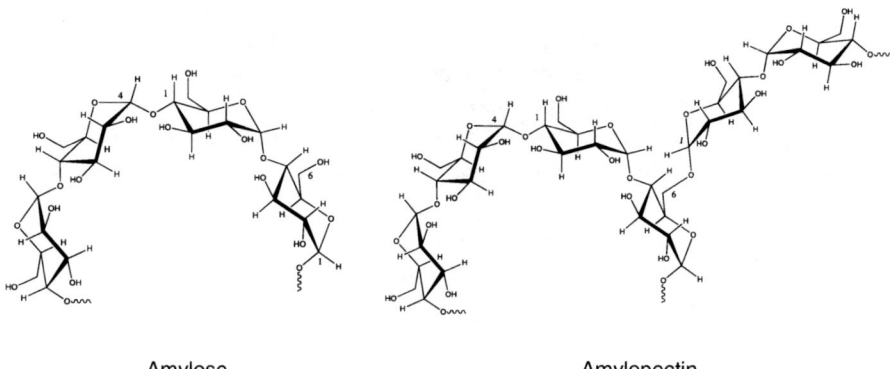

Amylose   Amylopectin

Figure 9.2.3   Amylose and amylopectin structures.

marked viscosity. Even after gelatinization, starch polymeric chains surrounded by water molecules have a strong tendency to form new hydrogen bonds among themselves. This results in a further increase in viscosity, which ultimately leads to a strong gel. The process is called retrogradation, which literally means "move backwards" and consists of a sort of re-crystallization of the polymeric chains that brings the polymer into a more crystalline state. Amylose is the main polysaccharide responsible for the retrogradation of starch paste as its structure facilitates the formation of pseudo-crystalline structures, which, linked to amylopectin, are widely responsible for the massive increase in viscosity and formation of strong gels. Conversely, the branched structure of amylopectin induces the formation of soft gels characterized by a relatively stable viscosity over time. The length of the chains helps to create an extended network of internal hydrogen bonds and, at the same time, the ramification of the polymer prevents a real crystallization of the amylopectin structure.

Retrogradation is strongly influenced by factors such as the amylose/amylopectin ratio, temperature and its variation, and the interaction between starch and other materials (proteins, lipids, *etc.*).

The functionalities of starch and its derived products that can be exploited in industrial applications are directly linked to its chemical structure and can be summarized as follows:

- Polymeric structure;
- Fermentation feedstock (glucose source);

- Chemical intermediate for synthesis;
- Protective colloid;
- Binder;
- Humectant;
- Carrier;
- Rheology modifier.

Modified Starches

Even though native starch can be used for some applications, chemical modifications of its structure are generally necessary to optimize a particular functionality and to obtain a product that can be handled on an industrial scale.

Even relatively low-level modifications in the chemical structure of starch have a significant impact on its properties such as:

- Viscosity;
- Paste viscosity stability;
- Retrogradation;
- Rheology;
- Water retention;
- Film formation capacity and film flexibility;
- Ionic charge;
- Affinity with hydrophobic molecules.

Chemical modifications are generally carried out on the starch granules dispersed into water as starch slurry. The reactants are dissolved into the slurry and the mixture is heated until the reaction is completed. After separation of the remaining reagents, water is removed. At the end of the process, modified starch is still in granular form and can be handled as a powder. The reaction conditions have to be carefully chosen to avoid the gelatinization of starch during the process. When a high degree of modification is required, gelatinization cannot be avoided and the final product is sold as a liquid.

Chemical modifications of starch include derivatization (also known as substitution), oxidation, cross-linking, hydrolysis (also known as thinning) and hydrolysis combined with internal rearrangements of the glucosidic bonds.

Derivatized starches are obtained by reaction with α-chloroacids, epoxy derivatives or anhydrides in alkaline conditions. In general, they have lower gelatinization temperature, better film-forming properties and tend to give softer gels. Typical products used in oil drilling fluids, adhesives and paper applications are carboxymethylated starches, hydroxypropylated starches, cationized starches and acetylated starches.

Oxidized products are normally prepared by treatment with an alkaline hypochlorite. This treatment results in two chemical reactions taking place simultaneously: a partial depolymerization that lowers molecular weight, and the introduction of some carbonyl and carboxyl groups.

After gelatinization, these starches form pastes with low viscosity at a relatively high concentration and with very stable viscosity. The resulting products are well suited for use in pigmented coating colours, adhesives and textile warp sizing.

Thinned products are the result of acid treatment of the granular starch that essentially produces depolymerized starches with lower molecular weight. Two straightforward consequences of this modification are: increased solubility compared to the native starch and a more pronounced retrogradation tendency, which results in a strong increase in viscosity of the gelatinized starch upon cooling. Thinned starch can be used at relatively high concentrations providing good binding properties as well as high film strength. Typical applications for such starches are as a binder in gypsum boards and in textile and paper-making applications.

Dextrins are produced by a severe thermohydrolytic action in conditions of low moisture (1–5%), acid pH and very specific ranges of temperature and time. Under appropriate conditions, two main chemical reactions occur: the hydrolysis of the 1,4 glucosidic bond and the rearrangement (or transglucosidation) of the 1,4 bond to 1,3 or 1,2 or 1,6 bonds. This rearrangement produces an amylopectin-like structure that is responsible for the superior viscosity stability of dextrins compared to the thinned products.

Dextrins' solubility in water varies from low (20%) to high (90%). The corresponding solutions are tacky and dry quickly, providing an excellent solution for adhesives.

Refinery Products: Maltodextrins, Glucose Syrups, Dextrose and Polyols

Hydrolysis of aqueous solutions of starch by acid or enzymatic reactions yields a mixture of different products, mainly consisting of D-glucose, maltose, maltotriose and higher oligosaccharides. Because of their complex composition, these hydrolyzed products are best characterized in terms of their "dextrose equivalent" (DE) value, which is a measure of the total reducing power of the saccharides relative to glucose taken as 100 and expressed on a dry weight basis.[8] The higher the DE, the higher the degree of hydrolysis of the product.

Starch hydrolysis products with a DE lower than 20 are defined by food legislation as maltodextrins, while those with a DE value higher than 20 are classified as glucose syrups.

A typical enzymatic production process for these products involves mixing enzymes and starch slurry, gelatinizing the starch by heating the mixture up to $\sim 75\,^{\circ}\mathrm{C}$, holding it at that temperature to allow for enzymatic cleavage and then heating it again or acidifying the mixture to inactivate the enzymes. After separation of the insoluble material, the product is stored as a liquid, or spray-dried to obtain a white powder. There are several versions of this process, and combinations of acid hydrolysis and enzymatic cleavage are also used. A good overview of such processes can be found in Howling's work.[9]

Table 9.2.2 Functional properties of glucose syrups as function of DE. Adapted from Howling.[9]

Property	Low DE High Average Molecular Weight	High DE Low Average Molecular Weight
Colour formation		⟶
Crystallisation control	⟵	
Emulsion stabilisation	⟵	
Fermentability		⟶
Foam stabiliser	⟵	
Hygroscopicity		⟶
Osmotic pressure		⟶
Viscosity	⟵	

Glucose syrups are produced *via* the same process described for maltodextrins but the hydrolysis reaction is run more extensively, resulting in products with higher DE. In general, the properties of syrups produced using the same process can be related to the DE value (Table 9.2.2).

Hydrolysate is a special type of glucose syrup in which starch has been hydrolyzed to obtain a D-glucose content of at least 90% on a dry basis, with a typical value ranging between 95 and 96%.

Hydrolysate is the starch-based product most widely used as a carbon source for fermentation processes. It is also the product used as feedstock to produce dextrose.

Dextrose is the common name of the monosaccharide D-glucose (CAS 50-99-7). It is an aldohexose that, when crystalline, assumes a hemiacetal form (α or β) of the six-membered ring (pyranose) while in solution it reaches equilibrium between the hemiacetal forms and the aldehyde open form (see Figure 9.2.4). The open form of D-glucose has the typical reducing property of the aldehyde group. It is worth mentioning that every amylose and amylopectin molecule at the end of the chain has one and only one glucose unit that is in equilibrium between the pyranose and the open form. This is normally called the reducing end.

Catalytic hydrogenation of an aqueous solution of glucose leads to the main polyol used in industrial applications: sorbitol. The aqueous solution of sorbitol is purified by ion exchange chromatography, decolourized and concentrated to give a 70% solution that is sold as such or further processed to give a crystalline powder (Figure 9.2.5).

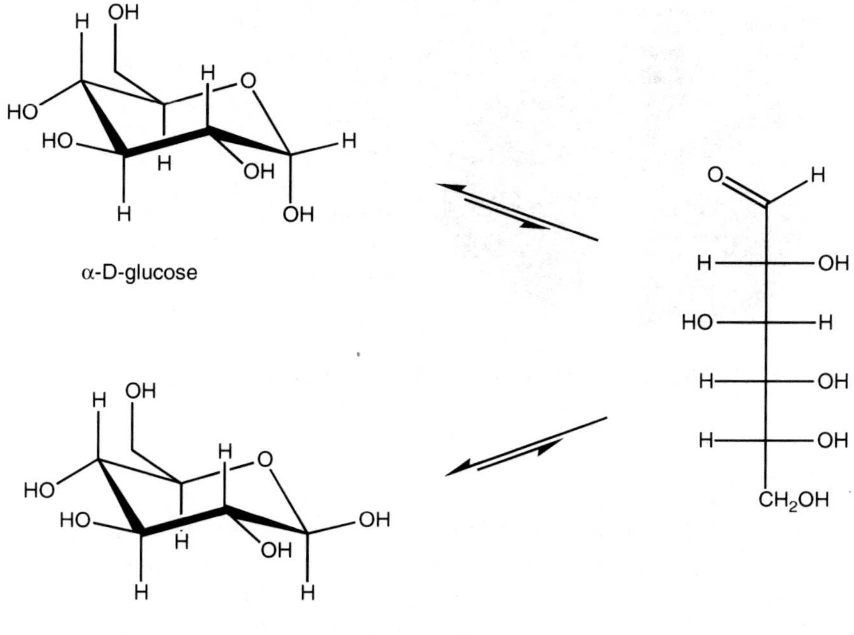

Figure 9.2.4   Equilibrium between hemiacetal and aldehyde forms of D-Glucose.

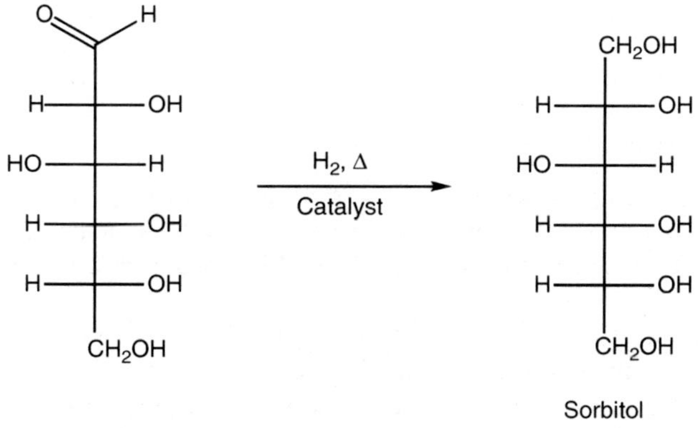

Figure 9.2.5   Sorbitol from D-glucose.

9.2.3   Food Applications

The market for starch-derived products in the food industry is generally seg-
mented into the following application areas: confectioneries, processed food,
beverages, dairy products and baked foods (see Figure 9.2.6).

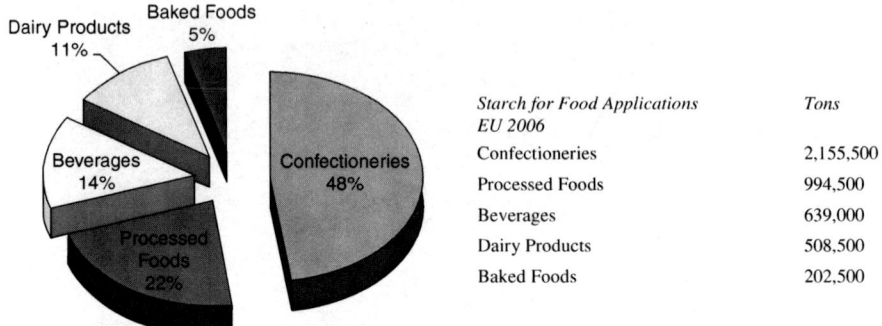

Starch for Food Applications EU 2006	Tons
Confectioneries	2,155,500
Processed Foods	994,500
Beverages	639,000
Dairy Products	508,500
Baked Foods	202,500

Figure 9.2.6 Starch products conumption in the food sector. Europe 2006.[10]

Confectioneries

Starches are widely used in confectioneries as binding or gelling agents. They provide quick gelling action upon cooling and texture-stabilizing properties.

Starch-containing gum candies are typically prepared with thinned starches that ensure a relatively low viscosity when cooked but then quickly form a gel upon cooling. This rapid gel formation is also appreciated in gelled confectioneries such as jelly beans and orange slices.

Maltodextrins are also used as a source of energy with sweetening activity in the so-called "energy bars".

Glucose syrups are mainly used to control crystal size and texture in various products.

Processed Foods

This market segment includes a wide variety of products with a very diverse set of requirements on starch functionality. In dressings and sauces, starches have to be stable at acidic pH and compatible with hydrophobic components such as oils and fats.

The use of harsh processing conditions, food safety and the long-shelf-life requirements are the major challenges for the starch product used in this segment.

In sauces and soups starch is used as a thickener that provides also a specific texture and mouthfeel. Those functionalities are particularly critical in low-fat products where starches are expected to provide the same textural and mouthfeel experience commonly associated with conventional products.

Beverages

A significant amount of glucose syrups are used in this segment as sweeteners. High Fructose Corn Syrup (HFCS), a glucose syrup with high concentration of fructose, is used in soft drinks as sugar (saccharose) replacement with very similar sweetness profile.

Glucose syrups are also used as fermentables in beer and alcohol production. Finally, modified starches are used to encapsulate vitamins that are frequently added to juices.

9.2.4   Pharmaceutical and Chemical Applications

The Pharmaceutical and Chemical Applications sector covers a wide range of different uses of starch and is one of the most promising development areas for carbohydrate chemistry in the future.

The following paragraphs give an overview of current industrial applications of starch-related green chemistry in this area.

Drug Formulations

Starch is used in many solid formulations as binder, diluent and disintegrant. A special grade suitable for direct compression has been developed for tablet formulations.

Polyols and dextrose are used to impart a sweet taste to chewable tablets and syrups.

Active Ingredients

Dextrose is an essential ingredient of intravenous and dialysis solutions. In addition to its nutritional value, it allows the control of the liquid osmotic pressure.

Maltodextrins develop lower osmotic pressure than low-molecular-weight carbohydrates; therefore they are used at higher concentration without causing stomach irritation.

Hydroxyethyl starches with specific molecular weights are commonly found in formulations of plasma expanders, typically required after an accident or surgery involving a large loss of blood. They can be used as a temporary measure to restore blood volume, water and salts while the patient's condition is stabilized.

Acarbose Precursor

In 1990 the first drug of a new class of oral antidiabetic was commercialized under the name of Acarbose (Glucobay[®])[11] and used to treat type-2 diabetes mellitus (Figure 9.2.7). Acarbose is an α-glucosidase inhibitor that acts by competitive and reversible inhibition of α-amylase from the pancreas and α-glucosidase. The inhibition of these enzymes results in a delayed absorption of glucose from the gut and a reduced post-prandial increase in plasma glucose and insulin.[12]

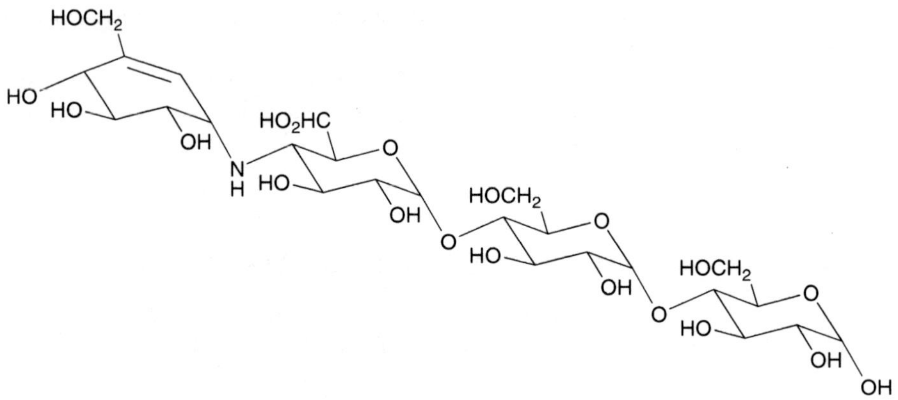

Figure 9.2.7 Acarbose structure.

The structure of this drug is built on a maltotriose skeleton that is further elaborated into the active product *via* a bioprocess. A special glucose syrup rich in maltotriose is the ideal substrate for this process.

Sorbitol as a Humectant

Sorbitol offers a broad range of functional properties including: humectancy, plasticizing ability, sweet taste, non-cariogenicity and good chemical stability in harsh conditions such as alkaline pH and heat. Furthermore, it is a *Generally Recognized as Safe* (GRAS) product by the United States Food and Drug Administration (FDA).

Toothpaste production is the second largest application of sorbitol, accounting for 50,000 mto a year in Western Europe alone. In the past, toothpaste formulations were glycerine-based but almost every producer went through a reformulation process to use sorbitol when the latter became available on an industrial scale.

With the recent drop in glycerine prices, the choice between sorbitol and glycerin is now primarily determined by their price.

Other important uses of sorbitol as a humectant include formulation of cough syrups, multivitamin preparations, emulsions and suspensions. Because of its humectancy properties, sorbitol reduces the tendency of liquid formulations to crystallize.

Glucose Syrups as Carbon Source for Fermentation Processes

Fermentation processes can be a valuable alternative to the conventional chemical synthesis, particularly when the finished product contains specific and complex stereochemistry.

Fermentation technology in the industrial synthesis of chemicals started to be used in the first decades of the twentieth century. Industrial production of citric

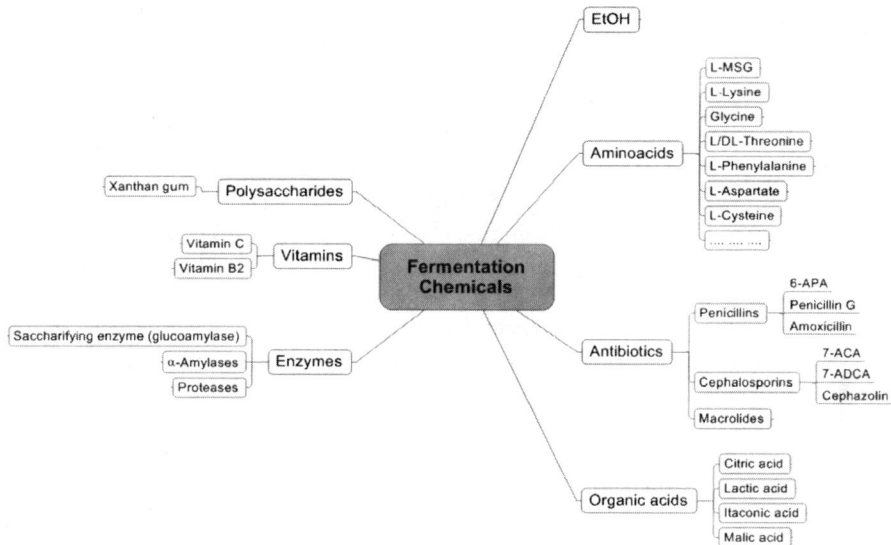

Figure 9.2.8 Main products produced by fermentation.

acid by fermentation, achieved by Pfizer in 1923, was an early success in this field. However, it was only with the production of penicillin during the Second World War that the whole sector took off. Today, the range of products that are produced by fermentation includes antibiotics, organic acids, aminoacids, polysaccharides, vitamins, enzymes and, more recently, ethanol for biofuel use (see Figure 9.2.8).

All the micro-organisms working in fermentation processes require the fundamental building blocks for their synthesis: a source of carbon, a source of nitrogen, salts and co-factors.

Saccharides and polysaccharides are the most common carbon sources used as feedstock; glycerol and oils can be used to a lesser extent.

The fermentation sector is one of the largest consumers of glucose syrup, accounting for some 600,000 tons of hydrolysate in Western Europe alone.

Over the last decade, mounting pressure on crude oil prices together with a desire to move to more sustainable processes pushed many companies to use the fermentation technology even for the synthesis of bulk chemicals. The large-scale productions of lactic acid for polymers and of ethanol as biofuel are two recent examples of this trend.

Biodegradable Plastics

Research efforts spanning eight decades have led to extensive improvements on the performances of conventional plastics such as polypropylene, polyethylene and polystyrene. On the other side, use of renewable materials in plastic production is still in an early stage, compared to traditional plastics, and still suffers from some performance limitations and high costs.

Starch has been one of the first materials extensively studied for its potential as polymer for biodegradable plastics. Impetus in this direction has come from shortage of municipal landfills and pressure to reduce the visual pollution caused by plastic residues abandoned in the landscape. Despite considerable interest from both the academic and industrial sectors, production of commercially viable materials has been delayed for technical and economic reasons. The former are inherent to the chemical structure of starch itself: the vast quantity of internal hydrogen bonds hampers thermoplastic behaviour, and the hydrophilic nature of the polysaccharide translates into sensitivity to the environmental humidity that, in its turn, may affect the mechanical properties of the material.

With extensive efforts, these technical challenges have now been effectively solved. Starch, in the presence of a suitable plasticizer, can be converted into a thermoplastic starch (TPS) that is suitable for conventional plastic processing technologies, such as extrusion, film blowing, injection moulding, *etc.* In addition, incorporation of polycapromolactone, polyvinylalcohol or other biodegradable and relatively hydrophobic polymers has made it possible to effectively reduce the strong hydrophilic character of the finished material.

The second factor hindering the development of starch-based material for plastics has been the high cost of the technology required to solve the technical issues, ultimately reflected in the cost of the finished products.

Today, there are products that satisfy the technical requirements and have an acceptable market price. One commercial success story is represented by Mater-bi[®],[13] a plastic material based on TPS and other polymers that is very versatile and suitable for applications such as composting bags, mulch film, disposable cutlery, *etc.*[14] However, due to the complexity of its production, its cost is higher than conventional polyethylene or polypropylene, and it is generally sold for applications where biodegradability is mandatory.

As legislation in several European countries is poised to limit the use of shopping bags made of conventional polyethylene, it is likely that starch-based materials will at least partially replace that type of plastic.

Expanded polystyrene (EPS) has been the sole packaging material for many years and its extensive use has led to environmental concerns, partly because of its visibility in landfills and partly because of the use of a chlorofluorocarbon, allegedly responsible for destroying the ozone layer, as blowing agent during its production. Today, starch-based foam materials are commonly produced and sold at a competitive price compared to EPS. They are biodegradable, have a very low density and, because of their better static electricity insulating properties, are considered even superior to EPS for use in electronic applications.

In 2007, about 73,640 metric tons of loose-fills made from starch were produced and consumed worldwide[15] (Table 9.2.3).

Table   9.2.3   Global biodegradable polymer market (mto) – adapted from Schlechter.[15]

Application	2006	2007	2012
Compost Bags	78,636	110,000	266,363
Loose-Fill Packaging	69,091	73,636	97,273
Other Packaging[a]	23,182	36,818	105,454
Miscellaneous[b]	15,000	25,455	77,727
Total	185,909	245,909	546,818

[a]Includes medical/hygiene products, agricultural, paper coating, *etc.*
[b]Unidentified biodegradable polymers.

Plasticizers

In general, polyols have good plasticizing properties, especially when it comes to plasticizing carbohydrate-based products such as TPS. They have a dual action: as primary plasticizers, altering the interaction between adjacent polysaccharide chains, and secondary plasticizers ensuring the presence of water – the most efficient plasticizer for carbohydrates.

All starch-based plastics on the market contain polyols such as sorbitol, glycerol or mixtures thereof ensuring flexibility and mechanical properties stability of the finished product.

Formaldehyde-based Resins

The presence of numerous hydroxyl groups able to react with formaldehyde makes starch-derived products suitable chemicals for formaldehyde-based resins. Research on this subject started many years ago and showed that in a number of applications it is possible to partially replace or "extend" urea formaldehyde, phenol formaldehyde and melamine formaldehyde resins without significantly affecting the finished product's performance.[16] In many applications, adhesive systems based on formaldehyde resins incorporate a polysaccharide component. More than 4.5 Mio mto of formaldehyde-based resins have been produced in Western Europe alone.[17] The use of carbohydrates allows lower consumption of oil-based resins and, consequently, reduced release of formaldehyde in the environment.

Chemical Synthesis

L-Ascorbic acid, known as vitamin C, occurs naturally in a number of sources such as citrus fruit, berries, kiwi fruit and green vegetables. Because of its pharmaceutical and nutritional value, ever since its discovery there has been a strong interest in obtaining it by chemical synthesis. In 1934 Haworth was awarded the Nobel Prize for establishing the first industrial process for the synthesis of vitamin C, but the process is named after the Polish chemist Tadeus

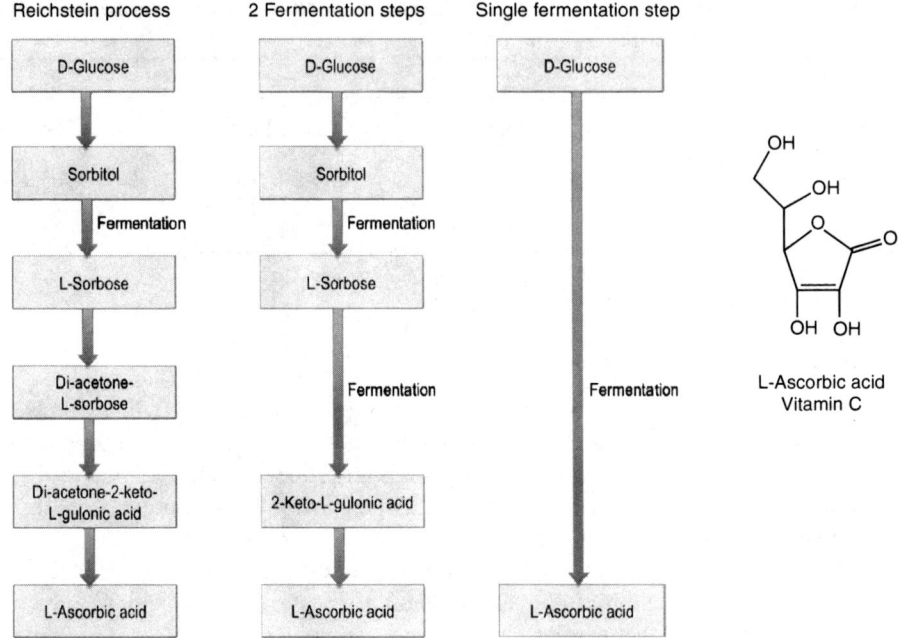

Figure 9.2.9   Processes for the synthesis of vitamin C.

Reichstein, who arrived independently at the same process somewhat earlier than Haworth.

Today, the bulk of vitamin C is produced *via* a "2 fermentation steps" reaction sequence in which L-sorbose is converted into 2-keto-L-gulonic acid (KGA) by fermentation (Figure 9.2.9).

The production of vitamin C without isolation of KGA has remained elusive, at least on an industrial scale, but studies in that direction are ongoing.

The current world production of vitamin C can be estimated to be as much as 80,000 tons per year consuming more than 140,000 tons of Sorbitol (100% dry substance).

The development of surfactants based on carbohydrates and oils is an interesting expression of "green chemistry" that led, among others, to two new classes of surfactants: the alkyl polyglycosides and the sorbitan esters.

After 10 years of development work, Rohm & Haas introduced on the market the first alkyl polyglycosides in the late 1970s. These were based on short hydrophobic chains (C8–C10) and had poor colour quality. Their application was limited to a few market segments such as the industrial and institutional sectors. In the 1980s other companies, among them Henkel KGaA, developed new APG® surfactants with different alkyl chains and with an improved colour quality that allowed the penetration of these products in very demanding market segments such as the cosmetic and the detergents industries.

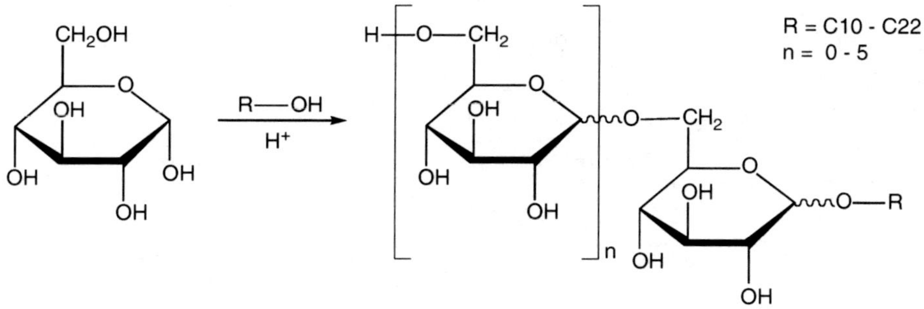

Figure 9.2.10 APG® synthesis scheme.

Despite many possible alternatives, the industrial production is still based on the "Fischer glycosidation" process that relies on the acid-catalyzed reaction of dextrose with fatty alcohols (see Figure 9.2.10). Dextrose monohydrate under the form of fine particle solids is suspended in a fatty alcohol acting as solvent and as reagent at the same time. The heterogeneous mixture is heated to evaporate immediately the crystallization water of dextrose firstly and the water coming from the condensation reaction as long as it is formed to drive the equilibrium towards the finished product. After cooling, separation of the by-products and a purification process, alkyl polyglycosides are isolated as a mixture of dextrose oligomers condensated with the fatty alcohols.[18] In general, commercial APG® surfactants contain different derivatized dextrose oligomers with an average degree of polymerization of 1.4 moles of glucose per mole of fatty alcohol.

Alkyl polyglycosides are known as very efficient wetting and dispersing agents that are readily biodegradable and have very low toxicity. They are stable in neutral and alkaline conditions; they tolerate a high salt concentration, and are compatible with many other surfactants.

Today alkyl polyglycosides are a typical ingredient in many cosmetics and in lightweight dishwashing formulations. Cognis (formerly Henkel) is the largest European producer with 25,000 tons capacity installed followed by Lamberti, SEPPIC and Akzo Nobel.

The esterification of a liquid solution of sorbitol with fatty acids leads to a variety of mild tensioactive agents commonly known as sorbitan esters. Under esterification conditions and high temperature (>200°C), sorbitol undergoes an internal dehydration reaction to give a tetrahydrofuran structure (sorbitan) that is esterified by the fatty acid mainly on the primary hydroxyl group (see Figure 9.2.11). When an excess of fatty acid is used, di-esters and tri-esters are obtained. These sorbitan esters are commonly referred to as Span®. They have low water solubility but are soluble in most cosmetic oils. They are typically used as water-in-oil (O/W) emulsifiers.

Sorbitan esters can be further reacted with ethylene oxide to introduce poly-oxyalkylene chains that result in a significant increase of the surfactant's water solubility. These products are commonly referred to as Tween® or polysorbates.

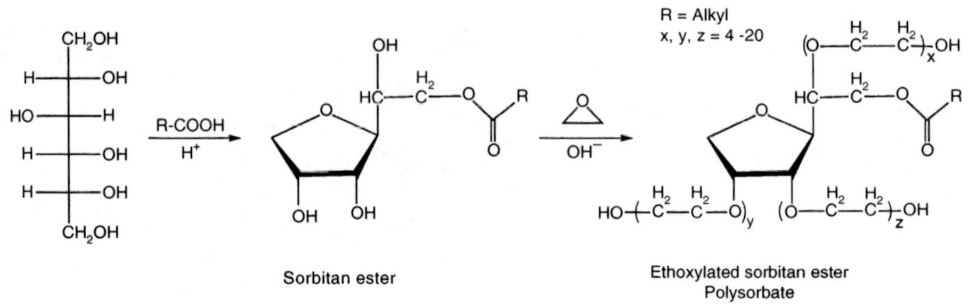

Figure 9.2.11 Sorbitan esters and polysorbates synthesis scheme.

Sorbitan esters are often used in the food industry, especially in chocolate and margarine production, where a high affinity for fats is desired.

Polysorbates are also commonly used as (O/W) emulsifiers in the food industry (salad dressing, ice cream), and in cosmetics, pharmaceuticals, detergents, and in a myriad of other industrial applications. Current Western Europe consumption of sorbitol for surfactants manufacture can be estimated at around 15,000 tons per year.

To produce polyether polyols (PEP), sorbitol is reacted in the presence of a catalyst with propylene or ethylene oxide (see Figure 9.2.12). The reaction is aimed at achieving polyols with a hydroxyl number[ii] of between 350 and 550. Sorbitol-based PEPs are then used to manufacture rigid or semi-rigid polyurethane foams by reaction with a suitable polyisocyanate. These foams are typically used in the construction industry and in packaging, industrial insulation, appliances and transport applications.

Sorbitol, sucrose, pentaerythritol and other polyalcohols or polyamines are used for the preparation of PEP for rigid foams where structures with high cross-linking density are required, while glycerol is mainly used for flexible foam production. Some PEP producers modulate the rigidity of the foam by using blends of sorbitol and glycerol in different ratios.

Simple carbohydrates and polyols have mild metal-complexing properties that are generally not good enough for industrial applications. However, it is sufficient to introduce a carboxylic group to significantly enhance this functionality. Today, two chemicals derived from glucose are used as sequestrant: gluconic acid and glucoheptonic acid.

Gluconic acid is commonly obtained by fermentation while glucoheptonic acid is synthesized from the cyanohydrin of glucose (see Figure 9.2.13).

These organic acids and their derivatives are mainly used as additives for concrete admixtures, in industrial cleaning formulations, metal finishing and agricultural applications. The functionality of gluconic and glucoheptonic acids in these applications, except for concrete, is based on their complexing ability towards polyvalent metal ions such as calcium, magnesium, iron, zinc, *etc.*

[ii] Hydroxyl number is defined as the number of milligrams of KOH that is chemically equivalent to the activity of one gram of the polyol.

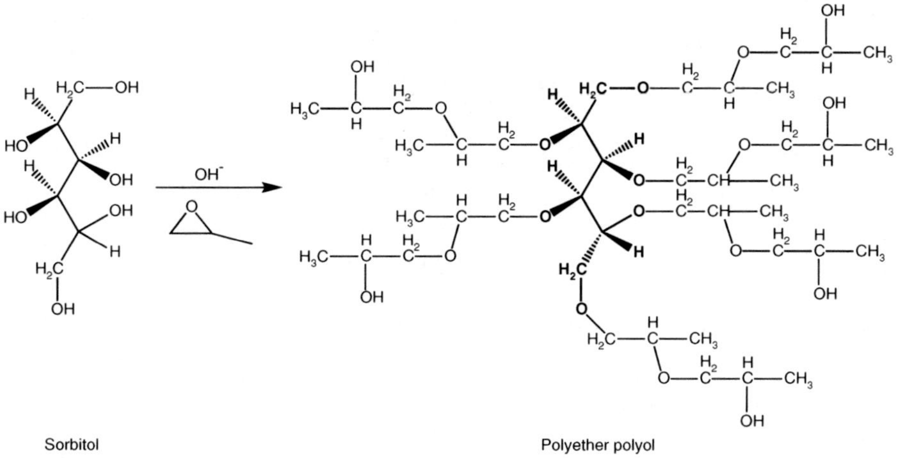

Figure 9.2.12 Polyether polyols synthesis scheme.

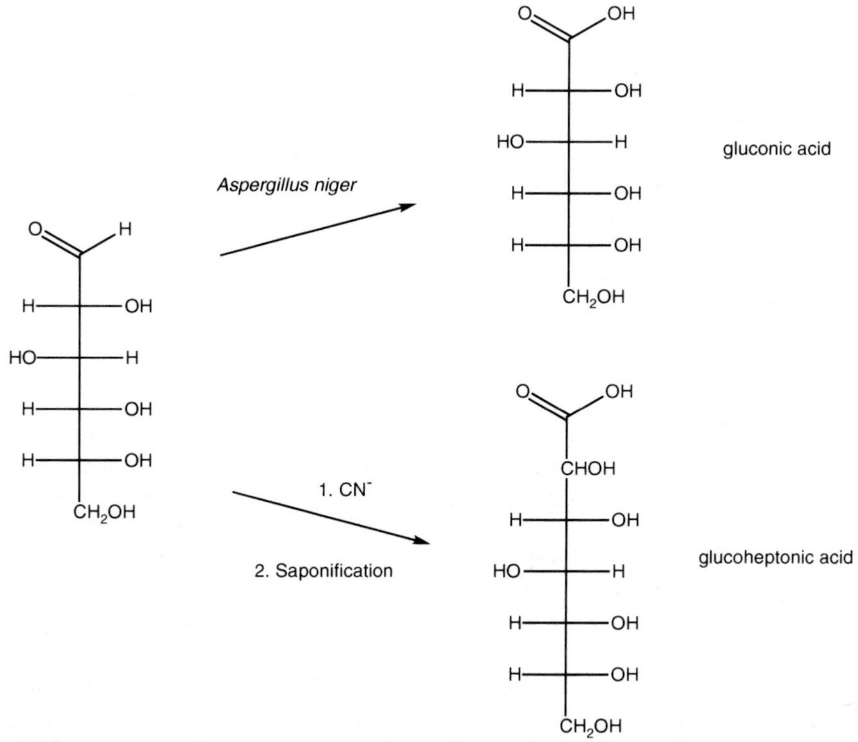

Figure 9.2.13 Gluconic and glucoheptonic acids from D-glucose.

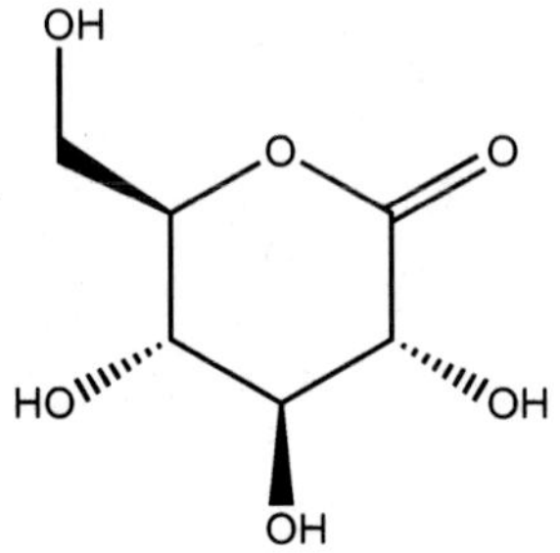

Figure 9.2.14   Glucono delta-lactone.

Differently from citrates and EDTA, they effectively chelate ions in strongly alkaline solutions and they are biodegradable.

Sodium gluconate is also the precursor of Glucono-delta-lactone (Figure 9.2.14), its cyclic internal ether, a product used in the food industry to control pH.

Calcium and ferrous gluconates are used in the treatment of calcium and iron deficiencies as well.

9.2.5   Industrial Binder Applications

One of the most widely known functionalities of starch is its capacity to bind materials together. Its polymeric structure combined with a strong tendency to form hydrogen bonds make starch an excellent adhesive, capable of binding many different materials, such as paper, wood, coal, *etc*.

Starch has been used as a natural adhesive for centuries, and even now in the presence of synthetic alternatives, it holds a significant market share in this application.

Coal Briquettes

During the production of coal briquettes, coal dust is mixed with starch and water before heating and compression into an ovoidal shape. Once gelatinized, starch provides structural support to the briquette shape and ensures that the structure is maintained during transport of the briquette to the oven on the conveyor belt. Once the briquette is dry, starch provides another interesting functionality that can be appreciated during the actual use of the briquette. When starch is exposed to high temperatures (as during combustion) a decomposition process starts. Among the possible decomposition pathways, one leads to the formation of a carbon skeleton that no longer burns but acts as a supporting structure for the briquette, increasing exposure of the coal to the air flow. This results in a better combustion and glowing of the briquette.

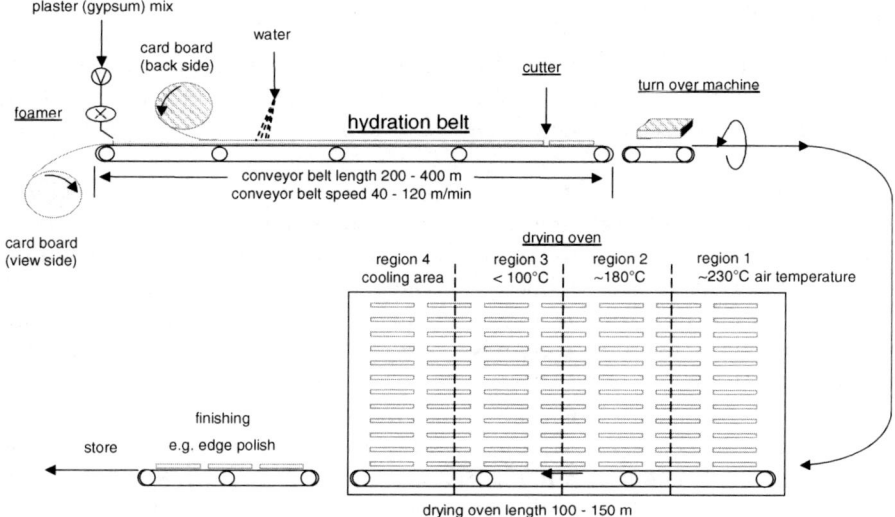

Figure 9.2.15 Gyspum board production process.

Gypsum Boards

Gypsum boards or plasterboards are typical materials used to build light walls. The production process is based on gypsum hydration technology. At the very beginning of the process de-hydrated calcium sulphate (gypsum) is converted into $CaSO_4 \cdot 0.5H_2O$ by heating. Once dehydrated, gypsum is mixed with starch, a foaming agent and other additives. Gypsum foam is then obtained and spread onto cardboard running on a conveyor belt (see Figure 9.2.15). $CaSO_4 \cdot 0.5H_2O$ starts the hydration and the crystallization process immediately after coming into contact with water to form $CaSO_4 \cdot 2H_2O$. The new crystals have a needle-like structure and interlock with each other creating a solid, firm structure. The crystallization of $CaSO_4 \cdot 2H_2O$ generates heat that helps starch gelatinize and migrate towards the cardboard.

During the drying phase, starch completes the gelatinization process and concentrates in the interface between gypsum and paper, where it functions as adhesive between the two materials.

9.2.6 Paper and Board Applications

The use of starch in papermaking is probably one of its oldest non-food applications. An old Egyptian paper document dating from 3500 BC shows that starch was already used for sizing. Starch as a sizing agent was also used in China, as documented by a paper document dating from AD 312 that was sized with starch.

Wet-end, Surface Sizing and Coating

After fibres and fillers, starch is still the third most important raw material for paper making. Various types of modified starches are applied throughout the papermaking process from the wet-end *via* the size- or film-press to coating and paper converting.

The papermaking process (Figure 9.2.16) consists essentially in continuously forming, dewatering, pressing and drying a web of cellulose fibres.

At the beginning of the process, a dilute (99% water) pulp slurry containing fibres, fillers, optical brightening agents and other additives is laid down to a wire screen to form a very loose bed of fibres. This is the wet-end part of the process in which cationic starch is used to impart the necessary strength properties for the paper web to be formed by associating with the anionic *i.e.* negatively charged cellulose fibres and inorganic fillers.

The next phase consists of the removal of the water by hydrodynamic and suction devices. When non-cationic starches were still used during this process, the majority of it (70% according to some authors) would be lost to the wastewater. On the other side, when a suitable cationic starch is used, then over 90% of it stays with the paper web.

The remaining water is then removed in a mechanical press device (press section) working like a mangle. After that step the paper has a water content of approximately 50%. This water cannot be mechanically removed but has to be evaporated by a system of steam-heated rolls (first drying section) to achieve a water content of 1–7%. At this point some paper grades are ready *e.g.* newsprint. Other grades of paper that need a further strengthening are impregnated with a solution of starch alone or in combination with other functional chemicals in a size- or film-press. As a result the surface of the paper becomes better suited for printing and the paper strength is increased. A variety of

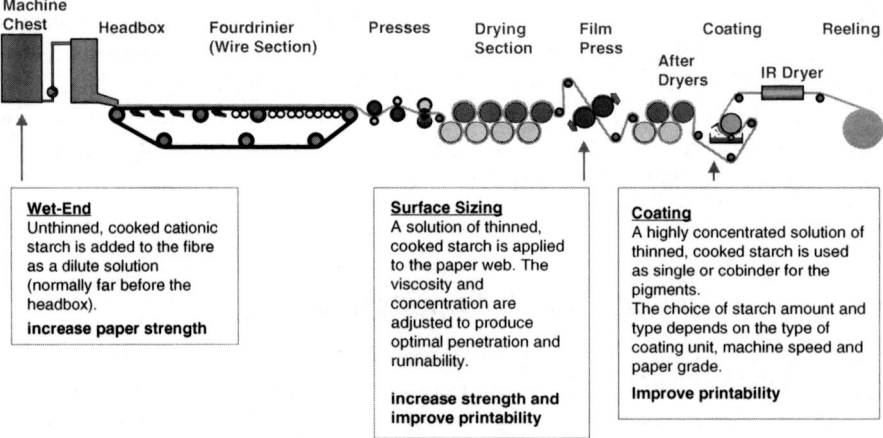

Figure 9.2.16 Papermaking process scheme.

starches can be used for this part of the process, the common denominator being the continuous film-forming ability to avoid cracks, a low viscosity to be able to operate at relatively high dry substance, and the whiteness of the product. Commonly used products here are: oxidized starches, hydroxypropylated starches and other stabilized starches.

After the surface-sizing section, the paper is dried again by an array of steam-heated rolls (after-drying section) to a moisture content of 6–8%. After an additional smoothing step calendar grades like *e.g.* office papers and packaging papers are ready for use.

For higher demands with respect to gloss and printability an additional finishing step is performed in which a coating colour consisting of inorganic pigments like calcium carbonates and clays together with binders and further functional additives is applied to the paper surface. This step can be performed only once on one or both sides of the paper or repeated several times in a sequence. After each application of a coating colour a drying step is necessary that mostly is done by a sequence of infrared, hot air and contact drying. At the end these papers are calendered to develop the desired gloss level. In coating formulations starch may be used alone or in combination with synthetic lattices to bind the pigments to each other and to the surface of the paper. The starches used for coating applications must fulfil more stringent requirements with respect to rheology, stability and paste solids than surface sizing starches.

Corrugated Board

A corrugated board is composed by a fluting medium (the wavy part of corrugated board) glued between two layers of linerboard to achieve a very strong structure. Very often, the paper used for the boards is recycled paper that has been surface sized with starch to reach a good strength. The glue used to bind the tips of the flutes to the linerboards is made of starches, caustic soda and borax. Special equipment is used to automatically prepare glues with the right viscosity and gel point for the variety of corrugated board qualities to be produced. This glue is applied to the tips of the flutes that are then contacted with the liner which is subsequently taken over a heated plate. Upon heating, the starch gelatinizes and forms a strong bond between the flutes and the linerboard. These steps are repeated on the other side of the fluting to create the known sandwich structure of corrugated board.

9.2.7 Outlook

Starch and its derivatives have proven to be very versatile materials for several different applications in many sectors. With relatively simple carbohydrate chemistry it has been possible to have access to carbohydrates with superior functionalities. There is still considerable potential to generate innovative solutions based on starch but this requires a better understanding of the starch chemistry and a strong cooperation between the agro-industry and the chemical industry. This cooperation is necessary for the starch industry to overcome the challenges associated with the complex nature of starch and is beneficial also to the chemical industry in its search for new and sustainable products to exploit.

Ambitious projects are already taking place and are expected to be successful in a few years. A few of them are mentioned below.

Isosorbide and Isosorbide Derivatives

Cyclization of sorbitol in suitable conditions results in the formation of the 4,8-dihydroxy-*cis*-2,6-dioxa-bicyclo[3.3.0]octane, also known as 1,4:3,6-dianhydro-D-sorbitol or Isosorbide (Figure 9.2.17).

This product is relatively expensive and is currently used in modest quantities mainly for pharmaceutical applications as a chemical intermediate.

However, Isosorbide has significant potential in the plastics sector as a new monomer for polymerization. For example, a modified poly-ethyleneterephthalate (PET) can be synthesized when Isosorbide is incorporated during polymerization.[19] Isosorbide's structural rigidity increases the stiffness of the PET chains and raises the glass transition temperature of this plastic giving better performances than conventional PET and requiring less material to produce an equivalent article.

In the field of plasticizers for PVC, some isosorbide esters give functional performance equivalent to some of the most widely criticized phthalates, such as DEHP.[20] As phthalates are used in a wide number of applications (PVC plasticizer, cosmetics, inks, lubricants, *etc.*) there is significant potential for isosorbide derivatives as an alternative to phthalates.

Dimethyl isosorbide (DMI) is a colourless liquid with excellent solvent properties. DMI has extremely low toxicity and facilitates the penetration of active ingredients through the epidermis. It has shown interesting properties in the delivery of substances like lactic acid, hydrocortisone, vitamin C and hyaluronic acid. It is a solvent with clear potential for cosmetic and pharmaceutical applications.

The challenge that is common to all these projects is the establishment of an industrial process that give access to a cost-competitive Isosorbide.

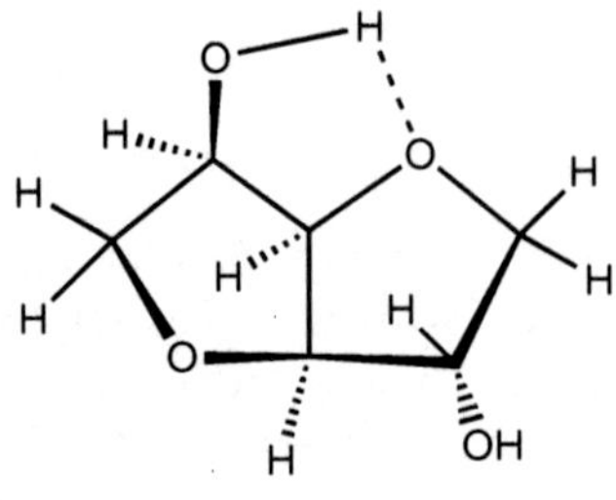

Figure 9.2.17   Isosorbide structure.

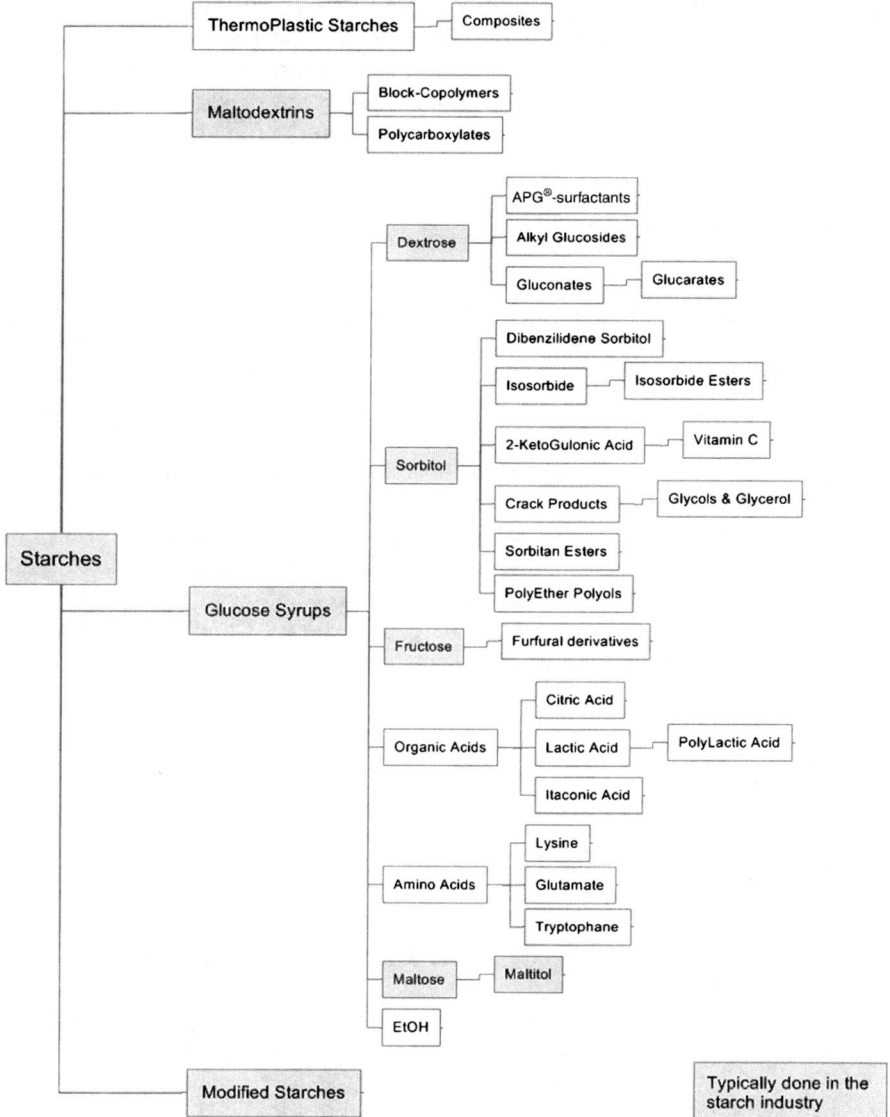

Figure 9.2.18   Starch-based product family tree.

New Molecules by Fermentation

Fast increasing crude oil prices and projected dwindling output mean that fermentation processes become more and more attractive. There are indeed a number of industrial projects aimed at producing relatively simple chemicals by fermentation, notably methionine, acrylic acid and succinic acid.

As these molecules do not require a complex chemical synthesis, one of the main challenges will consist in achieving fermentation processes that are economically sound. In addition to the economics, those processes may be preferable when they lead to substantial reductions in the production of CO_2 and the use of non-renewable raw materials, in line with the fundamental tenets of a "sustainable economy".

Starch as Part of the Biorefinery Concept

Renewable raw materials and their derivatives have been integrated in the industrial world for a long time. With a relatively small chemical effort, it has been possible to modify their structures and their functionalities in such a way that they became accepted even in the chemical industry as building blocks. Nevertheless, the volume of starch-derived products used for the technical industry is still very low when compared to the oil-based products.

In the last decade, the increasing cost of oil-based products has stimulated many efforts in finding sustainable alternatives. In that perspective, starch is a very promising material because it can be used as a source for several chemicals (see Figure 9.2.18). However, significant progress has still to be made in many disciplines including chemistry, biotechnology and microbiology, before we can handle the complexity of renewable raw material with the same efficiency that has been set as standard by the petrochemical industry.

References

1. R. K. Will, U. Loechner and K. Yokose, ed., *Hydrocolloids*, SRI Consulting, CEH Marketing Research Report, 2007.
2. Cargill internal data.
3. AAF data.
4. *Breaking the link between food and biofuels*, http://www.card.iastate.edu (retrieved 22.07.08).
5. European Commission Directorate-General for Agriculture and Rural Development, Prospects for agricultural markets and income in the European Union 2007–2014, 2008.
6. D. J. Manners, *Carbohydr. Res.*, 1981, **90**, 99–110.
7. A. Buléon, P. Colonna, V. Planchot and S. Ball, *Int. J. Biol. Macromol.*, **23**, 85–112.
8. J. S. Chronakis, *Crit. Rev. Food Sci.*, 1988, **38**(7), 599–637.
9. D. Howling, in *Starch Hydrolysis Products – Worldwide Technology, Production and Applications*, ed. F. W. Schenck and R. E. Hebeda, VCH, Weinheim, 1992, pp. 277–317.
10. Frost & Sullivan, Strategic Analysis of the European Food Starch Markets, 2007.
11. Glucobay® is a registered trademark of Bayer HealthCare AG.
12. W. Puls, U. Keup, H. P. Krause and G. Thomas, *Diabetologia*, 1977, **13**, 426.

13. Mater-bi® is a registered trademark of Novamont.
14. C. Bastioli, *Renewable Raw Materials For Industry*: Contribution to Sustainable Chemistry Symposium, Brussels, Oct 2007; C. Bastioli in *Degradable Polymers*, ed. G. Scott, Kluwer Academic Publ., Dordrecht, 2nd ed., 2002, p. 133–161.
15. M. Schlechter, BCC Research Report: PLS025C, 2007, Wellesley, Massachussets, USA.
16. H. Koch, F. Krause, R. Steffan and H. U. Woelk, *Die Stärke – Starch*, 1983, **35**(9), 304–313.
17. S. N. Bizzari, ed., *SRI Consulting*, CEH Marketing Research Report "Formaldehyde", 2004.
18. R. Eskuchen and M. Nitsche, in *Alkyl Polyglycosides: Technology, Properties, and Applications*, ed. K. Hill, W. Rybinski and G. Stoll, VCH, Weinheim, 1996, pp. 9–22.
19. D. J. Adelman, L. F. Charbonneau and S. Ung, WO 2003106531, 2003 (Du Pont de Nemours).
20. H. Luitjes and J. C. Jansen, WO9945060, 1998 (ATO-DLO).

CHAPTER 9.3
Industrial Sucrose

STEFAN FRENZEL, SIEGFRIED PETERS, THOMAS ROSE AND MARKWART KUNZ

Südzucker AG, Marktbreiter Straße 74, D-97199, Ochsenfurt, Germany

Sucrose, also called saccharose, or in general sugar, is produced on the industrial scale from sugar beet or sugar cane with a present annual production of 161 Mio mto (centrifugal sugar 2006/2007).[1] Thereby about a quarter of the total sugar produced is derived from sugar beet, three-quarters from sugar cane (Figure 9.3.1). Whilst the beet sugar amount has been quite constant during the last decades, the sugar production from cane has increased distinctly from the 1960s onwards.

Though the total amount of sugar produced increased dramatically during the past decades, today's world market price of sugar is comparable to that 20 years ago. Nevertheless, the price has been fluctuating dramatically (Figure 9.3.2) between 100 and 400 US\$ mto^{-1} (raw sugar) and 160–500 US\$ mto^{-1} (white sugar).

Sucrose is the most available low-molecular-weight carbohydrate but until now was mainly used for nutrition purposes and for the fermentation industry – above all for the production of bioethanol, citric acid and lactic acid – and only to a smaller extent as chemical feedstock.[2] Though the use of renewable resources in biotechnological processes has increased in the last years,[3,4] there is a strong price-driven competition between sucrose and starch as raw materials for fermentative processes being influenced by regional differences.

Due to the predicted shortage of fossil resources within the next few decades, a move from fossil resources towards renewables appears to become more and

RSC Green Chemistry No. 4
Sustainable Solutions for Modern Economies
Edited by Rainer Höfer
© The Royal Society of Chemistry 2009
Published by the Royal Society of Chemistry, www.rsc.org

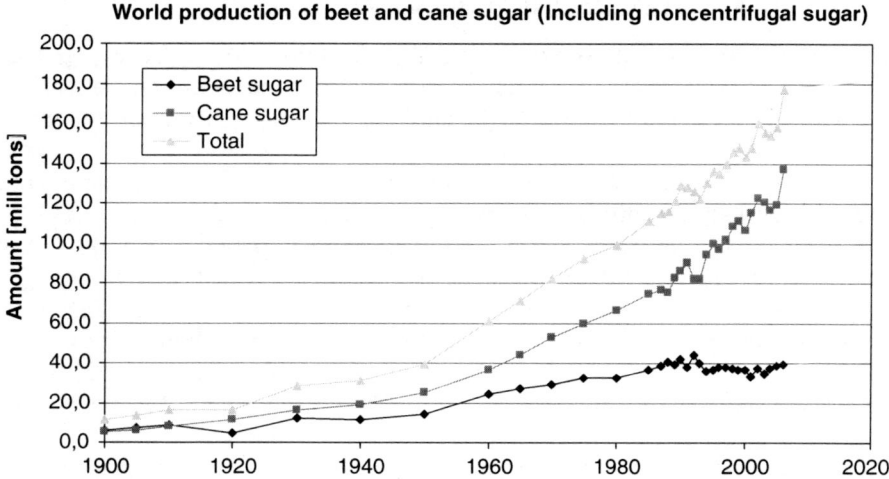

Figure 9.3.1 World production of beet and cane sugar since 1900.[1]

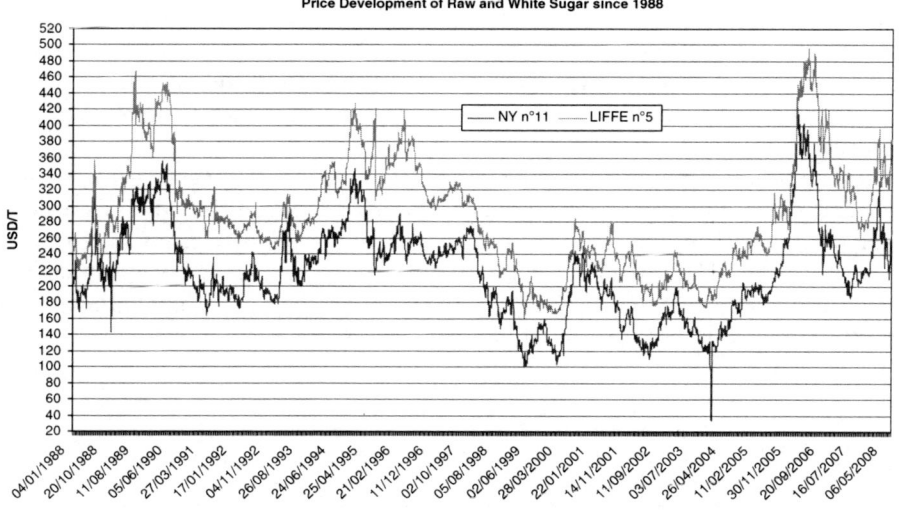

Figure 9.3.2 Price development of raw sugar and white sugar during the last two decades.

more imperative.[5] Thus, sucrose is depicted to be predestined, on renewable sources based organic compounds, for becoming an alternative feedstock for many up to now still petrochemically derived products or product applications. Many approaches to utilize sucrose as raw material for different organic chemicals have been attempted within the last decades but only a few are successful today (Section 9.3.2).[2]

9.3.1 Industrial Production of Sucrose[6-9]

The technology applied to extract, purify and crystallize sucrose and thus finally the overall economy of the process is governed to a large extent by the sucrose companion compounds and the physical texture of the plant material.

The differences in technology for processing sugar beet and sugar cane follow essentially the differences in the monosaccharides content, in the high fibre content of the cane and the physical characteristics (texture) of beet and cane (Table 9.3.1). Sucrose as an acetal is sensitive to acids and increasingly hydrolyzes into glucose and fructose with decreasing pH value. Therefore, for conservation of sucrose it is an advantage to apply alkaline processing conditions, whereas the monosaccharides (glucose and fructose) are stable under acidic conditions but react under alkaline conditions to form acids and colours.

Continuous efforts are made in the breeding and cultivation of sugar beet in order to reduce the non-sugar compounds in the crop, increasing the sugar yield and decreasing the environmental impact. Figure 9.3.3 shows the continuous reduction of the specific mineral fertilizer consumption in the last decades for the sugar beet cultivation.

The application of herbicides for the cultivation of sugar beet could also be reduced very significantly ($2.0 \, kg \, t^{-1}$ sugar in 1975–1980 to $0.7 \, kg \, t^{-1}$ sugar in 2007). Further improvements – *e.g.* reduction of the application of chemicals and primary energy – will be achieved by using new beet varieties including the potential of green biotechnology.

The residues from cane extraction (bagasse) are mainly used to fire steam-generators because their commercial energy value is higher than their feed value and thus a cane-sugar factory is operated independently from an external power supply. This means that certain process design principles, mainly regarding energy-saving technologies, applied in the beet sugar industry are even today less important for cane sugar processing.

Table 9.3.1 Contents relevant to the process engineering and properties of sugar beet and cane.

Content	Beet	Cane
Water content [g/100 g]	75–77	73–76
Dry mass (DM) content [g/100 g]	23–25	24–27
Non-water-soluble compounds [g/100 g]	Marc 4.5–5	Fibre 12–14
Sugar content [g/100 g]	14–20	12–14
Monosaccharides [g/100 g DM]	0.5–1	4–8
Raffinose [g/100 g DM]	0.2–0.5	
Amide content [g/100 g DM]	0.5	0.9
Oxalic acid [g/100 g DM]	0.2	0.1
Aconitic acid [g/100 g DM]		1.5
Other technologically-relevant compounds	Pectin	Dextran

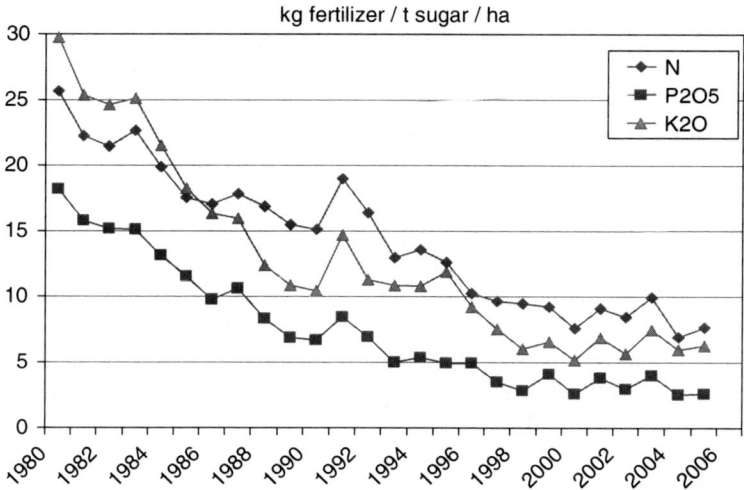

Figure 9.3.3 Trend of the consumption of mineral fertilizer for sugar beet cultivation in southern Germany.

Sugar Manufacture in the Beet Sugar Industry

In Central Europe, harvesting of the sugar beet and thus its processing normally begins in mid to end September and continues until December or January. The dissimilation of the sucrose in stored beets starts directly after harvesting. Therefore, to reach the best possible sugar yield, every effort is made to process the beets as soon as possible after harvesting.

After being washed (for total process overview *cf.* Figure 9.3.5) the beets are sliced into strips (cossettes), which afterwards are fed into the so-called diffusers where about 98% of the sucrose is extracted. High temperatures (65–70 °C) in the diffusion apparatus denature the plant cells and open the cell membranes for improved sucrose transfer. The entire extraction of the sucrose is carried out *via* a counter current flow of the extraction medium, *i.e.* water and the denatured cossettes.

The cossettes leave the diffusion plant after about 70 minutes residence time and are pressed by screw presses to a dry matter content of about 30% followed by a thermal drying in revolving drum driers. The dried cossettes compressed to pellets represent a valuable and storable animal feed.

The extract contains, in addition to the sucrose, further solved or colloidally dispersed components, which in the juice purification stage must be – as far as possible – precipitated or destroyed under alkaline conditions by adding calcium oxide. This is carried out by a multi-stage, continuous treatment with, first, calcium oxide and, second, carbon dioxide. Both processing aids are produced in a separate calcining process in parallel. The excess calcium oxide is precipitated in so-called carbonation vessels with carbon dioxide from the calcining process. All precipitates and deposits are separated in a multi-step filtration. The sludge produced contains lime, phosphate and other valuable compounds and is sold as fertilizer (carbonation lime). For technological, ecological or economic reasons the use of limited amounts of other processing aids, like anti-scalants or flocculants, may be necessary. They are used in accordance with good manufacturing practice and applicable food law regulations.

The clarified raw juice is called "thin juice" and contains between 14 and 16% of sucrose. Before crystallization, the sucrose content is enriched close to its solubility limit by evaporation of water in an energy efficient (6–7 effect) multi-stage evaporator station. Steam and thin juice run through a series of apparatus, each part of which is at a gradually lower pressure, until finally an intermediate but storable product, the so-called "thick juice", with 65 to 70% of dry matter content is produced. The beet sugar industry has improved the overall energy efficiency for white sugar production within the last decades significantly (Figure 9.3.4).

The thick juice is fed into evaporator crystallizers in which a further gentle concentrating under vacuum (0.2 bar abs.) is carried out to supersaturate the solution. After adding seed crystals, crystal growth proceeds *via* parallel further evaporation and reduction of progressing supersaturation through crystallization. For mechanical reasons, the crystallization is finished at about 50% crystal content.

The crystal magma is fed into a centrifuge in which syrup (mother liquor) is separated from the sucrose crystals by centrifugation. The resulting white sugar

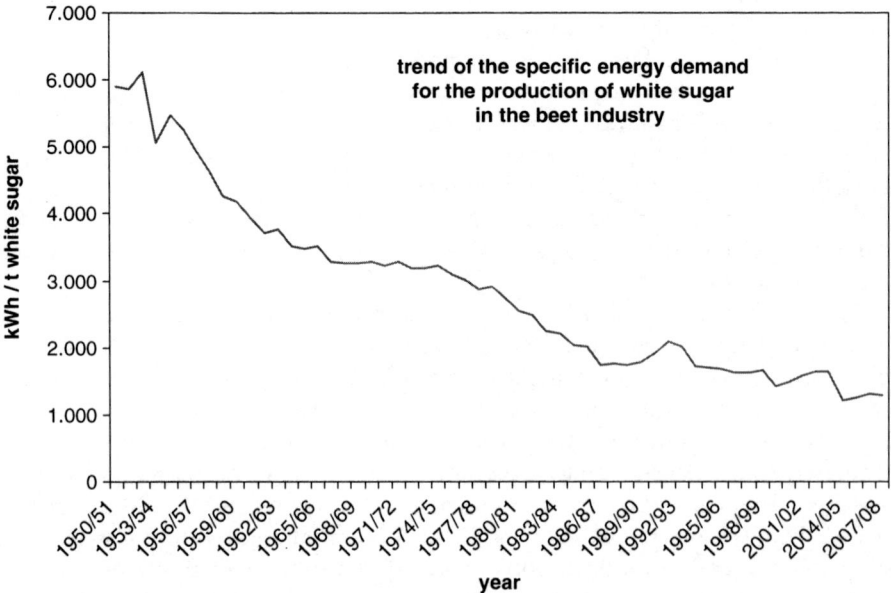

Figure 9.3.4 Specific energy demand for the production of white sugar in the beet industry.

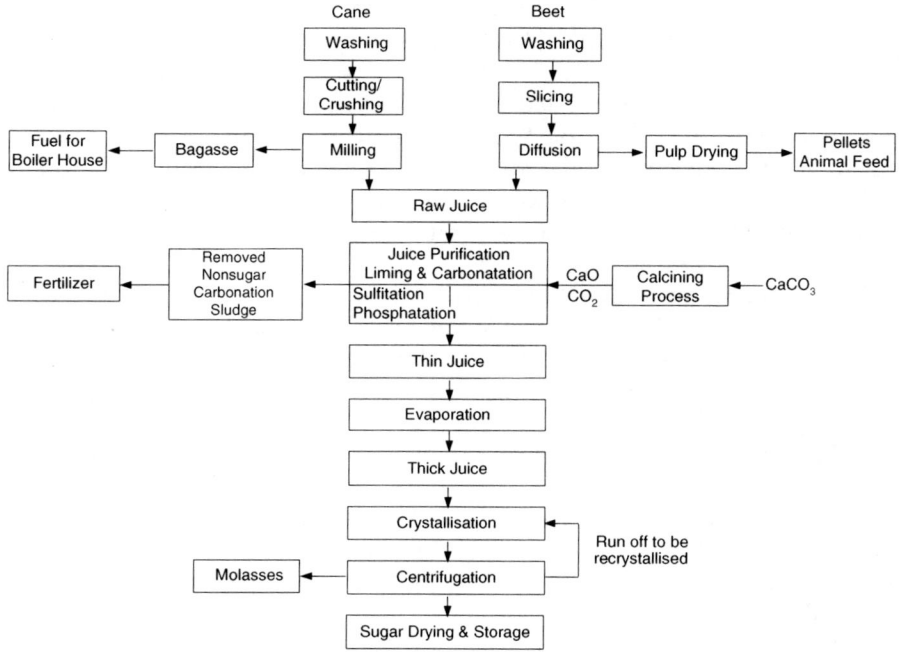

Figure 9.3.5   Process for cane and beet sugar production.

with a residual moisture content of about 0.5% is dried and cooled for the subsequent storage.

The syrup (mother liquor) separated in the centrifuge still contains sucrose which has to be recovered in further crystallization steps. In this way, there is a multi-stage process of crystallization yielding a final mother liquor (known as molasses) from which the sucrose cannot be economically further crystallized.

Depending on the economy – *i.e.* price of sugar crystallized *vs.* price of sugar in the molasses – in some regions of the world further extraction of sucrose using chromatography is applied.

Sugar Production in the Cane Sugar Industry

Storage of sugar cane is associated with a more intense hydrolysis of sucrose than in sugar beet. The cane must be processed with a minimum of delay after harvesting. The sucrose is stored in the innermost pith of the sugar cane and must be made accessible to extraction by destruction of the highly resistant outer skin. The cane is cut into short pieces by rotating knives and further crushed in so-called shredders (usually five). After shredding, the cane passes a multi-stage pressing and extracting – by adding water – process to separate the sucrose containing extract from the bagasse. The bagasse (25–30 kg per 100 kg cane) produced is normally adequately used to satisfy the energy demand of a cane sugar mill with bagasse-fired boilers.

With approximately 105–110% of the initial weight of the sugar cane, the extract is a slightly acidic raw juice which is characterized by a high content of invert sugar (some ten times higher than in sugar beet extracts) and high molecular companion compounds (*e.g.* dextran and starch). Unlike the classic juice purification process in the beet sugar industry, the invert sugar cannot be destroyed using strong alkaline conditions because high by-product and colour formation would be the consequence prohibiting commercial white sugar production. Thus high pH values and high temperatures have to be avoided in the following processing steps.

For producing raw cane sugar precipitating of non-sucrose juice constituents by the addition of calcium hydroxide while heating is widely practised. Various versions of this method are applied (cold and hot liming, fractionated liming with double heating). Phosphate content of the juices plays a significant role. The dehydrated and flocculated hydrocolloids in the raw juice are well wrapped by calcium phosphate, which improves their sedimentation ability as well as the filterability.

The production of juices for the direct manufacture of white sugar requires a more thorough elimination of non-sucrose compounds. This is achieved in the sulphitation process by the addition of milk of lime and sulphur dioxide and removal of calcium sulphite.

The purified juice (thin juice) with a dry mass content of some 15% is concentrated in a multi-stage evaporator station to 60–70% (thick juice).

Crystallization of the sucrose in the thick juice is carried out in the same way as in the beet sugar industry (see above). The high proportion of invert sugar has a marked effect on the residual sugar content in the cane molasses: the purity of the cane molasses, at about 40% (true purity), is distinctly lower than in the beet molasses, because the invert sugar content (up to 20%) reduces the solubility of sucrose and thus facilitates the crystallization of the sucrose.

Usually in sugar cane processing factories white commercial sugar – so-called plantation white – is only produced for domestic consumption; most often, raw sugar is produced. Raw sugar is traded on the world market and loaded and transported in bulk to refineries overseas. Raw sugar, because of its often critical hygienic status, is generally unfit for direct human consumption and therefore requires further processing into commercial sugar.

Today, raw sugar and bioethanol production using the mother liquors of sugar crystallization are executed very often in parallel.

9.3.2 Chemistry of the Sucrose Molecule

Sucrose (β-D-fructofuranosyl α-D-glucopyranoside, Figure 9.3.6) is a disaccharide of anomerically linked D-glucose and D-fructose, thus being a non-reducing sugar. Under acidic conditions hydrolysis of sucrose into these two building blocks occurs.

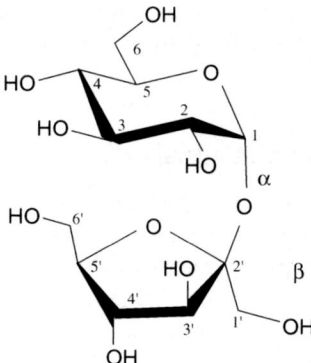

Figure 9.3.6 Sucrose (β-D-fructofuranosyl α-D-glucopyranoside).

Sucrose could be valorized to promising organic chemicals by three main material transformation pathways.

1. *Degradation of the sucrose framework:* Reactions or reaction cascades, including dehydration/rehydration, ring opening/closure, bond scission/formation lead to structures being of known value from petrochemically derived compounds, but are based on the renewable feedstock sucrose, the most prominent representatives being bioethanol and 5-hydroxymethyl furfural (HMF).

2. *Derivatization by maintaining the sucrose skeleton:* The eight OH groups of sucrose could be utilized for chemical reactions, thereby preserving the original sucrose structure as such. Three of those OH groups are primary (6, 1′, 6′) and five of them secondary ones. Nevertheless, their chemical differentiation and selectivity is not that simple like the different reactivity order of primary and secondary OH groups in other polyhydroxy compounds. In aqueous solutions hydrogen bond connections between OH-2 and OH-1′/OH-3′ clearly influence these OH groups by electron withdrawing effects. The OH-2 group is thereby the most acidic one. The type of reaction and the applied conditions – like the solvents – have a considerable impact on the selectivity, the degree of substitution and the regiochemistry.[2,10]

3. *Rearrangements/Reactions maintaining carbohydrate structure:* Inversion of sucrose to D-glucose and D-fructose is the simplest example, others – more complicated ones – are the enzymatic rearrangements of the sucrose building blocks (D-glucose and D-fructose) into more complex carbohydrates.

The following examples give an overview about possible material transformations of sucrose leading to valuable chemical intermediates for industrial bulk-scale production, some of them being industrially implemented today. The three mentioned reaction pathways, with and without retention of the sucrose skeleton and reactions leading to new carbohydrate structures, are thereby addressed.

9.3.2.1 Basic Organic Chemicals by Sucrose Degradation

Degradation of sucrose under elevated temperature and pressure conditions leads to a variety of chemical compounds either being known from petrochemicals or resembling petrochemicals in their structure having the potential of replacing non-renewables.[11] Also, fermentation of sucrose leads to "degradation" products. The most important example is bioethanol.

9.3.2.1.1 Bio-ethylene

One example for a successful effort to utilize sucrose as feedstock for the manufacture of bulk organic chemicals is the production of "green" ethylene from sugar *via* bio-ethanol. In Brazil, the company Braskem has announced "the first certified linear polyethylene in the world made from 100% renewable raw material" and planned its production on a 200,000 mto scale for 2010.[12] In a joint venture between SantelisaVale/Crystalsev and Dow Chemical the production of 350,000 mto of polyethylene is planned to start in 2011.[13]

Solvay Indupa announced the expansion of their manufacturing plant of polyvinyl chloride (PVC) in Santo Andre (Brazil) to a total amount of 360,000 mto per year. Thereby for this capacity extension the required monomer feedstock vinyl chloride should also be based on bio-ethylene (based on bio-ethanol sourced from Copersucar) and thus finally on sucrose and salt.[14]

The capacity of sugar-derived bio-ethylene seems low compared to the total annual ethylene production of 100 Mio mto (2000) but enough to show the capability of utilizing sucrose as feedstock for organic chemicals thereby opening the door for the whole ethylene-derived C2-chemistry. Bio-ethylene is particularly interesting for "decentralized" biorefinery concepts with good access to sucrose or other carbohydrate feedstock where crude oil-based integrated production facilities are not on-hand. There are far more syntheses possible for "petrochemicals" based on renewables. The long-term development of naphtha *vs.* sugars will decide the industrial implementation. Well-known examples are *e.g.* acrylic acid and its derivatives. A very promising new synthesis of methacrylic acid based on sucrose or other carbohydrates is the recently developed fermentation of sugars to 3-hydroxyisobutyric acid followed by catalytic dehydration.[207]

9.3.2.1.2 Access to 5-(Hydroxymethyl)furfural (HMF) and its
            Derivatives

HMF is an example for a potential bulk chemical obtainable from sucrose. The chemical structure of HMF, at first glance, does not immediately allow the

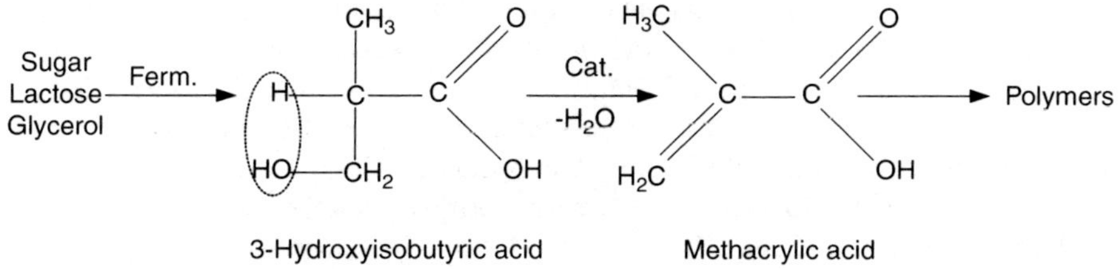

3-Hydroxyisobutyric acid   Methacrylic acid

Scheme 9.3.1 Synthesis of methacrylic acid via 3-hydroxyisobutyric acid by fermentation.

recognition of its carbohydrate origin. Due to being a furanic compound implemented with an alcoholic and an aldehyde-functionality, a petrochemical root seems to be much more feasible. But at least its sources are hexoses. At elevated temperature and under acidic catalysis, hexoses tend to eliminate three moles of water, thereby forming HMF (Scheme 9.3.1). In aqueous solution the dehydration reaction is subsequently followed by a re-hydration step leading to levulinic and formic acid. In addition, more intermediates, side products and the formation of coloured soluble or insoluble polymeric compounds accompany the dehydration reaction.[15,16]

The chemical structure of HMF places it into a unique position between petrochemical and carbohydrate chemistry.[17] Being regarded as a "secondary" base chemical derived from renewable recources,[18] HMF is evaluated widely as a potential replacer of existing petrochemically derived organic compounds. Many attempts for a successful HMF synthesis have been made in the last decades. Thereby, D-fructose has proven to be the feed of choice. In order to overcome the by-product formation and thus to optimize the HMF yield, many condition variations (catalyst, temperature, time, concentration and solvent) have been investigated and carefully balanced out for the D-fructose dehydration to HMF.[15–16,18] Higher, almost quantitative yields have been achieved by use of the solvent DMSO.[15,16,19–21] But DMSO is ecologically unfavourable and furthermore difficult to separate from HMF. The sub- and supercritical solvent system of acetone/water (9/1) proved to be advantageous over water as a sole solvent and was especially successful, reaching high HMF selectivity up to 77%; water/acetone is volatile, and therefore easy to separate from HMF.[22–25] Ionic liquids under acid or metal chloride catalysis allow HMF-yields higher than 80% – almost without any formation of by-products like levulinic acid[26] (Scheme 9.3.2). When using chromium(II) chloride as catalyst even D-glucose could be used as feedstock since this catalyst is effective for the *in situ* isomerization of D-glucose to D-fructose before dehydration takes place to yield HMF in 70%.[26]

The upscaling of HMF synthesis was successfully executed in pilot plants.[27]

However, due to dehydration the mass balance of this reaction appears quite unfavourable. Even when considering a favourable average molar yield of 70%, only 490 g of HMF could be obtained from 1 kg of fructose.

By basic chemical operations HMF could easily be transformed into organic chemicals with large industrial potential.[5,28] Especially the bi-functionalized derivatives, the 1,6-diol, the 1,6-diamin and the 2,5-dicarboxilic acid (FDCA) (Scheme 9.3.3) resemble petrochemically derived compounds like alkyldiols, hexamethylene diamine, adipic and terephtalic acid and have the potential to replace them for producing polyesters and polyamides.[10,29]

Despite the tremendous efforts in the elaboration of pathways for an optimized HMF production and the broad variety of already gained potential HMF-based chemical derivatives, HMF is not industrially manufactured until now. This seems to be driven by economic reasons since the costs for HMF production even at large scale are estimated at minimum 2 € kg^{-1} and thus still much more expensive than the production of petrochemical bulk chemicals.[3-4,23]

Besides HMF, the rehydration product levulinic acid is also regarded as a potential biomass-derived organic compound.[28,31,32] Large volume chemicals can be prepared from levulinic acid by chemical conversion, among them methyltetrahydrofuran and levulinate esters (fuel additives), delta-amino-levulinic acid

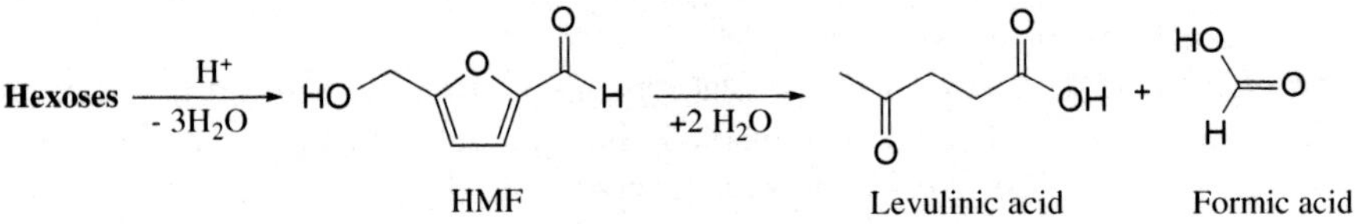

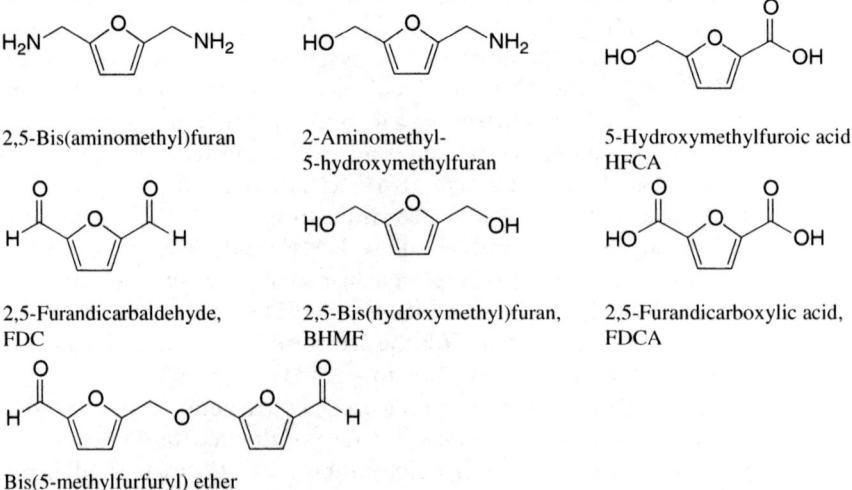

Scheme 9.3.2   Formation of HMF levulinic acid from hexoses.

2,5-Bis(aminomethyl)furan

2-Aminomethyl-
5-hydroxymethylfuran

5-Hydroxymethylfuroic acid,
HFCA

2,5-Furandicarbaldehyde,
FDC

2,5-Bis(hydroxymethyl)furan,
BHMF

2,5-Furandicarboxylic acid,
FDCA

Bis(5-methylfurfuryl) ether

Scheme 9.3.3   HMF-derived building blocks.[5,28,30]

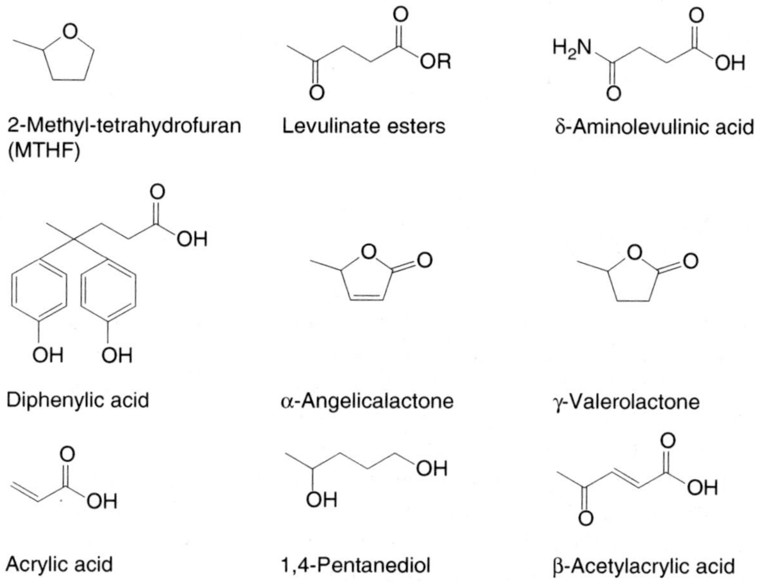

2-Methyl-tetrahydrofuran (MTHF)   Levulinate esters   δ-Aminolevulinic acid

Diphenylic acid   α-Angelicalactone   γ-Valerolactone

Acrylic acid   1,4-Pentanediol   β-Acetylacrylic acid

Scheme 9.3.4 Levulinic acid derivatives of industrial potential.[28,31,32]

(herbicide) and diphenolic acid (replacer for bisphenol A, polycarbonates). For these and other chemical compounds levulinic acid is suggested as potential starting material (Scheme 9.3.4). Like HMF, furfural is accessible by dehydration of pentosans. An important source therefore is the pentosan-rich bagasse of cane. Current world production for furfural is about $250{,}000\,t\,a^{-1}$. It is processed to important derivatives such as furfuryl alcohol, furoic acid and tetrahydrofuran.[33]

9.3.2.1.3 *1,2 Propylene Glycol and Other Polyhydric Alcohols*

Ethylene glycol, 1,2-propylene glycol, glycerol and other higher alcohols are accessible from sucrose either chemically by partial hydrogenation or by bio-catalysis.[18,34–36] Especially 1,2-propylene glycol is appreciated for the manu-facture of unsaturated polyesters de-icers, antifreeze and brake fluids as well as in cosmetics and pharmaceutical products.[37]

A process claimed by BASF describes the use of a cobalt/copper/manganese catalyst for the hydrogenation of an aqueous sucrose solution thereby showing the opportunity of influencing the composition of the product mixture by properly maintaining the reaction temperature (230–280 °C) and the hydrogen pressure (250 to 300 bar).[38,39] Under optimal conditions the reaction favours essentially the formation of 1,2-propylene glycol (50 to 65 weight%) and ethylene glycol (20 to 25 weight%) accompanied by lesser amounts of 1,2-butylene glycol (5 to 7 weight%) and hexane-1,2,5,6-tetraol (3 to 10 weight%) (Scheme 9.3.5).

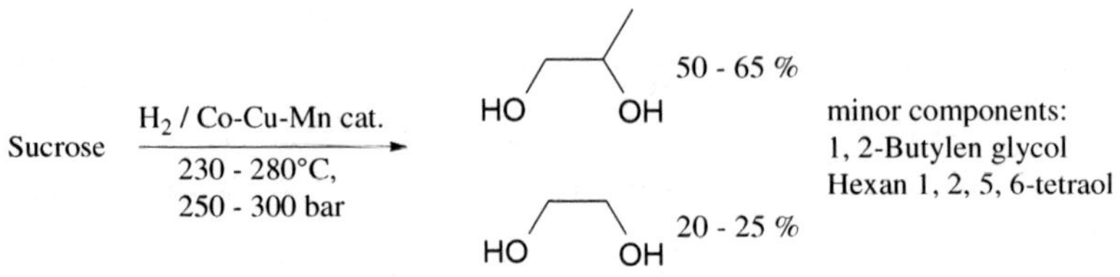

$$\text{Sucrose} \xrightarrow[\substack{230 - 280°C, \\ 250 - 300 \text{ bar}}]{\text{H}_2 \text{ / Co-Cu-Mn cat.}}$$

HO   OH   50 - 65 %

HO   OH   20 - 25 %

minor components:
1, 2-Butylen glycol
Hexan 1, 2, 5, 6-tetraol

Scheme 9.3.5 Conversion of sucrose to 1,2-propylene glycol and other di- and polyhydric alcohols.[38,39]

On depletion of the petrochemical feedstock, the mentioned manufacturing process based on the renewable feedstock sucrose might become an interesting alternative to the petrochemical pathway *via* hydration of propylene oxide.

In a subsequent reaction 1,2-propylene glycol and other polyols containing vicinal OH-groups could be reacted to aldehydes, respectively ketones, by acid-catalyzed dehydration and subsequent intramolecular rearrangement according to the pinacol-pinacolon-rearrangement (conc. sulphuric acid) or preferably under improved mild conditions using catalytic amounts of acetic acid in water by applying high temperature and pressure.[18]

9.3.2.2 Sucrose-derived Products of Industrial Relevance Maintaining the Sugar Skeleton

Chemical modifications of sucrose under preservation of the sugar skeleton fortunately have succeeded in several products with important food and non-food applications – among them the following examples:

9.3.2.2.1 Sucralose

Studies on the selective replacement of hydroxyl groups in the sucrose molecule by halogens led to a variety of halogenated sucrose derivatives which partly showed a remarkably enhanced sweetness compared to sucrose itself.[40–43] From these compounds finally sucralose (4,1',6'-Trichloro-4,1',6'-trideoxy-galactosucrose) was evaluated to show the most convenient physico-chemical properties concerning the quality and intensity of sweet taste, good solubility in water (for use in beverages) and attractive stability of the glycosidic linkage under acidic conditions. Dependant on the field of application, sucralose is about 400 to 800 times sweeter than sucrose and has been commercialized by Tate & Lyle under the Splenda® brand name[44] up to a scale nowadays of more than 1000 tons per year used in the food industry.

Since its first synthesis, optimizations of the detailed reaction sequence have been carried out during the past decades finally leading to a bundle of patents covering different possibilities to execute this reaction cascade.[45–64] Out of these patents a rather straightforward sucralose synthesis could be assembled (Scheme 9.3.6), the first step of which is a dibutyltin-oxide-supported selective

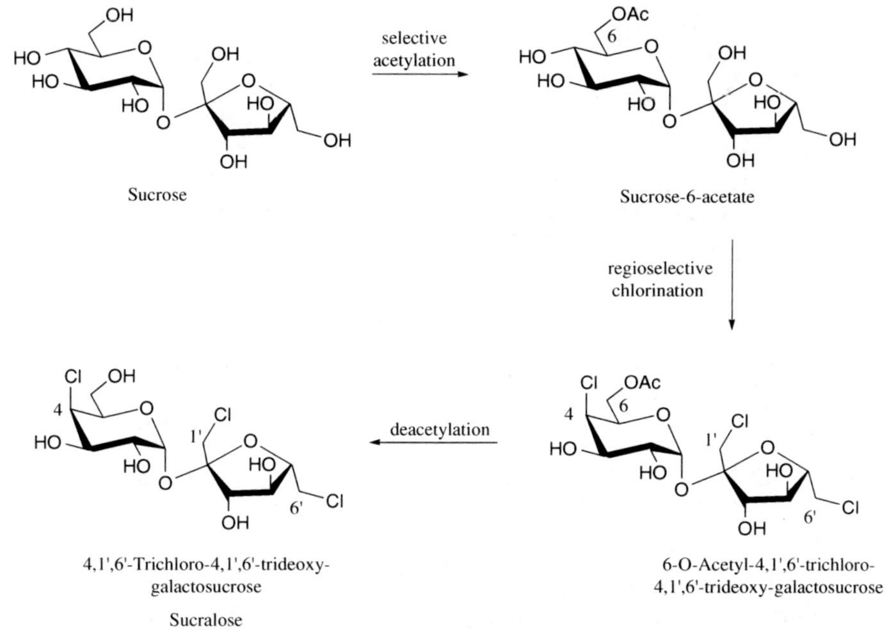

Scheme 9.3.6 Short and efficient reaction pathway from sucrose to sucralose.

acetylation of the 6-OH-group in the glucose part.[45–50] Other selective approaches are reported, like acetylation at low temperature,[51] using orthoester formation[52] or enzymatic techniques.[53–56] Secondly, the chlorination reaction performs with high efficiency and regioselectivity at the desired positions (4-OH, 1'-OH and 6'-OH) when using acid chlorides like $SOCl_2$, $POCl_3$, SO_2Cl_2 or PCl_5 in DMF according to a Vilsmeier reaction type.[57–59] The chlorination step takes place under inversion of configuration at C-4 leading to a chlorinated galactosucrose. In a third step, after removal of the solvent DMF, deacetylation is conducted under basic conditions to yield sucralose, which is purified by crystallization.[60–64] Using this reaction sequence sucralose could be obtained in overall yields of about 40%.

9.3.2.2.2 Sucrose Esters for Food and Non-food Application

Sucrose esters are non-ionic surface-active materials which could be utilized as emulsifiers and fat replacers.[65,66] They are synthesized by esterification linking lipophilic fatty acids to the hydrophilic sucrose. Especially the three primary OH groups of sucrose are more easily substituted through fatty acids compared to the secondary ones, thus leading preferably to mono-, di- and tri-esters. These lower-substituted sucrose esters are well suited for application as food emulsifiers especially in dressings and sauces, confectionery, icings and bakery fillings. Cosmetics are another important field of application.

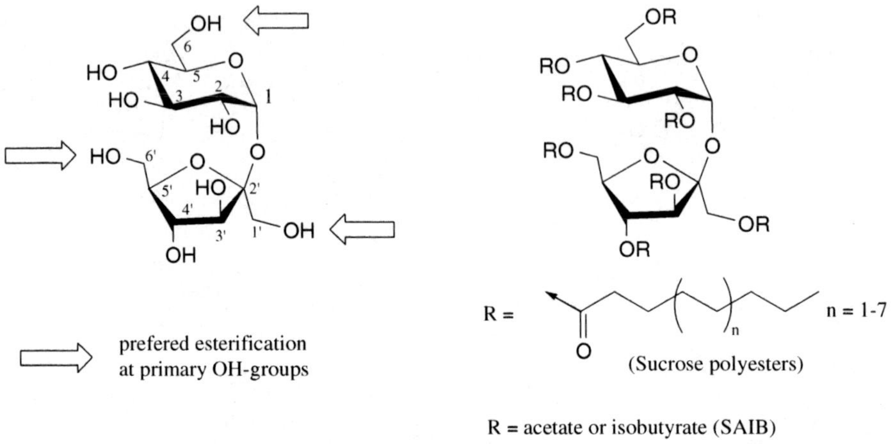

R = acetate or isobutyrate (SAIB)

Scheme 9.3.7 Preferred substitution positions for sucrose fatty esters and fully esterificated sucrose esters.

Sucrose polyesters (Scheme 9.3.7) have characteristic properties like fats and oils but are only partly or even totally non-digestible by the body's lipases and not absorbed by the human digestive system. In 1996, the US Food and Drug Administration (FDA) approved sucrose polyesters with 6 to 8 acyl moieties for certain uses as fat replacers[67] known under the brand names Olestra® and Olean®. They are used in food application *e.g.* in the preparation of dietary fried snacks.

Sucrose fatty esters could be prepared by reacting sucrose with fatty acid esters (interesterification) in a solvent (DMF, DMSO) or emulsion process (propylene glycol).[65–66,68–70] Most interesting – because it avoids hazardous solvents – is the melt process where sucrose and the fatty acid ester (*e.g.* methyl) are reacted together without any solvent.[65]

By these esterification reactions preferably the three primary OH groups of sucrose could be addressed, though the reactivity of the OH groups in sucrose decreases in the order 6-OH, 6'OH, 1'-OH and secondary-OH (Scheme 9.3.7) mixtures of different substitution patterns are obtained. Enzymatic esterifications became attractive due to the more specific reactions and thus better defined products.[65–66,71]

Sucrose acetate isobutyrate (SAIB, Scheme 9.3.7) is a mixed sucrose polyester prepared by reaction of sucrose with acetic and isobutyric anhydrides.[70] It is by far the largest industrially utilized sucrose ester and covers food (weighting agent in beverages) as well as non-food applications (gloss and adhesion improvement in lipstick and nail polish, pigment dispersion).[72,202]

9.3.2.2.3 Polyurethanes

Polyurethanes are formed by polyaddition of diisocyanates with dihydroxy compounds. The use of polyhydric alcohols instead of diols leads to cross-linked polyurethanes which proved to be more rigid: the higher the number of OH functionalities the bigger the impact on the rigidity of the polyurethane. Therefore, sucrose with its eight hydroxyl groups is of great importance in the preparation of rigid polyurethane foams.[73–76] Physico-chemical properties of the polyurethane foam are improved when sucrose is first alkoxylated with alkylene oxide (ethylene or propylene oxide) and the resulting polyether polyols are used as polyhydric components for the polyurethane formation reaction.[73–77]

A number of studies used partially protected sucrose esters to obtain well-defined polymers.[78] Investigations into the regioselectivity of the carbamoylation reaction of sucrose in DMF showed in the case of alkylisocyanates a clear preference for the reaction position at the 2-OH and that for phenylisocyanate the 6-OH group in the glucose-part is the favoured position.[79,80] These regioselectivities could also be proved in the reaction of the respective diisocyanates with sucrose.[80]

Besides sucrose itself, also short chain di- and polyols derived from partial hydrogenolysis of sucrose could be used for the formation of polyurethanes.

9.3.2.3 Sugar Derivatives While Maintaining Carbohydrate Structure

Based on the universally available renewable resource sucrose, a few technically interesting substances maintaining the carbohydrate structure can be produced. Thereby, the bound glucose and fructose units of the sucrose molecule are either split or rearranged (Figure 9.3.7).

The driving force of all these enzymatic reactions is the comparatively high free standard enthalpy of the glycosidic linkage between the glucose and the fructose unit ($\Delta G° = -27 \, kJ \, mol^{-1}$). This enthalpy is nearly in the same order as that of the phospho-anhydride bond within adenosine triphosphate (ATP, $\Delta G° = -31 \, kJ \, mol^{-1}$). ATP acts as an energy source in nearly all biological systems.[149] Thus, for these types of rearrangements sucrose is not only the material educt but also the energetic driving force that means no additional energy is required for the synthesis. This is completely different from most of the common chemical reactions.

Compared to classical chemistry, there are some great advantages of biotransformations especially concerning ecology: enzymatic reactions are normally carried out in water at ambient temperature and pH, without the need for high pressure and extreme conditions.[160] As a general rule, these reactions are very efficient without costly organic solvents and without producing many by-products. Many chemical reactions providing complex molecules require multi-step organic synthesis in different organic solvents often under the use of protecting groups. These groups are introduced into the molecule by chemical modification of a functional group in order to obtain chemoselectivity in a subsequent chemical reaction. Enzymes provide this chemoselectivity innately. This fact plays a key role in waste prevention.[86] The often-discussed disadvantage of low product concentrations for enzymatic reactions is not relevant here, because all enzymes mentioned in Figure 9.3.7 are active even at sucrose concentrations up to 50% (w/w). The low water activity of highly concentrated sucrose solutions also reduces the risk of microbial contamination.

9.3.2.3.1 D-Glucose and D-Fructose

D-glucose and D-fructose are the most easily obtainable carbohydrates from sucrose. Simple hydrolysis of the sucrose glycosidic linkage leads to these two building blocks. The reaction could be performed either under acidic conditions (homogeneous or heterogeneous) or enzymatically.[200] The resulting invert sugar syrup (the optical rotation changes ["inverts"] during the hydrolysis from + to –) contains, depending on the inversion degree, D-glucose, D-fructose and still remaining sucrose and is produced *e.g.* for the German market in an annual 350,000 to 400,000 tons scale. These syrups are mainly used in food, *e.g.* beverages and dairy products.

The heterogeneous acid-catalyzed process by an acidic ion exchange resin in a fixed bed column is established as the state-of-the-art process for the sucrose inversion. A sucrose solution (50–60 weight%) can be completely hydrolyzed continuously at a temperature of 30–45 °C. The inversion product, a mixture of D-glucose and D-fructose, could then be separated chromatographically. Thereby, the batch-wise process is replaced by semi-continuous or continuous simulated moving bed (SMB) technology yielding D-glucose and D-fructose fractions with a purity of minimum 95%.

By use of invertase from *Saccharomyces cerevisiae* (β-D-fructofuranosidase), a highly specific hydrolysis of sucrose could be achieved. However, high enzyme costs and the batch-wise operation mode proved economically disadvantageous over the ion exchange process.

9.3.2.3.2 Isomaltulose (Palatinose™) and Trehalulose

There are five other natural isomers of sucrose (α-D-glucopyranosyl-(1 → 2)-β-D-fructofuranoside, IUPAC name (2R,3R,4S,5R,6R)-2-[(2S,3S,4R,5R)-3, 4-dihydroxy-2,5-bis(hydroxymethyl)oxolan-2-yl]oxy-6-(hydroxymethyl)oxane-3,4,5-triol, CAS-No. 92004-84-7, Table 9.3.2), which, for example, are found in honey in small amounts.[91,130,151,153] These isomers are enzymatically provided by diverse bacteria using sucrose isomerases, also called α-D-glucosyl-transferases (EC 5.4.99.11).

Sucrose formally acts as α-D-glucose donor in the related enzymatic reaction. Because sucrose is an inexpensive and commonly available α-D-glucose donor, there is a high industrial potential for the enzymatic synthesis of the sucrose isomers and their follow-up products. The production of these follow-up products *via* functionalization of sucrose isomers by chemical modification can avoid the problem of the weak α-1,2-linkage of the glycosidic bond and the eight chemically very similar hydroxy groups of the sucrose molecule.[85,118,123,141]

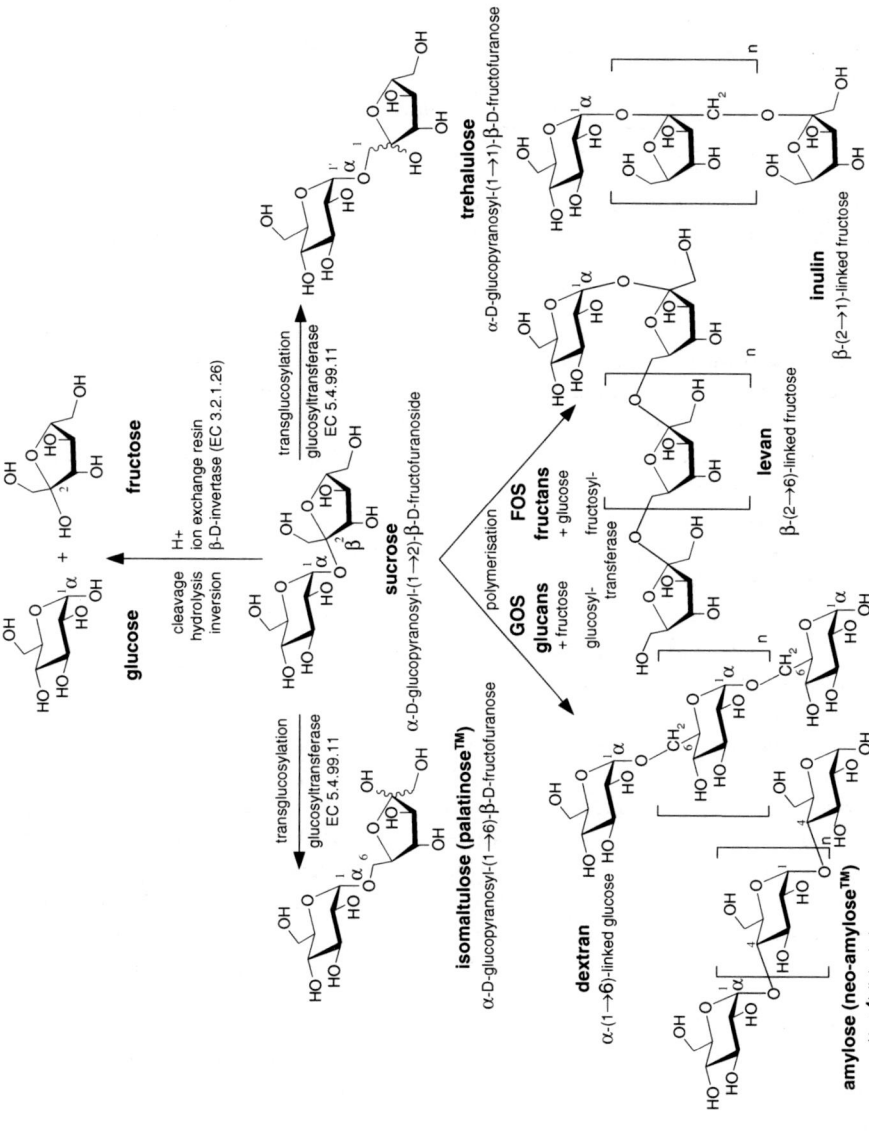

Figure 9.3.7 Reactions of sucrose maintaining the carbohydrate structure.

Table 9.3.2 Natural isomers of sucrose (α-D-glucopyranosyl-$(1 \rightarrow x)$-β-D-fructofuranose).

α-D-glu-$(1 \rightarrow x)$-β-D-fru	Isomer	CAS No.
1	Trehalulose	51411-23-5
2	Sucrose	92004-84-7
3	Turanose	5349-40-6
4	Maltulose	17606-72-3
5	Leucrose	7158-70-5
6	Isomaltulose (palatinose™)	13718-94-0, 15132-06-6

Up to now isomaltulose (Palatinose™) and the related trehalulose have obtained commercial relevance, whereas isomaltulose (Palatinose™) is produced on a large industrial scale.[84,112,113,124,137,142,146,154,157] Most of the isomaltulose (Palatinose™) produced is applied as a sugar replacer or used as a precursor for the production of the sugar replacer isomalt™ at present (see below). An additional advantage for "syntheses based on sucrose" is reached using isomaltulose (Palatinose™) as intermediate. In contrast to the hydroxyl groups of sucrose, the reactive carbonyl group of isomaltulose (Palatinose™) can selectively be functionalized. An example is the reductive amination of isomaltulose (Palatinose™) with hydrogen in the presence of n-dodecylamine, which leads to biodegradable surfactants.[88–89,196]

In addition to the particular main product the different sucrose isomerases also produce the other isomer as a by-product (Table 9.3.2). The ratio of the products obtained from the sucrose isomerase varies with the bacterial strain and the reaction conditions, particularly the temperature and pH. Furthermore small quantities of other products, such as isomaltose (glu-α-1,6-Glu, CAS No. 499-40-1), the trisaccharide isomelezitose (α-D-glucopyranosyl-$(1 \rightarrow 2)$-β-D-fructofuranosyl-$(6 \rightarrow 1)$-α-D-glucopyranoside), and the cleavage products glucose and fructose may be formed.[140,161,167–169]

It was shown for the sucrose isomerases of more than one strain that the enzyme catalyzes the reaction to both products, whereas isomaltulose (Palatinose™) is the kinetically and trehalulose the thermodynamically preferred product.[110–111,134,157,169] The difference in the reaction mechanism between the isomaltulose (Palatinose™) and trehalulose formation is an additional tautomerization step of the fructose from the fructofuranose to the fructopyranose form in the catalytic pocket of the enzyme, which finally results in the trehalulose formation.[169]

After clarifying the structure and the reaction mechanism of the sucrose isomerase specifically changes of relevant amino acid residues in the sequence, it seems to be possible to increase the product specificity and reaction rate.[81–82,140,167] These actual efforts and current classical optimization of the fermentation parameters, the strains and the related technical process partly including statistical methods like the response surface methodology[110,101–107,111,115–117,121,132–135] may further decrease the price of the bulk chemicals isomaltulose (Palatinose™) and trehalulose in future.

Isomaltulose, also known as Palatinose™ (IUPAC name (2R,3R,4S,5R,6S)-2-(hydroxymethyl)-6-[[(2R,3R,4S)-3,4,5-trihydroxy-5-(hydroxymethyl)oxolan-2-yl]methoxy] oxane-3,4,5-triol), and sucrose have similar physical and organoleptic properties. Isomaltulose (Palatinose™) is a reducing, free-flowing, non-hygroscopic crystalline compound which can be crystallized easily from aqueous solutions with one mole crystal water. Crystalline isomaltulose (Palatinose™) melts at 123–124 °C.[112,145,146,154] The crystal structure[197] and C[13]-NMR spectrum[97,150] were identified. The water solubility at 20 °C is about one-quarter compared to that of sucrose. The viscosity of aqueous isomaltulose (Palatinose™) solutions is in the same range as that of sucrose.[145,146] Isomaltulose (Palatinose™) shows a neutral sweetness like sucrose without any aftertaste, but it is only 42% as sweet as sucrose.[90,96,112,146] Apart from lower sweetness, isomaltulose (Palatinose™) is non-cariogenic.[87,112,131,138,139,143,155,156,158] Isomaltulose (Palatinose™) has a low glycaemic index therefore the insulin release is correspondingly reduced compared to other sugars.[108,109,112] In humans, isomaltulose (Palatinose™) is hydrolyzed by isomaltase and absorbed as glucose and fructose, thus its energetic value is 4 kcal g^{-1} as in sucrose.[112,146,156] It is suggested that isomaltulose (Palatinose™) has favourable effects on fat metabolism and mental performance in humans.[100] Because isomaltulose (Palatinose™) is less hygroscopic and more stable and acid resistant than sucrose, it is used more and more as a sweetener not only for diabetics but also for sports food and drinks.[112,115,156,166] After comprehensive verification of its safety, isomaltulose (Palatinose™) is approved as food for human nutrition.[83–84,98,125,126,166]

Most characteristics of trehalulose (IUPAC name (3S,4R,5R)-3,4,5,6-tetra-hydroxy-1-[(2S,3R,4S,5R,6R)-3,4,5-trihydroxy-6-(hydroxymethyl)oxan-2-yl]oxy-hexan-2-one) are similar compared to those of isomaltulose (Palatinose™). Trehalulose is almost as sweet as isomaltulose (Palatinose™), but its solubility in water is quite different. Trehalulose is a non-crystalline sugar which makes other applications possible.[112,134] But isomaltulose (Palatinose™) and trehalulose are not only used in the food and beverage industry. Isomaltulose (Palatinose™) and trehalulose are technically used as mild reducing agents because of their reducing properties. With isomaltulose (Palatinose™), trehalulose or mixtures thereof sulphur and vat dyes can be gently reduced in aqueous alkaline medium for example.[95,112]

Up to now, many other micro-organisms have been reported to be capable of transforming sucrose into isomaltulose (Palatinose™) but only the strain of *Protaminobacter rubrum den Dooren de Jong* (ATCC-No. 8457 and CBS 574.77, respectively) – which Lorenz et al.[128,129] originally isolated in 1957 from sugar beet raw juice – has been suitable for industrial scale use so far.[99,124–129,136,142,163–165] The molecular weight of *P. rubrum*'s α-D-glucosyltransferase is 69,000–70,000 and the isoelectric point is 9.9 (Figure 9.3.8).[137,198]

Technical processes using viable, free micro-organisms have the disadvantage of higher product purification costs and lower yield but techniques using immobilized non-viable whole cells[119] proved to be more cost effective.[92,93,120,142,146–148,159] The industrial process based on immobilized cells starts

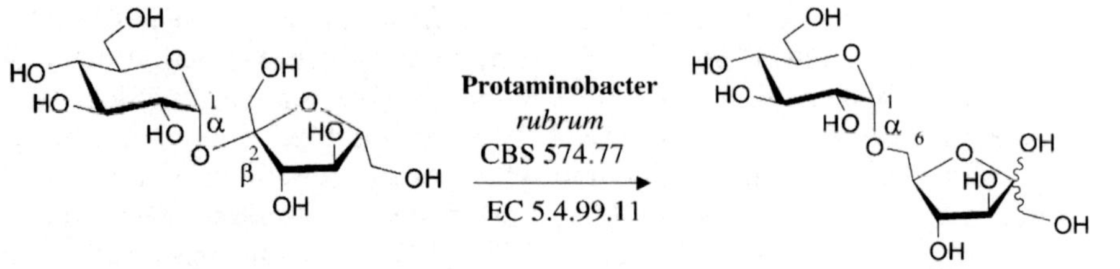

sucrose

α-D-glu-(1→2)-β-D-fru

isomaltulose (palatinose™)

α-D-glu-(1→6)-β-D-fru

Figure 9.3.8 Enzymatic conversion of sucrose into its isomer isomaltulose (Palatinose™).

Table 9.3.3 Product pattern of the glycosyltransferase of *P. rubrum*.

Product	% on total solids
isomaltulose (Palatinose™)	79.0–84.5
trehalulose	9.0–11.0
fructose	2.5–3.5
glucose	2.0–2.5
isomaltose	0.8–1.5
sucrose	0.5–1.0
isomelezitose	0.5
others	0.7–1.5

with the propagation of the cells. Because there is no need for isolation of the enzyme effecting the conversion of sucrose to isomaltulose (Palatinose™), the whole cells are immobilized. Almost all immobilization methods described in the literature are suitable for *P. rubrum*. The best technique so far is the immobilization in classical calcium alginate beads according to the method of Lim and Sun (1980).[142,146,147,152] The biocatalyst is transferred to a packed bed reactor. The sucrose used for conversion to isomaltulose (Palatinose™) is dissolved and the pH is adjusted to 6.0 before it is sterilized and pumped into the reactors. Downflow operation has been found to be advantageous. Flow rate is adjusted individually for each reactor to convert almost all sucrose being supplied. The product pattern of the solution leaving the reactors is shown in Table 9.3.3.[146,147]

Long-time stability of the immobilized enzyme is remarkably high even if high concentrations of sucrose up to $550\,g\,L^{-1}$ are used. A half-life of more than 5,000 hours is very common.[199]

The by-products can be easily removed from the product solution leaving the column reactors by multi-step crystallization. After evaporation and crystallization isomaltulose (Palatinose™) is separated by centrifugation and dried. The process yields isomaltulose (Palatinose™) as a free-flowing, non-hygroscopic crystalline substance with one mole crystal water. The mother liquor of the isomaltulose (Palatinose™) crystallization is the only by-product of the process. Because this liquid is used *e.g.* as feed stuff, the whole isomaltulose (Palatinose™) production process is free of waste products.[112,124,142,146,147]

Production of isomalt. Although isomaltulose (Palatinose™) is a sugar replacer itself, most of the product is currently used as a precursor for the production of the sugar replacer isomalt. If isomaltulose (Palatinose™) is processed to isomalt by hydrogenation wet crystals from crystallization are directly used.

Hydrogenation of isomaltulose (Palatinose™) solution can easily be done under moderate temperature and pressure conditions and at neutral pH or under mildly alkaline conditions using, for example, Raney-Nickel or supported *e.g.* noble metals in pelletized form in a fixed bed reactor. The reduction of isomaltulose (Palatinose™) is expected to yield an equimolar mixture of 1-O-α-D-glucopyranosyl-D-mannitol (1,1 GPM, CAS No. 20942-99-8 or 64519-82-0) and 6-O-α-D-glucopyranosyl-D-sorbitol (1,6 GPS) but depending on the conditions of hydrogenation the rate of each component can vary between 43 and 57% (Figure 9.3.9). The other sucrose isomers and saccharides can be hydrogenated in the same way leading to the corresponding sugar alcohols (Table 9.3.4).[114,142,146,147] After filtration and ion exchange (to remove traces of nickel) the isomalt solution is evaporated to appropriate concentration. The isomalt solution is either sold as such or evaporated to super saturation followed by crystallization.[114,142,146,147] The GPM part of isomalt crystallizes with two moles of crystal water,[127] whereas GPS crystals are anhydrous.[122,134] This results in an overall water content of about 5% in crystalline isomalt.[114,142,146,147]

Properties and applications of isomalt. Isomalt is an odourless, white, crystalline substance containing about 5% (w/w) water of crystallization. It tastes just as natural as sugar, is not sticky, is tooth-friendly (it helps to prevent cavities and plaque formation), is suitable for diabetics and has only half the

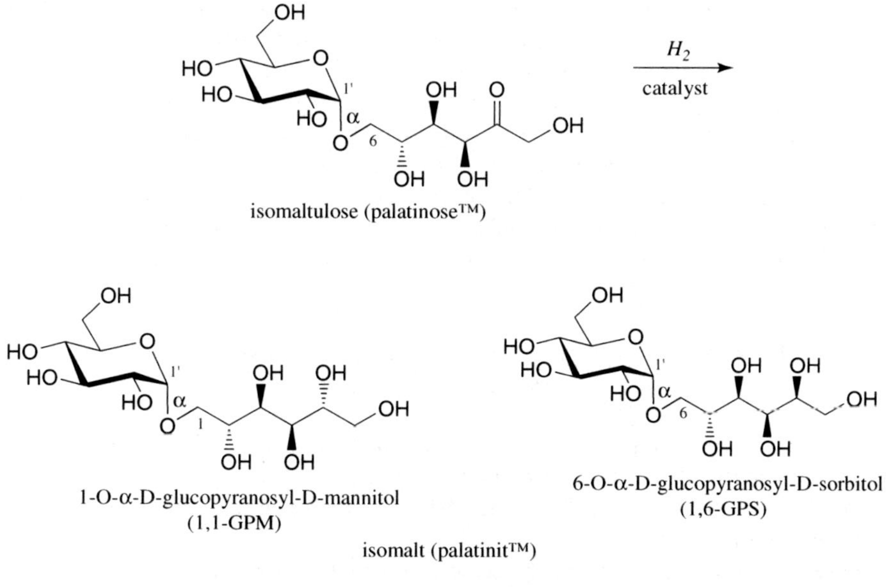

isomaltulose (palatinose™)

1-O-α-D-glucopyranosyl-D-mannitol
(1,1-GPM)

6-O-α-D-glucopyranosyl-D-sorbitol
(1,6-GPS)

isomalt (palatinit™)

Figure 9.3.9 Hydrogenation of isomaltulose (Palatinose™) to isomalt.

Table 9.3.4   Different saccharides and their corresponding saccharide alcohol
provided by hydrogenation.

Saccharide	*Corresponding saccharide alcohol*
trehalulose (1-O-glucosyl-fructose)	1-O-α-D-glucosyl-D-sorbitol (1,1 GPS)
	1-O-α-D-glucosyl-D-mannitol (1,1 GPM)
isomaltulose (6-O-glucosyl-fructose)	6-O-α-D-glucosyl-D-sorbitol (1,6 GPS)
	1-O-α-D-glucosyl-D-mannitol (1,1 GPM)
isomaltose (6-O-glucosyl-glucose)	6-O-α-D-glucosyl-D-sorbitol (1,6 GPS)
glucose	sorbitol
fructose	sorbitol
	mannitol
(glucosyl)$_2$-fructose	(glucosyl)$_2$-sorbitol
	(glucosyl)$_2$-mannitol
leucrose	others, *e.g.* 5-O-α-D-glucosyl-D-sorbitol
	(1,5 GPS)

calories of sugar because it cannot be completely metabolized. Isomalt is a bulk sweetener and can substitute sugar in a 1 : 1 mass ratio. It should not be mixed up with so-called intense sweeteners, which have much higher sweetening power (between a hundred and a thousand times as high) but no bulking properties. For this reason, these intense sweeteners are only applied in minute quantities.

Because isomalt is similar to sucrose, it is particularly suitable for making products such as candies, chewing gum, chocolate, compressed tablets or lozenges, baked goods, baking mixtures and pharmaceutical products using conventional equipment. Isomalt's nutritional and physiological benefits are ideal for use in sugar-free, low-calorie and diabetic products. It is particularly suitable for food and pharmaceutical applications (trade name galenIQ™). Isomalt also offers many advantages for technical uses.[114,142,146,166] Because the polyol isomalt is relatively thermally stable, it is used in the plastics industry for polymer stabilization in PVC.[162] In coatings based on melamine resins isomalt is used as polymer plasticizer.[114] The electrical industry uses isomalt as glue for temporary fixing of semi-conductor plates. The glue is washed out later.[144] In the production of ceramics isomalt can act as binder in the ceramic raw mixture. Later in the processing, isomalt causes no residues of combustion in the ceramics.[94]

9.3.2.3.3   Oligofructose, Inulin and Levan, Amylose (Neo-Amylose™) and Dextran

All these oligo- and polysaccharides in nature are synthesized (*in vivo*) *via* sucrose. Technical synthesis (*in vitro*) found industrial practices mainly for oligofructose, Neo-amylose™ and dextran following the reaction pathways as outlined in Figure 9.3.7.

Inulin. Fructans are natural polymers of fructose. In nature, there are two different types of fructans. One is named levan with a backbone of β-(2→6) linked fructose monomers which is a generally high molecular weight polysaccharide of micro-organisms. In contrast, inulin is characterized by a β-(2→1) linked backbone structure. Inulin is generally found in plants such as chicory (up to 20%), Jerusalem artichoke and onions as reserve carbohydrate, but also in some bacteria. The degree of polymerization (DP) of plant inulin is rather low (up to 200 maximum) compared to the DP of bacterial inulin which has a very high DP, ranging from 10,000 to over 100,000. Bacterial inulin is highly branched ($\geq 15\%$) which results in low intrinsic viscosity in spite of its high molecular weight. It has a compact, globular shape. In the case of plant inulin, the DP varies with plant species and with the life-cycle of the plant. The DP and the presence of branches are important properties since they influence the functionality of the molecule. For many possible applications it would be advantageous to produce high molecular weight inulin like the bacterial inulin, but without branches similar to plant inulin.

Both levan and inulin mostly have a glucose molecule at one terminus. A unique specific structural feature within the family of polysaccharides is that no bond of the fructose furanose ring is part of the macromolecular backbone. That means *e.g.* the linear polyfructan inulin has a comb-like structure with a "polyethylene"-like backbone and the furan ring being laterally fixed whereas in the analogue polyglucan amylase the pyran rings form the backbone. There are only a few natural polymers where the carbohydrate exists in the furanose form. Compared to the pyranose ring the furanose ring is less rigid which brings additional flexibility to the whole molecule as such. Inulin is produced from chicory in large scale for food applications including functional food. Oligofructose is produced semi-synthetically *via* sucrose (Figure 9.3.7) or *via* endo-inulinases from inulin. Oligofructose is used as prebiotic functional ingredient mainly in dairy products. Long-chain inulin (dp *ca.* 25) provides food with a fat-like structure and taste. The blend of oligofructose and long-chain inulin improves mineral, *e.g.* calcium, absorption. Fructans can also be used for the manufacturing of furan-based and other chemicals.[187,189,190,193,194, 203–206]

Bacterial and plant inulin are different not only in molecular weight and in the degree of branching but also in the synthesis and the enzymes applied. Both inulins are synthesized *via* sucrose. Glucose is generated as by-product.

The synthesis of plant inulin *in vivo* follows a two-step reaction, starting with a sucrose-1-fructosyltransferase (1-SST). One sucrose molecule acts as donor and another sucrose molecule as acceptor of a fructosyl unit. This leads to the formation of the trisaccharide 1-kestose. Catalyzed by a fructan-fructan-1-fructosyltransferase (1-FFT) fructosyl units are shuffled between the 1-kestose and higher polymeric β-(2→1) linked fructan molecules in the second step. Repetition of this step results in inulin with β-(2→1) linkages only.[186,187,190,191]

For the production of bacterial inulin only the enzyme fructosyltransferase (FTF, EC 2.4.1.9) is required, which shuffles fructosyl units from the sucrose donor to another sucrose molecule or 1-kestose or higher polymeric β-(2→1) linked fructan molecules acting as acceptor. The enzyme partly leads to β-(2 → 6) linkages, which results in branches within the inulin molecule.[187,195] This fructosyltransferase and the related fructan is interesting for many possible applications. Only based on sucrose as raw material high molecular weight inulin, preferably without branches similar to plant inulin, can be produced in aqueous solution. Scale-up is the same as the production design for dextran or neo-amylose. The utilization of the simultaneously formed glucose as food, feed or fermentation medium should be no problem.

Levan. Like dextran, microbial levan is often an undesirable by-product in the sugar industry. It was first reported by Lippmann in 1881 as a product which increases the viscosity of the processing sugar liquor.[192] In 1901, Greig-Smith found that a strain of *Bacillus* grown on sucrose produces a fructose polymer which was named levan by analogy to dextran.[188]

Levan is produced from sucrose under action of the enzyme levansucrase (sucrose 6-fructosyltransferase (FTF), EC 2.4.1.10) in a step-by-step addition of single fructofuranosyl units at the C-6 hydroxyl group of the non-reducing fructose terminal unit of the growing levan chain. It is possible to obtain levan either from the culture broth of bacterial strains like *Bacillus* or *Zymomonas* or from the enzymatic reaction of the cell-free supernatant with sucrose. Precipitation with alcohol and drying results in amorphous or microcrystalline levan. Levan has a varying solubility in cold water, is very soluble in hot water and is almost insoluble in ethyl alcohol. The solubility of levan is generally better than that of inulin. Levan can easily be hydrolyzed by acid. Molecular weight and viscosity of levan respectively vary with the strain and the reaction conditions used.[189,190] Even though levan has interesting properties, it has never had extensive industrial use up to now.[189]

Neo-amylose™. Neo-amylose™ is a biopolymer analogous to amylase, which can be synthesized *via* biocatalytical polymerization of sucrose to form a water-insoluble, unbranched poly-1,4-α-D-glucan with a chain length of up to 35–100 glucose units. As crystalline Neo-amylose™ cannot be hydrolyzed by human digestive enzymes, it can be claimed as resistant starch. The biocatalyst used for the synthesis of the glucan is the enzyme amylosucrase (AmSu; sucrose-glucan Glycosyltransferase/EC 2.4.1.4) derived from the non-pathogenic bacterium *Neisseria polysaccharea*, originally isolated from the throat of a healthy child.[170] Amylosucrase splits sucrose into glucose and fructose and adds subsequently the glucose moiety to the non-reducing end of an α-1,4-D-glucan. Fructose is the by-product of the enzymatic reaction. The advantage of the process is that no activated α-D-glycosyl nucleosid diphosphate substrate is needed.

$$Sucrose + (\alpha\text{-}1,4\text{-}Glucan)_n \overset{Amylosucrase}{\longleftrightarrow} (\alpha\text{-}1,4\text{-}Glucan)_{n+1} + Fructose$$

The enzyme amylosucrase can either be expressed in *Neisseria* itself or overexpressed in non-pathogenic *Escherichia coli* strains.[171] In 1999, the gene encoding the amylosucrase was sequenced and characterized in detail.[172] Briefly, the synthesis of the α-1,4-glucan is performed by fermenting the micro-organisms and extracting the amylosucrase from the biomass. Afterwards, the enzyme is brought into contact with a highly concentrated sucrose solution. As the chain of the glucan grows in length during biotransformation it crystallizes with other molecules and co-precipitates from the solution. Therefore, it can easily be isolated and dried.

The pure polymer powder derived from biotransformation can be processed *via* heat-moisture treatment to form a type-3 amylase resistant starch with a degree of resistance of >90% (RS$_3$).

Applications. Neo-amylose™ can be used for both food as well as non-food applications. As already mentioned at the beginning of this chapter, crystalline Neo-amylose™ cannot be hydrolyzed by human pancreatic α-amylases.[173] After consumption by human beings, as much as 90% of the heat-moisture treated polymer is transferred into the large bowel and is fermented by colonic bacteria to form short-chain fatty acids such as acetate, propionate and butyrate.

Besides the application as dietary fibre, Neo-amylose™ can be used for the production of smooth particles with a size of 10–100 μm by re-crystallization in dimethyl sulfoxide (DMSO). The application of these particles as a supplement to cosmetics such as creams, lotions or as UV-reflector has been described in WO 0038623.[174] Further hard or soft films can be produced from Neo-amylose™ in order to generate gelatine-free capsules for encapsulation purposes.[175]

Dextran. Dextran is a linear homopolysaccharide of glucose composed of α-(1→6)-linked α-D-glucose units with only occasional branches *via* O-2 (α-(1→2)), O-3 (α-(1→3)) or O-4 (α-(1→4)). The exact structure of each type of dextran depends on the producing dextransucrase and the microbial strain, respectively; several strains can produce more than one type of dextran.[178,179]

Because dextran is formed from sucrose it has been known for a long time as a source of trouble with slimy contaminations in food and especially in the sugar industry.[180] The corresponding enzyme dextransucrase (EC 2.4.1.5) is an extracellular glucosyltranferase (GTF), which catalyzes the cleavage of the sucrose molecule and the transfer of the D-glucopyranosyl residue to dextran while fructose is released. Commercial dextran is mainly produced by growing cultures of *Leuconostoc mesenteroides*. In addition to the sucrose (usually initial sucrose concentration is 2% (w/w)) the medium also contains an organic source of nitrogen such as peptone, growth factors, certain trace minerals and phosphate. After 24–48 h fermentation at 25 °C without aeration the cells are removed by centrifugation. The dextran is harvested from the cell-free supernatant by alcohol precipitation and purified by further precipitation after redissolution in water.

An alternative process uses the cell-free supernatant from fermentation for enzymatic synthesis of the dextran. This process allows fermentation and synthesis under different sets of conditions. The resulting product is more uniform and easier to purify. The elongation and branching of the dextran can moderately be influenced by the reaction conditions, especially by temperature and concentration.[176,178,179,182,183]

The biopolymer dextran produced on the basis of the renewable resource sucrose is a very interesting raw material for many applications. It was used first as a blood-plasma volume expander in the 1940s. Up to now, there are many other applications based on native dextran, partially degraded dextran and its derivatives such as esters or ethers.[176–179,182,185]

The reaction of dextran with epichlorohydrin results in a cross-linked gelatinous product that is used as a molecular sieve for separation and purification of biochemically important macromolecules like proteins, polysaccharides or nucleic acids. Its commercial name Sephadex® is derived from SEparation PHArmacia DEXtran and stands for many different successful products including further derivatized materials such as carboxymethyl, diethylaminoethyl, diethyl(2-hydroxypropyl)-aminoethyl, and sulfopropyl Sephadex®.[182]

Dextran and its derivates have found some applications in the medical field. Dextran sulphate has anticoagulant properties similar to those of heparin.[176] Recent studies show antiviral properties of the sulphate esters of dextran, particularly in the treatment of the human immunodeficiency virus.[181] Oral dextran sulphate has been used in Japan against arteriosclerosis for 20 years without harmful side-effects.[184,201] Mercaptodextran is discussed for therapy of acute heavy-metal poisoning because it has a higher affinity for heavy metal ions such as silver, mercuric, cupric and auric ions than most other thiols and chelating agents. The properties of mercaptodextran are of interest regarding environmental clean-up of heavy metal contaminations.[182]

By incorporating dextran into X-ray and other photographic emulsions, photographic companies could reduce silver usage without loss of grain fineness.[176]

Due to its excellent biocompatibility, moisturizing properties and proven stability under several conditions dextran can be found in several cosmetic formulations for the eye- and skin-care market. Furthermore, dextran improves the softness, crumb texture and loaf volume of bakery products and it is also known to be used in clinical nutrition and as an additive in products such as candies and ice-cream.[178,179]

9.3.3   Outlook

At first sight, the consumer's choice for products mostly depends on reasonable prices and convenience. But times have changed and the consumer's awareness focuses – besides price and application profiles – more and more on ecological aspects like greenhouse-gas emissions. There are far more products on the basis of sucrose developed in the past being able to replace ecologically unfavourable petrochemically based products of today.

Just to mention only two examples:

- Acetylated oxidized sucroses could replace nitrogen-containing ecologically unfavourable surfactant boosters in washing formulas both as booster and as sequestering agent.
- Alkyl amine derivatives of isomaltulose (Palatinose™) have very low critical micel concentrations (cmc) far better than today's petrochemically

based ones. They are very synergistic with classical anionic surfactants like sodium dodecyl sulphates and are easily biodegradable.

Both examples demonstrate that excellent application profiles combined with high ecological performance are today not sufficient to introduce products into the market. There are far more hurdles to overcome:

- Until today, petrochemicals often are very cheap products because they are produced on the basis of cheap by-products of oil refineries in a very integrated processing on large scales.
- Newly developed ecologically more favourable products based on renewables have to start on small scales but cannot compete price-wise with the cheap mass products based on non-renewables. Their production is not possible in the same integrated steam-cracker-based facilities of the chemical industry but very often in stand-alone production sites.

To conclude: even today – being faced with increasingly high oil prices – if society wants to have more ecologically favourable products it remains a political task to support the introduction of these products. But – bearing in mind regulations like "REACH" – the reality is exactly the contrary.

References

1. Bartens, *Zuckerwirtschaft Europa*, Verlag Dr. Albert Bartens KG, Berlin, 2007.
2. Y. Queneau, S. Jarosz, B. Lewandowski and J. Fitremann, *Adv. Carb. Chem. Biochem.*, 2007, **61**, 217.
3. D. Peters, *Nachwachsende Rohstoffe in der Industrie*, Fachagentur Nachwachsende Rohstoffe, 2. Aufl., Gülzow, 2007.
4. *Jahresbericht 2005/2006*, Fachagentur Nachwachsende Rohstoffe, Gülzow, 2006.
5. F. W. Lichtenthaler, in *Biorefineries – Industrial Processes and Products* , ed. B. Kamm, P. R. Gruber and M. Kamm, Wiley-VCH, Weinheim, 2006, **2**, p. 3.
6. P. W. van der Poel, H. Schiweck and T. Schwartz, *Sugar Technology – Beet and Cane Sugar Manufacture*, Dr. Albert Bartens KG, Berlin, 1998.
7. P. Rein, *Cane Sugar Engineering*, Dr. Albert Bartens KG, Berlin, 2007.
8. H. Hoffmann, W. Mauch and W. Untze, *Zucker und Zuckerwaren*, Paul Parey, Berlin, 1985.
9. E. Hugot, *Handbook of Cane Sugar Engineering*, Elsevier Publishing Company, Amsterdam/London/New York, 1972.
10. F. W. Lichtenthaler and S. Peters, *C. R. Chimie*, 2004, **7**, 65.
11. J. N. Chheda, G. W. Huber and J. A. Dumesic, *Angew. Chem., Int. Ed.*, 2007, **119**, 7298.

12. Press release from 28.03.2008, http://www.braskem.com.br (retrieved 31.07.2008).
13. Press release from 04.06.2008, http://www.marketwatch.com (retrieved 28.07.2008).
14. Press Release from 14.12.2007, http://www.solvinpvc.com (retrieved 25.02.2008); http://desarrolloydefensa.blogspot.com/2008/04/solvay-indupa-firma-contrato-con.html (retrieved 27.11.2008); N. Denis, A. Meiser, J. Riese and U. Weihe, *McKinsey on Chemicals*, 2008, 1, 22.
15. J. Lewkowski, *ArkiVoc*, 2001, **i**, 17.
16. B. F. M. Kuster, *Starch/Stärke*, 1990, **42**, 314.
17. H. Schiweck, M. Munir, K. Rapp and M. Vogel, in *Carbohydrates as Organic Raw Materials*, ed. F. W. Lichtenthaler, VCH, Weinheim, 1991, p. 57.
18. P. Claus and G. H. Vogel, *Chem. Ing. Tech.*, 2006, **78**, 991.
19. A. Gaset, L. Rigal, G. Paillassa, J.-P. Salome and G. Fleche, FR 2 551 754, 1985 (Roquette Freres).
20. D. W. Brown, A. J. Floyd, R. C. Kinsmann and Y. Roshan-Ali, *J. Chem. Technol. Biotechnol.*, 1982, **32**, 920.
21. R. M. Musau and R. M. Munawu, *Biomass*, 1987, **13**, 67.
22. J. Hirth, Dissertation, TU Darmstadt, 2002.
23. M. Bicker, T. Hirth and H. Vogel, *Green Chem.*, 2003, **5**, 280.
24. M. Bicker, *Stoffliche Nutzung von Biomasse mit Hilfe überkritischer Fluide*, Dissertation, TU Darmstadt, 2005.
25. M. Bicker, D. Kaiser, L. Ott and H. Vogel, *J. Supercrit. Fluids*, 2005, **36**, 118.
26. H. Zhao, J. E. Holladay, H. Brown and Z. C. Zhang, *Science*, 2007, **316**, 1597.
27. K. M. Rapp, EP 230 250, 1987 (Südzucker AG).
28. B. Kamm, M. Kamm, M. Schmidt, T. Hirth and M. Schulze, in *Biorefineries – Industrial Processes and Products*, ed. B. Kamm, P. R. Gruber and M. Kamm, Wiley-VCH, Weinheim, 2006, **2**, p. 97.
29. A. Gandini and M. N. Belgacem, *Prog. Polym. Sci.*, 1997, **22**, 1203.
30. F. W. Lichtenthaler, in *Ullmann's Encyclopedia of Industrial Chemistry*, Wiley-VCH, Weinheim, 6th edn, 2002, **6**, p. 237.
31. T. Werpy and G. Petersen, http://www1.eere.energy.gov/biomass/pdfs/35523.pdf (retrieved 01.08.2008).
32. L. E. Manzer, *A.C.S. Symp. Ser.*, 2006, **921**, 40.
33. D. T. Win, *Assumption University Journal of Technology*, 2005, **8**, 185.
34. R. Weidenhagen and H. Wegner, *Chem. Ber.*, 1938, **71**, 2712.
35. G. V. Ling, A. J. Driessen, A. C. Piet and J. C. Vligter, *Ind. Eng. Chem. Prod. Res. Dev.*, 1970, **9**, 210.
36. L. W. Wright, US 3 965 199, 1976 (CI America Inc.).
37. K. Weissermel and H. -J. Arpe, *Industrial Organic Chemistry*, Wiley-VCH, Weinheim, 2003.
38. L. Schuster and W. Himmele, DE 38 18 198, 1989 (BASF AG).
39. L. Schuster, DE 39 28 285, 1991 (BASF AG).

40. P. H. Fairclough, L. Hough and A. C. Richardson, *Carbohydr. Res.*, 1975, **40**, 285.
41. L. Hough, S. P. Phadnis, R. A. Khan and M. R. Jenner, GB 1 543 167, 1977 (Tate & Lyle).
42. S. V. Molinary and M. R. Jenner, *Food & Food Ingredients Journal of Japan*, 1999, **182**, 6.
43. S. V. Molinary and M. E. Quinlan, in *Sweeteners and Alternatives in Food Technology*, ed. H. Mitchell, Blackwell Publishing, Oxford UK, 2006, p. 130.
44. C. W. Sham, *Nutritional Bytes*, 2005, **10/2**, Article 5. (http://repositor-ies.cdlib.org/uclabiolchem/nutritionbytes/vol10/iss2/art5).
45. J. L. Navia, US 4 950 746, 1990 (Noramco Inc.).
46. D. S. Neiditch, N. M. Vernon and R. E. Wingard, US 5 023 329, 1991 (Noramco Inc.).
47. R. E. Walkup, N. M. Vernon and R. E. Wingard, US 5 089 608, 1992 (McNeill PPC Inc).
48. G. H. Sankey, N. M. Vernon and R. E. Wingard, US 5 470 969, 1995 (McNeill PPC Inc.).
49. J. E. White and B. A. Bradford, EP 776 903, 1997 (McNeill PPC Inc.).
50. J. D. Clark and R. R. LeMay, US 6 939 962, 2003 (Tate & Lyle).
51. K. S. Mufti and R. A. Khan, US 4 380 476, 1983 (Talres Dev.).
52. P. J. Simpson, US 4 889 928, 1989 (Tate & Lyle).
53. J. S. Dordick, A. J. Hacking and R. A. Khan, US 5 128 248, 1992 (Tate & Lyle).
54. J. S. Dordick, A. J. Hacking and R. A. Khan, US 5 270 460, 1993 (Tate & Lyle).
55. S. Bornemann, J. M. Cassells, L. Combes, J. S. Dordik and A. J. Hacking, US 5 141 860, 1992 (Tate & Lyle).
56. D. C. Palmer and F. Terradas, US 5 445 951, 1995 (McNeill PPC Inc.).
57. K. S. Mufti and R. A. Khan, US 4 380 476, 1983 (Talres Dev.).
58. R. E. Walkup, J. L. Navis and N. M. Vernon, US 4 980 463, 1990 (Noramco Inc.).
59. R. A. Khan, G. H. Sankey, P. J. Simpson and N. M. Vernon, US 5 136 031, 1992 (Tate & Lyle).
60. J. L. Navia, R. E. Walkup and D. S. Neiditch, US 5 530 106, 1996 (McNeill PPC Inc.).
61. J. L. Navia, R. E. Walkup, N. M. Vernon and D. S. Neiditch, US 5 498 709, 1996 (McNeill PPC Inc.).
62. J. L. Navia, R. E. Walkup, N. M. Vernon and R. E. Wingard, US 5 298 611, 1994 (McNeill PPC Inc.).
63. S. J. Catani, D. A. Leinhos and T. O'Connor, US 5 977 349, 1999 (McNeill PPC Inc.).
64. N. M. Vernon, E. Micinski, S. J. Catani and J. L. Navia, US 6 890 581, 2003 (McNeill PPC Inc.).
65. B. A. P. Nelen and J. M. Cooper, in *Emulsifiers in Food Technology*, ed. R. J. Whitehurst, Wiley-Blackwell, 2004, p. 131.

66. Y. -J. Wang, in *Chemical and Functional Properties of Food Saccharides*, ed. P. Tomasik, CRC Press, 2004, p. 35.
67. C. Bimal and Z. Guonong, *Food Rev. Int.*, 2006, **22**, 245.
68. H. B. Hass, in *Sugar Esters*, Noyes Development Corporation, Park Ridge NJ, 1968, p. 1.
69. L. Osipow, F. D. Snell, W. C. York and A. Finchler, *Ind. Eng. Chem.*, 1956, **48**, 1459.
70. L. Hough, in *Carbohydrates as Organic Raw Materials*, ed. F. W. Lichtenthaler, VCH, Weinheim, 1991, p. 33.
71. S. K. Karmee, *Biofuels, Bioprod. Bioref.*, 2008, **2**, 144.
72. B. E. Brockway, *Cosmetics & Toiletries Magazine*, 2005, **120**, 93.
73. J. W. Le Maistre and R. B. Seymour, *J. Org. Chem.*, 1948, 782.
74. V. Kollonitsch, *Sucrose Chemicals*, The International Sugar Research Foundation, Inc., Washington, 1970.
75. K. C. Frich and J. E. Kresta, in *Sucrochemistry*, ed. J. L. Hickson, ACS Symposium Series, ACS, Washington, **vol. 41**, 1977, p. 238.
76. K. Vohrspohl, W. Hinz and B. Güttes, in *Nachwachsende Rohstoffe – Perspektiven für die Chemie*, ed. M. Eggersdorfer, S. Warwel and G. Wulff, VCH, Weinheim/New York, 1993, p. 383.
77. P. Gupta, H.-J. Sandhagen, W. Betz, U. Leyrer and M. Hoppe, EP 618 251, 1994 (Bayer).
78. D. Jhurry and A. Deffieux, *Eur. Polym. J.*, 1997, **33**, 1577.
79. C. Chauvin, K. Aczko and D. Plusquellec, *J. Org. Chem.*, 1993, **58**, 2291.
80. R. Kohlstrung, *Polyurethan – Präpolymere auf der Basis von Sacchariden und Saccharid-Derivaten-Von der Selektivität zur anwendungstechnischen Prüfung*, Dissertation, TU Darmstadt, 2001.
81. S. J. Ahn, J. H. Yoo, H. C. Lee, S. Y. Kim, B. S. Noh, J. H. Kim and J. K. Lee, *Biotechnol. Lett.*, 2003, **25**, p. 1179.
82. A. Aroonnual, T. Nihira, T. Seki and W. Panbangred, *Enzyme Microb. Technol.*, 2007, **40**, 1221.
83. J. Asquith, K. Pickering, S. Trenchard-Morgan and S. Sangster, *Completion of Bacterial Reverse Mutation Test on Isomaltulose*, Toxicol. Lab. Ltd, Ledbury, Herefordshire, 1986.
84. J. Bruhns, *Zuckerind.*, 1991, **116**(8), 736.
85. K. Buchholz, E. Stoppock, K. Matalla, K. D. Reh and H. J. Joerdening, in *Carbohydrates as Organic Raw Materials*, ed. F. W. Lichtenthaler, VCH, Weinheim, 1991, p. 155.
86. K. Buchholz and V. Kasche, *Biokatalysatoren und Enzymtechnologie*, VCH, Weinheim, 1997, p. 11.
87. A. Baer, *Lebensm. Wiss. Technol.*, **22**, 46.
88. R. Cartarius, T. Krause and H. Vogel, *Chem. Ing. Tech.*, 2001, **73**, 1 + 2, 118.
89. R. Cartarius, T. Krause and H. Vogel, *Chem. Ing. Tech.*, 2002, **74**, 869.
90. P. S. J. Cheetham, C. E. Imber and J. Isherwood, *Nature*, 1982, **299**, 628.

91. H. Chick, H. S. Shin and Z. Ustunol, *J. Food Sci.*, 2001, **66**, 478.
92. P. Egerer, W. Haese, H. Perrey and G. Schmidt-Kastner, EP 160 253, 1985 (Bayer AG).
93. P. Egerer, W. Haese, H. Perrey and G. Schmidt-Kastner, EP 160 260, 1985 (Bayer AG).
94. W. Geis, P. Quirnbach, S. Schwartz and W. Weiland, DE 197 18 672, 1998 (Zschimmer & Schwarz).
95. D. R. Gruell, A. Haji Begli, N. Kubadinow, M. Kunz and M. Munir, EP 943 030, 1998 (Suedzucker AG).
96. J. H. Huang, L. H. Hsu and Y. C. Su, *J. Ind. Microbiol. Biotechnol.*, 1998, **21**, 22.
97. H. C. Jarell, T. F. Conwell, P. Moya and I. C. P. Smith, *Carbohydr. Res.*, 1979, **76**, 45.
98. D. Jonker, B. A. R. Lina and G. Kozianowski, *Food Chem. Toxicol.*, 2002, **40**, 1383.
99. T. Kaga and T. Mizutani, *Proc. Res. Soc. Jap. Sugar Refin. Technol.*, 1985, **34**, 45.
100. J. Kashimura, Y. Nagai and T. Ebashi, *J. Nutr. Sci. Vitamino. (Tokyo)*, 2003, **49**, 214.
101. H. Y. Kawaguti, E. Manrich, L. F. Fleuri and H. H. Sato, *Braz. J. Microbiol.*, 2005, **36**, 227.
102. H. Y. Kawaguti, E. Manrich and H. H. Sato, *Biochem. Eng. J.*, 2006, **29**, 270.
103. H. Y. Kawaguti, E. Manrich and H. H. Sato, *Electr. J. Biotechnol.*, 2006, **9**, 482.
104. H. Y. Kawaguti, M. F. Buzzato, D. C. Orsi, G. T. Suzuki and H. H. Sato, *Process Biochem.*, 2006, **41**, 2035.
105. H. Y. Kawaguti, M. F. Buzzato and H. H. Sato, *J. Ind. Biotechnol.*, 2007, **34**, 261.
106. H. Y. Kawaguti and H. H. Sato, *Biochem. Eng. J.*, 2007, **36**, 202.
107. H. Y. Kawaguti and H. H. Sato, *Process Biochem.*, 2007, **42**, 472.
108. K. Kawai, Y. Okuda and K. Yamashita, *Endocrinol. Jpn.*, 1985, **32**, 933.
109. K. Kawai, H. Yoshikawa, Y. Murayama, Y. Okuda and K. Yamashita, *Horm. Metab. Res.*, 1989, **21**, 338.
110. S. Kishihara, A. Ideno, M. Okuno, S. Fujii, T. Ebashi and Y. Nagai, *Proc. Res. Soc. Jap. Sugar Refin. Technol.*, 2002, **50**, 67.
111. S. Kishihara, H. Ikegami, M. Okuno, S. Fujii, T. Ebashi and Y. Nagai, *Proc. Res. Soc. Jap. Sugar Refin. Technol.*, 2002, **50**, 75.
112. J. Kowalczyk, in *Handbuch Süßungsmittel*, ed. K. Rosenplenter and U. Noehle, Behr's Verlag, Hamburg, 2007, p. 194.
113. J. Kowalczyk, in *Handbuch Süßungsmittel*, ed. K. Rosenplenter and U. Noehle, Behr's Verlag, Hamburg, 2007, p. 295.
114. J. Kowalczyk, in *Handbuch Süßungsmittel*, ed. K. Rosenplenter and U. Noehle, Behr's Verlag, Hamburg, 2007, p. 340.
115. A. Krastanov and T. Yoshida, *J. Ind. Microbiol. Biotechnol.*, 2003, **30**, 593.

116. A. Krastanov, D. Blazheva, I. Yanakieva and M. Kratchanova, *Enzyme Microb. Technol.*, 2006, **39**, 1306.
117. A. Krastanov, D. Blazheva and V. Stanchev, *Process Biochem.*, 2007, **42**, 1655.
118. M. Kunz, in *Carbohydrates as Organic Raw Materials*, ed. F. W. Lichtenthaler, VCH, Weinheim, 1991, p. 127.
119. C. Kutzbach, G. Schmidt-Kastner and H. Schutt, EP 049 801, 1982 (Bayer AG).
120. O. J. Lantero, US 4 390 627, 1983 (Miles Laboratories Inc.).
121. X. Li, C. Zhao, Q. An and D. Zhang, *J. Appl. Microbiol.*, 2003, **95**, 521.
122. F. W. Lichtenthaler and H. J. Lindner, *Liebigs Ann.*, 1981, 2372.
123. F. W. Lichtenthaler, P. Pokinskyj and S. Immel, *Zuckerind.*, 1996, **121**, 174.
124. A. Liese, K. Seelbach and C. Wandrey, in *Industrial Biotransformations*, Wiley-VCH, Weinheim, 2006, p. 512.
125. B. A. R. Lina, A. E. Smits-van Prooji and D. H. Waalkens-Berendsen, *Food Chem. Toxicol.*, 1997, **35**, 309.
126. B. A. R. Lina, D. Jonker and G. Kozianowski, *Food Chem. Toxicol*, 2002, **40**, 1375.
127. H. J. Lindner and F. W. Lichtenthaler, *Carbohydr. Res.*, 1981, **93**, 135.
128. S. Lorenz, *Zuckerind.*, 1958, **8**, 490.
129. S. Lorenz, *Zuckerind.*, 1958, **8**, 535.
130. N. Low and P. Sporns, *J. Food Sci.*, 1988, **53**, 558.
131. T. Minami, T. Fujiwara, T. Ooshima, Y. Nakajima and S. Hamada, *Oral Microbiol. Immunol.*, 1990, **5**, 189.
132. A. L. L. Moraes, C. Steckelberg, H. H. Sato and A. Pinheiro, *Ciénc. Tecnol. Aliment., Campinas*, 2005, **25**(1), 95.
133. P. Mundra, K. Desai and S. S. Lele, *Bioresource Technol.*, 2007, **98**, 2892.
134. Y. Nagai, T. Sugitani, K. Tsuyuki, K. Yamada, T. Ebashi and S. Kishihara, *Proc. Res. Soc. Jap. Sugar Refin. Technol.*, 2002, **50**, 57.
135. Y. Nagai, T. Sugitani, K. Yamada, T. Ebashi and S. Kishihara, *J. Jpn. Soc. Food Sci. Technol.*, 2003, **50**(10), 488.
136. Y. Nakajima, *Proc. Res. Soc. Jap. Sugar Refin. Technol.*, 1984, **33**, 55.
137. Y. Nakajima, *Denpun Kagaku*, 1988, **35**, 131.
138. T. Ooshima, A. Izumitani, S. Sobue, N. Okahashi and S. Hamada, *Infect. Immun.*, 1983, **39**, 43.
139. K. Ohata and I. Takazoe, *Bull. Tokyo Dent. Coll.*, 1983, **24**, 1.
140. S. Ravaud, X. Robert, H. Watzlawick, R. Haser, R. Mattes and N. Aghajari, *J. Biol. Chem.*, 2007, **282**(38), 28126.
141. E. Reinefeld, *Zuckerind.*, 1987, **112**(12), 1049.
142. T. Rose and M. Kunz, in *Practical Aspects of Encapsulation Technologies*, ed. U. Prüsse and K. D. Vorlop, Federal Agricultural Research Centre (FAL), Braunschweig, 2002, **241**, p. 75.
143. N. Sasaki, V. Topitsoglou, I. Takazoe and G. Frostell, *Swed. Dent. J.*, 1985, **9**, 149.
144. P. J. E. Schelwald, EP 601 615, 1993 (Philips Electronics N.V).

145. H. Schiweck, *Alimenta*, 1980, **19**, 5.
146. H. Schiweck, M. Munir, K. M. Rapp, B. Schneider and M. Vogel, *Zuckerind.*, 1990, **115**(7), 555.
147. H. Schiweck, M. Munir, K. M. Rapp, B. Schneider and M. Vogel, in *Carbohydrates as Organic Raw Materials*, ed. F. W. Lichtenthaler, VCH, Weinheim, 1991, p. 57.
148. I. Shimizu, K. Suzuki and Y. Nakajima, JP 55 113 982, 1980 (Mitsui Sugar Co.).
149. J. Seibel, R. Beine, R. Moraru, C. Behringer and K. Buchholz, *Biocat. Biotrans.*, 2006, **24**, 157.
150. F. R. Seymour, R. D. Knapp, J. E. Zweig and S. H. Bishop, *Carbohydr. Res.*, 1979, **72**, 57.
151. I. R. Siddiqui and B. Furgala, *J. Agric. Res.*, 1967, **6**, 139.
152. O. Smidsrød and G. Skjåk-Bræk, *Trends Biotechnol.*, 1990, **8**, 71.
153. P. Sporns, L. Plhak and J. Friedrich, *Food Res. Int.*, 1992, **25**, 93.
154. P. J. Straeter, *Zuckerind.*, 1987, **112**(10), 900.
155. I. Takazoe, *Int. Dent. J.*, 1985, **35**, 58.
156. I. Takazoe, G. Frostell, K. Ohta, V. Topitsoglou and N. Sasaki, *Swed. Dent. J.*, 1985, **9**, 81.
157. H. Taniguchi, in *Handbook of Industrial Catalysis*, ed. C. T. J. Hou, CRC Press, Taylor & Francis Group, BocaRaton, 2005, **20**, p. 1–25.
158. V. Topitsoglou, N. Sasaki and G. Frostell, *Caries Res.*, 1984, **18**, 47.
159. T. Tosa and T. Shibatani, *Ann. N. Y. Acad. Sci.*, 1995, **750**, 365.
160. D. Vasik-Racki, in *Industrial Biotransformations*, ed. A. Liese, K. Seelbach and C. Wandrey, Wiley-VCH, Weinheim, 2006, p. 29.
161. T. Verones and P. Perlot, *Enzyme Microb. Technol.*, 1999, **24**, 263.
162. W. Wehner, R. Drewes, K. J. Kuhn, H. J. Sander and M. Kolb, EP 677 549, 1995 (Ciba-Geigy AG).
163. R. Weidenhagen and S. Lorenz, *Angew. Chem.*, 1957, **69**, 641.
164. R. Weidenhagen and S. Lorenz, *Zuckerind.*, 1957, 533.
165. R. Weidenhagen and S. Lorenz, DE 10 49 800, 1959 (Süddeutsche Zucker AG).
166. R. Wilson, *Sweeteners*, Wiley-Blackwell, Oxford, UK, 2007.
167. L. Wu and R. G. Birch, *Appl. Environ. Microbiol.*, 2005, **71**(3), 1581.
168. D. Zhang, X. Li and L. H. Xhang, *Appl. Environ. Microbiol.*, 2002, **68**, 2676.
169. D. Zhang, N. Li, S. M. Lok, L. H. Zhang and K. Swaminathan, *J. Biol. Chem.*, 2003, **278**(37), 35428.
170. C. M. Anand, F. Ashton, H. Shaw and R. Gordon, *J. Clin. Microb.*, 1991, **29**, 22434.
171. V. Büttcher, T. Welsh, L. Willmitzer and J. Kossmann, *J. Bacteriol.*, 1997, **179**, 3324.
172. G. P. DeMontkalk, M. Remand-Simeon, R. M. Willemot, V. Planchot and P. Mousan, *J. Bacteriol.*, 1999, **181**, 375.
173. H. Bengs and A. Brunner, WO 0038537, 2000 (Aventis Res & Tech GmbH).

174. H. Bengs and A. Schneller, WO 0038623, 2000 (Aventis Res & Tech GmbH).
175. H. Bengs and A. Brunner, WO 0038537, 2000 (Aventis Res & Tech GmbH).
176. R. M. Alsop, in *Microbial Polysaccharides*, ed. M. E. Bushell, Elsevier, Amsterdam, 1983, p. 1.
177. A. N. de Belder, in *Polysaccharides in Medical Applications*, ed. S. Dumitriu, Marcel Dekker, New York, 1996, p. 505.
178. T. D. Leathers, in *Biopolymers, Vol. 5. Polysaccharides I Polysaccharides from Prokaryotes*, ed. E. J. Vandamme, S. de Baets and A. Steinbüchel, Wiley-VCH, Weinheim, Germany, 2002, p. 299.
179. M. Naessens, A. Cerdobbel, W. Soetaert and E. J. Vandamme, *J. Chem. Technol. Biotechnol.*, 2005, **80**, 845.
180. L. Pasteur, *Bull. Soc. Chim. Fr.*, 1861, 30.
181. J. Piret, J. Lamontagne, J. Bestman-Smith, S. Roy, P. Gourde, A. Desormeaux, R. F. Omar, J. Juhasz and M. G. Bergeron, *J. Clin. Microbiol.*, 2000, **38**, 110.
182. J. F. Robyt, in *Encyclopaedia of Polymer Science, Vol. 4*, ed. J. I. Kroschwitz, Wiley-VCH, New York, 1985, p. 753.
183. A. Tanriseven and J. F. Robyt, *Carbohydr. Res.*, 1993, **245**, 97.
184. R. Ueno and S. Kuno, *Lancet*, 1987, **3**, 796.
185. M. Yalpani, *CRC Critical Rev. Biotechnol.*, 1985, **3**(4), 375.
186. J. Edelman and T. G. Jefford, *New Phytologist*, 1968, **67**, 517.
187. A. Franck and L. de Leenheer, in *Biopolymers, Vol. 6. Polysaccharides II: Polysaccharides from Eukaryotes*, ed. A. Steinbüchel, Wiley-VCH, Weinheim, 2002, p. 439.
188. R. Greig-Smith, *Proc. Linn. Soc. N. S.*, 1901, **26**, 589.
189. Y. W. Han, *Adv. Appl. Microbiol.*, 1990, **25**, 171.
190. M. Iizuka, N. Minamiura and T. Ogura, in *Glycoenzymes*, ed. M. Ohnishi, Karger, Tokyo, 2000, **15**, p. 241.
191. A. van Laere and W. van den Ende, *Plant Cell Environ.*, 2002, **25**, 803.
192. E. O. Lippmann, *Chem. Ber.*, 1881, **14**, 1509.
193. H. G. Pontis, in *Methods in Plant Biochemistry Carbohydrates, Vol. 2*, ed. P. M. Dey, Academic Press, New York, 1990, p. 353.
194. S. K. Rhee, K. B. Song, C. H. Kim, B. S. Park, E. K. Jang and K. H. Jang, in *Biopolymers, Vol. 5. Polysaccharides I: Polysaccharides from Prokaryotes*, ed. E. J. Vandamme, S. de Baets and A. Steinbüchel, Wiley-VCH, Weinheim, 2002, p. 351.
195. J. Seibel, R. Moraru, S. Goetze, K. Buchholz, S. Na'amnieh, A. Pawlowski and H. J. Hecht, *Carbohydr. Res.*, 2006, **341**, 2335.
196. R. Cartarius, *Bioabbaubare Tenside durch reduktive Aminierung von Isomaltulose – Katalysator und Verfahrensentwicklung*, Dissertation, TU Darmstadt, 1999.
197. W. Dreissig and P. Luger, *Acta Crystallogr. Sect. B*, 173, **29**, 514.
198. Y. Nagai, T. Sugitani and K. Tsuyuki, *Biosci. Biotech. Biochem.*, 1994, **58**, 1789.

199. P. S. J. Cheetham, in *Methods in Enzymology*, ed. K. Mosbach, Academic Press, New York, 1987, **136**, p. 432.
200. W. Wach, in *Ullmann's Encyclopedia of Industrial Chemistry*, Wiley-VCH, Weinheim, 7th edn, 2004, DOI: 10.1002/14356007.a12.047.
201. J. S. James, *AIDS Treat. News (electronic journal)*, 1988, **50**.
202. M. A. Godshall, *Lut. Sug. J.*, 2001, **103**, 1233.
203. J. W. Yun, in *Recent Advances in Fructooligosaccharides Research*, ed. S. Norio, B. Noureddine and O. Shuichi, 2007, 19, p. 357.
204. M. J. Mabel, P. T. Sangeetha, K. Platel, K. Srinivasan and S. G. Prapulla, *Carbohydr. Res.*, 2008, **343**, 56.
205. J. Yoshikawa, S. Amachi, H. Shinoyama and T. Fujii, *Biotechnol. Lett.*, 2008, **30**, 535.
206. S. Kralj, K. Buchholz, L. Dijkhuizen and J. Seibel, *Biocatal. Biotrans.*, 2008, **26**, 32.
207. A. Marx, M. Poetter, S. Buchholz, A. May, H. Siegert, B. Albert, G. Fuchs and L. Eggeling, WO/2007/141208, 2006 (Evonik Röhm GmbH).

CHAPTER 9.4

Wood

ELISABETH WINDEISEN AND GERD WEGENER

Holzforschung München, Technische Universität München, Winzererstr. 45, D-80797 München, Germany

9.4.1 Introduction

The twenty-first century is expected to become the century of bio-based materials and biorefining processes,[1–5] which are now on the horizon to replace fossil-based materials gradually but continuously. Comparable developments are taking place in the energy sector where a multitude of renewable energy resources are complementing and substituting fossil-based energy and nuclear power.[6]

Driving forces are primarily the fact of dwindling fossil resources, along with uncertainties concerning availability and prices as well as the globally accepted certainty of society and politicians about climate change and the resulting effects on quality of life and living space for man, animals and plants. Keeping these aspects in mind, including all basics of sustainability, it is no exaggeration to say that forests and wood utilization play an important role in meeting global demands for materials and energy on the one hand and on the other hand contributing effectively to quality of life and climate protection.

In the context of "Green Chemistry" and against the background of sustainability wood utilization is demonstrated on the basis of the chemistry of wood in four sectors:

- Pulp and paper;
- Wood-based composite materials;
- Modified solid wood products;
- Wood as building material.

RSC Green Chemistry No. 4
Sustainable Solutions for Modern Economies
Edited by Rainer Höfer
© The Royal Society of Chemistry 2009
Published by the Royal Society of Chemistry, www.rsc.org

Wood and biomass are raw materials (Figure 9.4.1) for traditional building products like sawn timber, prefabricated timber elements and systems as well as wood-based materials in the form of boards (*e.g.* particle- and fibreboards), but also as construction elements from glued products like laminated beams, plywood, oriented structural boards or laminated veneer lumber.[7]

New developments give options in the form of hybrid materials combining wood with conventional plastics (*e.g.* Wood Plastic Composites, WPC) (Section 9.4.4) or as solid wood improved by heat treatment or chemical modification, *e.g.* acetylation, oil treatment *etc.* (Section 9.4.5).

Additionally, wood and biomass represent traditional and new energy products (such as pellets or briquettes) to be used both in small private heating systems and also in large biomass heat processing units, also producing electrical power by power-heat-coupling systems. In this sector wood and biomass compete with typical fossil energies such as oil, gas and coal but also with additional renewable energy sources like wind, water, geothermal heat and waste, besides the most promising natural energy coming from the sun and used by different techniques like sun collectors, photovoltaic systems, *etc.* For electricity, nuclear power is still an important factor for many countries worldwide. In the field of fuels for stationary and mobile systems biofuels based on forest and agricultural biomass (biodiesel, ethanol, *etc.*) are the option as alternatives to fossil fuels like petrol and diesel but also to future hydrogen power systems, *etc.* In the wide field of chemistry wood- and biomass-based and derived green chemicals compete with established oil-based chemicals and plastics as described in other chapters.

In the context of chemical utilization of wood, other parts of the tree, the so-called Non-Timber-Forest-Products (NTFP), have always been closely related

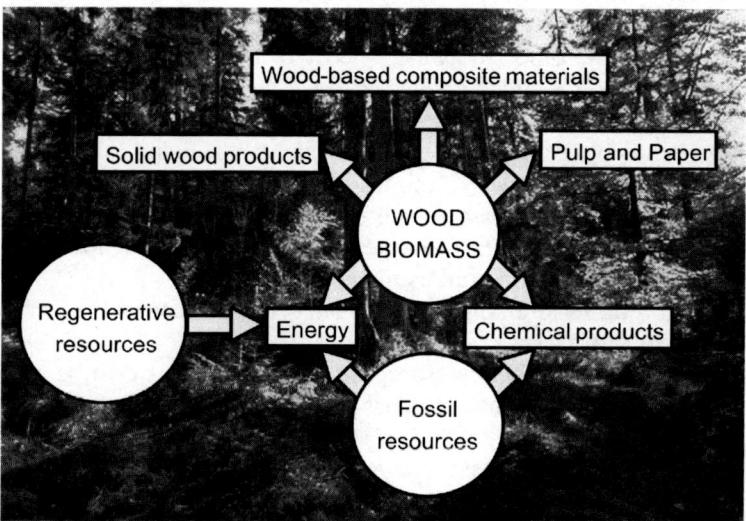

Figure 9.4.1 Wood utilization in competitive markets.

to the cultural development of mankind. The Egyptians already used myrrh for embalming mummies and for the preparation of salves in 2000 BC; the hand-written books from the High Middle Ages could not have been written without ink from oak galls and gum arabic. Besides the utilization as foods (fruits, nuts, spices like cinnamon), raw materials (exudates, *e.g.* maple syrup, caoutchouc, chicle for the production of chewing gum), odorous substances for perfumes and cosmetics (*e.g.* eucalyptus oil), dyestuffs, tanning agents and many others have been used. The progress in organic chemistry at the beginning of the twentieth century changed the market for NTFP. More and more, it was possible to produce these products synthetically. Nevertheless a few products, such as nat-ural caoutchouc, still remain indispensable due to their excellent properties.

9.4.1.1 Perspectives of Sustainability

Sustainability intrinsically means planning ahead for the future. Apart from economic action, ecological criteria as well as social aspects and responsibility need to characterize the life and economic activities of a continually growing world population. Only then will the planet be able to provide a humane foun-dation for life on Earth. Today, there are 6.5 billion people living on Earth, in 15 years there will be 1.5 billion more. Especially in industrial countries this presents increasing pressure on politics, economy and society to rapidly and effectively review economic processes and consumer behaviour. At global, regional and local levels, suitable strategies will have to be developed or made use of. This will not be possible by way of a more efficient technology alone, but can only succeed if economic, social and cultural rethinking result in joint efforts of redirected acting. Sustainability in its sense of planning for the future is multi-faceted: thus the climate is to be protected, resources are to be conserved, energy should be employed efficiently, renewable resources and renewable energies are to be uti-lized and at the same time biodiversity needs to be sustained and, on a global scale, fair trade is to be safeguarded. Not least, individuals find themselves in a position where there is a need to change their hitherto existing concept of prosperity from living standard to quality of life. While living standard describes personal materialistic prosperity in terms of gross national income or *per-capita* income, quality of life comprises much more. The latter also implies other ele-ments and values such as education, health, social status, occupational career, working environment quality and availability of nature.

Apart from long-term changes to be aimed for, measures that swiftly take effect and setting courses is required in order to fulfil the climatic objectives of the European Union by 2020: carbon dioxide emissions are to be reduced by 20%, renewable energies are to be part of the energy mix and energy is to be employed more efficiently by 20%. In this context, the former director of the UN Environmental Programme, Klaus Töpfer, recently wrote on the topic of "Sustainable Construction and Reconstruction" "that it would be out of all reason, even inexcusable, not to pay regard to the contribution of the forests and the usage of wood to its full extent".[8]

9.4.1.2 Forests as Ecosystem and Resource

The forests of Earth are multi-faceted and constitute vital ecosystems. With their 39 Mio km² they cover about one-quarter of the land area worldwide. About half of the global forest area is shared by five countries – Russia, Brazil, Canada, USA and China. This ought to be considered with regard to forestry and timber strategies, when it comes to third-party certification with respect to sustainable management. *Per annum*, still 130,000 km² of forests are cleared – which is more than the entire forest area of Germany. This contrasts with an increase in the forest area in other regions by 57,000 km². However, this mostly does not concern real forest, but plantations, mainly in the southern hemisphere.

In many cultures forests play an important role. They fulfil many tasks for water sources, soil and air. They are habitats for plants, animals and humans alike. Via photosynthesis, they produce wood, the quantitatively most important raw material throughout the world. The forest industry supplies industry with wood, which in turn processes it into products like pulp and paper, composite materials, construction material and energy. A very large part of humanity is dependent on firewood as energy source. With 52%, more than half of the globally utilized 3.4 billion m³ (= solid cubic metre) wood is used as fuel wood.[9]

Since forests serve as wildlife habitat, biotope and recreation refuge, and supply wood as raw material and a source of energy, their cultivation and afforestation effectively contribute to the objectives of sustainability and climate-protection: since forests take up and store carbon dioxide, they help to attenuate climate change, especially generated by the usage of fossil fuels like carbon, crude oil and natural gas. Wood products retain this carbon even after the felling of a tree. At the end of their life span they can be burnt, still providing similar amounts of energy as direct burning of wood does. Presently, forests in Germany take up and store temporarily about 8% of the around 900 Mill. t carbon dioxide annually emitted in Germany. Subsequently, this carbon is retained in the forests themselves or in wood products. Moreover, wood products replace products that, like plastics, are made from crude oil as well as products like steel, concrete or aluminium, which have to be produced and disposed of in energy-intensive processes. In order to meet the growing world population's increasing demands for products and energy without continuing to endanger the climate, not only does energy need to be saved as well as energy and resources to be employed more efficiently, but also biomass and thus especially forests need to be utilized in sustainable ways.

9.4.1.3 From Wood Resources to Wood Products

The fact that not only forests but also wood products retain carbon is an important strategic element concerning climate protection politics. From the nearly 1.7 billion m³ industrial roundwood harvested annually, 420 Mio m³

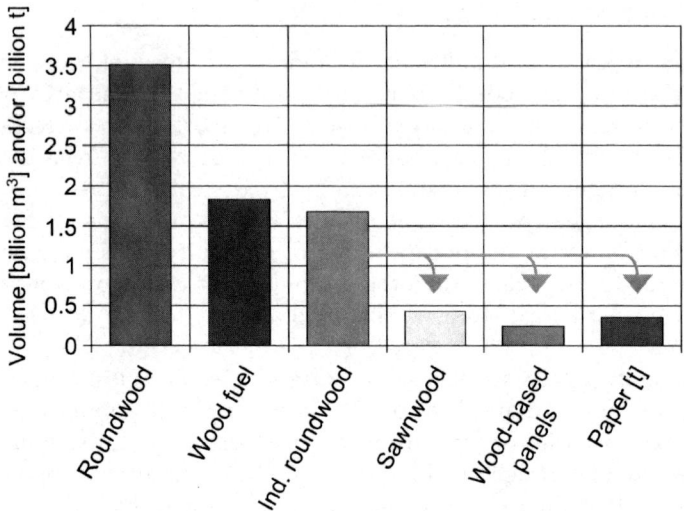

Figure 9.4.2 Global annually harvested amount of round wood and its utilization.

timber, 220 Mio m^3 wood-based materials and 370 Mio mto paper are produced (Figure 9.4.2).[9]

Even though Europe only has 5% of the global forest area, it produces nearly 30% of all wood products worldwide due to its large stocks of wood, efficient manufacturing technologies and a high rate in net product.

Today, in a global comparison, Germany, ahead of Sweden (2.9 billion m^3), holds the largest stocks of wood in its forests (3.4 billion m^3). Eight percent of the entire European net product is contributed by the forest, wood and paper industry. Europe also holds the globally leading market for wood and wood-based products. 60% of all exports and nearly 50% of all imports are transacted in the European market.

In view of economy, and from the viewpoint of ecology, forest, wood and paper constitute "heavyweights". According to the European definition also downstream economic sectors like the printing and publishing industry are assigned to the forest and wood domain. Ecological advantages arise already as wood is being produced in the forest. In addition wood products replace fossil and other non-renewable resources and their subsequently manufactured products.

Since the residues accruing during wood processing make good raw material for further products, the wood industry hardly produces any refuse. When wood products like paper are no longer wanted, they can either be recycled or burnt, subsequently supplying new products or utilizable energy. Therefore, already today, the forest and wood domain contribute substantially to meeting important environmental demands. Thus, a renewable resource is employed snatching greenhouse gas from the air; its production processes use up little energy and generate few pollutants. Not only does the bioproduct wood supply sustainable energy, but it also facilitates building sustainability.

For climate protection construction material too needs to be selected according to ecological criteria. This is particularly important, because construction represents an essential economic element, also reflecting the intellectual and cultural value benchmark of a society.

In Germany, for example, more than two-thirds of the wood utilized "ends up" in buildings as construction material, insulating material or furniture. Therefore, the forest and wood industry plays an important part where environmental objectives are concerned. Its task is to offer suitable wood species characterized by the required properties like density, strength, stability, hardness, durability, surface shape and colour. From glulam (glue-laminated timber) to ultra-light structural particle board, high-capacity raw material and suitable building parts as well as structural systems need to be supplied, fulfilling all the technological, aesthetic and hygienic requirements of future construction.

Building has always been of significant importance for mankind's culture and development. It determines the appearance of our villages and cities, reveals the social structures of the society and is an important economic factor itself.

Architecture and construction are essential characteristics of all types of buildings in general. Under ecological and environmental aspects, however, the choice of the building materials is additionally gaining more and more importance. Taking these aspects into account a renaissance of wood as a building material is taking place. There are good reasons for that, *e.g.*:

- Mankind was always familiar with the utilization of wood;
- Wood is produced in an extraordinary factory under exceptional environmentally friendly conditions; forests additionally fulfil a multitude of environmental beneficial functions;
- Wood is produced by the basic process for all life on Earth: photosynthesis;
- Photosynthesis implies capturing CO_2 from the atmosphere and fixation of carbon in the material, thus acting as a carbon sink. Any use of wood as building material therefore represents an important contribution to one of the biggest challenges of mankind today: climate change as a result of an exaggerated release of CO_2 resulting from the use of fossil resources;
- Photosynthesis generates a material for multiple use and a wide range of products due to its chemical composition and structure ($\rightarrow$ green chemistry, pulp and paper) and anatomical and morphological structures and properties as well;
- This makes wood the most important and extremely variable renewable raw material and a universal construction material with many unique characteristics and properties such as:

  - high strength and stiffness and low mass at the same time;
  - it can be processed easily and with low energy input;
  - it has excellent heat and sound insulation properties;
  - it provides a healthy living climate in houses due to the ability of absorbing and releasing moisture;

    o it reveals important fire safety characteristics;

    o thousands of different wood species with different properties and appearance offer a wide range of options for the use of wood as construction, equipment and decoration material;

    o wood and wood products keep their energy content over the whole lifetime of a building and the energy can be used at the end of the life span.

Wood for building purposes can be used in the form of roundwood, but mostly as sawn timber such as beams, planks, boards or slats as traditional components of building elements such as floors, walls, ceilings, roofs, *etc*. Mostly softwoods, preferably spruce or pine wood, but also hardwoods such as oak, are used *e.g.* for windows, doors or parquet.

The solid wood products are supplemented by a large number of wood-based composites (see Section 9.4.4).

9.4.2   Chemistry of Wood

9.4.2.1   Survey

Wood is a material with very differing properties: different wood species, the site, the tree age, stemwood or knotwood, heartwood or sapwood offer an extremely broad range of biological, chemical, technological and appearance properties.

The vast diversity of nature can be a disadvantage for industrial purposes as all the various wood properties must be taken into account during processing. On the other hand, many advantages emerge as well. There is always at least one wood species fitting for the intended use.

Out of about 30,000 woody plant species worldwide some 500 are regarded as timber available in a significant amount and are being traded internationally. 20–30 commercial wood species are commonly and typically used in Europe. The most important softwood species are spruce, pine, larch, fir and Douglas fir, and the most important hardwood species are beech, oak, aspen, birch, poplar, maple, lime, locust along with lesser available species, such as cherry or walnut.[10,11]

It is a matter of common knowledge that from a chemical point of view, wood is composed of mainly organic compounds, which include, on the one hand, macromolecular components which are important for pulp production particularly the polysaccharides cellulose and polyoses (hemicelluloses) as well as lignin, and on the other hand the low molecular components of wood. In the case of the latter being of organic origin they are referred to as accessory components or more frequently as extractives according to their isolation method while the inorganic components are usually identified as ash (see Figure 9.4.3).[12]

In nearly all wood species of the temperate zones 90–99% of the wood substance is made up from polysaccharides (cellulose, polyoses) and lignin

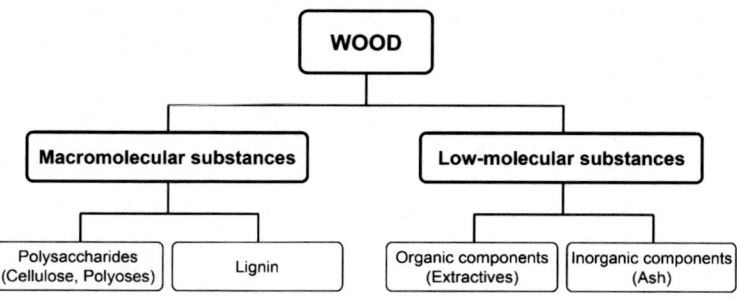

Figure 9.4.3 Chemical composition of wood.[12]

Table 9.4.1 General chemical composition of wood.

Constituent	Hardwoods [%]	Softwoods [%]
Cellulose	40–50	40–50
Polyoses (hemicelluloses)	25–35	20–30
Lignin	20–30	25–35
Extractives	0.5–10	0.5–10
Inorganics (ash)	0.1–1	0.1–1

forming the cell wall; the extractives can often be found in comparatively low percentages (Tables 9.4.1 and 9.4.2).

Even if, in the first instance, it might seem that this material does not vary a lot, the different amounts and differing chemical composition of the polyoses (see Section 9.4.2.3) and the lignins (see Section 9.4.2.4) in hardwood and softwood strongly affect the manufacturing processes. This applies for the extractives too, which due to their large chemical diversity influence some of the properties of wood and behave respectively different with regard to the chemical utilization of the wood.

As previously mentioned the chemical diversity of the wood species developed by nature most often is a utilization benefit; however, with regard to the chemical utilization, methods and processes must often be optimized for each wood species separately, as due to the specific chemistry of the wood species it is not always possible to transfer them.

In the following, the chemical components cellulose, polyoses, lignin, extractives and ash are briefly described. Further information can be found in various books and literature cited therein.[12–17]

9.4.2.2 Cellulose

With an estimated vegetable production of 10^{11} mto per annum, cellulose is on a quantity basis the most important organic natural material worldwide.[12] Since cellulose is a homopolymer it is composed of a single component, namely the D-glucopyranose, which is linked by ß-1-4-glycosidic bonds. The cellulose

Table 9.4.2 Comparison of the chemistry of various wood species excluding inorganic components[18] (with permission of Wiley-VCH Verlag GmbH & Co. KGaA).

Wood species	Common name	Cellulose [%]	Polyoses [%]	Lignin [%]	Extractives [%]
Softwood					
Abies balsamea	Balsam fir	38.8	28.5	29.1	2.7
Pseudotsuga menziesii	Douglas fir	38.8	26.3	29.3	5.3
Tsuga canadensis	Eastern hemlock	37.7	27.9	30.5	3.4
Juniperus communis	Common juniper	33.0	30.3	32.1	3.2
Pinus radiata	Monterey pine	37.4	33.2	27.2	1.8
Pinus sylvestris	Scots pine	40.0	28.5	27.7	3.5
Picea abies	Norway spruce	41.7	28.3	27.4	1.7
Picea glauca	White spruce	39.5	30.6	27.5	2.1
Larix sibirica	Siberian larch	41.4	29.6	26.8	1.8
Hardwood					
Acer rubrum	Red maple	42.0	28.9	25.4	3.2
Fagus sylvatica	Common beech	39.4	33.3	24.8	1.2
Betula verrucosa	Silver birch	41.0	32.4	22.0	3.2
Betula papyrifera	Paper birch	39.4	34.5	21.4	2.6
Alnus incana	Gray alder	38.3	30.9	24.8	4.6
Eucalyptus globulus	Blue gum	51.0	25.2	21.9	1.3
Acacia mollissima	Black wattle	42.9	33.6	20.8	1.8

content in wood varies between 40 and 50%. The mean degree of polymerization of native cellulose ranges on average between 8000 and 10,000. Cellulose forms linear, unbranched chains. Intramolecular hydrogen bridges, which may predominately occur in native cellulose (cellulose I) between the hydroxyl groups on C2 and C6 and the hydroxyl groups on C3 and the C5 oxygen (Scheme 9.4.1), prevent free rotability and account for the relatively high stiffness of the straight-chained molecular chains. Due to the formation of intermolecular hydrogen bridges further degrees of order are formed. Thus, in solid state, crystalline regions alternate with the ones of a lower degree of order (amorphous regions).

Thirty to one hundred strands of cellulose aggregate into so-called elemental fibrils (cross section approximately 2–4 nm) *via* H-bridges, which in turn aggregate into microfibrils (cross section approximately 10–30 nm) and these into macrofibrils (cross section approximately 500 nm). The orientation of the cellulose fibrils in the wood cell wall differs. The strongest cell wall layer, for example, shows a constitution parallel to the axis which accounts for the high tensile strength of wood.[12]

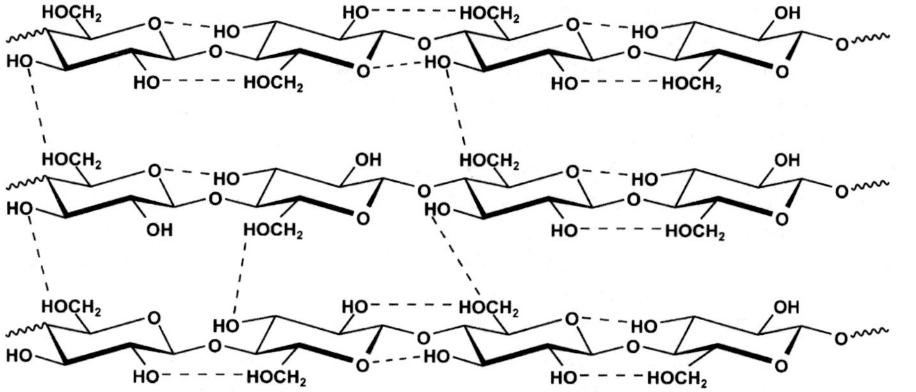

Scheme 9.4.1 Structure and molecular chain constitution of cellulose I (native cellulose) with hydrogen bonds.

Due to its structure, cellulose does not dissolve in common solvents nor in water or in dilute acids, even though non-crystalline structural regions can swell in water and in doing so adsorb 8–14% of its volume to water. In alkaline solutions, cellulose strongly swells at first and the short-chained parts with DP < 200 are dissolved (*e.g.* α-cellulose can be obtained after treatment with 17.5% NaOH or 24% KOH). Only strong acids cause a hydrolytic degradation. In addition to several solvent systems containing heavy metal salts and ammonia or aliphatic di- or triamines (Schweizers reagent respectively cuoxam (Cu/NH₃), cuen (Cu/ethylene diamine), nitren (Ni/diethylene triamine), 4-methylmorpholine-4-oxide (NMMO), dimethylsulfoxide blends and ionic liquids (ILs), besides others, are employed as organic solvents for cellulose.[12,18–23]

The chemical reactivity of cellulose generally depends on the percentage of accessible hydroxyl groups, which are not occupied by hydrogen bonds. At first, chemical reactions occur in amorphous regions of a lower degree of order until crystalline regions take part. A further decisive factor for the succession and the extent of the interaction is the molecular size and character of the interacting agent in relation to the capillaries and interstices allowing its penetration into the cellulose substrate.[24]

The possible uses of cellulose of various forms is manifold (fibres, cellulose derivatives as well as new products such as cellulose aero gels), however a chemical synthesis of cellulose, as demonstrated by nature, has not yet been realized.[20,25,26] In order to meet the high requirements for products made of cellulose, it is usually necessary to obtain cellulose from plant material by selective removal of non-cellulose components. In the case of wood, mechanical, mainly mechanical-chemical, as well as chemical pulping is employed (see Section 9.4.3).

9.4.2.3 Polyoses (Hemicelluloses)

Polyoses (hemicelluloses) are irregular, amorphous polymers, which, in contrast to cellulose, are composed of various sugar components like hexoses (glucose, mannose, galactose, rhamnose) and pentoses (xylose, arabinose) as well as uronic acids. The molecular chains are significantly shorter than cellulose and have a significantly lower DP of 50–250. Furthermore, polyoses have side groups (*e.g.* parts of the OH groups are substituted for acetyl groups) as well as side chains. Due to their structure and chemical composition polyoses can serve as bonding components between lignin and cellulose in the cell wall.

As already mentioned, there are differences in the content and the composition of polyoses in hardwood and softwood. While xylans (pentosans) are dominant in hardwood, mannans (hexosans) predominate in softwood. The xylans are so-called homopolymers, as the main chain consists of xylose only (Scheme 9.4.2), whereas the mannans present a heteropolymer consisting of glucose and mannose (Scheme 9.4.3). Furthermore, softwood has a higher content of polyoses as well as acetyl groups compared to softwood. The higher content of acetyl groups in hardwood can cause elimination of acetic acid in certain procedures (*e.g.* steaming of beech or thermal modification, Section 9.4.5).

In contrast to cellulose, polyoses are alkaline-soluble and partly water-soluble (*e.g.* arabinogalactan of larch).

At present, only a few polyoses from wood are chemically used. An example is the utilization of xylose (Figure 9.4.4), which can be obtained as a by-product in the pulping process of beech or birch wood and which is used in the form of the sugar alcohol xylitol as a sugar substitute (E 967) as well as arabinogalactan as a thickening agent (E 414) in the food industry.

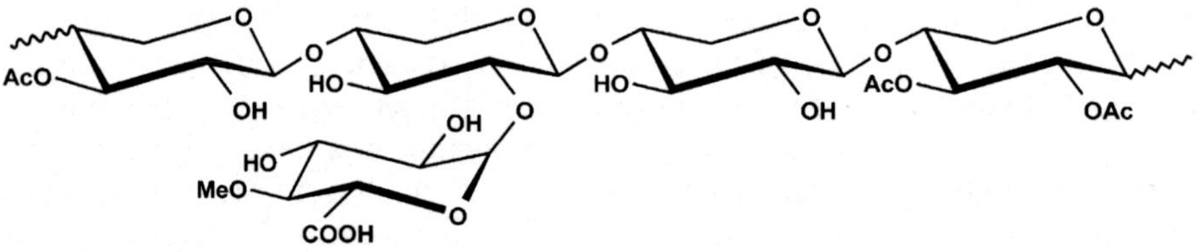

Scheme 9.4.2   Hardwood xylan.

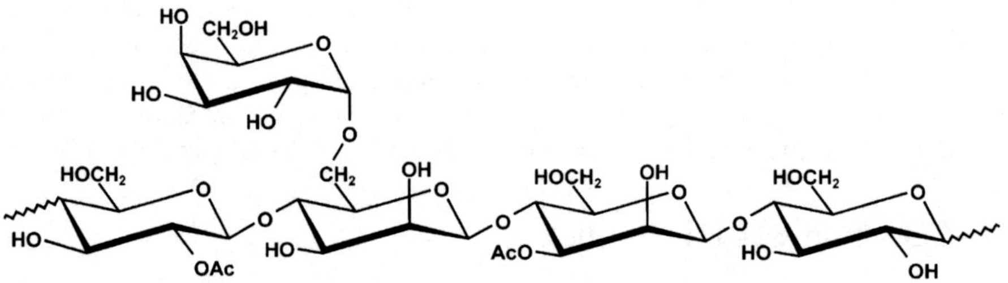

Scheme 9.4.3   Softwood glucomannan.

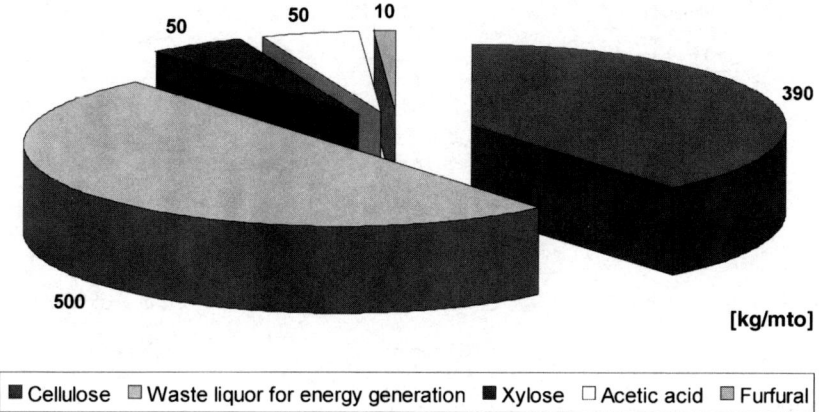

50   50   10

390

500

[kg/mto]

■ Cellulose   ▢ Waste liquor for energy generation   ■ Xylose   ▢ Acetic acid   ▢ Furfural

Figure 9.4.4   Example of mass balance of a beech sulphite pulping.

9.4.2.4   Lignin

Lignin (lat. lignum = wood) is the second most common organic natural material and the real lignification substance. By integration of lignin in the cell wall and by association with the cellulose fibrils *via* the polyoses, high material strength develops, whereby the tree can stand upright and absorb static and dynamic forces. Hence, the presence of lignin in the cell wall of the plant is a pre-condition for the strength of the wood.

The aromatic lignin is formed by way of the so-called Shikimate biosynthesis from glucose. Lignin components form phenylpropan units with differing amounts of methoxy groups (Scheme 9.4.4). Depending on the plant group, the percentage of the basic units of coumaryl- (H), coniferyl- (G-) and sinapyl alcohol (S-) varies. Thus, softwood mainly contains G-lignin, whereas hardwood is composed of G- and S-units. The lignin of softwood lignin is more strongly condensed than the lignin of hardwood; it features a higher carbon content due to the lower amount of methoxy groups as well as less acid-soluble parts.[12]

Lignin is an amorphous, three-dimensional heteropolymer without any defined chemical structure (Scheme 9.4.5). According to present knowledge, the polymerization not only takes place by reaction of monomers with polymer stages but also by reaction of already existing aggregates (di-, tri-, *etc.* lignans). The described molar masses, *e.g.* approximately 10,000 in the case of spruce MWL (milled-wood-lignin) produced in the laboratory, of isolated lignins are not uniform or representative, as they depend on the isolation method.

Lignin shows thermoplastic properties at approximately 160 °C (glass transition), which are used for the manufacture of certain composites (Section 9.4.4).

Regarding lignin, it always has to be considered that it is not a uniform and defined product, but it can vary a lot depending on the origin and the isolation

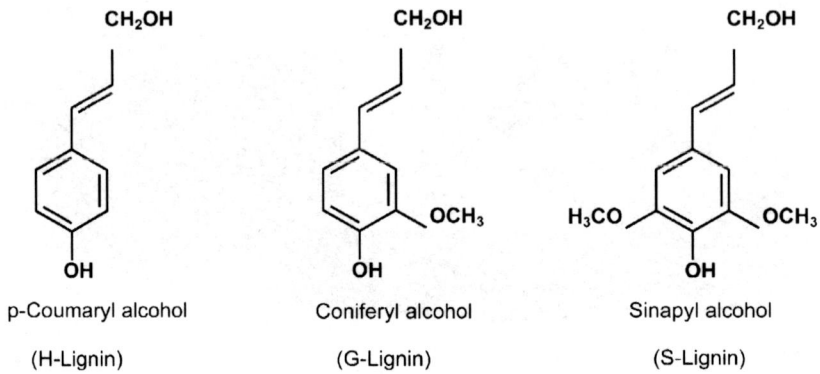

Scheme 9.4.4   Basic units of lignin.

method.[12,27] This is one of the reasons why lignin is materially utilized only in specific cases, although it is, along with the fossil resources, the only natural resource for aromatic compounds.

Most of the 70 Mio mto of lignin annually accruing as a by-product in pulping is used for energy generation in the course of sulphate pulping (Kraft pulping). Only about 1.5% of lignin and its by-products are materially utilized. In particular sulphurous lignin variations like lignin sulphonate obtained from the sulphite pulping are commercially utilized in quantities of 1,000,000 mto a^{-1} or Kraft lignins (alkali lignin) obtained from Kraft pulping in quantities of 100,000 mto a^{-1}.[28-30]

The existing markets of lignin are either for low-value products as lignosulphonates mainly in dispersing and binding applications or limited to very narrow market segments as high-quality dispersants from chemically modified Kraft lignin as well as speciality chemicals like vanillin or agrochemicals.[31] Nevertheless, numerous potential applications have been studied, e.g. as tanning agent, resin component for wood-based boards, dispersing agent, antioxidants and binder for different purposes.[12,29,32] Lignin has a relatively high application potential in the sector of wood-based products, where wood adhesives, phenol and epoxy resins or carbon fibres can be made from lignin. Approximately 300,000 mto a^{-1} of lignin are required for the resin market. Lignin with a high degree of purity is employed as a UV-stabilizer for the polyolefin production or as a filler for the production of polyurethane.[31,33]

In the form of thermoplastically processed products (e.g. ARBOFORM® and FASAL®), lignin is used as a composite material for various applications (see Section 9.4.4). However, these are mainly niche products shown by the production volume of ARBOFORM® of approximately 300 mto a^{-1}.[34] A significantly higher material utilization of lignin would, on the one hand, increase the added value of wood and on the other hand, crude oil-based substances like plastics could partly be replaced.

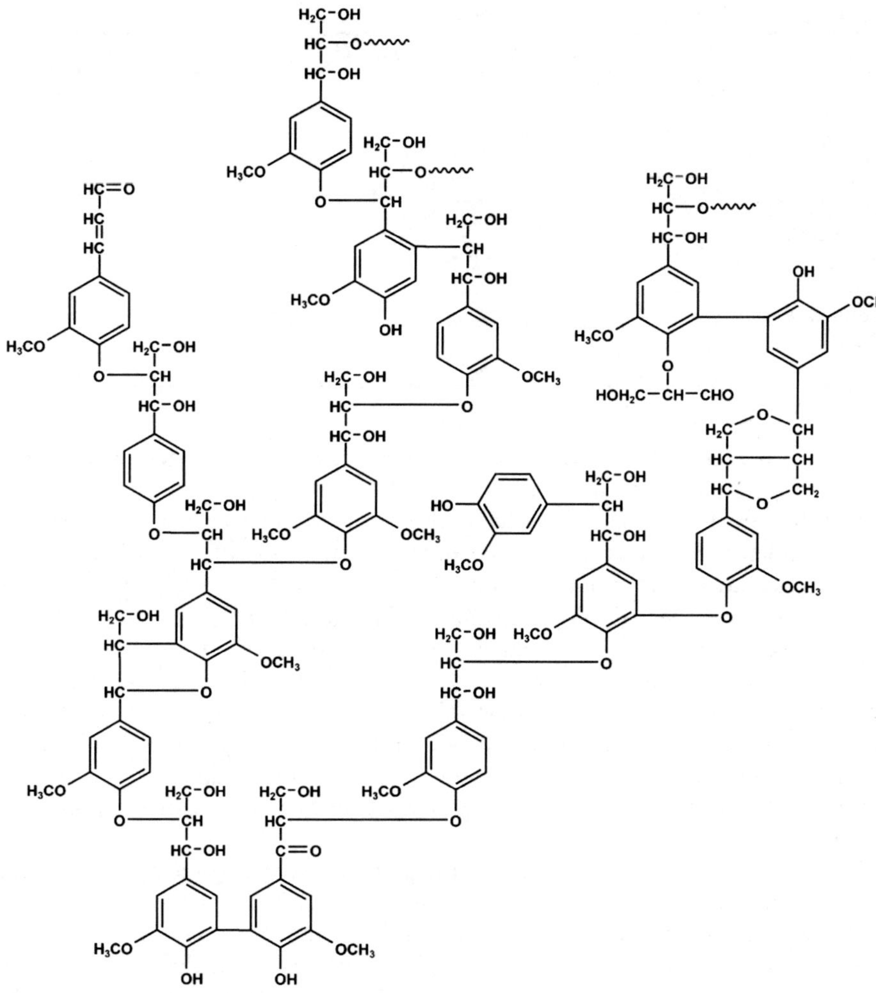

Scheme 9.4.5 Section of softwood lignin.

9.4.2.5 Extractives

Table 9.4.3 and Scheme 9.4.6 show that contents as well as structure of the extractives can vary greatly between the wood species. Furthermore, there are large differences between wood parts of, for example, sapwood and heartwood (see below) or between wood and bark. Therefore, it is very difficult to briefly describe extractives and the reader is referred to the literature.[12,14,15,16,17]

Since the name of the extractives refers to their isolation method, it follows that the term extractives includes non-specifically all substances which can be extracted from the wood using solvents with most different polarities. Hence the term *extractives* must be considered in a much differentiated manner, as

Table 9.4.3 Typical extractive contents in wood.[35]

Solvent	Cyclohexane/Ethanol [%]	Water [%]
Softwood		
Spruce	1–2	0.5–2
Pine	4–7	0.5–2
Larch	3–6	8–10
Douglas fir	3–6	3–6
Hardwood		
Beech	0.5–1.5	<1
Oak	3–8	6–10
Tropical wood		
Teak	5–7	2–3
Quebracho	up to 40	20–25

most probably only part of the low molecular organic compounds of a number of wood species have yet been described.

The classification of extractives is rarely made strictly according to substance classes. Therefore it can usually be divided into lipophilic and hydrophilic extractives due to the polarity (Table 9.4.4) or into primary and secondary extractives due to their biological function (Table 9.4.5).

The first differentiation is used in various sectors of the wood-processing industry like the pulp and paper industry, as its chemical behaviour is defined by the polarity. Depending on the chemical pulping, certain extractives can show negative effects on the production, *e.g.* pinosylvin and its mono-methyl ether during acid sulphite pulping.[13] In contrast, other pine extractives (resin acids, terpenes, *etc.*) are useful by-products in Kraft pulping (see Section 9.4.3).

Looking at the biological function, the primary extractives are necessary for maintaining the metabolism and supply of the components for the biosynthesis of the secondary substances. Typical primary substances include carbohydrates (sugars), which can be found for example as monosaccharides, disaccharides and as starch as well as fats and proteins. As sapwood is responsible for the constant water and nutrient supply of the whole tree, mainly primary extractives are involved, which are also used by various micro-organisms as a source of food.

Heartwood contains very few or no primary extractives, but usually significantly more secondary extractives than sapwood. The secondary substances include almost all the other extractive classes, hence they have enormous structural variety including most different substance classes like alkaloids, quinones, fats, oils, saponins, terpenes, tropolones, waxes and a number of aromatic or phenolic compounds (stilbenes, lignans, flavonoids, tannins, *etc.*). The secondary substances play an essential role in the use of wood, for example the heartwood of most of the wood species is more durable than the sapwood.

Whether and to what extent the chemical, biological, physical and optical properties of the wood species are influenced by the extractives depends generally on the quantity and the substance class, which is in turn specific to the

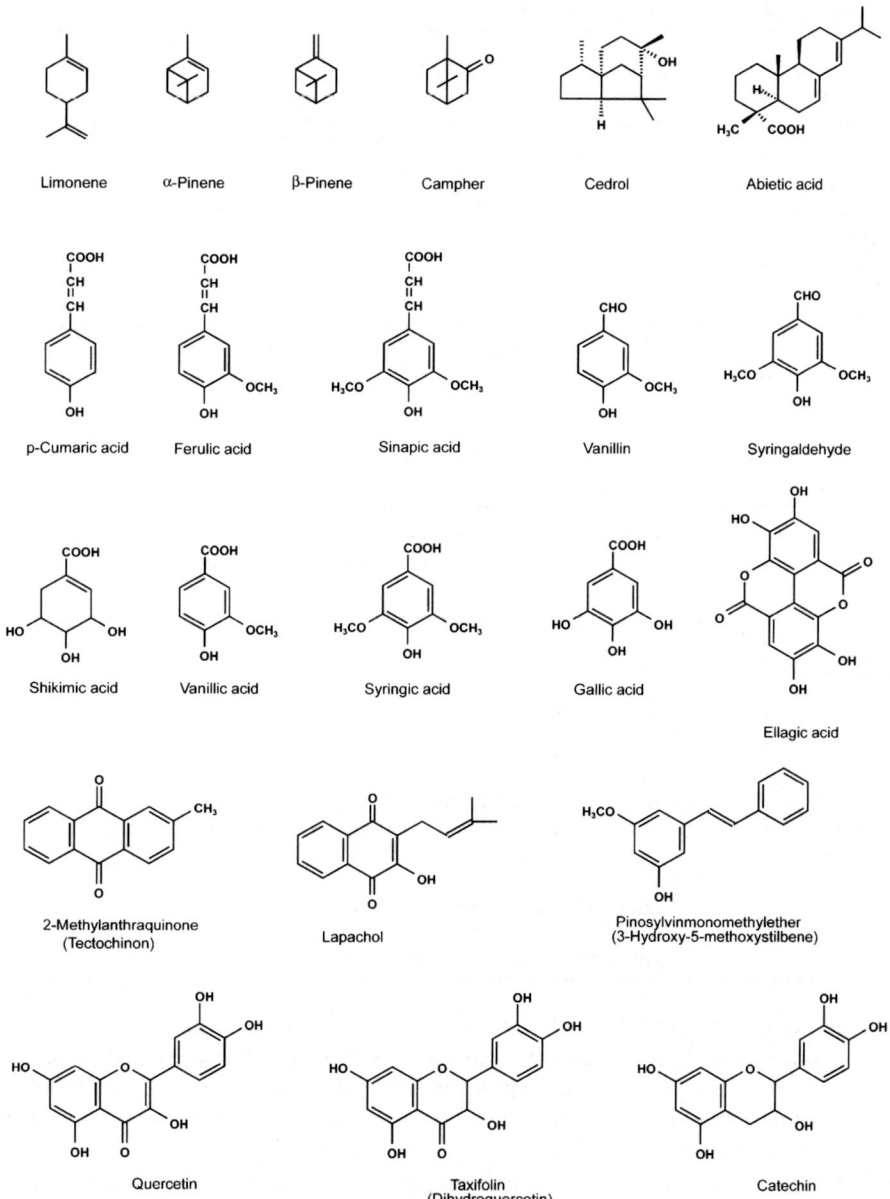

Scheme 9.4.6 Examples of extractives.

wood species. In Table 9.4.5, the effects on properties as well as the processability are summarized.[35]

A very important criterion determined by the extractives is the natural durability of wood. Furthermore, colour, odour and taste are mainly caused by

Table 9.4.4 Solubility of wood extractives.[14]

Major compound classes	Resin acids, monoterpenoids, other terpenoids	Fats, fatty acids, steryl esters, sterols	Phenolic substances	Glycosides, sugars, starch, proteins	Inorganics
Solubility in					
Alkanes	+ + +	+ + +	0	0	0
Diethyl ether or dichloromethane	+ + +	+ + +	+ +	0	0
Acetone	+ + +	+ + +	+ + +	+ +	+
Ethanol	+ +	+ +	+ + +	+	+
Water	0	0	+	+ + +	+ +

+ + +: easily soluble, + +: soluble, +: slightly soluble, 0: insoluble

Table 9.4.5 Extractives in wood: functions, effect on properties and processability.

Function	Primary extractives: substances for maintaining the metabolism; formation of components for the biosynthesis of secondary substances; Secondary extractives: substances for the protection against biological decay *etc.*
Effect on properties	Colour, Odour, Taste, pH value, Natural durability, Light resistance, Dimensional stability, Calorific value, Combustibility *etc.*
Effect on processability	Reactions with additives or metals (adhesion properties, staining, corrosion *etc.*); Surface treatment; Reactions during pulping (e.g., pitch problem); Toxic or allergic effects; Volatile organic compounds (VOC)

the extractives present. In addition to the decorative aspects in various areas of wood utilization, extractives are distinguished, for example, for the production of wine, whiskey and cognac, where mainly white oak and chestnut wood is used for barrels. The colourless distillates absorb the phenolic compounds of the wood as well as wood dye, where the typical more or less golden colour of, for example, cognac comes from. Concurrent oxidation and esterification processes improve the aroma and the taste.[36]

Furthermore, extractives determine the pH value due to their solubility, which can often play a role in wood processing. Certain extractives, which can be found for example in larch, oak or various tropical woods, can cause staining or even corrosion when in contact with ferrous materials. In the same way, light resistance or swelling and dimensional stability of the wood can be affected by lipophilic extractives. Furthermore, a relation between certain extractives and the calorific value or even combustion has been reported.[15]

In general, reactions with additives might occur in such a way that adhesion properties or coatings and surface treatment processes are affected or staining is caused. Furthermore, pitch problems in pulping as well as toxic or allergic reactions might be induced by components of certain wood species during processing.[14,37] In recent years, parts of the lightly volatile extractives have been discussed with regard to the problem of volatile organic compounds (VOCs), which concerns mainly softwood or products thereof.[38]

Since time immemorial, humans have used nature as a pharmacy and plant parts of the tree other than wood have also been used. *Hippocrates* recommended using the bark of poplar and willow trees against rheumatic pains. Although the willow bark does not contain salicylic acid but only salicylic alcohol derivatives, it was not only an eponym (Latin salix = willow), but also a paradigm of one of the most famous drugs namely Aspirin®. Furthermore, in a Chinese book concerning agriculture and medicinal plants, probably composed between 300 BC and AD 200, the use of extracts of the foliage of *Gingko biloba* as medicament is described, which is still traded as a natural remedy for the purpose of increased mental efficiency by presumed improved blood circulation of the brain.[20]

Even today, researchers are looking for new and highly efficient substances in plants and animals. In recent years, the so-called phytomedicine showed an increasing interest in biologically active substances.

Although wood and barks still play an inferior role, there might be a large potential.[39,40] The need for further research is demonstrated on a number of native wood species where there is still little known about the biological activity of their substances.[41] Another aspect is that the secondary products of pulp production are insufficiently used from an economic point of view. Some well-known as well as new and promising aspects are presented below.

A few years ago, a Finnish research group at Abo Academy discovered a new potential for the utilization of knotwood extractives.[42] It was found that knotwood contains a very high percentage of bioactive phenolic extractives, mainly lignans, which are on their way to being technically utilized. Thereby, knotwood, which usually is not desired in pulp production, has turned out to be an interesting raw material for the pharmaceutical industry. Knotwood contains up to 24% lignans of which 0–85% is 7-hydroxymatairesinol (Scheme 9.4.7), which is already in possession of the Investigational New Drug

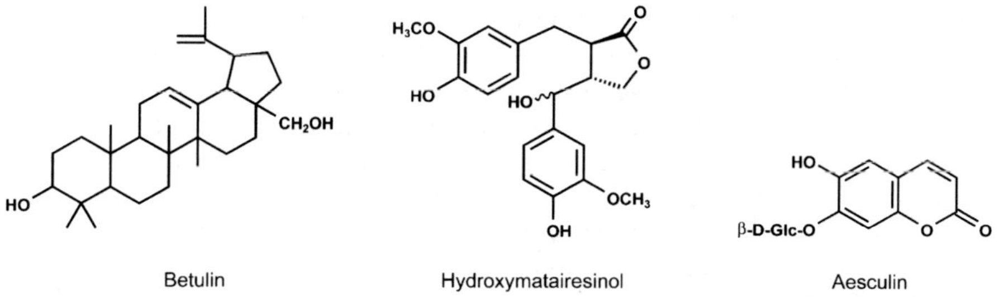

Scheme 9.4.7 Chemical structures of betulin, hydroxymatairesinol and aesculin.

Approval by the FDA, as it seems to be effective against previously incurable cancer diseases. In Finland, a process was developed to separate and extract knotwood during the production of wood chips.[42–45]

Phytopharmaca have been developed, for example, due to the cytostatically effective (-)-taxol of the bark of the yew tree, even if today it is predominantly obtained from plant cell cultures or partially synthetic from taxanes existing in the needles of yew trees.[20] Further examples for biologically active substances include the analgetic salicin of various willow species as well as avicine from the Australian acacia or Argan oil from the Moroccan argan tree, which can be used in cosmetics (see Chapter 9.7).[46,47] Pycnogenol®, a water extract of the bark of lodgepole pine (*Pinus maritima*), which consists of >90% flavonoids, shows a high antioxidant effect.[48]

Furthermore, betulin, which is found in birch bark at up to 25–35%, is the subject matter of many investigations. The pharmacological spectrum efficacy of betulin (Scheme 9.4.7) is very broad. In the treatment of skin diseases and skin disorders, betulin proved to be anti-inflammatory, anti-bacterial and beneficial to wound healing. In recent research, further potential and very interesting properties of the medicine emerged. Antiviral, hepatoprotective and antitumorigenic effects were found. This is currently the subject matter of worldwide betulin research.[17,49]

The reason why the horse chestnut was voted medicinal plant of the year 2008 in Germany is the fact that modern medicine ascribes medically valuable aspects with regard to humans to the awarded tree. In addition to the abatement of diarrhoea or itching, the ingredients of the horse chestnut are especially efficient against venous diseases, in particular vein weakness. The bark of the horse chestnut is rich in tanning agents, which can abate diarrhoea and haemorrhoidal disorders like weeping and itching. Furthermore, it contains an especially high amount of aesculin (Scheme 9.4.7), which is isolated from the plant and then included in sunscreens. It also contributes to the abatement of chronic venous insufficiency.[50]

There is certainly still potential regarding the material utilization of extractives including in particular other parts of the tree like the bark which have been barely used or not used at all industrially up to now.

9.4.2.6 Inorganic Components (Ash)

The term ash reflects the determination method. Herewith all inorganic substances are described that can be quantitatively detected after incineration. The ash is composed of important nutritional elements (*e.g.* potassium, calcium, magnesium, manganese, sodium, phosphorus) and trace elements (iron, zinc and many more). While the ash content of wood species in temperate zones ranges between 0.2 and 0.5%, it can be significantly higher in tropical wood species (>1%) due to silicon incorporation.[12] This also applies to the bark, which often contains ten times more than the wood. Usually, inorganic components are irrelevant in the processing of wood. However, during energetic utilization, certain elements influence the processes (see Chapter 7), *e.g.* potassium together with an acidic pH value may cause corrosive effects.[6] Silica compounds influence tools used in mechanical machining processes (saw blades, knives, *etc.*).

9.4.3 Pulp and Paper

9.4.3.1 Production and Environmental Aspects

Currently applied processes for the chemical utilization of wood are almost exclusively geared towards chemical pulping. Although, to some extent, the utilization of by-products like vanillin, xylose, furfural, acetic acid or resins (tall oil, rosin and products thereof (*e.g.* sitosterol); see Chapter 9.7) is also included, very often not more than 50% of material utilization is obtained (*e.g.* sulphite mass balance in Figure 9.4.4).

Pulping of lignocellulosic materials results in fibres termed pulps. Pulping of wood can generally be divided into mechanical and chemical pulping or mixtures of both types. In chemical pulping, nowadays the alkaline Kraft sulphate pulp process is predominant compared to the sulphite process.

In principle, all pulps are intermediates in paper production obtained by different treatments of wood or other non-wood fibrous raw materials such as different grasses, bagasse (sugar cane, stalks after sugar removal) or cotton fibres and secondary or recovered fibres from used paper.

About three thousand paper assortments and grades are finally composed of different portions of selected specific pulps and varying additives to yield the final paper properties with respect to strength, surface quality, printability, colour, *etc.*

The treatment may exclusively be a mechanical treatment by grinding or refining the raw materials in the presence of hot water or vapour in specialized processes, yielding pulp products like groundwood or thermo mechanical pulps (TMP). Both types of mechanical pulping can be improved and supplemented by addition of small amounts of chemicals to the aqueous systems, or working under pressure, *etc.* Since in these types of processes lignin is not removed the yields are high with values between 90 and 98%.

Another group of pulps is produced by a large number of so-called semi-chemical or high-yield processes in which mechanical and chemical treatments are combined. Since, in this case, part of the lignin is removed the yields differ between 60 and 90%.

The most important pulps are chemical pulps produced by a cooking treatment with selective lignin dissolving chemicals in a wide pH range from strong acidic to strong alkaline conditions. The prominent and important pulping process applied worldwide is the sulphate or Kraft process (Table 9.4.7) using Na_2S, $NaOH$ and Na_2CO_3 as leading chemicals for the removal of most of the lignin. Thus, the typical yields for chemical pulps drop to about 50% of the starting material as most of the polyoses and part of the cellulose is also dissolved along with lignin and extractive components.

While mechanical pulps are used mostly in low-quality paper grades and assortments, *e.g.* for newsprint or packing papers, Kraft pulps are the strongest pulps used in all-quality papers and paperboards where high strength properties are required.[12,18,21,51,52]

To attack residual portions of lignin which were not dissolved during the cooking process in chemical pulping a series of different bleaching sequences are applied. In principle there are two types of bleaching concepts: reductive bleaching and oxidative bleaching. Reductive bleaching with reducing chemicals (*e.g.* $NaHSO_3$, $Na_2S_2O_4$ or formamidinesulfinic acid [FAS]) does not remove lignin but only converts chromophoric groups in more oxygen and light stable functionalities. To prevent the paper from yellowing, oxidative chemicals used in bleaching, today mostly oxygen, ozone, hydrogen peroxide, *etc.*, remove residual lignin portions to yield pulps with high brightness and brightness stability.

In all cases in modern pulp bleaching, elemental chlorine is excluded and only chlorine dioxide is applied to yield elemental chlorine-free pulps (ECF pulps). If any chlorine-containing compound is avoided totally chlorine-free pulps (TCF pulps) are achieved. Concerning the bleaching, Kraft pulp mills have been converted predominantly to elemental chlorine-free (ECF) bleaching. Totally chlorine-free (TCF) bleaching is the method of choice for sulphite mills. Therefore, the extent of TCF bleaching remains at around 5% of the pulp production.

Many efforts regarding the research on alternative pulping processes have not yet been realized on an industrial scale. This is true for organosolv pulping, using organic solvents to support the removal of lignin from wood or acid-catalyzed organosolv pulping processes such as Formacell and Milox or ASAM (alkaline sulphite with anthraquinone and methanol) pulping.

Although the operation of a pulping process requires energy, modern mills do not require fossil fuel as the source, because all the energy comes from combustion of the compounds (mainly lignin) dissolved during the pulping process.[18]

Recovery of the pulping chemicals has been increased significantly and substantially contributes to the economy of pulp manufacture.

The pulp and paper industry has been regarded as a "dirty" industry, neglecting environmental aspects and matters of sustainability up until the 1980s and 1990s due to huge amounts of required energy and water, and the release of volatile and solid emissions and contaminants to the atmosphere, to the aquatic environment and to the ground. In addition, the industry was claimed to be partly responsible for diminishing valuable forests by the use of ever-increasing amounts of suitable wood species and assortments.[53]

In contrast, most of the modern pulp mills nowadays have reduced their environmental impacts applying "accepted modern technology" (AMT) or "best available technology" (BAT). Improvements in technology are supplemented by the use of wood which is certified by means of internationally accepted and applied forest certification standards *e.g.* FSC and PEFC.

The essential steps towards clean technologies in pulping were achieved by preferring Kraft pulping to sulphite pulping (see Table 9.4.6), which reduced the negative environmental impacts by better recoveries of involved chemicals and dissolved organic wood components. In bleaching, the removal of elemental chlorine and the extended use of oxygen and oxygen-based chemicals in bleaching sequences as well as high standards in the sewage plants, including sophisticated cleaning and biodegradation technologies also contribute to environmental improvements.[18,53] Overall, modern and up-to-date pulp and paper production can be characterized as an extremely complex and difficult conversion technology and business which succeeded over the last 20 years in reducing the negative impacts of pollution, effluents, solid wastes, odour, water, energy consumption and CO_2 emissions. In addition, the ecological standards of raw materials and final products on an international level are fulfilled by the pulp and paper industry. An example for the reduction of energy in the German pulp and paper industry is given in Figure 9.4.5.

An overview of the worldwide pulp production, divided into the respective categories, is presented in Table 9.4.6. Figure 9.4.6 shows the countries with more than 1 Mio mto annual production.[9] Figure 9.4.7 reflects the major trade flow of wood pulp worldwide. The percentage of mechanical pulp is approximately 10% and that of chemical pulp is approximately 40%. Thus, it can be clearly seen that, by now, approximately 44% of the pulping raw materials is composed of recycled fibres. The use of waste paper strongly depends on the pulp percentage produced inland. Thus, in Central Europe, for example in Germany (58%), significantly more waste paper fibres are used than in Scandinavia. Moreover, 5% of the total pulp production is made from non-wood plant fibres, including stalk, bast, leaf and seed fibres.

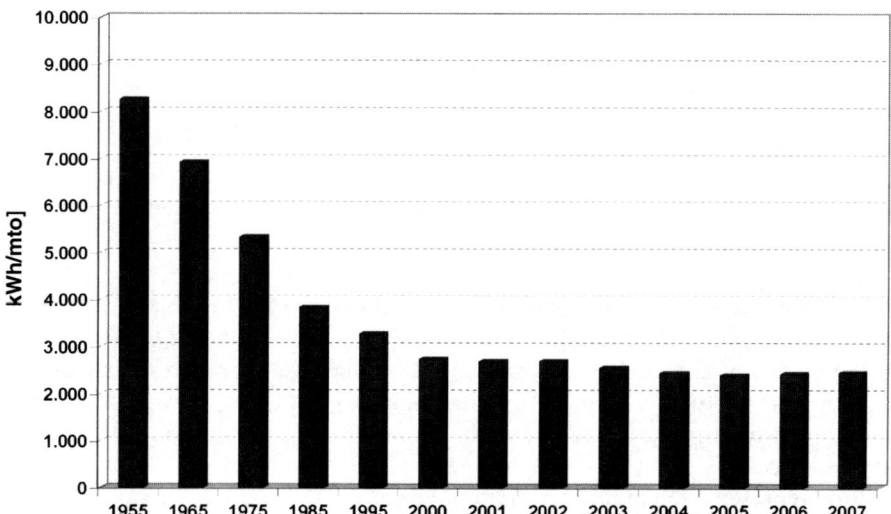

Figure 9.4.5 Energy input per ton of pulp and paper production in Germany.[54]

Table 9.4.6   Global pulp production by category (2000)[18] (with permission of Wiley-VCH Verlag GmbH & Co. KGaA).

Pulp category	Pulp production [Mio mto]
Chemical	131.2
Kraft	117.0
Sulfite	7.0
Semichemical	7.2
Mechanical	37.8
Nonwood	18.0
Total fibres	334.0
Total virgin fiber	187.0
Recovered fiber	147.0

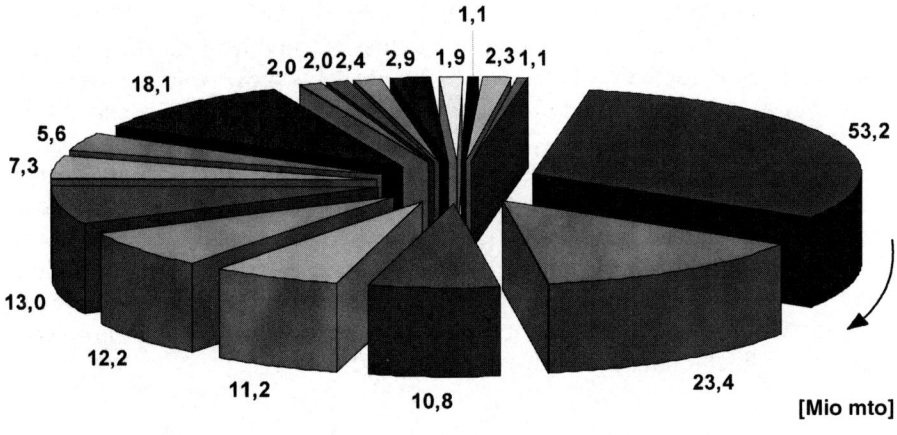

Figure   9.4.6   Worldwide production of total (mechanical and chemical) pulp; countries with more than 1 Mio mto annual production.

The pulp and paper industry is a branch where large, multinational companies dominate pulp production on a global scale, often combined with mass paper production. On the other hand, medium-sized and small companies are producing a large number and variety of speciality papers. In Germany, for example, four to five international companies ("global players") produce pulp and paper in big tonnages, while about 160 paper mills act in the field of speciality papers.

Pulp represents the major raw material basis for the two main applications: a) about 95% for paper and board production, where the pulp fibres are additionally modified to give a coherent paper sheet and b) about 5% for chemical

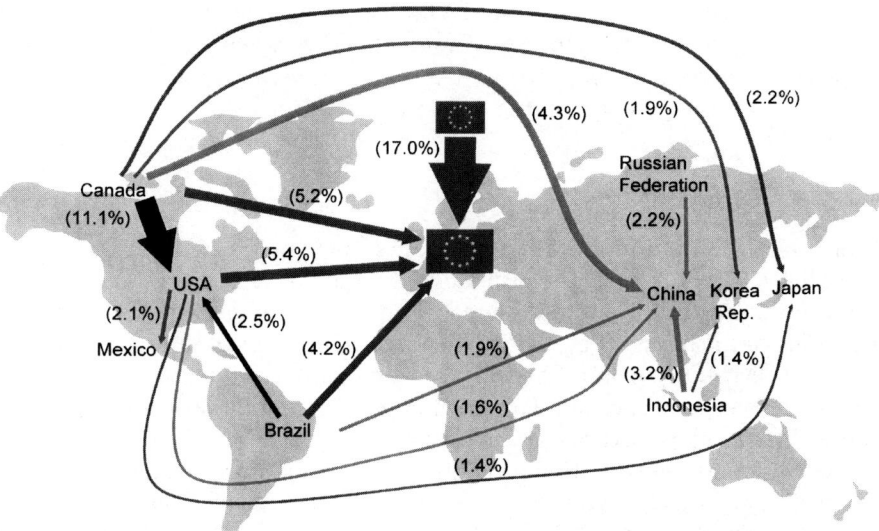

Figure 9.4.7 Major trade flow of wood pulp worldwide (2004). (Thickness of arrows corresponds to trade volume.)

conversion to products such as regenerated fibres and cellulose derivatives. The former is denoted as paper grade pulp, the latter as dissolving grade pulp.[18,21]

The paper grades produced by the paper industry are multitudinous and run into thousands. However, the paper grades can be summarized into groups, like graphic paper including printing paper and writing paper, as well as packaging paper and board, sanitary paper and technical speciality paper, as can be seen in Figure 9.4.8.[21]

Dissolving pulps are high-quality, intensely bleached pulps. As they are used as chemical intermediate product or source material for viscose rayon and

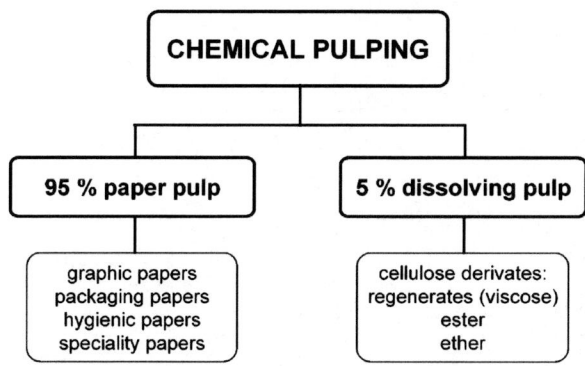

Figure 9.4.8 Applications of chemical pulping.

besides cotton linters for cellulose derivatives (*e.g.* cellulose ether and ester), they feature a significantly higher degree of purity when compared to paper pulp. Thus, the contents of polyoses (0.5–10%), lignin and extractives (<0.2%) are much lower depending on the application.[55]

9.4.3.2 Products

There are manifold applications for dissolved pulp, as can be seen in Table 9.4.7. Today, regenerated cellulose fibres are of utmost importance. They are produced either through the viscose process by derivatization with CS_2 or according to the lyocell process through direct solution in 4-methylmorpholine-4-oxide (NMMO). In addition, there are a number of cellulose derivatives with very specific properties. Thus, cellulose ethers and esters like methyl cellulose (MC), methyl hydroxypropyl cellulose (MHPC), carboxymethyl cellulose (CMC) and hydroxyethyl cellulose (HEC) are used, for example, as glue for wall paper (MC), as protective colloids in suspension polymerization of vinyl chloride (MC, MHPC) and in emulsion polymerization of vinyl acetate (HEC) as entire deposition agent for laundry detergents (CMC) or as thickening agents in aqueous paints, cosmetics and food (*e.g.* E 461–E 466) due to their ability to influence the rheology of liquid systems. Cellulose esters such as acetates are used for cigarette filters due to their specific filtration and adsorption properties and they are also used as raw material in photographic films due to their excellent optical properties. Properties like surface activity, impact strength, transparency and brilliancy present an advantage when nitrocellulose is used as a coating raw material.[56]

As already mentioned, 95% of the pulp is used for the production of paper. In Figure 9.4.9 it can be seen that, despite the development of electronic media, the worldwide production of paper is still increasing. Thus, the paper demand in developing countries is especially increasing.[54]

In Table 9.4.8 the largest paper and paperboard producing countries worldwide and their share in the world production are presented.

Table 9.4.7 World production of dissolving pulp by end-use (2003)[18] (with permission of Wiley-VCH Verlag GmbH & Co. KGaA).

Cellulose product	*Examples of end-use*	*[Mio mto]*
Regenerated fibres	Staple, filaments	2.20
Cellophane	Sponges, casings	0.10
Cellulose acetate	Tow, filament, plastic	0.53
Cellulose ether	Protective colloids, wall paper glue thickening agents	0.47
Microcrystalline cellulose	Moulding powder	0.09
Others (e.g. nitrocellulose)	Varnishes, speciality papers, explosives	0.26
Total		3.65

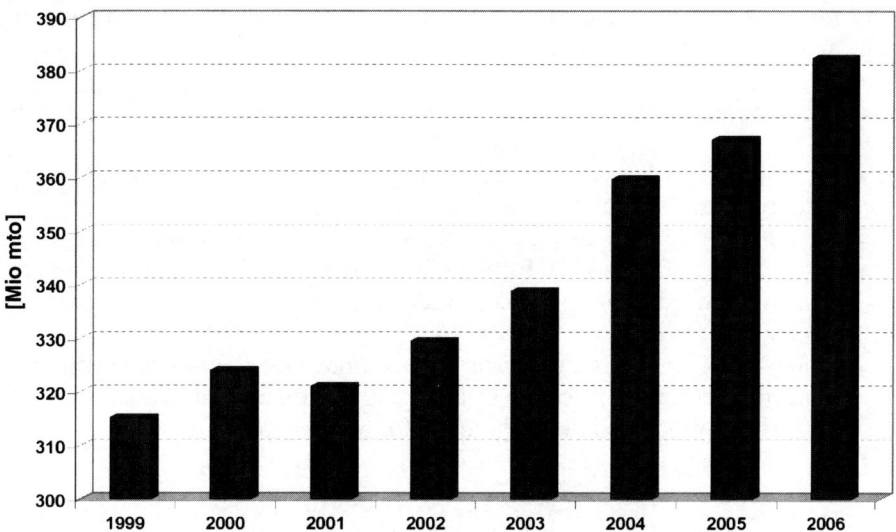

Figure 9.4.9 Development of paper production worldwide.

Table 9.4.8 Top Ten of worldwide paper manufacturing countries (2006).

Country	[Mio mto]	[%]
USA	84.0	22
China	65.0	17
Japan	31.1	8
Germany	22.6	6
Canada	18.1	5
Finland	14.1	4
Sweden	12.0	3
South Korea	10.7	3
France	10.0	3
Italy	10.0	3
Total of top ten	277.6	73
World	382.6	100

9.4.4 Wood-based Composites

9.4.4.1 Conventional Concepts and Products

The concept of wood-based products is to glue together wood parts or particles in a specific way. The wooden part may be boards, veneers, strands, particles or fibres thus yielding *e.g.* laminated beams, plywood, oriented strand boards (OSB), particle boards, fibreboards (*e.g.* medium-density fibreboards, MDF) in a multitude of specifications. The gluing components are predominantly oil-based glues, but can also be inorganic adhesives like gypsum or cement.

The advantageous characteristics of wood-based products in comparison with solid wood products are *e.g.*:

- Wood-based products overcome the limited dimensions determined by height and diameter of the tree;
- Wood-based products are exclusively processed in industrial plants yielding reproducible characteristics and quality;
- Parts of wood-based products, *e.g.* particle boards, can afford the use of low quality and mixtures from different raw materials and assortments (*e.g.* residues containing bark and branches, used timber and used products);
- Wood-based products are much more homogenous compared with solid wood. Therefore, dimensional stability, surface quality, individual strength, properties, *etc.* can be perfectly made for special applications and needs.

9.4.4.2 New Concepts and Products

The materials referred to as Wood-Plastic-Composites (WPC) mainly consist of the renewable raw material wood, synthetic resins, recycled and conditioned plastic chips and additives. The composites are mixed and processed according to established plastics-processing technologies. For lack of suitable standards, the terminology for WPCs is quite vague. Synthetic resins impart bonding; additives like UV absorbers, heat stabilizers or plasticizers and lubricants improve processing and performance properties, pigments are added for colouring. WPCs can be processed by thermal-forming technologies like extrusion, injection moulding or moulding techniques.[57,58]

Suitable plastics for the production of WPCs are relatively cheap and can be processed at temperatures below 200 °C. Further increase in temperature would cause damage of the wood components (lignin, wood polyoses). Suitable plastics are, for example, polyethylene, polypropylene or polyvinyl chloride.[57]

The wood components are mainly obtained from the by-products of the saw mill industry. The subsequent mechanical properties of WPCs are mainly influenced by the fineness and the slenderness ratio. It depends on the slenderness ratio whether the wood is used as filler or product reinforcement.[57]

Bonding agents like maleic anhydride are used for improved bonding between wood, the plastic, calcium carbonate, talcum or mica. The strength and the heat resistance are increased; creeping is compensated. Additional additives will be added depending on the application of the WPC in order to reduce combustibility or to improve the resistance and the durability.[57]

Compared to wood-based products, WPCs are freely mouldable, moisture resistant and fairly resistant to rot. After-treatment of the products is not required.

Furthermore, they show higher strength, stiffness and a significantly lower coefficient of expansion compared to purely used plastics. An increase in the wood percentage leads to efficient cost savings. Since, however, WPC is mainly for exterior use, resistance of the product against fungi is important.[57] Therefore, the addition of fungicidal acting additives is required.

WPCs are mainly used for interior and exterior floor covering, but also for the construction of patios, stairs and terraces (Figure 9.4.10). Furthermore, WPCs are used for promenades, noise barriers, docks, fences, railings, window frames and doors.[59,60] Future markets could be found in the furniture and automobile industry.[60]

In the United States, approximately 75% of the WPCs are used in the building sector and approximately 50% of the entire market volume is used for terraces and patios.[60,61] In the year 2003, the market share of WPC products for exterior use was 14%, comprising a financial volume of approximately 3.5 billion US$. These so-called deckings consist of polyethylene with a wood percentage of 50–70%.

The annual growth of the markets is considerable. Worldwide, over 1 Mio mto of WPC are produced per year. While it is just under 1 Mio mto in Northern America, Europe has, with an estimated 105,000 mto (2006), still quite low values, followed by Asia with approximately 100,000 mto.[60]

The mechanical properties of Wood-Plastic-Composites substantially depend on the quality of the raw material, the composite and the manufacturing process. The density of WPC lies between 0.9 and 1.4 g cm^{-3} and is higher than that of one of the two components, plastic and wood, alone. The porous wood structure is densified during the production process and plastics and additives partly penetrate into the cavities. Thus, water absorption is made harder, and the material swells less and more slowly.[57]

In general, the stiffness of a WPC increases linearly with increasing wood percentage. The stiffness of WPCs with a wood percentage of 50% is five times

Figure 9.4.10 WPC for exterior use.

higher than the one of pure polyproylene.[62] Mean bending strengths of up to 40 N mm^{-2} are common for WPC.[57]

In principle, WPC products are recyclable. They are melted down and processed into granules. A second extrusion or injection moulding is technically feasible. However, the quality of recycled WPCs and the reaction of the wood component is still unclear.[57] Furthermore, the blending of various plastic components often makes the recycling even more complex.[63]

ARBOFORM® and FASAL® are examples for the development of composites made of wood and other materials corresponding to the idea of "green chemistry".

ARBOFORM® consists of the three basic components, lignin, plant fibres and additives on the basis of renewable raw materials. For the first time, this mixture was successfully developed as a material capable of injection moulding at the Fraunhofer-Institute for Chemical Technology Pfinztal, Germany, in 1996. Further development of the material was carried out by TECNARO, a company founded by the Fraunhofer-Institute in 1998.[64] The product name is derived from the Latin term for tree (*arbor*).[34] At present, ARBOFORM® is successfully used for the manufacture of marketable products like precision parts, toy figures or loudspeaker housings by means of extrusion and injection moulding.[34] Ongoing research projects deal with the development of circuit boards, the manufacture of bipolar panels in batteries and building boards (Figure 9.4.11).[60,64–66]

Qualitative and quantitative modification of the raw materials significantly influences the product properties. The properties of the lignin are influenced by the origin, since the proportion of the elements (p-hydroxyphenyl-, guaiacyl-, syringyl units) in softwood, hardwood or annual plants varies and, hence,

Figure 9.4.11   Boards made of ARBOFORM®.

differences in the chemical reaction behaviour result. Particularly, the kind of the technical lignin (from either the Kraft or the sulphite process) influences the material properties of the lignin.

Resource and manufacturing processes affect the chemical composition and reactivity as well as the geometry of the fibres in plant fibres. Lignin, reactive substances like tannins, resins and tanning agents as well as reactive wood polyoses or smooth surfaces of the elements significantly influence the quality of the bonding and, hence, the later product properties. Furthermore, particle size and slenderness ratio play a role.

The additives have different functions within the product. They afford the bonding of the lignin on the plant fibres or particles and improve the rheology. Other additives act as flame retardants or as fungicides.

Similarities between FASAL® and ARBOFORM® can be found. The fibre components of FASAL® are wood particles, rice husks or coconut fibres. Natural starch, such as for example from maize, and partly proteins are used as binders. Resins or pigments from natural raw materials are used as additives. Processing is done by processing techniques used in the plastics industry, *e.g.* by extrusion. The application is comparable to the one of injection moulded ARBOFORM®.[67]

Since the two products consist of renewable raw materials, they can be recycled or energetically used after their service life.

9.4.5 Modified Solid Wood Products

Although wood is used for many applications because of its excellent material properties, it can show disadvantages with undesirable effects in specific application areas too. Prominent examples are the variation in dimensional stability as a result of altering atmospheric conditions or the low natural durability of many domestic wood species. Wood modification has also become a topic of renewed interest because of increasing environmental considerations, *e.g.* some modified woods can substitute tropical hardwoods due to their darker colours and higher durability. Wood modification represents a process that is used to improve the material properties of wood, but produces a material that does not show at the end of a product life-cycle a higher environmental hazard than the unmodified wood during the disposal.

Driving forces for wood modification are:

- Improvement of important application properties (higher dimensional stability, higher durability);
- Increasing demand for outdoor applications of wood (higher durability);
- Substitution of tropical wood species (higher durability and/or darker colours);
- Alternatives for chemically preserved wood (durable and low-maintenance assortments of domestic wood).

This chapter can give only a short insight into some important aspects. A comprehensive overview with further and detailed information can be found in the literature.[68,69]

9.4.5.1 Chemical Modification

"Chemical modification of wood is defined as the reaction of a chemical reagent with the wood polymeric constituents, resulting in the formation of a covalent bond between the reagent and the wood substrate."[69]
"Impregnation modification of wood is defined as any method that results in the filling of the wood substance with an inert material (impregnant) in order to bring about a desired performance change."[69]

In this chapter no distinction is made between a real chemical modification (acetylation), where covalent bonds between the reagent and the chemical constituents of wood are formed, and the impregnation modification, where monomers diffuse into the wooden cell wall and become polymerized there, like Hill suggested in his definitions (see above), because it was shown that DMDHEU and furfuryl alcohol not only swell the cell wall and polymerize there but also have the potential to form some covalent bonds.[70,71]

This chapter deals mainly with commercially used processes, which are at the moment:

- Acetylation with acetic anhydride (*e.g.* Accoya® of TITAN WOOD);
- DMDHEU = Dimethylol dihydroxy ethylene urea (Belmadur® of BASF);
- Furfurylation with furfuryl alcohol (*e.g.* Kebony®, VisorWood®).

During the chemical modification of wood with acetic anhydride (Scheme 9.4.8), the reaction takes place with the cell wall polymeric hydroxyl groups to form ester bonds. Acetic acid is produced as by-product.

A commercial plant (Titan Wood) has started to produce acetylated wood and has given the acetylated wood the name of a "new wood species": Accoya®. Improved properties are high durability, high dimensional stability and high UV-resistance.[68]

Dimethylol dihydroxy ethylene urea (DMDHEU), a resin primarily developed for the textile industry, was used on wood (Scheme 9.4.9), *e.g.* beech

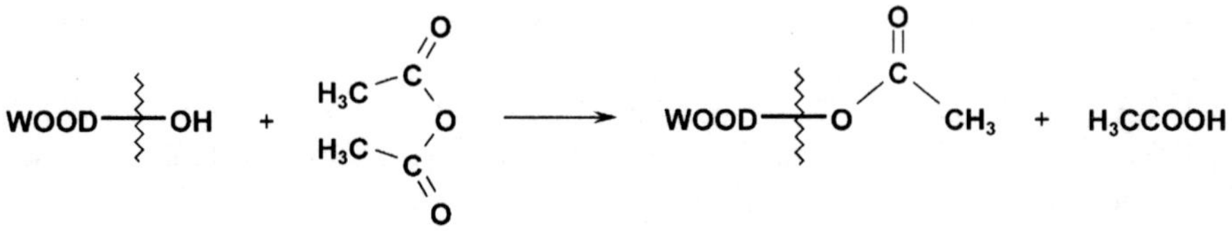

Scheme 9.4.8   Reaction of acetic anhydride with wood.

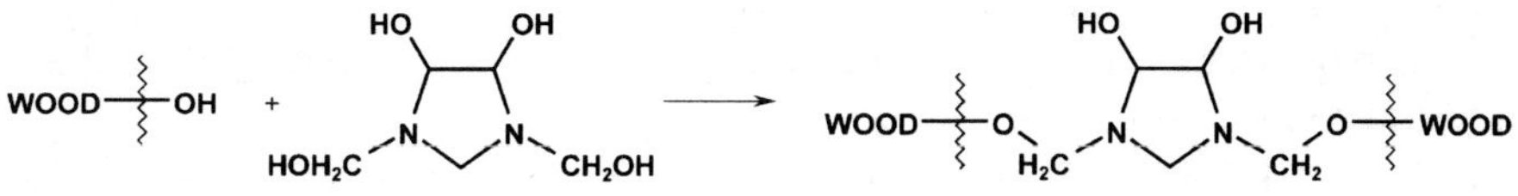

Scheme 9.4.9 Reaction of DMDHEU with wood.

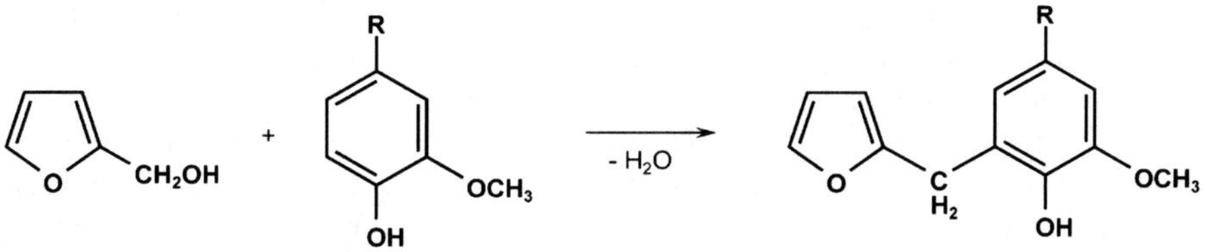

Scheme 9.4.10 Reactions of furfuryl alcohol.

veneers. After impregnating the wood with DMDHEU solution, drying at a temperature of more than 100 °C causes cross-linking and polycondensation reactions.

A commercial product is Belmadur® from BASF, which is harder, more durable and more dimensionally stable than the respective untreated wood.[72]

The principles of forming furan polymers from the condensation of furfuryl alcohol have become attractive for modifying wood, when ways of controlling the polymerization in dilute systems were discovered. There are several competing reactions: condensation and cross-linking reactions of furfuryl alcohol with itself and also the reaction between furfuryl alcohol and lignin units is supposed (Scheme 9.4.10).[70,71]

Sweden is the largest producer with nearly 1,000,000 m³ impregnated wood in 2003. Finland and Norway were treating approximately 360,000 m³ each, while Denmark treated 151,000 m³ in 2003.

VisorWood® is furfurylated Scots pine (sapwood and heartwood) and is used for outdoor applications like cladding and decking roof boards, where durability is required.[68]

Kebony® is sapwood of various pines or temperate hardwoods like beech and ash and is marketed as an alternative for tropical hardwoods and is suited for floors, interior panels and other applications where tropical appearance, hardness and dimensional stability are important criteria.[68]

9.4.5.2 Thermal Modification

"The thermal modification of wood is defined as the application of heat to wood in order to bring about a desired improvement in the performance of the material."[69]

The thermal modification of wood has become a topic of great interest since the first heat-treatment plant was built in Finland in the early 1990s. The production volume of thermally modified wood in Europe increased from about 100 m^3 in 1995 to about 130,800 m^3 in 2007 (Figure 9.4.12).[73] However, also in North America there are some developments concerning the thermal modification of wood.

One consequence of increased market relevance is that in November 2007 the European Committee for Standardization (CEN) drew up the first technical specification (TS) about thermally treated wood (TMT).

Without application of chemicals as wood preservative, the thermal treatment can be seen as a contribution to the increasing environmental protection, even though an energy input is necessary for the production of TMT.[69] There are different types of thermal modifications (Table 9.4.9).

Of greatest importance still is the classical thermal modification in an atmosphere of water vapour and volatiles from wood without additional pressure. Most of the existing plants are based on this process, which has been optimized in different ways regarding the chamber system or the process sequence. The technical details in which the processes and accordingly the plants differ are company secrets.

The other processes apply partly different atmospheres. In the oil-heat-treatment (OHT) vegetable oil is used; in the French systems *e.g.* nitrogen is used (Retified[®]Wood [bois rétifié], NOW [new option wood]). The multi-stage PLATO (Providing Lasting Advanced Timber Option) process uses in the first

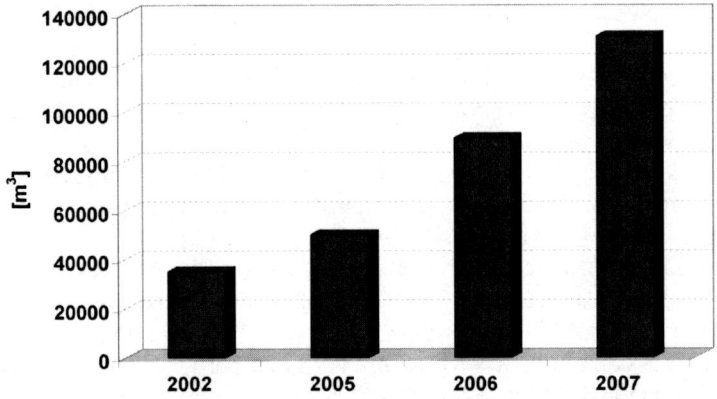

Figure 9.4.12 European production of TMT.

Table 9.4.9 Thermal treatment processes.[73]

Process	Process conditions
ThermoWood[®], Stellac[®]Wood, Thermoholz, Perdure *etc.*	Water vapour and volatiles from wood without additional pressure
PLATO	Thermo hydrolysis (autoclav) and high temperature treatment
Retification[®] (Baladur, NOW),	Nitrogen
OHT	Vegetable oil

step water vapour under pressure, which is named hydro thermolysis followed by drying, curing and conditioning.

Typical applications of thermally treated wood are *e.g.* garden furniture (Figure 9.4.13), deckings for balconies, terraces or swimming pool borders and for wet areas.

The application of heat to wood results in chemical changes in the material.[69] The properties of the modified products depend on the one hand on the specific chemical composition of the different wood species, but, most notably between hardwoods and softwoods, on the other hand on the process conditions. Maximum temperatures, atmosphere and the duration of the thermal load are among the main parameters. Further influencing factors are the sample dimensions and whether the system is open or closed.

The results of several published chemical investigations of thermally treated wood are not comparable in each case, because the changes and their extent are due to treatment conditions, such as temperature and atmosphere during the treatment or duration of the treatment.[69,74–77]

In general, properties such as colour, natural durability and dimensional stability of thermal treated wood differ from those of untreated wood. Depending on the duration and the power of the process, thermally treated wood shows improved rot-resistance properties and also demonstrates increased resistance to both fungi and climatic influences.[69,74] To warrant the natural durability it must be ensured, however, that the TMT products have no contact to the ground.[78] Furthermore, its dimensional stability can be improved as well as its adhesion property against hydrophobic materials.[79,80]

Figure 9.4.13 Furniture at the German Horticultural Show in München 2005

Moreover, darker tints, which are normally characteristic only for tropical wood species, can be achieved.[79,81]

It is well known that the chemical changes are caused by autocatalytic reactions and that changes starting with hydrolysis of polyoses mainly depend on the applied temperature. Afterwards amorphous cellulose and lignin are affected by the treatment as reported in numerous papers.[12,69,75–77] Due to the various chemical natures of wood species it is nevertheless difficult to make prevailing predictions of the treatment results in detail. In general, there are still several deficits. Above all, there are varying properties due to the multiplicity of different thermally modified assortments. Missing standardization and quality control for market products leads sometimes to a lack of confidence of potential customers. Furthermore, there is still a lack of knowledge on the mechanisms of modification and application restrictions due to varying strength properties.

9.4.6 Outlook

The principles of green chemistry may be summarized in the simple statement "applying fundamental knowledge of chemical processes and products to achieve elegant solutions with the ultimate goal of hazard-free, waste free, energy-efficient synthesis of non-toxic products without sacrificing efficacy of function".[82]

As the worldwide shortage of exhaustible resources can be predicted quite reliably and is accompanied by an increase in price of crude-oil-based products, this will in turn promote the marketability of substitution products on the basis of renewable raw materials, which is more sustainable than the production of oil-based materials and products.

The trees and the forests as well as any other biomass are a renewable resource, the potential of which has not yet been exploited. Hence, different compounds like flavonoids, lignans, terpenes, phenols, alkaloids, sterols, waxes, fats, tannins, sugar, gum, suberins and resins from different woods and plant components can be extracted and converted to useful market products. The utilization of wood as a provider of components for niche and premium products is certainly able to grow. In this regard, various research fields have been opened and extended, focused on raw materials, products, processes and development of methods as well. This needs efforts and contributions by politics, industry and research. Even if science will not eventually invent or detect another acetylsalicylic acid, there are a number of open questions like identification of unknown, biologically active compounds or the regulation of the biosynthesis of various extractives. Furthermore, a large added value could be achieved by the production of pharmaceutically active or health beneficial ingredients, which include for example phytohormones, polyphenols or triterpenes.

In future, by using wood, or one of its components, as a renewable resource, the entire process chain or an entire reaction system can be regarded under "green" aspects including energy efficiency.

An established instrument for ecological evaluation of processes or products is life-cycle-assessment, which is particularly useful for ecological comparison and optimization of processes and products. A consequent realization of the mentioned aspects and challenges will lead to a better and more sustainable situation on our planet for both nature and mankind.

References

1. J. E. Winandy, A. W. Rudie, R. S. Williams and T. H. Wegner, *Forest. Prod. J.*, 2008, **58**, 6–16.
2. G. Gellerstedt, in *Proceedings 10th European Workshop on Lignocellulosics and Pulp*, Stockholm, 2008.
3. C. Konijnendijk, ed., *Forest-Based Sector Technology Platform – FTP*, 2008.
4. J. H. Clark, V. Budarin, F. E. I. Deswarte, J. E. Hardy, F. M. Kerton, A. J. Hunt, R. Luque, D. J. Macquarrie, K. Milkowski, A. Rodriguez, O. Samuel, S. J. Tavener, R. J. White and A. J. Wilson, *Green Chem.*, 2006, **8**, 853–860.
5. B. Kamm, P. R. Gruber and M. Kamm, ed., *Biorefineries – Industrial Processes and Products*, Wiley-VCH, Weinheim, 2006.
6. M. Kaltschmitt and H. Hartmann, *Energie aus Biomasse*, Springer Verlag, Berlin, 2001.
7. T. Herzog, J. Natterer, R. Schweitzer, M. Volz and W. Winter, ed., *Holzbau Atlas*, Institut für internationale Architektur Dokumentation, München, 4. Auflage, 2003.
8. K. Töpfer, in *Nachhaltig bauen und modernisieren*, ed. Holzabsatzfonds, Bonn, 2006, p. 2.
9. FAO, http://faostat.fao.org/site/381/default.aspx (retrieved 01.07.08).
10. R. Wagenführ, *Holzatlas*, Carl Hanser Verlag, 6. Auflage, 2007.
11. J. Sell, *Eigenschaften und Kenngrößen von Holzarten*, Baufachverlag Lignum, Dietikon, 4. Auflage, 1997.
12. D. Fengel and G. Wegener, *Wood. Chemistry, Ultrastructure, Reactions*. Reprint Kessel, http://www.forstbuch.de, Remagen, 2003.
13. R. Alén, I. Forsskahl, B. Holmbom and P. Stenius, *Forest Products Chemistry*, Fapet Oy, Helsinki, 2000.
14. E. Sjöström and R. Alén, ed., *Analytical Methods in Wood Chemistry, Pulping, and Papermaking*, Springer, 1999.
15. J. W. Rowe, (Ed.), *Natural Products of Woody Plants I and II*, Springer, 1989.
16. R. M. Rowell, (Ed.), *Handbook of Wood Chemistry and Wood Composites*, Taylor & Francis, Boca Raton, London, NJ, Singapore, 2005.
17. M. Ek, G. Gellerstedt and G. Henriksson, ed., *Ljungberg Textbook, Wood Chemistry and Wood Biotechnology*, Stockholm, 2006.
18. H. Sixta, ed., *Handbook of Pulp* Volume 1 and 2, Wiley-VCH, Weinheim, 2006, p. 9, 11, 23, 1–20, 997–1004.

19. D. Klemm, B. Philipp, T. Heinze, U. Heinze and W. Wagenknecht, *Comprehensive Cellulose Chemistry*, Wiley-VCH, Weinheim, 1998.
20. J. Falbe and M. Regitz, ed., *Römpp Chemielexikon*, Thieme Verlag, 10. Auflage, 1996.
21. L. Göttsching and C. Katz, ed., *Papier Lexikon*, Deutscher Betriebswirte-Verlag, Gernsbach, 1999.
22. R. P. Swatloski, R. D. Rogers and J. D. Holbrey, US 2003/0157351, 2002 (Univers. of Alabama).
23. D. A. Fort, R. C. Rensing, R. P. Swatloski, P. Moyna, G. Moyna and R. D. Rogers, *Green Chem.*, 2007, **9**, 63.
24. H. A Krässig, *Cellulose. Structure, Accessibility and Reactivity*, Gordon and Breach Science Publishers SA, Yverdon, 1993.
25. J. F. Kennedy, G. O. Philips, P. O. Williams and L. Piculell, (Ed.), *Cellulose and Cellulose Derivatives*, Woodhead Publishing, 1995.
26. T. Rosenau, http://bfw.ac.at/nawaros/documents/02_Praes_Rosenau.pdf (retrieved 14.08.08).
27. S. Y. Lin and C. W. Dence, ed., *Methods of Lignin Chemistry*, Springer Verlag, Berlin, 1992.
28. R. J. A. Gosselink, E. de Jong, A. Abächerli and B. Guran, *Ind. Crops Prod.*, 2004, **20**, 121–129.
29. International Lignin Institute, http://www.ili-lignin.com (retrieved 13.04.2006).
30. M. Hofrichter and A. Steinbüchel, *Biopolymers: Lignin, Humic Substances and Coal*, Wiley-VCH, 2001.
31. R. J. A. Gosselink, E. de Jong, A. Abächerli and B. Guran, in *The 7th ILI Forum: Bringing lignin back to the headlines*, Proceedings, Barcelona, 2005.
32. W. G. Glasser and S. Sarkanen, ed., *Lignin: Properties and Materials*, American Chemical Society, Washington, 1989.
33. B. A. Acha, C. Capurro, N. E. Marcovich, N. I. Aranguren and M. M. Reboredo, in *The 7th ILI Forum: Bringing lignin back to the headlines*, Proceedings, Barcelona, 2005.
34. H. Nägele and J. Pfitzer, http://www.tecnaro.de (retrieved 17.01.2006).
35. E. Windeisen and G. Wegener, *Mat. Sci. Forum*, 2009, **599**, 79–106.
36. H.-D. Belitz, W. Grosch and P. Schieberle, *Lehrbuch der Lebensmittelindustrie*, Springer, Berlin, 2001.
37. B. Hausen, *Woods Injurious to Human Health*, Walter de Gruyter, Berlin, New York, 1981.
38. M. Ohlmeyer, M. Makowski and D. Meier, http://www.bfafh.de/iud/projekte/4-2004/6-2.pdf (retrieved 23.01.2008).
39. D. C. Ayres and J. D. Loike, *Lignans: Chemical, Biological and Clinical Properties*, Cambridge University Press, 1990.
40. P. M. Dewick, *Medicinal Natural Products*, Wiley, 2001.
41. Ch. Schwager and W. Lange, *Biologischer Holzschutz. Literaturstudie über akzessorische Bestandteile dauerhafter Holzarten mit resistenzwirksamer Aktivität*, Landwirtschaftsverlag, Münster, 1998.

42. B. Holmbom, C. Eckermann, P. Eklund, J. Hemming, L. Nisula, M. Reunanen, R. Sjöholm, A. Sundberg, K. Sundberg and S. Willför, *Phytochem. Rev.*, 2003, **2**, 331–340.
43. http://web.abo.fi/meddelanden/english/2007_01/2007_01_trees.sht (retrieved 23.01.2008).
44. http://www.finnfacts.com/pdf/FF_4_06_saksa_web.pdf (retrieved 23.01.2008).
45. http://www.tekes.fi/eng/publications/Tekes_Foods.pdf (retrieved 29.11. 2007).
46. http://www.pte.at/pte.mc?pte=010925008 (retrieved 23.01.2008).
47. G. S. Jayatilake, D. R. Freeberg, Z. Liu, S. L. Richheimer, M. E. Blake, D. T. Bailey, V. Haridas and J. U. Gutterman, *J. Nat. Prod.*, 2003, **66**, 779–783.
48. L. Packer, G. Rimbach and F. Virgili, *Free Radical Biol. Med.*, **27**, 704–724.
49. http://www.betulin.de/ (retrieved 23.01.2008).
50. http://www.uni-wuerzburg.de/sonstiges/meldungen/single/artikel/rosskastan/ (retrieved 23.01.2008).
51. J. Gullichsen and C.-J. Fogetholm, (Ed.), *Chemical Pulping 6A, 6B*, Fapet Oy, Helsinki, 1999.
52. M. Ek, G. Gellerstedt and G. Henriksson, ed., *Ljungberg Textbook, Pulping Chemistry and Technology*, Stockholm, 2006.
53. W. Sandermann, *Papier. Eine Kulturgeschichte*, Springer, Berlin, 3. Auflage, 1997.
54. Verband Deutscher Papierfabriken e.V. (VDP), ed., *Papier: Ein Leistungsbericht*, 2008.
55. R. Janzon, B. Saake and J. Puls, http://www.bfafh.de/iud/projekte/3-2006/5-1.pdf (retrieved 23.01.2008).
56. H. Harms, in *Lignovisionen, High-Performance Materials from Nature*, University of Natural Resources and Applied Life Sciences, Vienna, 2006, pp. 27–34.
57. Anonymous, http://www.nova-institut.de/pdf/06-01_WPC-Studie.pdf (retrieved 13.04.2006).
58. A. Teischinger, U. Müller and H. Korte, *Holztechnologie*, 2005, **46**, 30–34.
59. M. P. Wolcott and P. M. Smith, in *Nachrichtenportal für Nachwachsende Rohstoffe*, 2004: Opportunities and challenges for Wood-Plastic Composites in structural applications (retrieved 28.05.2004).
60. Fachagentur Nachwachsende Rohstoffe e.V., ed., *Naturfaserverstärkte Kunststoffe*, 2008, p. 29.
61. P. Schwarzbauer and A. Eder, in *4. Internationales Seminar für die Holzindustrie in der Schweiz*, Wien, 2003.
62. D. Kaczmarek and J. Wortberg, *Kunststoffe*, 2003, **93**, 18–23.
63. Anonymous, *Holz-Zentralblatt*, 2004, **130**(51), 661–666.
64. G. Wegener, E. Windeisen, G. Scholz, C. Schrader, J. Pfitzer and H. Nägele, in *9th European Workshop on Lignocellulosics and Pulp*, Proceedings, Wien, 2006, pp. 126–129.

65. S. Krischok, M. Himmerlich, S. I. U. Ahmed, A. Kauffmann, J. Pfitzer, M. Huber, W. Bengel and J. A. Schaefer, in *Biannual Report 2005/2006*, Technische Universität Ilmenau, 2006, p. 136–137.

66. E. Windeisen, G. Wegener, J. Pfitzer and H. Nägele, in *Zukunftsmärkte für das Bauen mit Holz*, DRW Verlag, 2008, pp. 86–99.

67. http://www.austel.at/deutsch/home.html (retrieved 28.02.2007).

68. C. A. S. Hill, D. Jones, H. Militz and G. A. Ormondroyd, ed., *The Third European Conference on Wood Modification*, Cardiff, 15–16th October 2007.

69. C. A. S. Hill, *Wood Modification. Chemical, Thermal and Other Processes*, Wiley, 2006.

70. M. Westin and S. Lande, http://www.bfafh.de/inst4/45/ppt/3furfury.pdf (retrieved 04.03.08).

71. S. Lande, Doctoral Thesis, Norwegian University of Life Sciences, As, 2008.

72. A. Nurmi, http://www2.spm.slu.se/bitwon/seminars/mikkeli_060527_28/Nurmi_Chemical%20modification_Mikkeli_06.pdf, (retrieved 04.07.08).

73. W. Scheiding, in *Tagungsband 5. Europäischer TMT-Workshop*, Dresden, 24–25 April 2008, p. 1–6.

74. M. Boonstra, J. van Acker, E. Kegel and M. Stevens, *Wood Sci. Technol.*, 2007, **41**, 31–57.

75. B. Esteves, J. Graca and H. Pereira, *Holzforschung*, 2008, 62, online DOI 10.1515/HF.2008.057.

76. E. Windeisen, C. Strobel and G. Wegener, *Wood Sci. Technol.*, 2007, **41**, 523–536.

77. E. Windeisen and G. Wegener, *Mat. Sci. Forum*, 2009, **599**, 143–158.

78. S. Hofer, C. R. Welzbacher, A. O. Rapp and C. Brischke, in *Tagungsband der 25. Holzschutz-Tagung der DGfH*, Biberach 20–21 September, 2007, pp. 143–161.

79. D. Johansson, Thesis, Lulea University of Technology, 2005.

80. B. Sundquist, Doctoral Thesis, Lulea University of Technology, 2004.

81. J. Follrich, U. Müller and W. Gindl, *Holz Roh Werkst.*, 2006, **64**, 373–376.

82. http://www.chem.monash.edu.au/green-chem/docs/cgc-report-2001.pdf (retrieved 04.07.08).

CHAPTER 9.5
Natural Rubber

LAURENT VAYSSE,[a] FRÉDÉRIC BONFILS,[b] PHILIPPE THALER,[c] AND JÉRÔME SAINTE-BEUVE[b]

[a] CIRAD, UMR 1208, Joint Research Unit "Agropolymers Engineering and Emerging Technologies" CIRAD, Montpellier SupAgro, INRA, Université de Montpellier 2; Rubber Technology Laboratory Agro-Industry Building 3, 8th floor Kasetsart University, Bangkok 10900, Thailand; [b] CIRAD, UMR 1208, Joint Research Unit "Agropolymers Engineering and Emerging Technologies", CIRAD, INRA, Montpellier SupAgro, Université de Montpellier 2, 73 rue JF Breton, F-34398 Montpellier Cedex 5, France; [c] CIRAD, UPR80, Functioning and management of tree based cropping systems, Campus international de Baillarguet, F-34398, Montpellier Cedex 5, France

9.5.1 Introduction

Natural rubber, India rubber and caoutchouc are all names for the solid elastic material isolated, one way or another, from the 'milk' or latex of *Hevea brasiliensis*, and various other tropical plants like *Castilloa elastica*. Natural rubber and natural rubber products have been known to the ancient cultures of Amazonia and Mesoamerica since time immemorial. According to the Spanish historian Antonio de Herrera Tordesillas, Christopher Columbus returned from his second voyage, bringing back the first rubber balls from the West Indies to Europe. The term *cahuchu* or *caoutchouc* was used by the Maninas Indians according to the French explorer Charles de la Condamine. It is generally taken to be based on the Indian *cao ochu*, which could be translated as

RSC Green Chemistry No. 4
Sustainable Solutions for Modern Economies
Edited by Rainer Höfer
© The Royal Society of Chemistry 2009
Published by the Royal Society of Chemistry, www.rsc.org

teardrops of the weeping tree, an interpretation which was publicized by Vicki Baum's eponymous novel *The Weeping Wood*.[124]

Rubber is one of the most important products to come out of the rainforest. The history of natural rubber since the beginning of the industrial age is a fascinating sequence of inventions (vulcanization, Charles Goodyear and Thomas Hancock, 1843/1844; pneumatic automobile rubber tyres, Robert William Thomson, 1846; synthetic rubbers: polyisoprene, Fritz Hofmann, 1909; polydimethylbutadiene, IG Farben 1910; BUNA, IG Farben, 1910; SBR emulsion polymerization, Walter Bock, 1929) as well as a trade embargo (the Brazilian export embargo for *Hevea* seeds enforced by the death penalty and broken by Henry Wickham's "rubber theft", which laid the ground for the Southeast Asian rubber plantations), the opulence of social life and the brutality of the rubber barons in Manaus (*e.g.* Alberto Vazquez-Figueroa's novel, *Manaos* and Werner Herzog's epic movie *Fitzcarraldo*, starring Klaus Kinski).

Rubber is one of the few examples where chemical synthesis succeeded in a nearly identical performance copy of a natural polymer (polyisoprene) – albeit with a completely different chemical composition (styrene-butadiene-rubber, SBR). Regarding sustainable development, the complete imbalance of the early rubber history has emanated during recent years into equilibrium between natural and synthetic rubber.

Natural rubber from *Hevea brasiliensis* is a natural polymer composed of an association of poly(*cis*-1,4-isoprene) [poly(2-methyl-1,3-butadiene)] and biological elements,[1] giving it highly specific properties. Originating from the Amazon Basin, *Hevea* was already booming in Asia at the turn of the twentieth century.

9.5.2 Challenges Facing the Supply Chain

Natural rubber (NR) production – 9.7 Mio mto in 2007[2] – is mainly provided by estates (usually >40 ha) and smallholdings (of around 2 ha) primarily located in Asia between the Tropics; Thailand and Indonesia are the two main producers. Some tens of millions of people earn a living from this crop, making it a very important development factor in the region. Among the other stakeholders operating in the production system, agro-industry is present but does not hold the place that it occupies in the palm oil supply chain, which might explain the difference in behaviour between these two supply chains in terms of responsiveness (weaker in the case of NR, both economically and environmentally). In terms of consumption, the scheme is different, with an oligopoly-type situation where three tyre manufacturers[3] share 50% of the market.

Both NR and SR are traded in a dry and in a liquid form. The elastomer market (Table 9.5.1) is divided into three major zones (USA, Europe and Southeast Asia) each of which has its own dynamics very closely linked to its internal growth. In the years 2006–2007 China increased its synthetic rubber consumption (+19.3%) more quickly than its natural rubber consumption

Table 9.5.1 Elastomer market (1000 mto) – Source: International Rubber Study Group (IRSG); NR = natural rubber; SR = synthetic rubber.

	NR production 2007	SR production 2007	NR consumption 2007	SR consumption 2007
North America	0	2790	1157	2129
Latin America	228	684	565	861
Africa	454	71	117	105
Europe	0	2811	1377	2623
Russia and others	0	1267	273	1057
Thailand	3056	156	374	217
Indonesia	2755	48	391	133
Malaysia	1200	67	446	123
China	600	2215	2550	3435
India	811	103	851	290
Japan	0	1655	887	1162
Others, Asia	622	1671	896	1129
Total	9726	13538	9884	13264

(+6.3%), whilst over the same period North America reduced its synthetic rubber consumption (–3.4%) and stabilized its natural rubber consumption (+0.8%). The effect of globalization has been to relocate the means of tyre production to Southeast Asia, particularly to China. The rising price of energy is likely to modify that situation profoundly as transport costs escalate. In addition, the impending arrival of new stakeholders – especially in the Middle East – in the synthetic rubber production sector risks destabilizing the elastomer market.

Newly planted rubber trees take 5 to 7 years before they can be tapped and reach their peak production in 10 to 20 years. Of the major global challenges, the environment occupies a dominant place for rubber trees and natural rubber, even on smallholdings where *Hevea* is usually intercropped with other tree crops or food crops, thereby constituting agro-forests, which nowadays have important secondary functions (maintaining biodiversity, environmental conservation, rehabilitation of degraded zones, *etc.*).[4]

As a tree, *Hevea* plays a role in carbon sequestration and may therefore be eligible for the Clean Development Mechanism (CDM).[5]

Lastly, *Hevea* produces wood, which is effectively exploited today[6] and produces natural rubber in its laticifer rings, which man has learned perfectly to use to his advantage. The car industry exerted such demand that supplies fell short, opening up the way for synthetic rubbers, which have become the ideal substitute capable of satisfying all the requirements of the industry. Substitution of natural rubber by synthetic rubber peaked in 1979 (29.5% of elastomers consumed worldwide were made from natural rubber). Since then, the trend has reversed, mainly due to industrial development in producing countries, which favoured their own national production of natural rubber; in 2006 43% of elastomers consumed worldwide were made from natural rubber. Since 2001,

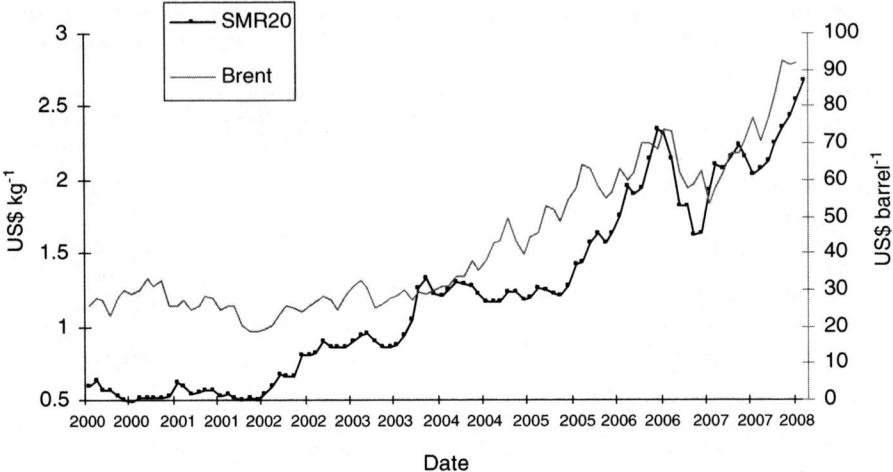

Figure 9.5.1 Natural rubber prices – TSR 20 – FOB Kuala Lumpur. SMR = Standard Malaysian Rubber, SMR 20 is a medium grade

China has amplified the phenomenon, creating a supply/demand imbalance fuelling a spectacular rise in natural rubber prices (+540% in 6 years between 2002 and 2007 – FOB price Kuala Lumpur in US$ – TSR 20) – Figure 9.5.1 – further amplified by speculation. The price of natural rubber currently exceeds that of most synthetic rubbers (around 20% higher). Might that inversion risk encouraging further substitution by synthetic rubbers produced from fossil carbon? In some areas of industry, a readjustment is likely to occur, but it will not clearly alter the share of natural rubber insofar as the global context has evolved, especially as regards energy, the existence of greenhouse gases and European regulations (REACH):

- in the coming years, the new energy context – permanently expensive petroleum oil and climate change – will modify transport, but also the cost of producing synthetic rubbers; oil consumption will therefore have to be reduced;[7] it takes 0.4 mto of oil equivalent (TOE) to produce one tonne of natural rubber, whereas 3.7 to 5 TOE are required to make one tonne of synthetic rubber.[8] Natural rubber consumption therefore allows savings of 36 million TOE, amounting to 40% of total primary energy consumption in France;[9]
- it will be necessary to store or use greenhouse gases. Recycling greenhouse gases into natural polymers through the green factory – *Hevea* – is a tremendous opportunity; using polluting and toxic products during elastomer processing will be profoundly modified in the next 10 years, in order to achieve the REACH standard. Substitution by preferably natural non-toxic products will be necessary; better knowledge of products – apart from isoprene – existing naturally in latex and remaining unexploited for the moment should provide some initial solutions (particularly antioxidants).

9.5.3 Water and Carbon Budget of the Rubber Tree

9.5.3.1 Carbon and Water in Plants

Like any plant, the rubber tree relies on photosynthesis for carbon acquisition and on root uptake for water and nutrient acquisition. Through photosynthesis, solar energy is intercepted and converted to synthesize energy-producing carbohydrates $(CH_2O)_n$, which is the basis for all subsequent metabolic pathways in plants. These carbohydrates are monosaccharides (glucose, fructose, ribose), oligosaccharides (sucrose, maltose, raffinose) and polysaccharides (starch). Following synthesis in leaves (sources), carbohydrates are partitioned among the different sinks (leaves, shoots, trunk, flowers, fruits, roots) through phloem vessels. The major form of carbohydrate transported is sucrose, whereas the major non-structural reserve form is starch, mainly stored in the parenchyma. Cellulose, hemi cellulose and pectin constitute structural carbohydrates. In sink organs, carbohydrates are catabolized to provide energy for cell functions, releasing CO_2 and consuming O_2 (respiration). Consequently, the respiration rate of any organ is directly related to its activity, be it growth, maintenance or specific functions.

Water taken up together with mineral nutrients by roots is transported *via* the xylem towards the leaves, where it is transpired through stomata. According to the cohesion theory of sap ascent in plants, transpiration creates a tension (negative pressure) in xylem conduits, driving water from soil to root and throughout the shoot. Atmospheric water vapour pressure deficit (VPD) is therefore the main climatic factor determining water demand in plants, whereas soil water availability limits actual uptake. However, transpiration is not a passive process; it is regulated by plants by controlling stomatal opening. Under water stress, stomatal conduction drops, in order to limit water loss. On the other hand, as CO_2 uptake takes place in stomata too, stomatal closure also limits photosynthesis. Water use efficiency (WUE) for carbon, the amount of absorbed carbon to the amount of transpired water, is a major parameter in plant ecophysiology.

9.5.3.2 Photosynthesis and Water in the Rubber Tree

Rubber clones differ in their photosynthetic activity, not only in terms of maximum net photosynthesis, P_{max}, or the light response curve,[10] but also in their ability to cope with changing light conditions.[11] In a rubber plantation, a large proportion of the leaves in the canopy are found under limiting light conditions; however, in leaves exposed to sunlight, stomatal conductance declines rapidly in the afternoon, limiting photosynthesis activity. Thus, shaded leaves in rubber plantations may contribute significantly to total photosynthesis.[12]

Photosynthesis in rubber is highly temperature-dependent. It decreases sharply below 20 °C and above 36 °C.[13] The response of photosynthesis to temperature is a key area concerning both global climate change[14] and adaptation of crops to new planting areas. Temperature may directly affect the properties of key enzymes such as ribulose-1,5-bisphosphate carboxylase oxygenase (Rubisco) and the regeneration of ribulose-1,5-bisphosphate (RuBP).[15] Moreover, photosynthesis rates can be affected by temperature not only through biochemical processes but also by impact on stomatal conductance for CO_2.[16]

As rubber plantations are currently expanding to new areas under drier climatic conditions than traditional ones (north-eastern Thailand, Mato-Grosso in Brazil, north-western India), the performance of rubber trees under water stress has become a major issue. Two main traits are to be considered: performance (the ability to transfer water and grow under water stress) and resistance (the ability to avoid damage and survive under water stress). Performance mainly relies on the conductivity of the hydraulic system from root to leaf,[17] whereas resistance mainly relies on susceptibility to cavitation, leading to embolism, within xylem vessels.[18] Stomatal closure occurs to avoid severe cavitation. Hence, differences in susceptibility to cavitation among rubber clones[19] are likely to induce differences in resistance to severe water stress.

9.5.3.3   Tapping, Latex Yield and Carbon Budget of the Rubber Tree

Tapping brings about major changes in rubber tree functioning. Both rubber production and growth require assimilates from photosynthesis, mainly in the form of sucrose. Sucrose is both the source of metabolic energy and the basic molecule for metabolic pathways to isoprene biosynthesis (see Section 9.5.4). A fundamental feature is that an untapped rubber tree synthesizes very little rubber, as latex is not naturally exuded by the tree. However, in response to tapping (cutting a thin slice of trunk bark each day to every five days) the laticifer vessels included in soft bark tissue are severed and latex flows out. A secondary metabolism is artificially induced for regeneration of exported latex, which uses assimilates derived from the other sinks,[20–22] particularly from trunk growth. The negative relation between latex production and wood biomass creation is well documented.[21–23] Silpi *et al.*[24] showed an 80% growth deficit for tapped trees as compared to untapped trees, over a period of 5 months of tapping. Preserving a balance between these two sinks is a key to high and sustainable productivity in rubber plantations[21,25] particularly when wood production is considered, too. However, the amount of carbon stored in the wood biomass of untapped rubber trees is always higher than the amount of carbon in the wood biomass of tapped trees, plus the equivalent in carbon exported by latex exploitation (taking into account the higher carbon cost of latex than wood). It therefore looks as though some carbon is missing in tapped trees. This may be partly explained by increased respiration activity in trunk bark following repeated wounding by tapping.[26]

Inside laticifer vessels, the sucrose content drops sharply following tapping.[27] Latex sucrose content indicates the balance between sugar uptake and utilization (catabolism to rubber). Crossing sucrose laticifer content with inorganic phosphorus content (Pi, indicator of latex metabolic activity) is the basis of

Latex Diagnosis (LD)[28] methods used to evaluate tapping systems and characterize rubber clones. Rubber clonal typology is mainly based on sucrose and Pi content under low intensity tapping and changes in these parameters following an increase in tapping intensity.[29] Some clones are considered "quick starters" as they exhibit high latex yield together with high Pi and low sucrose under low tapping intensity. On the other hand, some clones are "slow starters" but exhibit high potential for intensification as their metabolic activity is able to increase and their sucrose resources are high.

However, the range in yield-related sucrose changes is clone-dependent. For a given amount of exported latex, hence a given amount of sucrose consumed, some clones, such as PB 217, are able to maintain a higher sucrose balance than others.[30] As isoprene biosynthesis pathways are not likely to differ among clones, this shows that clones differ in their ability to upload sucrose into laticifer cells. Regulation of active sucrose transport from phloem and parenchyma to laticifer vessels through medullar rays may explain such differences that could also be related to differences in carbohydrate reserves in the vicinity of the tapping cut. Despite increased competition for carbohydrate induced by tapping, there is more reserve accumulation in the trunk wood of tapped trees than of untapped trees.[31] Moreover starch accumulation is positively correlated to yield when different tapping systems are compared.[32] Such surprising findings show that trees tend to adapt their level of carbohydrate reserves to metabolic demand, at the possible expense of radial growth. The trunk area affected by tapping is not limited to the vicinity of the cut, but covers almost all the trunk area up to at least 2 m high.[26] Latex metabolic activity, latex sucrose content, wood and bark starch content, as well as radial growth, is affected throughout both sides of the trunk, even in areas which are too far from the cut to be directly involved in latex regeneration following tapping. Latex diagnosis mapping is used to determine the latex regeneration area (high Pi, low sucrose). Latex yield per tapping day is positively correlated to the size of that area.[33]

9.5.3.4 Tapping and Water Budget of the Rubber Tree

Latex contains approximately 65% water. Consequently, human activity artificially removes water from the tree by tapping. However, the volumes of water flowing out of the trunk following tapping (a few dl per tapping) are negligible when compared to the amount ascending along the xylem and transpired through leaves (between 60 and 200 l per day in 12-year-old trees).[34] However, tapping may locally and temporarily disrupt bark water status.[35] In the first minutes after tapping, latex flow – driven by turgor pressure – is high and water has to flow from the parenchyma surrounding the laticifer tissues to compensate for water loss from latex. Radial contraction of trunk bark was measured after tapping, indicating interference between water mobilization towards laticifer tissues and normal changes in the water content of parenchyma throughout the day.[36] A transient imbalance between latex flow and water recharge may be a triggering factor in bark necrosis symptoms under unsuitable soil and climatic conditions.[37]

As turgor pressure is responsible for latex flow, water availability for laticifer tissues is a major factor determining flow duration and consequently rubber yield. It determines the more or less rapid blocking or "plugging" of the open ends of latex vessels. A plugging index has been defined and is widely used to characterize the decrease in latex flow dynamics.[38] Seasonality of this index is well related to rainfall or soil water content. Moreover, very early observations showed that tapping at different times of the day influenced latex yield. This is related to a marked drop in turgor pressure at the tapping site during the day, when tension within the xylem increases following the onset of transpiration. Tapping is therefore carried out early in the morning in most places and even at night in Thailand.

Carbon, assimilated through photosynthesis into carbohydrates, is partitioned among the main sinks: vertical growth, radial growth, respiration and reserves (Figure 9.5.2). When trees are tapped, some of the carbohydrates are diverted to regenerate the exported latex, a sink that is artificially created and maintained by the exploitation process.

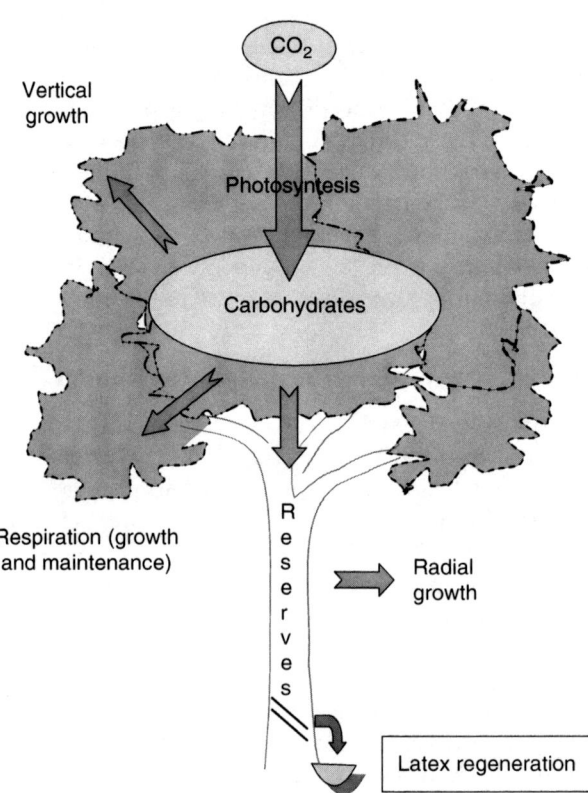

Figure 9.5.2 Effect of tapping on carbohydrate partitioning in the rubber tree.

9.5.4 Biosynthesis of poly(*cis*-1,4-isoprene)

Poly(*cis*-1,4-isoprene) is the polymer that gives natural rubber a very competitive share of the elastomer market. In *Hevea brasiliensis* latex, this polymer is stored in so-called "rubber particles", which make up 20 to 45% of the latex volume. Much work has been undertaken to understand the biosynthesis pathway of this very-long-chain polymer, which stands out through having almost 100% *cis* double bonds.

9.5.4.1 Polyisoprenoids

Poly(*cis*-1,4-isoprene) belongs to the family of polyisoprenoids, which are the most structurally diverse and abundant natural products known, with more than 23,000 primary and secondary metabolites. This huge family comprises, for example, sterols which display not only structural functions (control of biological membrane fluidity) but also hormonal functions (steroid hormones). Key phyto-hormones, such as abscisic acid, gibberellins and cytokinins, are isoprenoids too. Moreover, isoprenoids are used in protein prenylation, which is a key step in the activation and the localization of metabolic enzymes in many organisms. The first common step of all isoprenoid biosynthesis pathways is the formation of isopentenyl diphosphate (IPP).[39]

9.5.4.2 Biosynthetic Pathway

The biosynthetic pathway of linear isoprenoids can be divided into three steps: initiation, elongation, and termination.[39-42] A synthetic pathway is presented in Figure 9.5.3.

9.5.4.2.1 Initiation

The initiation step consists in the formation of IPP **(12)**, and its isomer DMAPP **(13)**. The conventional metabolic pathway to form these two molecules is called the acetate/mevalonate pathway (MVA pathway) in which three molecules of acetyl-CoA **(3)** condense successively to form 3-hydroxy-3-methylglutaryl-CoA (HMG-CoA) **(5)**, which leads to a key intermediate molecule, namely mevalonic acid (MVA) **(6)**. The latter is further phosphorylated and decarboxylated to form the IPP molecule **(12)**. In *Hevea brasiliensis*, this cytosolic pathway was described by Lynen[43] and Lebras[44] in the early 1960s and reviewed more recently by Kekwick[45] and Ohya.[46] Most experimental validations were obtained by observing the incorporation of radioactive tracers, such as [2-^{14}C] MVA and [3-^{14}C]HMG-CoA. The incorporation of [^{14}C]IPP into rubber was found to be much faster than that of [2-^{14}C] MVA. This was assumed to be due to slow conversion of MVA into IPP.[43-44,46] Another explanation might be that the MVA pathway was not exclusive for IPP biosynthesis. Indeed less than 10 years ago, a new, mevalonate-independent, IPP biosynthesis pathway was discovered by Rohmer.[47] This plastidic DXP-MEP pathway initiates with a

From glucose to natural rubber polyisoprene

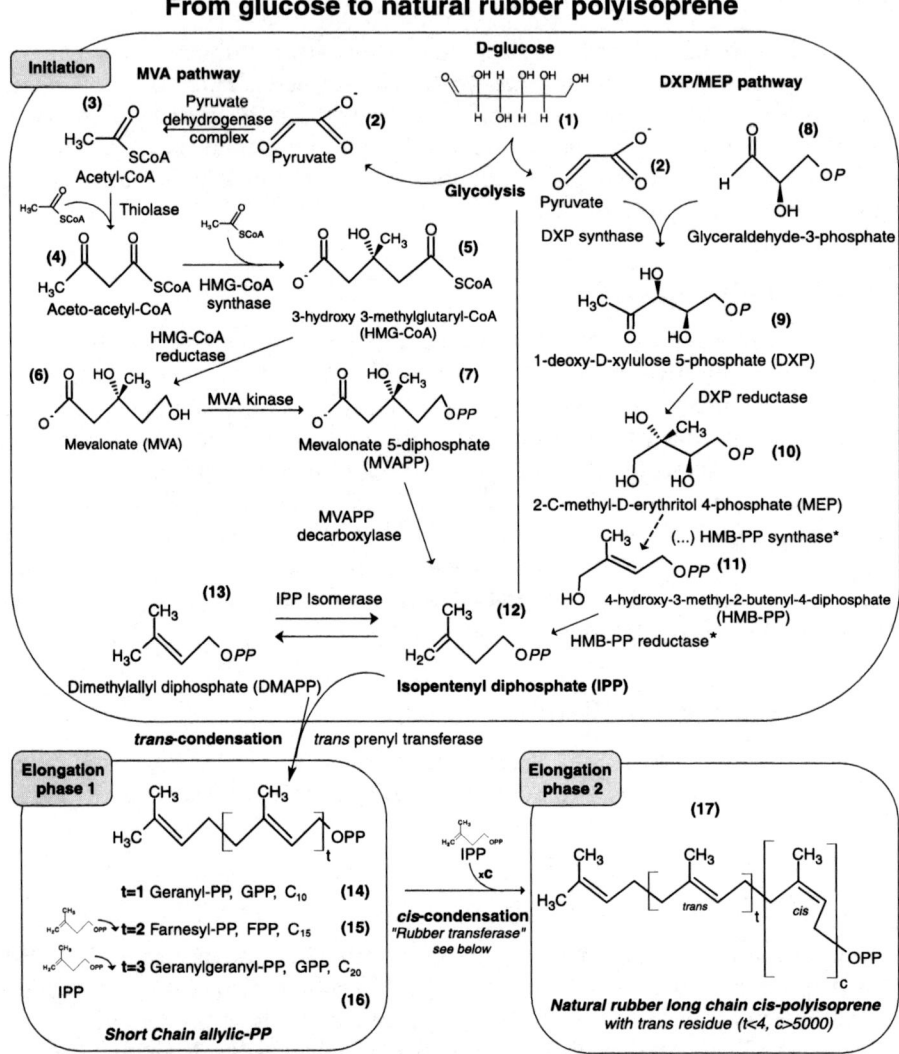

Figure 9.5.3 Pathway of poly(*cis*-1,4-isoprene) biosynthesis.

condensation of pyruvate (**2**) and D-glyceraldehyde 3-phosphate (**8**) to form 1-deoxy-D-xylulose 5-phosphate (DXP) (**9**) catalyzed by DXP-synthase. DXP is converted to 2-C-methyl-D-erythritol 4-phosphate (MEP) (**10**) by the DXP reducto-isomerase. The last step of this pathway is the reduction of

4-hydroxy-3-methyl-2-butenyl-4-diphosphate (HMBPP) **(11)** catalyzed by HMBPP reductase, whose structure and reaction mechanisms are still unknown.[48] This new pathway has been found in bacteria, algae and higher plants.[47] Concerning *Hevea brasiliensis*, the expression of DXP synthase has been shown in latex and leaves.[49] The IPP derived from this second pathway may diffuse from the plastids and join the cytosolic pool.[40]

Given that acetylCoA **(3)** is produced from pyruvate **(2)** by the pyruvate dehydrogenase complex, both pathways (MVA or DXP-MEP) leading to the formation of IPP start with a pyruvate which is a product of glucose **(1)** catabolism by glycolysis.

In addition to the formation of IPP **(12)**, initiation includes as well the formation of dimethylallyl diphosphate (DMAPP) **(13)** by isomerization of IPP, catalyzed by IPP isomerase.

9.5.4.2.2 Elongation

Elongation consists of the sequential condensation of IPP and allylic diphosphate through the action of the prenyl chain elongating enzyme, commonly called prenyl transferase. The reactions catalyzed by prenyltransferases start with the formation of allylic cations after the elimination of pyrophosphate ions, to form allylic prenyl diphosphate. This is followed by addition of an IPP with stereo-specific removal of the proton at the 2-position. This is the key reaction which determines *cis*- or *trans*-configuration of the double bonds contained in the linear isoprenoid chain. *Cis-trans* isomerism is dependent on the nature of the prenyltransferase involved in elongation catalysis. A comprehensive review of *cis*- and *trans*-prenyltransferases was recently undertaken by Takahashi and Koyama.[39] *Cis*-prenyltransferases are much less well known than their *trans*-homologues.

For linear isoprenoid biosynthesis, elongation of the polymer chain can be described in two main phases as shown in Figure 9.5.3:

- Phase 1: the formation of short-chain allylic diphosphates where geranyl-PP **(14)**, farnesyl-PP **(15)** and geranylgeranyl-PP **(16)** are formed by the action of *trans*-prenyltransferases.
- Phase 2: the *trans*-short-chain allylic diphosphates obtained are then employed as allylic primer substrates for additional IPP condensation with *trans*- or *cis*-configuration.[39]

In the case of natural rubber from *Hevea brasiliensis*, which is almost entirely made of *cis*-isoprene units, it could be assumed that *cis*-prenyltransferase is exclusively involved in phase 2. Unfortunately it is more complex. Authors agree on phase 1, which is *trans*-condensation catalyzed by a cytosolic soluble *trans*-prenyl transferase[40,50–52] Farnesyl diphosphate synthase has been cloned from rubber latex.[53] The most probable prenyldiphosphate used as a primer for phase 2 is farnesyl-PP **(15)** or geranylgeranyl-PP **(16)**.[46,54,55] Phase 2 is catalyzed by a still not clearly identified "rubber transferase" system. Different proteins have been

described as being involved in *cis*-condensation of IPP by acting as prenyltransferase itself, or as co-factors. A dimeric protein with a monomeric molecular mass
of 38 kD was purified by Light[56] and showed prenyltransferase activity leading to
chain elongation. Nevertheless, in the presence of DMAPP, that enzyme only
catalyzed *trans*-condensation of IPP leading to FPP and GGPP.[46] Many authors
have stated that this reaction is catalyzed by the association of a prenyltransferase
with several co-factors. The two most cited are Rubber Elongation Factor (REF),
which is a 14 kD protein tightly bound with large rubber particle membrane, and
Small Rubber Particle Protein (SRPP), which is a 22 kD protein.[50,57,58] More
recently, two *Hevea brasiliensis* genes encoding *Hevea* rubber transferases (HRT1
and HRT2) were isolated by homology with other *cis*-prenyltransferases cloned in
other organisms.[52] The expressed protein HRT2 had a molecular mass of 33 kD
and showed *cis*-prenyltransferase activity in the presence of washed bottom
fraction particles (WBFP). It is interesting to note that in most of the reported *in
vitro* studies, the addition of some part of initial latex, such as washed rubber
particles or WBFP, is required to achieve *cis*-prenyltransferase activity. This
shows that the rubber synthesis machinery is complex and probably involves the
participation of several proteins. In addition to proteins, the presence of metal
divalent cations such as magnesium or manganese ions is required as a co-factor
of rubber transferase activity.[54]

9.5.4.2.3   Termination

Very little is known about the final biochemical termination event. Different
alpha end groups have been identified by NMR indicating the presence of
phosphate and fatty acid. These could be attributed to the presence of a
phospholipid molecule.

   Although it is not part of the biosynthesis pathway, *stricto sensu*, it is worth
mentioning that after termination the linear isoprene chain may undergo further chemical modifications. Indeed, abnormal groups such as aldehyde, epoxy
or lactone have been detected on the polymer chain.[59]

9.5.4.2.4   Polyisoprene Chain Length Control

The control of polyisoprene chain length and therefore molecular weight has
been attributed to several factors:

- The concentration of substrates such as IPP and FPP. High IPP concentration leading to higher molecular weight, while a contrary phenomenon occurs with FPP concentration;
- The magnesium ion, which is a necessary co-factor for *Hevea brasiliensis*
  prenyltransferase, has also been described as a molecular weight
  regulator;[54]
- Tangpakdee[60] also assumed that molecular weight could be regulated by
  the activity of IPP isomerase, which is necessary to produce DMAPP, and

then to initiate rubber biosynthesis. A lower IPP isomerase activity would lead to a higher molecular weight.

9.5.4.3 Localization of Rubber Biosynthesis

The synthesis site for allylic diphosphate primers and *cis*-polyisoprene is largely assumed to occur on the surface of pre-existing rubber particles,[51,61,62] but rubber biosynthesis activity has also been localized in the membrane of non-rubber particles from the bottom fraction after ultracentrifugation of latex.[52,63,64] The latter authors[64] presumed that previous localization of rubber biosynthesis on rubber particles was due to an artefact resulting from the rapid deterioration of bottom fraction (BF) particles after tapping, which led to the migration of rubber synthesis machinery from BF particles to rubber particles.

9.5.4.4 Conclusion

The details and understanding of rubber biosynthesis are still ambiguous and clear evidence has not yet been convincingly presented. Modern proteomic techniques should provide a clearer understanding of reaction mechanisms, especially the *cis*-condensation of IPP, as well as the biosynthesis location. This knowledge would be very useful for considering genetic improvement of the biosynthesis machinery. This work should be done in close collaboration with rubber tree physiologists who are working on sugar availability in laticiferous cells: indeed, it is pointless optimizing engine power if there is a lack of fuel.

9.5.5 Natural Rubber Structure

9.5.5.1 Introduction

As for all materials, a study of natural rubber and of the relations existing between its structure and properties calls for different structural levels to be defined. We have opted to define them in terms of microstructure, mesostructure and macrostructure. The **macrostructure** of natural rubber concerns the raw material in its entirety, which is assumed to be homogeneous with little or no destructuring. We shall not be looking at this structural level in detail here as it primarily involves rheological characterizations. For a polymer, definition of the **microstructure** varies depending on the authors. Koenig[65] and Sperling[66] consider the microstructure to involve the (chemical) composition, configuration and conformation of the chain. Mirau *et al.*[67] propose: *the precise way in which monomers are linked to each other to form a polymer (chemical structure)*. For natural rubber, the microstructure concerns the chemical structure of the poly(*cis*-1,4-isoprene) macromolecules, but also the composition of non-isoprene constituents (lipids, proteins, *etc.*), their chemical structure and their interactions with poly(*cis*-1,4-isoprene) macromolecules. These interactions play a paramount role in the **mesostructure** of natural rubber and its specific properties. For natural rubber, the mesostructure corresponds to the macromolecular structure and to the aggregates formed by the association of poly(cis-1,4-isoprene) molecules (gel).

9.5.5.2   Microstructure

For synthetic rubbers, analysis problems boil down to two determinations: identification of the constitutive groups and determination of spatial arrangement. For natural rubber, those two determinations have undergone a great deal of research already mentioned.[68] The poly(cis-1,4-isoprene) chain in the plants studied comprises three parts (Figure 9.5.4): a (ω) end group containing 1 dimethylallyl group and 2 or 3 *trans* groups, a long chain of *cis* configuration and an (α) end group (Figure 9.5.4a). For natural rubber from *Hevea*, the chemical structure of the two end groups has yet to be completely identified. Neither the hydroxyl group nor the dimethylallyl group have been revealed. Studies on deproteinized natural rubber from which lipids had been removed by methanolysis led Tanaka *et al.* [69,70] to suggest the existence of a protein (or polypeptide) on the (ω) group and a phospholipid (or fatty acid) on the (α) end group (Figure 9.5.4b). Those two non-isoprene compounds attached by covalent bonds to the polyisoprene chain would seem to lie behind the interactions responsible for the associative structure of natural rubber.

Non-isoprene compounds (lipids, proteins and sugars) will be covered in Section 9.5.6.

9.5.5.3   Mesostructure

The mesostructure covers both the macromolecular scale (dimension, conformation and architecture of the macromolecules) and also the supramolecular scale (complex aggregates between macromolecules). It is this last point that is the key to the specific properties of natural rubber. This associative

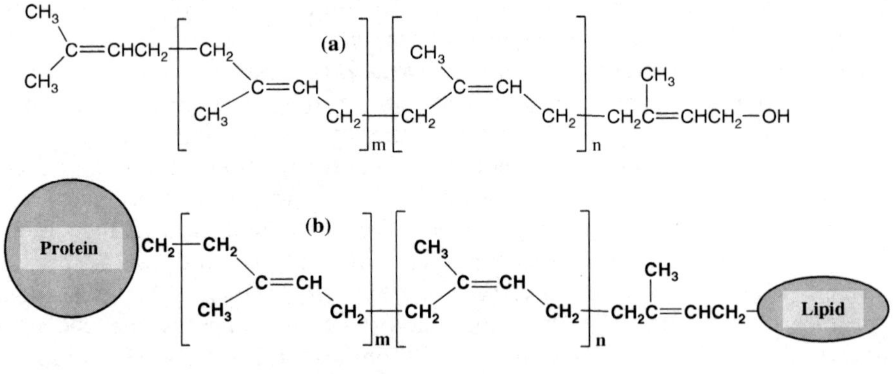

Figure 9.5.4   Microstructure of poly(cis-1,4-isoprene) (a) for latex plants other than *Hevea* and (b) for *Hevea*.

mesostructure is gradually and partly destroyed when dissolved in a conventional solvent of polyisoprene (cyclohexane, tetrahydrofuran). However, in many cases, a proportion of natural rubber remains insoluble in those solvents, a fraction commonly referred to as the gel phase or **macro gel**. The soluble fraction contains polyisoprene macromolecules and a variable quantity of micro-aggregates making up the **micro gel**.

9.5.5.3.1 *Macromolecular Structure*

In 1972, Subramaniam[71] investigated inherent molar mass distribution (MMD_0) in natural rubber. MMD_0 is the MMD of natural rubber on leaving the tree without undergoing any modification linked to post-harvest processing (coagulation, maturation, drying, *etc.*). The MMD_0 of natural rubber is an important criterion for forecasting a certain number of natural rubber properties that will be obtained after processing.[72] Depending on the clone, MMD_0 will lie between two extreme types: bimodal or unimodal with a shoulder (Figure 9.5.5). MMD evolves after processing depending on the process used.[72] Natural rubber from *Hevea* is characterized by polyisoprene chains of very variable size ranging from a few thousand to some tens of millions of $g\,mol^{-1}$. It is therefore characterized by a very wide molar mass distribution, with a polymolecularity index (I) sometimes exceeding 8, and a very high average number molar mass ($150,000 < Mn < 300,000\,g\,mol^{-1}$). Among around twenty

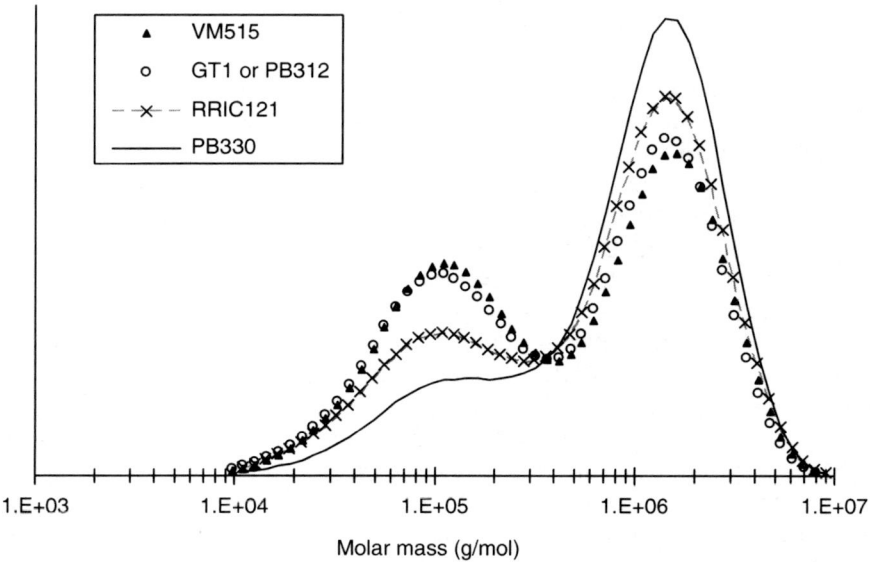

Figure 9.5.5 Examples of inherent molar mass distribution for several *Hevea* clones (Size Exclusion Chromatography in cyclohexane according to Bonfils *et al.*[80]).

latex-bearing plants studied by Swanson *et al.*[73] only rubber from guayule contained such long macromolecules.

9.5.5.3.2   Gel

Non-isoprene mechanisms and compounds responsible for gel formation have been intensely studied. Prior to the recent proposals made by Tanaka *et al.*,[69] so-called abnormal compounds were incriminated in the interactions responsible for gel. Those compounds (aldehydes,[74] epoxy,[75] esters[76]) were linked to the polyisoprene chain and reacted with some non-isoprene compounds (amino acids[77] or proteins,[76] metal ions[78]).

Many methods can be used to determine the quantities of macrogel and microgel in a polymer.[79] For natural rubber, the proportion of macrogel is determined by gravimetry after centrifugation. The quantity of microgel can be measured by filtration, but that quite laborious method can be replaced by size exclusion chromatography (SEC).[80] Macrogel and microgel exist in both dry rubber (TSR or RSS) and stabilized latex (field latex or concentrate) that has matured.[81,82]

The proportion of macrogel can vary enormously, reaching rates of almost 60% in extreme cases depending on the authors. Allen and Bristow[83] showed that the proportion of gel depended on the solvent used, and the dissolution time and temperature. It is therefore unlikely that difficulties in dissolving dry rubber are solely linked to chemical cross-linking problems.

In many studies, the authors settled for measuring the amount of macrogel whilst microgel is clearly preponderant in many cases.[84] Bloomfield[85] defined the microgel found in latex harvested from *Hevea* left to rest for several months as: *internally cross-linked molecules similar in size to latex particles.* The quantity of microgel, determined by filtration, in a natural rubber sample (5 to 30%) would seem to vary and greatly depend on clonal origin.[86] Larger quantities can be found depending on the technique used.[87] Campbell and Fuller[88] associated the same definition with the term microgel to describe the cloudy solutions obtained when dissolving a sample of SMR L in dichloromethane. Allen and Bristow[83] presumed that microgel is due to chemical bridges generated inside latex particles. Those bridges would seem to form between leaving the tree and coagulation, maybe even inside the tree. Voznyakovskii *et al.*[89] used dynamic light scattering (DLS) to study natural rubber solutions (RSS and DPNR,[i] [NR] ≈ 1 to $3\,\mathrm{mg\,ml^{-1}}$). They showed that microgel displayed two populations, one centred on $1.1\,\mu\mathrm{m}$ (P1, $0.9 < d < 5\,\mu\mathrm{m}$) and the other around $10\,\mu\mathrm{m}$ (P2, $5 < d < 12\,\mu\mathrm{m}$). The deproteinized sample (DPNR) only contained small-sized micro-aggregates (P1), whereas for smoked sheet (RSS1) they were mostly around $10\,\mu\mathrm{m}$ (P2). Steam treatment, which eliminates a proportion of the proteins, reduced the quantity of large micro-aggregates (P2) to the benefit of small micro-aggregates and the soluble portion. Although protein removal did not eliminate all the microgel,

---

[i] RSS: Ribbed Smoked Sheet; DPNR: Deproteinized Natural Rubber.

population P1 persisted. They concluded that the interactions lying behind micro-aggregate formation were linked to the existence of proteins. They also discovered the thermo-reversibility of the interactions leading to microgels. Heating solutions (20 to 80 °C) did not eliminate microgels but reduced the average size ($d \approx 10\,\mu$m at 20 °C, $d \approx 5\,\mu$m at 60 °C, for RSS). The concentration and size of micro-aggregates also depend on the concentration in the solution.

9.5.6 Non-isoprene Components of Natural Rubber

Non-isoprene components of *Hevea brasiliensis* latex account for around 10% of its dry matter while they account for about 5% of the raw dry rubber derived from latex (Table 9.5.2). They comprise proteins, carbohydrates, lipids and inorganic constituents and represent the main composition difference between natural rubber (NR) and its synthetic counterpart, namely poly(*cis*-1,4-isoprene) synthetic rubber (IR). They could therefore be involved in the irreplaceable specific qualities of NR. The nature and quantity of these non-isoprene components can vary greatly depending on the clones, the exploitation system and environmental conditions.[90] Some of these components are either dissolved or suspended in the aqueous medium of the latex while the others are adsorbed on the surface of rubber particles.

9.5.6.1 Non-isoprene in the Different Compartments of *Hevea brasiliensis* Latex

Many hydrophilic non-isoprene components of latex are leached by the numerous washing operations with water during the process leading to dry rubber. Nevertheless, it is interesting to observe the initial allocation of non-isoprene in latex compartments.

Table 9.5.2   Composition of natural rubber latex and raw dry rubber.

	Latex[a]		*Dry rubber*[c]
	% w/v fresh latex[a]	*% w/w dry matter of latex*[b]	*% w/w dry matter*
Rubber hydrocarbon	35.0	87.0	94.0
Proteins	1.5	3.7	2.2
Carbohydrates	1.5	3.7	0.4
Lipids	1.3	3.2	3.4
Organic solutes	0.5	1.1	0.1
Inorganic substances	0.5	1.2	0.2

Approximate values only (highly dependent on clone, season and physiological status of the tree).
[a]averaged from data published by Wititsuwannakul, 2001.[100]
[b]calculated.
[c]Sainte-Beuve, 2006.[115]

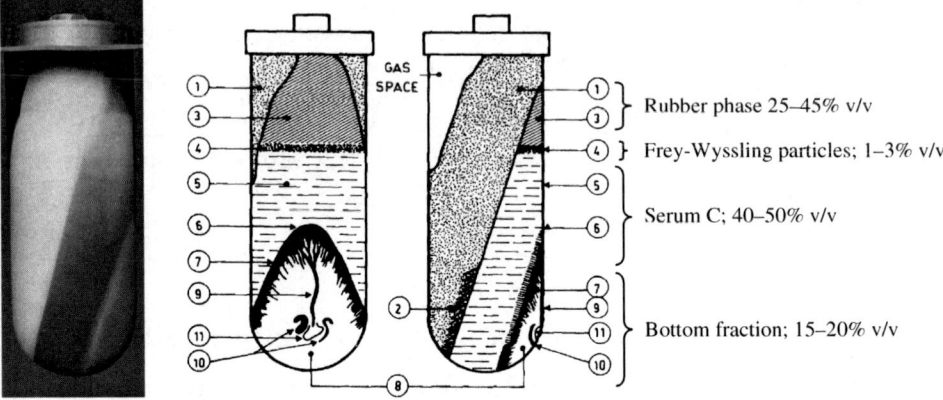

Figure 9.5.6   Various fractions of centrifuged latex. From Moir.[91]

Fresh latex can be separated into various layers by ultracentrifugation at low temperature (Figure 9.5.6). Four main layers are identified:[91]

- A white upper layer or rubber cream consisting of rubber particles amounting to 25 to 45% of latex volume. The particles are strongly protected in suspension by a film of adsorbed proteins and phospholipids. This protein and phospholipid layer imparts a net negative charge to the rubber particle, contributing to colloidal stability.[92,93]
- An orange or yellow layer containing Frey-Wyssling particles which form 1 to 3% of latex volume. These particles are surrounded by a typical double membrane and accumulate lipid globules, isoprenic compounds, carotenoids, plastochromanols and plastoquinones.[94]
- An aqueous layer, called serum or cytosol, amounting to 40 to 50% of latex volume. It contains carbohydrates, minerals, organic acids, amino acids, nitrogenous bases, reducing agents and high-molecular-weight compounds such as proteins, lipids and nucleic acids.
- A bottom layer consisting mainly of lutoid particles and also including other particulate components with a density greater than that of serum.[95–98] Lutoid particles amount to 15 to 20% of latex volume. Their negatively charged tonoplast membrane, rich in phosphatidic acid, encloses a fluid serum known as B-serum. The vacuolar medium is acidic and positively charged. It includes cations, basic amino acids and a whole range of hydrolases.[96]

9.5.6.2   Non-isoprene Families

Non-isoprene can be classified into four families: proteins, carbohydrates, lipids and inorganic compounds. Table 9.5.2 presents an approximate composition expressed *versus* either fresh or dry matter. Around half of non-isoprenes are leached or degraded during dry rubber processing, those remaining being logically the most hydrophobic, *i.e.* lipids.

9.5.6.2.1 Proteins

Fresh latex contains roughly 1.5% of proteins of which about 25–30% are bound to the rubber phase, another 25% being associated with the bottom fraction (lutoid) and the remainder (45–50%) being reported as cytosolic.[99,100] Two major particle surface-bound proteins of 14 kD (rubber elongation factor) and 22 kD (small rubber particle protein) have been identified. In the bottom fraction, the two major proteins are hevein (> 50%), an anionic protein, and hevamine. The molecular weight of the proteins in this fraction is narrow, ranging from 14 to 45 kD. In C-serum, proteins are numerous and range from 14 kD to 133 kD.[94,98,100]

Proteins play an important role in metabolism, in the structure of rubber particles and in the quality of rubber products. Enzymes involved in the glycolysis as well as the rubber biosynthesis pathways are obviously part of the protein pool of latex. Furthermore, rubber particle proteins confer, with phospholipids, a negative charge to the particle surface, which ensures colloidal stability. Proteins have been declared to influence the technological properties of rubber. For example, they are involved in the water retention process during rubber drying[101] and in the stress relaxation of unfilled NR vulcanizate.[102] Lastly, an unwanted property of latex proteins is their allergenic character in latex-dipped goods. However, proper washing processes can reduce allergenicity to a reasonable level. Nevertheless, this remains an issue for a minor, overexposed population such as healthcare workers.[99]

9.5.6.2.2 Carbohydrates and Inositols

Latex cytosol contains sucrose (0.3–0.4% of latex weight), which is the sugar supply for the rubber biosynthesis pathway as the source of pyruvate (see rubber biosynthesis section). Sucrose concentration is a key parameter in latex diagnosis used to assess the physiological status of the rubber tree enabling monitoring of the tapping intensity.[103–105] In addition to sucrose, a smaller amount of glucose, fructose and raffinose are present in latex.

But the most important water-soluble non-isoprene is another carbohydrate called quebrachitol, or 2-O-methyl-L-inositol (see chemical structure in Figure 9.5.7), which can account for more than 1.2% of fresh weight. Its physiological function remains unknown.[100] Quebrachitol can be used as a starting material to synthesize chiral drugs with therapeutic effects for the treatment of cancer, early ageing, diabetes and AIDS. An international patent for the extraction process has therefore been filed by a Malaysian team.[106]

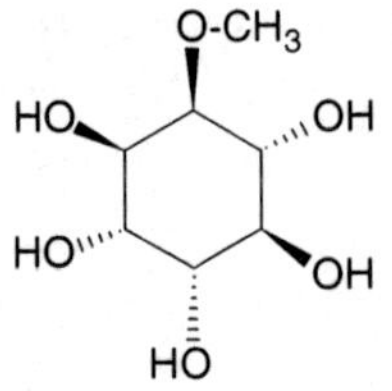

L-quebrachitol   **Furan fatty acid**
>1% of fresh latex weight   >1% of dry rubber weight (for specific clone only)

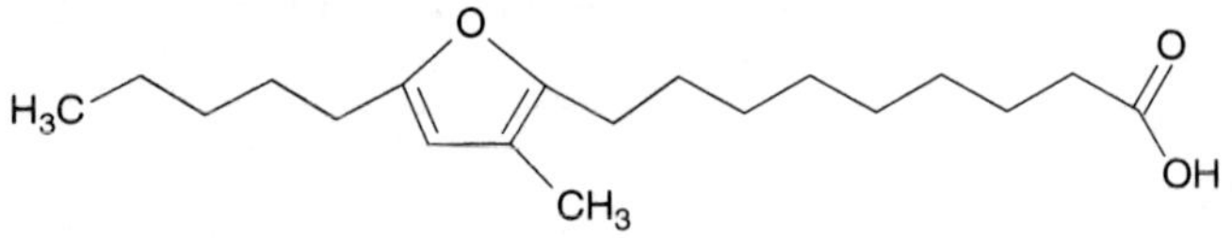

Figure 9.5.7   Two examples of potentially high value-added by-products from *Hevea* latex.

9.5.6.2.3   Lipids

As mentioned before, lipids are the family of non-isoprenes that is the most retained with the rubber during processing. Their extraction for analytical purposes has been optimized recently by Liengprayoon.[107] Depending on the clone, lipids amount to 2.5 to 3.8% *versus* rubber in latex, and 1.8 to 3.3% in dry natural rubber. Three families of lipids can be distinguished:[108]

- Phospholipids: these account for 18–25% of lipids in fresh latex but only 4–5% of lipids in dry rubber. The main phospholipid is phosphatidyl choline, amounting to more than 60% of latex phospholipids. As amphiphilic molecules they are involved in the structure and charge of particle membranes, but could also be linked to polyisoprene chains.
- Glycolipids: these account for 25–35% of lipids in fresh latex and 10–14% of lipids in dry rubber. Like phospholipids, they are degraded and/or leached during dry rubber processing. The two main glycolipids are digalactosyldiglyceride (DGDG, approximately 45% of latex glycolipids) and steryl glucoside (SG, approximately 35%). They also have a structural role in membranes.
- Neutral lipids: these account for 40–60% of lipids in fresh latex but 82–86% of lipids in dry rubber. This share increase is due to the leaching of hydrophilic lipids (phospho- and glycolipids) as well as to the hydrolysis of fatty acid moieties from those hydrophilic lipids: newly released free fatty acids thus join the neutral lipid family. Indeed, free fatty acids account for less than 2% of lipids in latex while they account for more than 20% of those extracted from dry rubber. Neutral lipids can be divided into fatty acids and unsaponifiables.

  - Fatty acid composition: linoleic acid (C18:2) is generally the most important with 40 to 50% of the fatty acids, followed by oleic acid (C18:1), stearic acid (C18:0) and furan fatty acid. These latter fatty acids each account for approximately 10–15% of the fatty acids. For a specific clone, furan fatty acid (see structure in Figure 9.5.7) can amount to 80% of the fatty acid pool. This uncommon fatty acid has been reported to have a biological activity in the treatment of cardiovascular diseases.[109] As for quebrachitol, this co-product, present in not insubstantial quantities, could be put to use. Fatty acids have been described as having a technological role in some properties of rubber, such as green strength[110] and tack property[111] and are known to be activators of vulcanization.

  ○ Unsaponifiables: the major unsaponifiable is β-sitosterol (37–45% of unsaponifiables) followed by γ-tocotrienol (17–22%), Δ5-avenasterol (16–20%), stigmasterol (6–8%), octadecanol (4–7%) and α-tocotrienol (4–6%). Some unsaponifiables may have an antioxidant activity, which is of great importance in protecting the double bonds of isoprenic chains.[112,113]

9.5.6.2.4 Inorganic Compounds

The total concentration of inorganic ions in fresh latex is about 0.5%, the major ions being potassium, magnesium, copper, iron, sodium, calcium and phosphate.[114] Even in small quantities they are key co-factors in metabolic pathways, and are also involved in the technological properties of rubber, such as susceptibility to thermal oxidation.[113]

9.5.6.3 Conclusion

Non-isoprenes of natural rubber are various and play important physiological roles, especially in metabolism (carbon source and enzymes for polyisoprene biosynthesis), and in the structural organization of latex (membranes of rubber, lutoid or Frey-Wyssling particles, colloidal stability). They are also involved in some technological properties of rubber, which give natural rubber its unique properties. Finally, uncommon molecules with high value-added potential, such as quebrachitol or furan fatty acid, are present in large quantities ($> 1\%$), which confirms that *Hevea brasiliensis* is a green chemical factory of great interest.

9.5.7 Specific Properties *versus* Synthetic Counterparts

Through its partly associative structure (molar mass, molar mass distribution, gel and overlap density) natural rubber from *Hevea brasiliensis* exhibits a set of properties unequalled by its synthetic rivals, but also certain drawbacks. The main advantages are:

• Higher elasticity;
• Strain-induced crystallization;
• Low heat build-up at high temperature;
• Higher tack and high green strength.

The main disadvantages of natural rubber still remain today its variable processability and allergy risks for latex-based articles for uses involving human contact.

9.5.7.1 Elasticity

Natural rubber displays elastic properties – the ability to return to its initial shape after initial stress is halted – which are remarkable in raw rubber but also when it is vulcanized (around 700% elongation at break) due to its ability to crystallize under strain (100% *cis*- and associative structure). Its properties are used in elastic bands but also in dipped items such as gloves.

9.5.7.2 Strain-induced Crystallization

Unlike synthetic elastomers such as polybutadiene with a high level of *cis* configuration or polychloroprene, natural rubber crystallizes very slightly at cold temperature without strain. Kawahara *et al.*[116] suggested that the presence of free fatty acids in natural rubber might be conducive to cold crystallization.

But under important strain (>400% of elongation) natural rubber crystallizes at high temperature (>100 °C), undergoing crystallization that is partly perpendicular to the strain depending on the specificities of the network.[117] Such crystallization is reversible, with crystallites disappearing when stretching stops.[118] This property is also found in synthetic elastomers, but at lower deformation rates.[119] Strain-induced crystallization is largely due to the stereochemical regularity of macromolecules (NR: 100% *cis*-), but for natural rubber it is probably linked to the associative structure, which gives rise to branching.[119] Even recently, studies were conducted to elucidate the structure and form of crystallites,[120] or the influence of cross-linking density on crystallization kinetics.[117]

9.5.7.3 Heat Build-up

Heat build-up is the temperature rise in a tyre resulting from the dissipation in heat form of part of the energy (viscous or loss part) due to an applied strain (rolling+weight). As rubbers have low thermal conductivity, heat is not easily dissipated. Energy loss in the medium under strain is linked to the existence of chain extremities that can move freely. The more chains are linked by bridges as, for example, during vulcanization, or even by the presence of gel involving non-isoprene compounds, the less there will be heat dissipation. Too great an increase in tyre temperature, particularly the sidewalls, will lead to a certain number of problems and may cause the tyre to burst under certain conditions of use. Natural rubber is the elastomer with the lowest heat build-up, making it unavoidable for dynamical applications with high strain and high frequencies such as tyres subjected to heavy strains (trucks, planes, *etc.*).

9.5.7.4 Tack and Green Strength

Tack and green strength are two important properties of many elastomer compounds. In manufacturing, these two properties enable the many components of a non-vulcanized tyre to hold together until moulding and curing. Tack is the ability of two materials to resist separation after bringing their surfaces into contact for a short time under light pressure. The green strength of an elastomer is its resistance to deformation and fracture before vulcanization. Natural rubber has good tack (very mobile low molecular weight chains) and good resistance to separation; the high molar mass chains are able to deform without slipping before the process of crystallization under stress starts.

9.5.7.5 Vulcanization

Vulcanization remains a prime stage in the manufacture of rubber articles. It consists of linking (cross-linking) the polymer chains with each other with sulphur, sometimes with peroxides. This essential stage eliminates flow phenomena by adding crosslinking points. It makes it possible to reduce hysteresis by maintaining an acceptable level in terms of elongation/break. Numerous chemicals are involved with the sulphur to catalyze the reaction (stearic acid, zinc oxide and vulcanization accelerators).[121] Unlike its synthetic counterparts, natural rubber withstands much higher vulcanization rates without losing its properties and displays a shorter scorch time. This phenomenon has been attributed to the existence of natural vulcanization activators, notably free fatty acids (1.5 to 2.5% w/w NR). Some studies in this field have also identified the involvement of certain amino compounds: neutral and basic amino acids[122] and the amino bases arising from the hydrolysis of certain phospholipids.[123] Whilst there is abundant literature on vulcanization chemistry[121,123] very few studies have been undertaken to understand the action mechanisms of natural activators in natural rubber.

9.5.8 Conclusion

Rubber and *Hevea* offer a tremendous opportunity for offering some solutions to major global challenges. Indeed, rubber growing is relatively easy and not very sophisticated, making it accessible to the largest number of people; it does not require rich soils. Current prices are advantageous for smallholders, which encourages them to plant rubber. It is a tremendous machine for recycling atmospheric carbon, enabling the production of polymers that will become increasingly rare in the future given the expected shortages in petroleum oil production. Lastly, the richness of natural elements occurring in latex – apart from isoprene – has yet to be exploited and suggests uses in some very different fields.

Acknowledgement

Thanks to Mr Philippe Schill for his shrewd advice, which enabled us to gain a clearer understanding of the specificities of natural rubber.

References

1. P. Compagnon, *Le caoutchouc naturel*, ed. G. P. Maisonneuve et Larose, Paris, France, 1986; M. J. R. Loadman, *The Exploitation of Natural Rubber*, Publ. 1531, Malaysia Rubber Producer's Res. Association, Brickendonbury, Engl. version from U. Giersch and U. Kubisch, *Gummi – die elastische Faszination*, Nicolai, Berlin, 1995.
2. H. Smit, *Tire Technology International*, 2007, 100–104.
3. F. Nkoa and B. Daviron, *Recherche, Développement*, 1995, **2**(4), 27–31.
4. S. Diaz-Novellon, E. Penot and M. Arnaud, Characterisation of biodiversity in improved rubber agroforests in West-Kalimantan, Indonesia: Real potential uses for spontaneous plants. *International Symposium Land-use, nature conservation and the stability of rainforest margins in Southeast Asia*, Bogor, Indonesia, 2002.
5. O. Hamel and J. M. Eschbach, *Oleagineux-Corps-gras-Lipides, Nov.-Dec.*, 2001, **8**(6), 599–610.
6. J. Balsiger, J. Bahdon and A. Whiteman, *The utilization, processing and demand for rubberwood as a source of wood supply*, in Paper Series FAO, Rome, 2000, **50**, 75.
7. P. R. Bauquis, *Quel avenir pour les hydrocarbures à l'approche des pics pétrolier et gazier?* – Conférence-débat "Les pics pétrolier et gazier: conséquences et enjeux", IFP, 11 Mai 2006.
8. K. P. Jones, *Kautschuk Gummi Kunststoffe*, 2000, **53**(12), 735–742.
9. Observatoire de l'Energie, 19 Oct. 2007.
10. R. Ceulemans, R. Gabriels, I. Impens, P. K. Yoon, W. Leong and A. P. Ng, *Trop. Agr.*, 1984, **61**(4), 273–275.
11. A. Nugawela, S. P. Long and R. K. Aluthhewage, *J. Nat. Rubber Res.*, 1995, **10**(4), 266–275.
12. A. Nugawela, D. C. A. Abeysinghe and R. K. Samarasekera, *Journal of the Rubber Research Institute of Sri Lanka*, 1995, **75**, 1–12.
13. B. Kositsup, P. Kasemsap, P. Thaler and T. Ameglio, *Effect of temperature constraints on photosynthesis of rubber (Hevea brasiliensis)*, Proceedings IRRDB International Natural Rubber Conference, Siam Reap, Cambodia, 2007, p. 161–166.
14. C. J. Bernacchi, E. L. Singsaas, C. Pimentel, A. R. Portis Jr and S. P. Long, *Plant Cell Environ.*, 2001, **24**(2), 253–259.
15. G. D. Farquhar, S. von Caemmerer and J. A. Berry, *Planta*, 1980, **149**, 78–90.
16. B. E. Medlyn, E. Dreyer, D. Ellsworth, M. Forstreuter, P. C. Harley, M. U. F. Kirschbaum, X. Leroux, P. Montpied, J. Strassemeyer, A. Walcroft, K. Wang and D. Loustau, *Plant Cell Environ.*, 2002, **25**, 1167–1179.

17. J. S. Sperry, U. G. Hacke, R. Oren and J. P. Costock, *Plant Cell Environ.*, 2002, **25**, 251–263.
18. M. T. Tyree and J. S. Sperry, *Annu. Rev. Plant Phys. Mol. Bio.*, 1989, **40**, 19–38.
19. K. Sangsing, P. Kasemsap, S. Thanisawanyangkura, K. Sangkhasila, E. Gohet, P. Thaler and H. Cochard, *Trees Struct. Funct.*, 2004, **18**, 109–114.
20. J. K. Templeton, *J. Rubb. Res. Inst. Malaya*, 1969, **21**, 259–273.
21. P. R. Wycherley, *J. Rubb. Res. Inst. Malaya*, 1976, **24**, 169–194.
22. M. R. Sethuraj, *Plant Cell Environ.*, 1981, **4**, 81–83.
23. E. Gohet, J. C. Prévôt, J. M. Eschbach, A. Clément and J. L. Jacob, *Plantations, recherche, développement*, 1996, **3**(1), 30–38.
24. U. Silpi, P. Thaler, P. Kasemsap, A. Lacointe, A. Chantuma, B. Adam, E. Gohet, S. Thanisawanyangkura and T. Améglio, *Tree Physiology*, 2006, **26**, 1579–1587.
25. E. Gohet, *La production de latex par Hevea brasiliensis. Relations avec la croissance. Influence de différents facteurs: origine clonale, stimulation hormonale, réserves hydrocarbonées*, Thèse de Doctorat, Université Montpellier II – Sciences et Techniques du Languedoc, 1996.
26. U. Silpi, *Carbon partitioning in Hevea brasiliensis Muell. Arg.: Dynamics among functional sinks (latex regeneration, respiration, growth and reserves) at trunk scale*, PhD thesis, Graduate school, Kasetsart University, 2006.
27. J. Tupy, *Physiol. Vég.*, 1973, **11**(1), 1–11.
28. J. L. Jacob, J. M. Eschbach, J. C. Prévôt, D. Roussel, R. Lacrotte, H. Chrestin and J. d'Auzac, in *Proceedings International Rubber Conference*, Kuala Lumpur, ed. J. C. Rajarao and L. L. Amin, Kuala Lumpur, *J. Rubb. Res. Inst. Malaya*, 1985, 43–65.
29. J. M. Eschbach, D. Roussel and H. Van De Sype, *Physiol. Vég.*, 1984, **22**, 295.
30. J. L. Jacob, J. C. Prévôt, R. Lacrotte, A. Clément, E. Serres and E. Gohet, *Plantations, recherche, développement*, 1995, **2**(5), 43–49.
31. U. Silpi, A. Lacointe, P. Kasemsap, S. Thanisawanyangkura, P. Chantuma, E. Gohet, N. Musigamart, A. Clément, T. Améglio and P. Thaler, *Tree Physiology*, 2007, **27**, 881–889.
32. P. Chantuma, P. Thaler, E. Gohet, S. Thanisawaniangkura and P. Kasemsap, *Carbohydrate distribution at trunk level of Hevea brasiliensis*, International Natural Rubber Conference, Ho Chi Minh City, Vietnam, 2006, 175–182.
33. U. Silpi, P. Chantuma, P. Kasemsap, P. Thaler, S. Thanisawanyangkura, A. Lacointe, T. Améglio and E. Gohet, *J. Rubber Res.*, 2006, **9**(2), 115–131.
34. P. Thaler, M. Verdier, O. Roupsard, A. Chantuma, P. Siripornpakdeekul, P. Kasemsap and K. Sangkhasila, Partitioning of water flux in rubber plantations. Estimation of tree transpiration by sap flow measurement. *International Workshop on Flux Estimation over Diverse Terrestrial Ecosystems in Asia* – AsiaFlux Workshop, Chiang Mai, Thailand, 2006, p. 30.
35. S. W. Pakianathan, G. Harridas and J. d'Auzac, in *Physiology of Rubber Tree Latex*, ed. J. d'Auzac, J. L. Jacob and H. Chrestin, CRC Press, Boca Raton, 1989, p. 233–256.

36. F. Ninane, *Les aspects écophysiologiques de la productivité chez Hevea brasiliensis au Cambodge*, Thèse, Université Catholique de Louvain, 1970.
37. S. Sargnkool Na Ayutthaya, J. Junjittakarn, F. C. Do, K. Pannengpetch, J. L. Maeght, A. Rocheteau and D. Nandris, *Drought and trunk phloem necrosis (TPN) effects on water status and xylem sap flow of Hevea brasiliensis*, RRDB International Natural Rubber tree latex, Siam Reap, Cambodia, 2007, p. 75–84.
38. G. F. J. Milford, E. C. Paardekooper and C. Y. Ho, *J. Rubb. Res. Inst. Malaya*, 1969, **21**, 274.
39. S. Takahashi and T. Koyama, *Chem. Rec.*, 2006, **6**, 194–205.
40. K. Cornish, *Nat. Prod. Rep.*, 2001, **18**, 182–189.
41. Y. Tanaka, *Prog. Polym. Sci.*, 1989, **14**, 339–371.
42. J. E. Puskas, E. Gautriaud, A. Deffieux and J. P. Kennedy, *Prog. Polym. Sci.*, 2006, **31**, 533–548.
43. F. Lynen, *Revue Générale des Caoutchoucs*, 1963, **40**, 83–97.
44. J. Lebras, *Revue Générale des Caoutchoucs*, 1963, **40**, 1501–1526.
45. R. G. O. Kekwick, in *Physiology of Rubber Tree Latex*, ed. J. d'Auzac, J. L. Jacob and H. Chrestin, CRC Press, Boca Raton, 1989, 145–164.
46. N. Ohya and T. Koyama, in *Biopolymers. Vol. 2: Polyisoprenoids*, ed. T. Koyama and A. Steinbüchel, Wiley-VCH, Weinheim, 2001, p. 73.
47. M. Rohmer, *Nat. Prod. Rep.*, 1999, **16**, 565–574.
48. W. N. Hunter, *J. Biol. Chem.*, 2007, **282**, 21573–21577.
49. J. H. Ko, K. S. Chow and K. H. Han, *Plant Mol. Biol.*, 2003, **53**, 479–492.
50. P. Priya, P. Venkatachalam and A. Thulaseedharan, *Plant Sci.*, 2006, **171**, 470.
51. B. L. Archer and E. G. Cockbain, *Meth. Enzymol.*, 1969, **15**, 476–480.
52. K. Asawatreratanakul, Y. W. Zhang, D. Wititsuwannakul, R. Wititsuwannakul, S. Takahashi, A. Rattanapittayaporn and T. Koyama, *Eur. J. Biochem.*, 2003, **270**, 4671.
53. K. Adiwilaga and A. Kush, *Plant Mol. Biol.*, 1996, **30**, 935–946.
54. B. M. T. Da Costa, J. D. Keasling and K. Cornish, 2005, **6**, 279–289.
55. Y. Tanaka, A. H. Eng, N. Ohya, N. Nishiyama, J. Tangpakdee, S. Kawahara and R. Wititsuwannakul, *Phythochemistry*, 1996, **41**, 1501–1505.
56. D. R. Light and M. S. Dennis, *J. Biol. Chem.*, 1989, **264**, 18589–18597.
57. H. Kang, Y. S. Kwak, J. Y. Lee, I. J. Kim, H. J. Ryu, E. J. Lim, Y. S. Na and S. B. Ryu, *FASEB J.*, 2001, **15**, 188.
58. K. S. Chow, K. L. Wan, M. N. M. Isa, A. Bahari, S. H. Tan, K. Harikrishna and H. Y. Yeang, *J. Exp. Bot.*, 2007, **58**, 2429–2440.
59. A. H. Eng, J. Tangpakdee, S. Kawahara and Y. Tanaka, *J. Nat. Rubber Res.*, 1997, **12**, 11–20.
60. J. Tangpakdee, N. Nishiyama, S. Kawahara and Y. Tanaka, *Molecular weight distribution and branching in natural rubber*, Conference proceedings, IRC International Rubber Conference, Manchester, 1996, 96.
61. J. d'Auzac, J. L. Jacob, J. C. Prevot, A. Clement, R. Gallois, H. Crestin, R. Lacote, V. Pujade-Renaud and E. Gohet, *Recent Res. Devel. Plant Physiol.*, 1997, **1**, 273–331.

62. S. K. Oh, H. Kang, D. H. Shin, J. Yang, K. S. Chow, H. Y. Yeang, B. Wagner, H. Breiteneder and K. H. Han, *J. Biol. Chem.*, 1999, **274**, 17132–17138.
63. J. Tangpakdee, Y. Tanaka, K. Ogura, T. Koyama, R. Wititsuwannakul, D. Wititsuwannakul and K. Asawatreratanakul, *Phythochemistry*, 1997, **45**, 261–267.
64. D. Wititsuwannakul, A. Rattanapittayoporn, T. Koyama and R. Wititsuwannakul, *Macromol. Biosci.*, 2004, **4**, 314–323.
65. J. L. Koenig, *Chemical Microstructure of Polymer Chains*, Wiley-Interscience, New York, 1980, ch. 6–8.
66. L. Sperling, *Introduction to Physical Polymer Science*, John Wiley & Sons, New York, 4th edn, 2006.
67. P. Mirau, F. Bovey and L. Jelinski, in *Encyclopedia of Physical Science and Technology*, ed. R. Meyers, Elsevier, Amsterdam, 2004, 857–901.
68. A. H. Eng and Y. Tanaka, *Trends Polymer Sci.*, 1993, **3**, 493–513.
69. Y. Tanaka, *Rubber Chem. Tech.*, 2001, **74**, 355–375.
70. L. Tarachiwin, J. Sakdapipanich, K. Ute, T. Kitayama and Y. Tanaka, *Biomacromolecules*, 2005, **6**, 1858–1863.
71. A. Subramaniam, *Rubber Chem. Tech.*, 1972, **45**(1), 346–358.
72. F. Bonfils, C. Char, Y. Garnier, A. Sanago and J. Sainte-Beuve, *J. Rubber Res.*, 2000, **3**(3), 164.
73. C. L. Swanson, R. A. Buchanan and F. H. Othey, *J. Appl. Polymer Sci.*, 1979, **23**, 743.
74. B. Sekhar, *Abnormal groups in rubber and microgel*, Proc. 4th Rubber Technol. Conf., London, 1960.
75. D. R. Burfield and S. N. Gan, *J. Polymer Sci., Polymer Chem. Ed.*, 1975, **13**, 2725–2734.
76. T. Shiibashi, *Int. Polymer Sci. Tech.*, 1987, **14**(12), T/33.
77. M. J. Gregory and A. S. Tan, *Some observations on storage hardening of Natural Rubber*, International Rubber Conference, Kuala Lumpur, 1975.
78. S. N. Gan and K. F. Ting, *Polymer*, 1993, **34**(10), 2142–2147.
79. S. Lee, *Trends Polymer Sci.*, 1993, **1**(10), 303–309.
80. F. Bonfils and C. Char, in *Encyclopedia of chromatography*, ed. J. Cazes, Marcel Dekker, 2nd edn, 2005, pp. 1101–1104.
81. F. Bonfils, J. Sainte Beuve, S. Sylla, A. Allet Don and J. C. Laigneau, *J. Nat. Rubber Res.*, 1995, **10**(3), 143–153.
82. J. Tangpakdee and Y. Tanaka, *Rubber Chem. Tech.*, 1997, **70**, 707.
83. P. W. Allen and G. M. Bristow, *Rubber Chem. Tech.*, 1963, **36**, 1024.
84. F. Ngolemasango, E. Ehabe, C. Aymard, J. Sainte-Beuve, B. Nkouonkam and F. Bonfils, *Polymer Int.*, 2003, **52**, 1365–1369.
85. G. F. Bloomfield, *J. Rubb. Res. Inst. Malaya*, 1951, **13**, 18–24.
86. R. Freeman, Microgel in latex and sheet rubber, *3rd Rubber Technology Conference*, London, 1954.
87. E. Ehabe, F. Bonfils, C. Aymard, A. Akinlabi and J. Sainte-Beuve, *Polymer Test.*, 2005, **24**, 620–627.
88. D. S. Campbell and K. N. G. Fuller, *Rubber Chem. Tech.*, 1984, **57**, 104.

89. A. P. Voznyakovskii, I. P. Dmitrieva, V. V. Klyubin and S. A. Tumanova, *Polymer Sci.*, 1995, **38**(10), 1153–1157.

90. S. Sylla, J. C. Laigneau, A. Koman Achi and A. Allet Don, *A clonal typology approach to natural rubber technology, IRRDB Symposium on the Technology and End Uses of Natural Rubber*, Beruwela, Sri Lanka, 1996, 7–20.

91. G. F. J. Moir, *Nature (London)*, 1959, **184**, 1626–1628.

92. E. Yip and J. B. Gomez, *J. Rubb. Res. Inst. Malaya*, 1980, **28**, 86.

93. C. C. Ho, T. Kondo, N. Muramatsu and H. Ohshima, *J. Colloid Interface Sci.*, 1996, **178**, 442–445.

94. J. L. Jacob, J. C. Prevot, D. Roussel, R. Lacrotte, E. Serres, J. d'Auzac, J. M. Eschbach and H. Omont, in *Physiology of Rubber Tree Latex*, ed. J. d'Auzac, J. L. Jacob and H. Chrestin, CRC Press, Boca Raton, 1989, 345–382.

95. B. L. Ancher, B. G. Audley, G. P. McSweeney and T. C. Hong, *J. Rubb. Res. Inst. Malaya*, 1969, **21**, 560–569.

96. J. d'Auzac, J. L. Jacob, J. C. Prevot, A. Clement, R. Gallois, H. Crestin, R. Lacote, V. Pujade-Renaud and E. Gohet, *Recent Res. Devel. Plant Physiol.*, 1997, **1**, 273–331.

97. J. L. Jacob, J. d'Auzac and J. C. Prevot, *Clin. Rev. Allergy*, 1993, **11**, 325.

98. N. U. Nair, in *Natural Rubber Agromanagement and Crop Processing*, ed. P. J. George and C. K. Jacob, Rubber Research Institute of India, Kerala, 2000, 249–260.

99. E. Yip and P. Cacioli, *J. Allergy Clin. Immunol.*, 2002, **110**, 3–14.

100. D. Wititsuwannakul and R. Wititsuwannakul, in *Biopolymers.Vol. 2: Polyisoprenoids*, ed. T. Koyama and A. Steinbüchel, Wiley-VCH, Weinheim, 2001, p. 151.

101. J. Sainte-Beuve, S. Sylla and J. C. Laigneau, *J. Rubber Res.*, 2000, **3**, 14–24.

102. A. B. Othman, G. A. W. Murray and A. W. Birley, *J. Nat. Rubber Res.*, 1996, **11**, 183–199.

103. Y. Le Roux, E. Ehabe, J. Sainte-Beuve, J. Nkengafac, J. Nkeng, F. Ngolemasango and S. Gobina, *J. Rubber Res.*, 2000, **3**, 142–156.

104. E. Gohet, *The production of latex by Hevea brasiliensis: relation with growth. Influence of different factors: clonal origin, hormonal stimulation, carbohydrate reserves*, PhD thesis, Montpellier II University, 1996.

105. U. Silpi, A. Lacointe, P. Kasemsap, S. Thanysawanyangkura, P. Chantuma, N. Musigamart, A. Clement, T. Ameglio and P. Thaler, *Tree Physiology*, 2007, **27**, 881–889.

106. Y. Udagawa, M. Machida and S. Ogawa, US 5041689, 1991 (Yokohama Rubber, The Board of the Rubber Research Institute of Malaysia).

107. S. Liengprayoon, F. Bonfils, J. Sainte-Beuve, K. Sriroth, E. Dubreucq and L. Vaysse, *Eur. J. Lipid Sci. Technol.*, 2008, **110**, forthcoming.

108. S. Liengprayoon, *Characterization of lipid composition of sheet rubber from Hevea brasiliensis and relations with its structure and properties,*

PhD thesis, Montpellier Sup Agro, France – Kasetsart University, Thailand, 2008.

109. G. Spiteller, *Lipids*, 2005, **40**, 755–771.
110. S. Kawahara, I. Isono, T. Kakubo, Y. Tanaka and A. H. Eng, *Rubber Chem. Tech.*, 1999, **73**, 39–73.
111. M. O. David, T. Nipithakul, M. Nardin, J. Schultz and K. Suchiva, *J. Appl. Polymer Sci.*, 2000, **78**, 1486–1494.
112. N. Na-Ranong, H. de Livonnière and J. L. Jacob, *Plantations, Recherche, Développement*, 1995, **2**, 44–54.
113. H. Hasma and A. B. Othman, *J. Nat. Rubber Res.*, 1990, **5**, 1–8.
114. J. L. Jacob, J. d'Auzac and J. P. Prévot, *Clin. Rev. Allergy*, 1993, **11**, 325.
115. J. Sainte-Beuve, L. Vaysse and F. Bonfils, *in La Chimie Verte*, Coord. P. Colonna, Editions Tec&Doc Lavoisier, Paris, 2006, 215–237.
116. S. Kawahara, T. Kakubo, Y. Isono and Y. Tanaka, *Polymer*, 2000, **41**, 7483–7488.
117. M. Tosaka, S. Murakami, S. Poompradub, S. Kohjiya, Y. Ikeda, S. Toki, I. Sics and B. S. Hsiao, *Macromolecules*, 2004, **37**, 3299–3309.
118. S. Toki, B. Hsiao, S. Kohjiya, M. Tosaka, H. Tsou and S. Datta, *Rubber Chem. Tech.*, 2006, **79**, 460–488.
119. S. Toki, I. Sics, S. Ran, L. Liu and B. S. Hsiao, *Polymer*, 2003, **44**, 6003–6011.
120. J. M. Chenal, L. Chazeau, L. Guy, Y. Bomal and C. Gauthier, *Polymer*, 2007, **48**, 1042–1046.
121. G. Heideman, R. Datta and J. Noordermeer, *Rubber Chem. Tech.*, 2004, **77**, 513.
122. A. B. H. C. Othman and H. Hasma, *Plast. Rubber Compos. Proc. Appl.*, 1993, **19**(3), 185–194.
123. A. Chapman and M. Porter, in *Natural Rubber Science and Technology*, ed. A. Roberts, Oxford University Press, Oxford, 1988, p. 593–595.
124. V. Baum, *The Weeping Wood*, Doubleday, Garden City, 1943, *Cahuchu, Strom der Tränen*, Kiepenheuer & Witsch, Köln, 1943, 1952.

CHAPTER 9.6
Natural Fibres

MARTIN MÖLLER[a] AND CRISAN POPESCU[b, c]

[a] DWI an der RWTH Aachen e.V., Pauwelsstraße 8, D-52056 Aachen, Germany; [b] DWI an der RWTH Aachen e.V., Pauwelsstraße 8, D-52056 Aachen, Germany; [c] University "Aurel Vlaicu", Bd. Revolutiei 77, RO-310130 Arad, Romania

9.6.1 Generalities

Natural fibres are raw materials directly obtainable from an animal, vegetable or mineral source for which the diameter is negligible in comparison with the length. Along the centuries until comparatively recently, natural fibres were the basis for producing clothes, paper, ships' sails and insulation and building materials.

It is not certain when people first started wearing clothes. Scientists estimate that this happened more than 100,000 years ago, because the body louse (*pediculus humanus humanus*) apparently diverged from the head louse (*pediculus humanus capitis*) at that time. The first clothes were made from natural materials: animal skin and furs, grasses and leaves and the first needles are recorded about 30,000 years ago.

The use of natural fibres, both plant and animal, to meet our needs plays a significant role throughout history. Wool, as a representative of animal fibres, has been used since the dawn of mankind. One of the oldest recorded uses of plant fibre for fabrics is that of hemp, already being cultivated in China in 2800 BC. The common nettle is probably the best example of how society's interest in natural fibres fluctuates along the years. It was used for cloth in Neolithic times, and then became important for nets for fishing. Nettle cloth was manufactured in Scandinavia and Scotland until the nineteenth century and shortages of

RSC Green Chemistry No. 4
Sustainable Solutions for Modern Economies
Edited by Rainer Höfer
© The Royal Society of Chemistry 2009
Published by the Royal Society of Chemistry, www.rsc.org

cotton during the First World War forced the Germans to use it again for making fabrics. Now it has almost vanished from textile classifications.

With the industrial revolution, textile manufacturing turned into industry and the requirements for a fibre to be used for textiles became more exactly defined. For use in textiles a fibre should have properties which render it spinnable and wearable, enumerated below:

- geometrical properties: minimum length of 6–7 mm, diameter to length ratio of 1 : 2–5000
- mechanical properties: strength, extensibility and flexibility
- comfort properties: moisture management, dyeability, fast to sun and heat
- economic properties: cheap and abundant

The minimal values of the properties are very much related to those of natural fibres, which were the only fibres when textile manufacturing emerged as an industry.

During the last century there has been a turn away from natural fibres towards chemically obtained materials, mostly derived from petrochemicals. This change was a result of the technological revolution and the short-term economic advantages of synthetics.

As a result of the developments there is a large variety of fibres available on the market, classified into two large classes: natural and chemical fibres. Each of them may be further developed into subclasses, as detailed in Figure 9.6.1.

Out of the group of natural fibres, those under mineral fibres (asbestos fibres) are of diminishing interest after being related to lung cancer.[1]

The newly inserted group of synthetic fibres based on renewable sources is just beginning, with poly-lactic fibres (PLA) as the prominent representative. One may also consider that, due to the move away from petrol sources, this group will become increasingly interesting in the coming decades.

9.6.2 Demands and Restraints for Sustainable Fibres

Sustainability is a characteristic of a process or state that can be maintained at a certain level indefinitely. The term, in its environmental usage, refers to the potential longevity of vital human ecological support systems, such as the planet's climatic system, systems of agriculture, industry, forestry and fisheries, systems on which they depend in balance with the impacts of our unsustainable or sustainable design.

We are now witnessing a growing movement away from petrochemical-based fibres back to natural fibres. There are some reasons for this, *e.g.*:

- petrochemical-based fibre production has undergone continuing rising costs;
- synthetic fibres rely on precious non-renewable resources and incur environmental costs in their production;

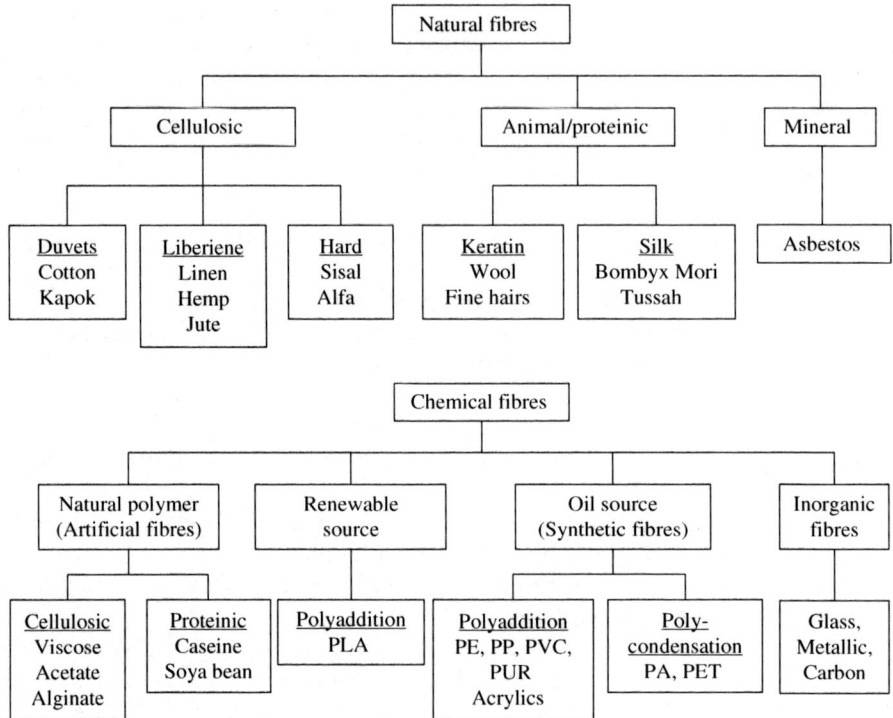

Legend: PLA polylactic acid, PE polyethylene, PP polypropylene, PVC polyvinyl chloride, PUR polyurethane, PA polyamide, PET polyethyleneterephthalat

Figure 9.6.1 Classification of fibres

- petrochemical-based products pose a health risk in most applications, both from direct exposure and also from secondary exposure through soil, water and air pollution.

Although environmentally and health conscious consumers have traditionally favoured natural fibres over synthetic clothing, there are several factors which pledge against their use, such as:

- due to the amount of insecticides required, cotton is one of the most environmentally unfriendly crops grown;
- during the processing of the fibres, cotton and wool produce the largest amount of polluted effluents per kilogram of fibre.

The new "organic" fibre industry that is rapidly developing across North America and Europe aims at answering these problems. However, there is still the need to address the steadily increasing demand of fibres by an increasing population (now at 7.94 kg *per capita* for 6.3 billion inhabitants, 2006; see Chapter 5, Figure 5.2).

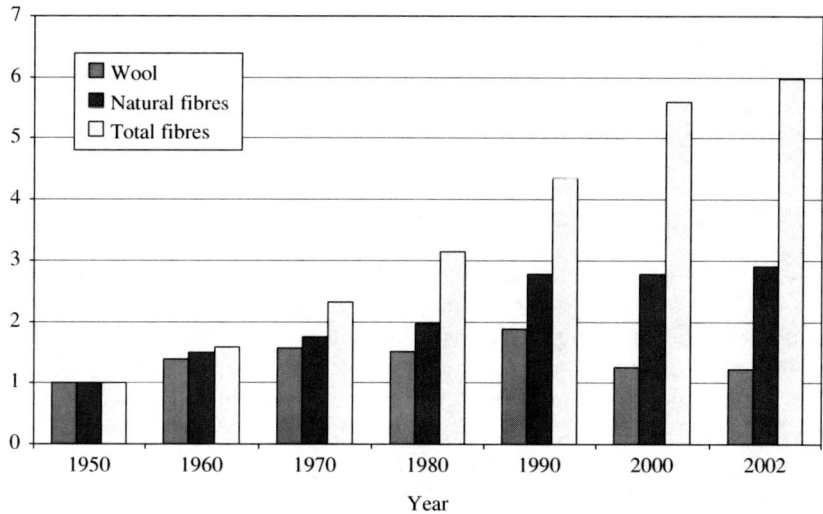

Figure 9.6.2 The world fibre production growth (1950 = 1).

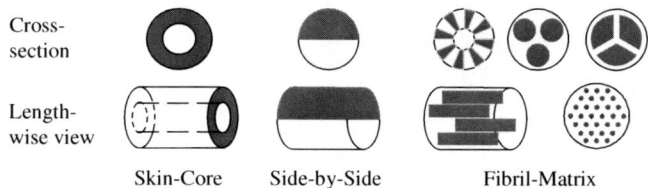

Figure 9.6.3 Schematic fibre structures.

As shown in Figure 9.6.2, the demand has been responded to, so far, by a continuously increasing contribution of chemical fibres to the total, with natural fibres seeming to reach a plateau in production. The limits of natural fibres are imposed by the amount of arable land required for producing them: the total arable land available is estimated at $20\,\text{Mio}\,\text{km}^2$ and for producing the $55\,\text{Gkg}$ fibres used in 2006[2] one has a maximum available of some $3.5 \times 10^{-4}\,\text{km}^2$ per kilogram of fibre. When comparing this with the demands of roughly $0.02\,\text{km}^2$ for 1 kg clean wool, and $0.01\,\text{km}^2$ land for 1000 kg cotton, respectively, it appears that developing a sustainable chemical fibres industry is the only solution to keeping most arable land for food production.

9.6.3 Fibre Structure

Fibres have a complex structure. A simplified view of the general types of structure is given in Figure 9.6.3.

While any of the chemical fibres is easily described by one of the structures in Figure 9.6.3, natural fibres look more complex, with a defined arrangement at each level of organization. This reflects, ultimately, the fact that natural fibres are produced by living organisms (in the cells) and their apparent homogeneity comes from the supra-organization of similar elements.

9.6.3.1   Chemistry and Structure of Cellulose Fibres

Cellulose fibres originate from the seed of the plant (cotton), its stem (bast fibres) or its leaves (sisal, alfa), having, as a consequence, different percentages of cellulose, lignin and hemicellulose. Basically, any plant may be used as a source of cellulose fibres, and it is a matter of historical development, availability and abundance that cotton, hemp and linen (flax) are today the most used cellulosic fibres.

Cellulose, the base of all cellulosic fibres, is a complex composite material which structurally comprises three hierarchical levels:

   (i) the molecular level of the single molecule;
  (ii) the super-molecular level concerning the packing and aggregation of the molecules in crystals called micro-fibrils; and
 (iii) the morphological level, *i.e.* the arrangement of micro-fibrils and interstitial voids in relation to the cell wall.

On the molecular level, cellulose is a linear polymer of β-$(1 \rightarrow 4)$-D-glucopyranose units in $^4C_1$ conformation. The fully equatorial conformation of β-linked glucopyranose residues stabilizes the chair structure, minimizing its flexibility.

The chains consist of 2000–14,000 residues which form crystals (cellulose I_α) where intra-molecular (O3-H $\rightarrow$ O5$'$ and O6 $\rightarrow$ H-O2$'$) and intra-strand (O6-H $\rightarrow$ O3$'$) hydrogen bonds hold the network flat allowing the more hydrophobic ribbon faces to stack. Each residue is oriented 180° to the next with the chain synthesized two residues at a time. This tendency to form crystals utilizing extensive intra- and intermolecular hydrogen bonding makes it completely insoluble in normal aqueous solutions (although it is soluble in special solvents such as aqueous N-methylmorpholine-N-oxide, NMNO). It is thought that water molecules catalyze the formation of the natural cellulose crystals by helping to align the chains through hydrogen-bonded bridging, as shown subsequently.

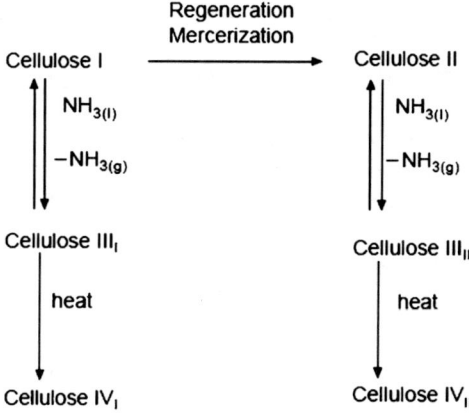

The cellulose molecules in cotton are organized into parallel arrangements called crystallites, and subsequently into larger aggregates called fibrils. The elementary fibril is made up of 36 cellulose chains[3] and is widely regarded as the basic crystalline unit of cotton cellulose.[4] In nature, elementary fibrils associate into aggregates to form micro-fibrils and larger aggregates called macro-fibrils. Different papers report slightly different dimensions for these fibrils, but typically the elementary fibrils measure 3.5–10 nm wide, the micro-fibrils 10–40 nm and the macro-fibrils 60–300 nm.[5]

The spatial arrangement of elementary fibrils and larger aggregates determines the morphology and properties of a fibre and in maturing cotton the arrangement of these structural units differs, depending on the state of fibre development.

There are six polymorphs of cellulose (I, II, III_1, III_{11}, IV_1 and IV_{11}), which can be inter-converted, as shown in Figure 9.6.4.[6,7] More recently, evidence for two polymorphs of cellulose I has been offered;[8] that is to say what was previously thought to be one polymorph (I) has now been found to be a mixture of two polymorphs (I_α and I_β). Proof of the polymorphy of cellulose comes from nuclear magnetic resonance (NMR), infrared and atomic force

Figure 9.6.4 Inter-conversion of the polymorphs of cellulose.

microscopy (AFM) studies.[9] Cellulose I, or native cellulose, is the form found in nature. Cellulose II, the second most extensively studied form, may be obtained from cellulose I by either of two processes: a) regeneration, which is the solubilization of cellulose I in a solvent followed by re-precipitation by dilution in water to give cellulose II, or b) mercerization, which is the process of swelling native fibres in concentrated sodium hydroxide, to yield cellulose II on removal of the swelling agent. Celluloses III_1 and III_{11}[10] are formed, in a reversible process, from celluloses I and II, respectively, by treatment with liquid ammonia or some amines, and the subsequent evaporation of excess ammonia.[11] Polymorphs IV_1 and IV_{11}[12] may be prepared by heating celluloses III_1 and III_{11}, respectively, to 206 °C, in glycerol.

The natural crystal is made up from meta-stable Cellulose I with all the cellulose strands parallel and no inter-sheet hydrogen bonding. The reason for this meta-stable crystalline form is that the conversion of glucose into cellulose takes place on the surface of an enzyme having about 30 active sites.[13] As a consequence polymer molecules grow together in the same direction, producing a parallel faces crystal.

Cellulose I (natural cellulose) contains two allomorphs, namely cellulose I_α (triclinic) and cellulose I_β (monoclinic). They are found in varying proportions dependent on its origin, cellulose I_α being found more in algae and bacteria, whilst cellulose I_β is the major form in plants.

Cellulose I_α and cellulose I_β have the same fibre repeat distance (1.043 nm for the repeat dimer interior to the crystal, 1.029 nm on the surface[14]) but different displacements of the sheets relative to one another. The neighbouring sheets of cellulose I_α (consisting of identical chains with two alternating glucose conformers) are regularly displaced from each other in the same direction whereas sheets of cellulose I_β (consisting of two conformationally distinct alternating sheets) are staggered.

It has been found that cellulose I_β significantly alters the water structuring at its surface out to about 10 Å, which may affect its enzymatic digestion.

Cellulose I_α and cellulose I_β are inter-converted by bending during microfibril formation and meta-stable cellulose I_α converts to cellulose I_β on annealing.

When re-crystallized (for example, from base or CS_2), cellulose I gives the thermodynamically more stable Cellulose II structure with an anti-parallel arrangement of the strands and some inter-sheet hydrogen-bonding. Cellulose II contains two different types of anhydroglucose (**A** and **B**) with different backbone structures; the chains consisting of -**A-A**- or -**B-B**- repeat units. Cellulose III is formed from cellulose mercerized in ammonia and is similar to cellulose II but with the chains parallel, as in cellulose I_α and cellulose I_β.

9.6.3.2 Chemistry and Structure of the Protein Fibres

Protein fibres are grouped into animal fibres (wool and other animal hairs, known also as α-keratin fibres) and insect fibres (silk, a fibroin fibre). Both α-keratin and fibroin are fibrous proteins. Their elemental analysis shows carbon, hydrogen, oxygen, nitrogen and sulphur. The high sulphur percentage (around 5 wt.%) found in wool and other animal hairs results from the high cystine content of these fibres and distinguishes them from silk.

Total hydrolysis of the peptide bonds in proteins yields the 20 common natural α-amino acids given in Table 9.6.1 (cystine, thiocysteine and cysteine are considered to be faces of the same amino acid).[15] More than 100 amino-acids bind to each other to form the protein chains, schematically shown subsequently.

Table 9.6.1 The 20 common natural α-amino acids found in protein fibres.

Group	Name	Side chain
"Acidic" amino acids and their ω-amides		
	Aspartic acid	$-CH_2-COOH$
	Glutamic acid	$-(CH_2)_2-COOH$
	Asparagine	$-CH_2-CONH_2$
	Glutamine	$-(CH_2)_2-CONH_2$
"Basic" amino acids and tryptophan		
	Arginine	$-(CH_2)_3-NH-C(NH_2)=NH$
	Lysine	$-(CH_2)_4-NH_2$
	Histidine	
	Tryptophan	
Amino acids with hydroxyl groups in the side chain		
	Serine	$-CH_2-OH$
	Threonine	$-CH(CH_2)-OH$
	Tyrosine	$-CH_2-C_6H_4-OH$
Sulphur-containing amino acids		
	Cysteine	$-CH_2-SH$
	Thiocysteine	$-CH_2-S-SH$
	Cystine	$-CH_2-S-S-CH_2-$
	Methionine	$-(CH_2)_2-S-CH_3$
Amino acids without reactive groups in the side chain		
	Glycine	$-H$
	Alanine	$-CH_3$
	Valine	$-CH(CH_3)_2$
	Proline	
	Leucine	$-CH_2-CH(CH_2)_2$
	Isoleucine	$-CH(CH_2)-CH_2-CH_3$
	Phenylalanine	$-CH_2-C_6H_5$

Because they contain both cationic and anionic groups, protein fibres are amphoteric. The cationic character is due to the protonated side groups of arginine, lysine and histidine, and free terminal amino groups. Anionic groups are present as dissociated side groups of aspartic and glutamic acid residues and as carboxyl end groups.

The amino acid amounts differ in α-keratin fibres from silk, as revealed by data in Table 9.6.2.

The peptide arrangement in protein fibre has been investigated since the first half of the twentieth century. Astbury[18,19] used X-rays to demonstrate the nature of a crystalline phase in hair. The X-ray diffraction pattern of animal hairs shows a meridian reflection at 0.51 nm and an equatorial reflection at 0.98 nm. Interpreting these results Pauling *et al.*[20] proposed the α-helix structure to give account of the secondary structure of the keratin fibre, shown in Figure 9.6.5.

The α-helix contains 18 amino acid residues in five turns, *i.e.* 3.6 amino acid residues per turn. To obtain the distance between successive turns of the helix that leads to the observed meridian reflection (0.51 nm), the helical chain must itself be slightly coiled (super-helix, coiled coil[21]). Two super-helices combine to form a left-handed two-stranded rope-like assembly in which the super-helices are arranged in such a way that the hydrophobic side groups at the outside of the helices interlink to form a stable "buttonhole" structure.[22] These dimers are the actual structural subunits of the micro-fibrils, and can be termed "molecular twins". The force that keeps two α-helices together in the coiled-coil dimer (the "brick" of the intermediate filament rod, IF) is given by the geometry of the arrangement of amino acid residues in the polypeptide chain and by the hydrophobic effect. The geometry requires a repeating sequence of seven

Table 9.6.2 Amino acids composition of α-keratin[16] and of fibroin and sericin.[17]

Amino-acid (mol %)	α-Keratin (wool)	Fibroin (silk)	Sericin (silk)
Glycine	8.1	42.8	16.6
Alanine	5.0	30.0	4.7
Serine	10.2	12.2	31.3
Glutamine + Glutamic acid	12.1	1.4	4.1
Cystine	11.2	0.1	0.8
Proline	7.5	0.5	0.7
Arginine	7.2	0.5	3.2
Leucine	6.9	0.6	1.1
Threonine	6.5	0.9	7.9
Asparagine + Aspartic acid	6.0	1.9	19.0
Valine	5.1	2.5	3.6
Tyrosine	4.2	4.8	1.9
Isoleucine	2.8	0.6	0.7
Phenylalanine	2.5	0.7	0.6
Lysine	2.3	0.4	2.3
Tryptophan	1.2	0.2	1.2
Histidine	0.7	0.2	1.4
Methionine	0.5	0.1	0

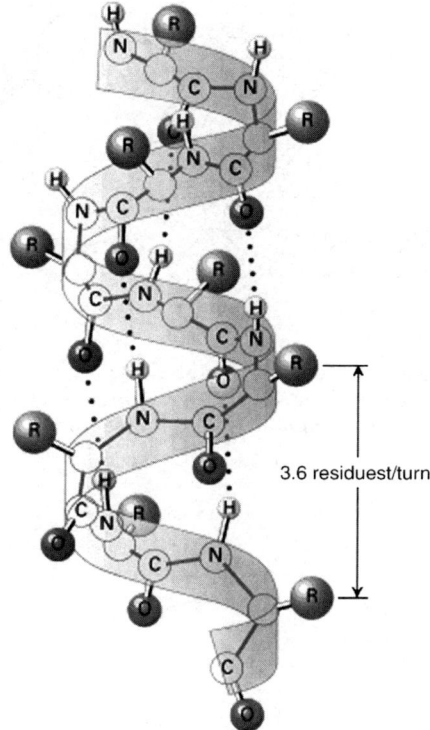

3.6 residuest/turn

Figure 9.6.5   Structure of α-helix as proposed by Pauling, Corey and Branson.[20]

amino acids (**abcdefg**), a heptade, with the residues **a** and **d** representing hydrophobic ones, as shown in Figure 9.6.6.[15]

The further organization of the α-helices in proto-filaments, proto-fibrils to micro-fibrils or intermediary filaments follows a pairing rule illustrated in Figure 9.6.7.

Morphologically, the fibres are composed of the cortex and the cuticle. Each of the two components is formed of various other morphological components (Table 9.6.3). The cortex contains cortical cells and the cell membrane complex. The cortical cell is further composed of macro-fibrils and intermacro-fibrillar material. The macro-fibrils consist of micro-fibrils and intermicro-fibrillar matrix. In summary, the cortex is formed of micro-fibrils (intermediate filament, IF, or keratin proteins, KP) and keratin associated proteins (IFAP or KAP), which compose the intermicrofibrillar matrix containing cytoplasmatic and nuclear remnants. This ensemble is wrapped up in the cuticle, as an external sheath which also has its own architecture, being formed of four layers: the epicuticle, the a-layer, the exocuticle and the endocuticle.

Thus, the α-keratin fibre is the best example of a natural composite system, having a complex dual structure at all levels.

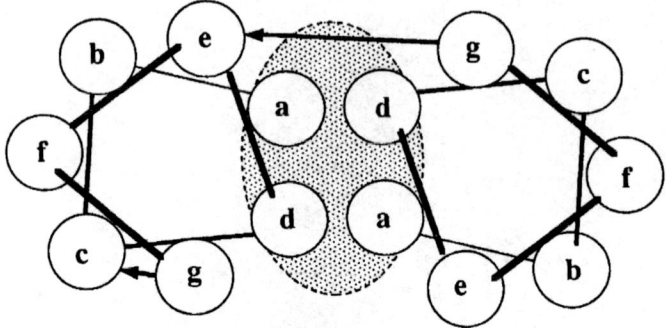

Figure 9.6.6 The coiled-coil dimer structure. The letters may be replaced by any amino acid residue from Table 9.6.1, with the only requirement that amino acids **a** and **d** are hydrophobic ones.[15]

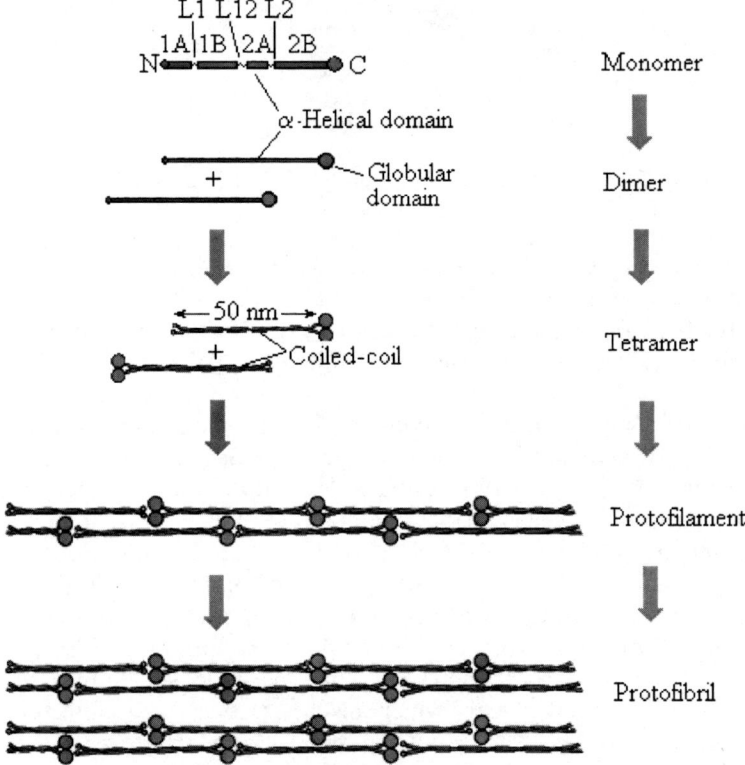

Figure 9.6.7 Structure of the intermediate filaments. A and B: helical domains; L: non-helical linkers; N, C: N and C termini, respectively.[16]

Table 9.6.3 Animal fibre structure.

Composite	Type	Component 1	Component 2
Hair fibre	Ring/core	Cuticle	Cortex
Cortex	Filament in matrix	Cortex cells (spindle shape)	Cell membrane complex
Cortex cell	Filament in matrix	5–8 macro-fibrils	Intermacrofibrillar matrix
Macro-fibril	Filament in matrix	500–800 micro-fibrils (IFs)	Intermicrofibrillar matrix

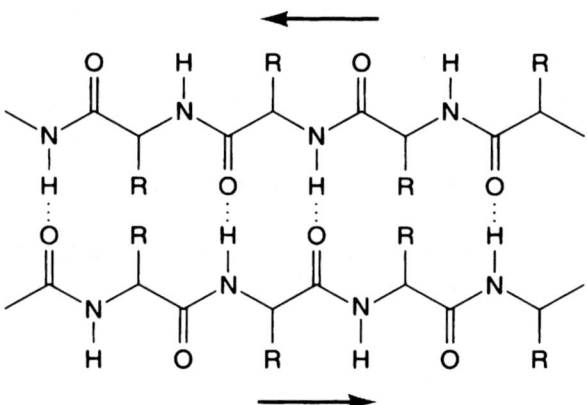

Figure 9.6.8 Arrangement of hydrogen bonds in the anti-parallel chain pleated sheet β-structure described by Pauling and Corey.[23]

The different production history and composition in terms of amino acid percentages found in silk reflects into a different X-ray pattern, described by an arrangement of the chains in what is termed as anti-parallel β-sheet and shown in Figure 9.6.8.[23] The same configuration, which is stable for silk fibroin, is also achieved metastably by α-keratins when stretched in a wet environment.[24]

Fibroin contains long regions of anti-parallel β-sheets. Other parts of fibroin are not in the form of β-sheets. These contain bulky amino acids that interrupt the β-sheet structures and may account for the "stretchiness" of silk fibres. The β-sheet regions contain, almost exclusively, multiple repetitions of the sequence:

[Gly-Ala-Gly-Ala-Gly-Ser-Gly-Ala-Ala-Gly-(Ser-Gly-Ala-Gly-Ala-Gly)$_8$]

Notice that almost every other residue is a glycine and that between them lie either alanine or serine residues. This repeating structure results in simple, tightly organized structures, such as the structure of silk fibroin shown in Figure 9.6.9.

In raw form silk consists of some 75% fibroin, 20% sericin and 5% of various salts. The former is crystalline fibrous proteins and the latter is amorphous and soluble in hot water or dilute alkali solutions.[15] The degree at

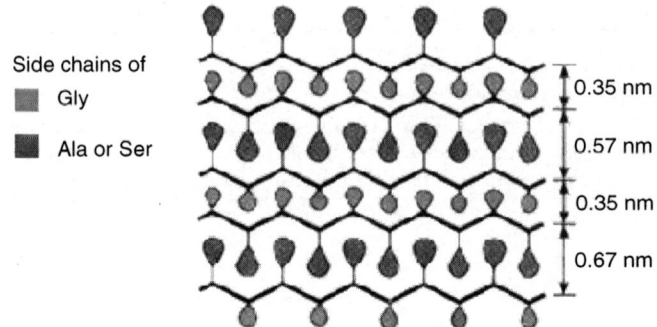

Figure 9.6.9   Packing of the amino acid chains in fibroin.

which sericin is removed leads to the various qualities of silk materials, with the highest value for the highest percentage of fibroin.

The amino acid composition of the two proteins forming the raw material differs, as shown in the last two columns in Table 9.6.2.

9.6.4   Fibre Sourcing

9.6.4.1   Cotton

Cotton fibres are the seed hairs of the plant *Gossypium hisutum*. The fibre consists typically of between 88 and 96% cellulose with the rest being protein, hemicellulose, ash and wax. When the fibre begins growing inside the seed, it emerges like a hollow cylinder. The hole in the middle is the lumen and, around it, the fibrils are laid down in a helical form on the inner wall under a fairly constant angle of 21° throughout the thickness of the fibre. The sense of the helix reverses periodically up to 50 times along the length of the fibre and this is supposed to provide a torsional balance.

The fully developed cotton fibre consists of a waxy cuticle that envelopes it, a cell wall that is differentiated into primary (outer) and secondary (inner) layers and residual protoplasm called the lumen. Although this concept of the fibre structure persists, more recent ideas do not differentiate between the cuticle and the primary wall, which is less than half a micrometer thick and consists of around 50% cellulose, with pectin, waxes and proteins making up the remainder. The secondary wall, which differs considerably in chemical composition and structure from the primary wall, consists of up to 95% cellulose.[3]

The cellulose in the secondary wall is laid down in a series of concentric growth rings or lamellae that reflect diurnal temperature fluctuations during fibre development.[25] Lord[26] reported up to 50 lamellae; other workers[27] have estimated between 25 and 40 lamellae in a "mature" fibre. The first rings deposited on the inside of the primary wall are known as the transition lamellae.

When a cotton boll opens, the fibres are cut off from the transpiration system of the plant. The fibres then dehydrate and become flattened and twisted. The cross section of the cotton fibre after dehydration is generally kidney-shaped, although the shapes range from near circular in mature fibres to flat in immature fibres. The kidney shape is an inherent phenomenon due to zones of different density in the secondary wall layers.[28]

Cotton grows on every continent, in a strip, the Cotton Belt, between 38° north latitude and 28° south latitude.[29] Cotton plants need heat (10–25°C for 150 days), a lot of sunshine and water, particularly during flowering. It is one of the crops most severely affected by pests and diseases worldwide. Consequently, cotton is one of the most environmentally expensive fibres to produce. Cultivation of cotton accounts for 25% of global insecticide sales. It has the second largest agricultural use of pesticides in the world, with five of the nine top "nasty" pesticides used. Cyanide, dicofol, naled and propargite are commonly used in cotton production and these chemicals are known cancer-causing chemicals. This prompted research for genetically modified (GM) plants. The best known and most widely grown GM cotton varieties produce a protein that kills bollworms, enabling farmers to cut their chemical insecticide use.

Cotton plantations are intrinsically tied to the history of the erstwhile Confederate States of the USA. Cotton is a crop that, besides fibre (cotton lint), delivers food (cottonseed oil) and feed (meal). For each 100 kg of cotton fibre produced, the plant produces about 150 kg of cottonseed. About 50 cottonseeds are contained in a pod, which opens when it is ripe. After removal of the cotton, the seeds are still covered with fine hairs (linters), which are removed with delintering machines. The linters are used as a chemical cellulose source in personal care products, in batting for upholstered furniture and mattresses, in high-quality paper and for celluloseether production.

The collecting of cotton fibres that surround the seeds in cotton bolls is known as picking. The picked product, "seed cotton", comprises roughly 55% seeds, 40% fibre and 5% trash and requires cleaning by machines for removing firstly the large impurities (leaves, twigs or bolls) and for separating it from seeds (ginning). By the end of the line, the clean cotton fibre is compacted and compressed for being delivered to further processing.

The production of cottonseed varies directly with cotton fibre production and cotton growers generally consider seed and oil as by-products. Cottonseed oil is the sixth most widely consumed oil worldwide (see Chapter 9.1). It is high quality and rich in polyunsaturated fatty acids and vitamin E (Figure 9.6.10.).

9.6.4.2 Bast Fibres (Flax, Hemp)

Bast fibres are obtained from the stems of the corresponding plants: hemp is a variety of *Cannabis sativa L* (differing from marijuana by having a considerably low content of Δ-9-tetrahydrocannabinol, the narcotic substance), and flax is obtained from the plant *Linium usitatissimum*. The fibres contain 70–75% cellulose, with about 4% lignin, 17% hemicellulose and 6% pectin in hemp and

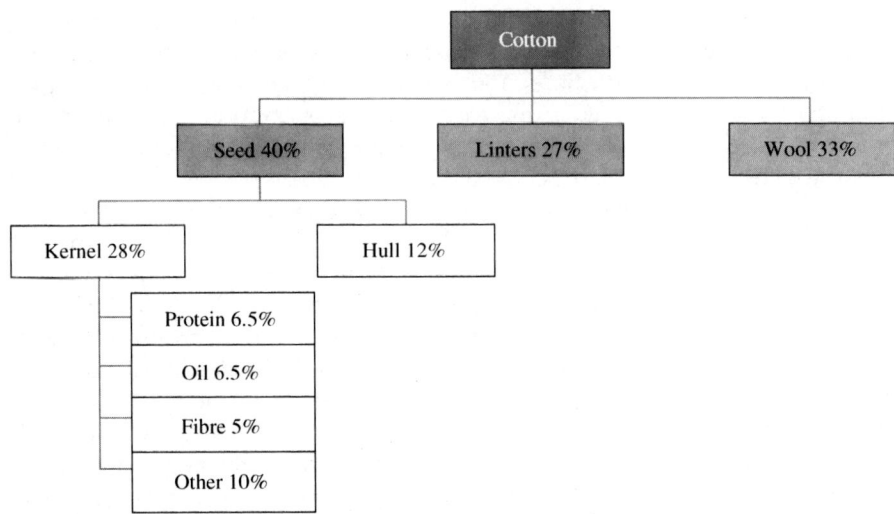

Figure 9.6.10 Composition of cottonseed.

2% lignin, 20% hemicellulose and 3% pectin in flax.[30] The helical arrangement of the fibrils found in cotton exists also here, but the angle of the helix and its sense differ at each fibre: the helix makes 4° and follows a Z sense in hemp fibre, and makes 10° with an S sense in flax.

Within the stem there are a number of fibre bundles, each containing individual fibre cells or filaments. The filaments are made of cellulose and hemicellulose, bonded together by a matrix of lignin or pectin.

The pectin and lignin also surrounds the bundle for holding it on to the stem. For separating the bundles of fibres from the rest of the stem and using them industrially the pectin and other impurities have to be removed. The first stage for achieving this is retting, a microbial process that breaks the chemical bonds that hold the stem together and allows separation of the bast fibres from the woody core. The two traditional types of retting are field and water retting, differing by the amount of water required (dew only for field and water basins for water retting, respectively) and the obtained quality. High-quality fibre results from water retting, but the process is very labour- and capital-intensive, requiring knowledgeable workers and uses large volumes of clean water that must be treated before being discharged. Research is carried out using micro-organisms or enzymes for producing textile-quality bast fibres. Scutching is the next step to be done. This process mechanically extracts the fibres from the retted straw. The resultant fibres are further processed by industry.

The fibres are only part of the total plant culture; they represent 4–5% of hemp, or 35–40% of flax plant, and 33–40% of cotton seeds (Figure 9.6.11). The rest of the plant is also used for various other purposes, more particularly fodder and oil (see Chapter 9.1). The fibre output per hectare of cultivated land differs also from one plant to another: cotton reaches 1000 kg fibres ha^{-1}, hemp may yield 1800 kg fibres ha^{-1}, and flax up to 2000 kg fibres ha^{-1}.

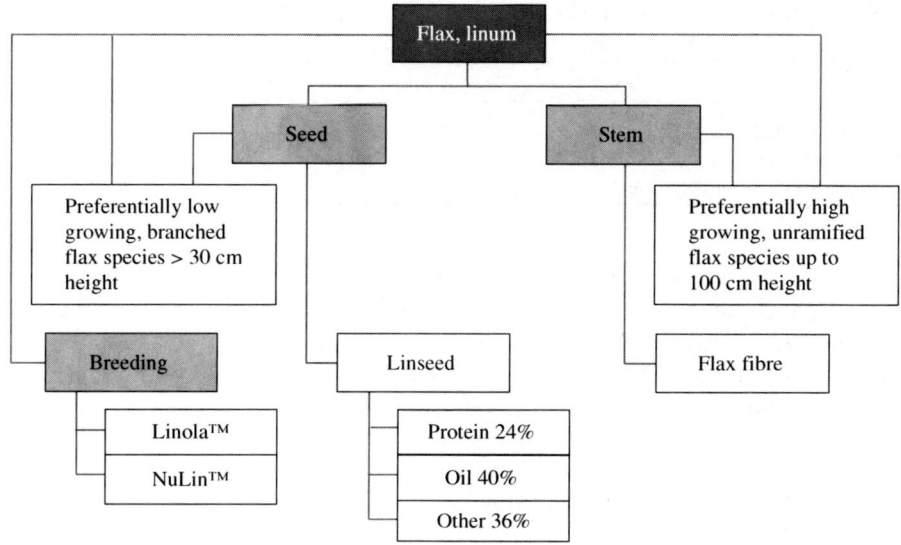

Figure 9.6.11 Composition and industrialization of flax.

Compared to cotton the cultivation of bast fibres is much more sustainable. Although not botanically related, both plants grow well in a moderately cool, temperate climate and can be grown in almost any country in the world. The plants are resistant to pests and do not require pesticides.

9.6.4.3 Animal Fibres

Animal fibres have a slightly elliptic cross section and are protected by the scales arranged on their surface as tiles on a roof.

Keratin fibres are available almost everywhere in the world. Wool, produced by sheep, is by far the most widely used keratin fibre, while shatoosh produces the most expensive one and is found only on the high peaks of Himalaya.

Among all keratin fibre producers, the sheep is the most important one. Economically the sheep is a fibre factory without wastes. The by-products of the factory are defined from the point of view of the down processing industry. For example, the textile industry regards milk and meat (lamb) as secondary products.

Compared to other keratin producers, sheep give the largest fibre amount from a given pasture surface. Due to the way sheep eat grass the pasture regenerates quickly after a sheep passage – to compare with the Cashmere goat after whose passing at least 2–3 seasons are required for regenerating the pasture. Very roughly, a sheep produces 1 kg greasy wool (or some 0.6 kg clean wool) annually from 1 ha of average pasture.

Sheep have several pests and there are insecticides that are used for fighting them. For fighting flystrike, occurring notably in Australia, farmers developed a process called mulesing. Mulesing is a surgical procedure by which the wrinkled skin in the animal's breech area is cut away from the perianal region down to the

Table 9.6.4 Composition (%) of greasy wool. The micron limits for the three wool types are only orientative.

Wool type	Grease and suint	Sand and dirt	Vegetable matters	Fibre
Merino ($<25\,\mu m$)	15–30	5–40	0.5–10	30–60
Crossbred (25–33 μm)	15–30	5–20	1–5	40–65
Long wool ($>33\,\mu m$)	5–15	5–10	0–2	60–75

top of the hind limbs. This is thought to be painful for sheep and for this reason the animal protection organization PETA (People for the Ethical Treatment of Animals) spoke against the practice. There is research targeting the development of an alternative for this, since by 2010 mulesing will be phased out.

The bringing of the fibres to the textile industry is a labour-intensive process. The shearing and collecting of greasy wool are manual operations and so are the skirting (selecting the parts of the fleece) and classing. The collected greasy (raw) wool comprises various amounts of different impurities, as detailed in Table 9.6.4.

To deliver clean fibres to industry the greasy wool goes through scouring. This consists of washing the raw fibres with about 1% surfactant in a continuous 5–6 bowl line for removing the grease, suint, sand and dirt from the fibres. The discharged waters from the first three basins from washing fibres with more than 10% grease (usually Merino type) are further used for extracting lanolin by an Alfa-Laval process. In spite of lanolin separation the waste waters from wool scouring are heavily polluted with organic matter and much effort is dedicated to cleaning and reusing them for saving water consumption. The recent development of biodegradable surfactants[31] helps toward building a more environmentally friendly wool scouring process. Some 5–7 litres of water are required for obtaining 1 kg clean wool under severe control of the parameters, but 20 litres is more common.

Vegetable matter, if more than 2–3%, has also to be removed before going to further processing. The operation, known as "carbonization", makes use of the good resistance of wool to strong acids (particularly sulphuric acid) and of the hydrolysis of cellulose in the same environment. The fibres with vegetable matter (grass, burrs) are soaked in a 20% sulphuric acid solution, dried and baked and then crushed for separating the carbonized cellulosic matters from the rest. The fibres are then neutralized and dried. The process produces acid-polluted waters and carbon dust in air, besides weakening the wool fibres,[32] for which reason alternatives, like the use of enzymes, are sought.

9.6.4.4 Silk

Silk is produced by silkworms (caterpillars) fed with fresh mulberry leaves, while spinning their cocoons. Liquid silk produced by spinnerets is coated in sericin, and solidifies on contact with the air. Within 2–3 days, the caterpillar spins about 1500 m of filament and is completely encased in a cocoon. Most caterpillars are then killed by heat (boiling) and some are allowed to meta-morphose into moths to breed the next generation of caterpillars.

Stifled cocoons are sorted, and then brushed to find filaments. Several fila-ments are gathered together and wound onto a wheel (reeling). Each cocoon yields approximately 1000 m of silk filament, known as raw silk, or silk-in-the gum, fibre.

Wild silk (Tussah) production cannot be reeled, because cocoons are har-vested after the moth has matured and left. As an average yield one may take 30 kg mulberry leaves for 1 kg cocoon. As the harvesting of the silk from the cocoon requires the killing of the larvae, silk-culture has been criticized by animal rights activists.

9.6.5   Summary of the Properties of Natural Fibres

The "classical" two-phase model assuming a composite arrangement of distinct crystalline and extended amorphous regions to describe the superstructure of natural fibres apparently has to be revised. Concepts like crystallinity and amorphicity are well adapted to describe homogeneous states of matter. They are, however, rather ill-defined when it comes to treating dense composite materials like cellulose, or proteins, given that intermolecular correlations do not build up or die off abruptly at some fictitious interfaces.

The mechanical characteristics of the fibres well reflect the complexity of their structures. While wool shows a decrease of E-modulus and an increase of elongation at a break in the wet stage compared to the dry one, cellulosic fibres behave inversely. A summary of the mechanical properties of natural fibres is given in Table 9.6.5.

9.6.6   Processing of Natural Fibres

The numerous chemical and mechanical processes through which the fibres are turned into end-product can be grouped into:

* operations which transform the fibres into fabric;
* operations of cleaning the fabrics;

Table 9.6.5 Properties of glass and natural fibres.

	Cotton	*Bast*	*Wool*	*Silk*	*E-Glass*
Density (g.cm^{-3})	1.5–1.55	1.4–1.52	1.32	1.25	2.55
E-modulus dry[a] (Gpa)	8–12	60–75	4–6	20–25	73
E-modulus wet[b] (Gpa)	3–4	20–30	2–4	5–6	73
Elongation dry[a] (%)	6–10	1.5–4	35–50	20–30	3
Elongation wet[b] (%)	8–12	7–10	45–65	40–50	3
Moisture absorption[a]	7–11	8–10	15–18	9–11	–

[a]measured at 65% relative humidity and 22 °C.
[b]measured at 100% relative humidity and 22 °C.

- operations of stabilizing the dimensions of fabrics;
- operations of coating-infiltrating for colouring and finishing the fabrics;
- operations of surface treatment.

These operations are similar for both natural and chemical fibres. The details of the operations differ with the fibre chemistry and origin, and with the way the fabric was obtained: by weaving or by knitting, respectively. In all cases chemical products are used for assisting, or for producing the required effect, and in all cases pollutants are produced, either in effluents, or in the atmosphere.

Most operations make extensive use of surfactants for assisting mechanical processes, for washing/scouring, for improving the quality of bleaching, of dyeing, or for finishing (softening, stiffening) the materials. The surfactants are subsequently discharged into the waste waters and pose serious problems for cleaning. Besides, the surfactants interact with the skin/eyes of the operators in the mills, or of those laundering at home.

The actual trend of natural surfactants based on renewable raw materials[33] combine a high emulsifying power with good electrolyte stability and extremely good skin compatibility.

The biodegradability of chemical substances depends on their chemical structure. Branched hydrocarbon chains such as tetrapropylene alkylbenzene sulphonates, alkylphenol polyglycol ethers on the one side and quaternary ammonium compounds on the other show poor biodegradability, more particularly under anaerobic conditions.[33,34]

The nature of renewable raw materials matches the requirement for biodegradability under aerobic and anaerobic conditions (Table 9.6.6). The linear hydrocarbon chains facilitate biodegradability without the formation of ecotoxic metabolites as is the case with alkylphenol polyglycol ethers.

Textile softeners form a special group among surfactants. They should remain on the textile in order to make the fabric ultra smooth and soft without affecting its texture. Depending on the washing conditions (temperature, *etc.*)

Table 9.6.6 Anaerobic biodegradability of various surfactants.[33,34]

Surfactant type	*Biodegradability*
Sulphonated anionic surfactants (LAS, SAS)	biodegrades poorly
Sulphated anionic surfactants (fatty alcohol sulphates, alcohol ethoxysulphates)	biodegrades well
Fatty acids and soaps	biodegrades well
Fatty alcohol ethoxylates	biodegrades well
Sugar-based surfactants (alkyl polyglucosides, glucamides)	biodegrades well
Alkylphenol ethoxylates	partially degradable, leaving alkylphenol residues
Mono-or di-alkyl quaternary compounds (TMAC, DTDMAC)	biodegrades poorly
Esterified mono-or di-alkyl quaternary surfactants (esterquats)	biodegrades well

the softeners will be washed off after 1–5 washing cycles at the earlist, which also means the product stays in contact with the human skin during wear. Besides they form the largest part of finishing agents, which share about 65% of textile auxiliary sales.[35] Because of this, potential allergenic problems are regarded more closely. For that purpose the German Skin and Allergy Relief Fund (DHA *Deutsche Haut-und Allergiehilfe*) has conducted a survey of dermatologists, paediatricians and allergists regarding the use of consumer market laundry softeners (*Weichspüler* in German). Seven out of ten of those surveyed considered the use of fabric conditioners by people with sensitive skin as useful and convenient, and one in four even recommended fabric conditioners explicitly, since the smoothened fibre surfaces reduce the potential of mechanical friction, which is advantageous for sensitive and damaged skin.[36]

Textile softeners are mainly cationic or pseudo-cationic surfactants. They attach by electrostatic forces to textile fibres with the polar (charged) end of the cation oriented toward the fibre and the fatty tail exposed which imparts the feeling of softness to the fabric.

The main ingredients of fabric softening formulations used by the main manufacturers in Europe and since proliferated around the world are the so-called *Esterquats*.[37]

Their synthesis is performed by the condensation of fatty acids and ethanolamines, followed by the quaternizing of the tertiary nitrogen with a methylchlorine, or dimethyl-sulphate:

$$2RCOOH + CH_3N(CH_2CH_2OH)_2 \rightarrow CH_3N(CH_2CH_2OOCR)_2 + 2H_2O$$
$$CH_3N(CH_2CH_2OOCR)_2 + CH_3Cl \rightarrow [(CH_3)_2N(CH_2CH_2OOCR)_2] + Cl^-$$
$$\text{Esterquat}$$

Esterquats based on triethanol amine, diethanol amine or N,N-dimethyl-3-aminopropane-1,2-diol are usually abbreviated to TEAQ (triethanol amine quat), DEEDMAC (diethyloxyester dimethylammonium chloride) or HEQ (Hamburg Esterquat), respectively. Esterquats as cationic surfactants were introduced in the European market in the early 1990s when concerns were raised about the environmental profile of the cationic surfactants used up to then (Di-Hardened Tallow Di-Methyl Ammonium Chloride, or DHTDMAC) and initiated their decline in use and international initiatives (*e.g.* PARCOM recommendation 93/4) calling for them to be phased out. Esterquats are similar to what was used before except that ester links have been introduced into the head-group of the molecules, which have the quality that they are subject to degradation by hydrolysis and greatly facilitate biodegradation. Most, if not all, fabric conditioners marketed in Europe are now based on these three Esterquat materials: TEAQ, DEEDMAC and HEQ. They combine good environmental profile – especially ready and ultimate biodegradability – with the structural features required for an effective fabric conditioner.

Also being sold are the amphoteric softeners, the betaines (general structure: $(CH_3)_3N^+CHR_1\text{-}COO^-$). Amphoteric/betaine surfactants are generally mild with low eye and skin irritation. They produce a good handle and greatly increase the hydrophilic properties. Their mildness makes them suitable for baby clothing laundry detergents and their compatibility with quats and esterquats for "detergent and fabric softener in one" formulations. Betaines and non-ionic products, such as ethoxylated fatty alcohols, $RO(CH_2CH_2O)_nH$, ethoxylated fatty acids, $RCOO(CH_2CH_2O)_nH$, ethoxylated fatty amides, $RCONH$-$(CH_2CH_2O)_nH$, or ethoxylated fatty amines $RN(CH_2CH_2O)_nH$ and acetylated sugar ethers increase hydrophilicity and have found applications as co-softeners.

Also polydimethylsiloxane (PDMS) emulsions and micro-emulsions are well known as conditioners in textile finishing imparting softness, dimensional stability and wrinkle and stretch recovery. When combined with quats and esterquats they improve the water absorbency of softened cotton fabric. Long-chain PDMS with amino- or amido-functional side groups anchor the silicone to the fibre by way of attractive electrostatic forces. Interfibre friction is reduced in this way, producing a distinct and substantive softener effect.[38]

9.6.6.1   Operations Which Transform Fibres into Fabric

This group gathers several mechanical operations which help transform the fibres into a yarn (carding, combing and spinning) and further a yarn into a knitted or woven fabric (knitting and weaving, respectively). The making of the yarn requires the grouping of fibres in a roving of fairly constant density, which is eventually spun. The fibres are kept together in the yarn by fibre–fibre friction and the twisting movement during spinning ensures a compact packing. Providing the fibres in the yarn are parallel or randomly arranged the yarn is known to be worsted (its formation includes a combing for parallelizing the fibres and the yarn is fine and smooth) or woollen (carded yarns, as combing is not used). The yarn is then used for producing the 3D structure (fabric) by knitting or weaving.

Alternatively, non-woven fabrics are produced by directly transforming fibres after carding (formation of the web) into a fabric, without the intermediation of the yarn. Except in the production of felts, which makes use of the felting property of wool (and other animal fibres), fibre–fibre friction plays no role here and there are resins used for bonding.

These operations do not make a distinction between fibres, which is to say that all fibres are processed alike.

9.6.6.2   Cleaning Operations

Washing/scouring and bleaching are the two operations of cleaning the fabrics.

Scouring aims to wash out the chemical auxiliaries used in assisting the spinning, knitting or weaving.

While animal fibres go through a scouring process before being industrially processed into end-products, cotton fibres meet water only as yarns or fabrics. Consequently the scouring of cotton fabrics is more severe (boiling temperature, enzymes for desizing and hypochlorites or chlorites for cleaning), as it is required in eliminating the rest of the seeds and other waxes from the plant.

The bleaching process also has peculiarities depending on the type of fibre which is treated. The most common and environmentally friendly bleaching agent used for fibres is hydrogen peroxide. Cotton is also partly bleached while scouring, under the action of sodium chlorite. Some yellow wools or bast fibres may require a harsher bleaching, for which reason sodium dithionite or sodium formaldehyde sulphoxylate (for wool) and sodium chlorite (for bast fibres) are also used.

Due to the fact that chemical fibres are produced under controlled parameters, their cleaning requires fewer chemicals and even the bleaching is, sometimes, unnecessary.

A common bleaching agent used domestically (more particularly in Mediterranean countries and North America) is Javel water, which is sodium hypochlorite in water.

9.6.6.3 Stabilizing the Dimensions

This group of operations is the most diverse depending on the nature of the fibre.

Wool fabric requires crabbing, wet and dry decatizing for setting the dimensions. The chemical background of wool fabric setting is the breaking and reforming of interchain cystine bonds in the desired places by using a wet environment, mechanical stretch and heat-cold shock. A chemical setting is also possible by using treatment with thio-glycols under stretching, followed by hydrogen peroxide reformation of the disulphide bonds, as in permanent waving of hair, but it is considered too expensive for industrial purposes.

Cotton and bast fibres fabric dimensions are set by mercerizing and sanforizing.

The mercerizing is based on the action of cold strong alkali (usually NaOH) solution and concomitant application of a stretching force, followed by neutralization. As is shown in Figure 9.6.4, this allows cellulose fibre to pass from the meta-stable natural crystalline structure I to the stable cellulose II form. The mercerization produces effluents with very high pH, requiring further acidic treatment for discarding.

Sanforizing, a complementary operation for stabilizing the dimensions of cellulose fabrics, is an induced shrinkage of cotton fabric achieved by controlled dampening and drying of the relaxed fabric to remove any tensions and distortions which have appeared as a result of other processing. It does not produce pollutants.

The stabilizing of fabrics made of chemical fibres requires a thermosetting operation, performed in the stenter. The pollutants produced are volatile organic compounds (VOCs) arisen from the oligomers volatilizing at 180–200 °C required by thermosetting.

9.6.6.4 Coating and Infiltrating

The coating and infiltrating operations cover most of the dyeing and wet finishing processes.

The colouring process renders most of the commercial value of a fabric. It is achieved either by coating (pigment printing) or by infiltrating (printing and dyeing) the fibres. While pigment printing binds by a coating resin colouring pigments which have no affinity for the macromolecules of the fibre, the other printing systems and the dyeing makes use of molecules which diffuse and eventually bind on active sites inside the fibre, or aggregate to form difficult to removable clusters. The dyes used for the natural fibres belong to practically any dyeing classes; it is only the disperse dyes which are of low interest, due to their poor affinity, and this class remains dedicated exclusively to polyester fibres. Historical textile dyeing handicraft operations, because of the multi-colour aspect, are highly attractive for tourism on one side, but on the other the colouring operations, either dyeing or printing, produce polluted (coloured) waters because the process yield is usually only up to 80%.[39]

The pigment printing issues also volatile organic compounds (VOCs) and fumes from resin curing.

The wet finishing operations comprise flame-proofing treatment, softening or easy-care. Among them the easy-care finish is probably the most important one.

Easy-care finishing reduces the tendency of the fabric to crease in wear and makes it much easier to iron after laundering. It may also prevent shrinkage during washing. These types of finish have greatly reduced the maintenance which cotton and wool in everyday use require, and have increased their suitability for apparel of all types. However, there is a price for these advantages: pollution.

The easy-care treatment of wool fabrics has to handle the felting property, which affects also the dimensional stability of wool fabrics. The felting process makes use of the particular surface of wool fibre, whose scales enable the directional frictional effect. Consequently, any mechanic action in a wet environment on a wool fabric translates to a shrinking of dimensions by increasing the compactness. The felt-proofing treatment aims at shaving the scales on fibre surface and/or at fixing the fibre-to-fibre by chemical resins. The cleaning of the scales is achieved by Alwörden's reaction,[40] which involves a controlled cold chlorination followed by neutralization and a resin application. The process produces large amounts of halogenated organic compounds (AOX) in the effluent, besides gaseous chlorine in the atmosphere, for which reasons new methods are under investigation. Treatments with enzymes,[41] with ozone[42] or with plasma,[43] followed by resin application, are some of the available alternatives.

The easy-care treatment of cellulose fabrics uses padding with methylol-urea resins and curing them to cross-link inside the fibres. Consequently for-maldehyde is the most important pollutant accompanying the easy-care finish. Alternatives were developed with resins without formaldehyde (reactive elas-tomere silicones, or dicarboxylic acids able to react with the OH groups of cellulose) but the results need still to be improved. Ammonia may also be used for improving the easy-care treatment in a mercerization-like process. This produces, on the other side, ammonia pollutants.

The mechanics of the chemical fibres make easy-care treatments unnecessary for fabrics made out of these fibres. The blends of natural fibres with chemical fibres are based on this feature for producing easy-maintainable fabrics for apparels at low cost of pollution.

9.6.6.5 Surface Treatments

Surface treatments are, generally, mechanical operations giving the final aspect to fabric surface. Pressing (ironing), raising, suing and shearing are, probably, the most important ones.

These operations, particularly raising, suing and shearing, produce certain amounts of dust and fuzzy fibres as pollutants.

9.6.7 Conclusions

While artificial fibres came onto the market with the 1889 invention of rayon by Hilaire de Chardonnet, and synthetic fibres emerged with the 1931 DuPont patent for Nylon, natural fibres have been around since the dawn of mankind, flax having a documented history (burial shrouds for the Egyptian pharaohs) of more than 7000 years. No matter in which climatic zone humans settled, they were able to find and utilize the fibres of native species to make products such as clothes, cloths, buildings and cordage.

The first composite material known was made with clay and straw to build ziggurats in the ancient world. About 3000 years later there is a strong belief that the field of natural fibre composites is set to expand enormously in the near future. The environmental and health problems associated with synthetic fibres mean that they should be replaced by natural alternatives in a vast number of different applications. Their moderate mechanical properties restrain the natural fibres from being used in high-tech applications, but for many reasons they can compete with glass fibres. The natural fibres are lightweight, which makes them particularly attractive to the automotive industry. Statistics show that already in 2002 Germany and Austria together used around 17,000 mto of natural fibres in the automotive industry, out of which about 9,000 mto are flax and 2,500 mto are hemp fibres, and the yearly growing rate is 22%.

The modern composite materials, which merge the advantages of natural fibres with synthetic matrices and environmental goals, aim to go beyond the fibre competition during the last 100 years and to answer the new challenges.

References

1. M. Albin, F. D. Pooley, U. Strömberg, R. Attewell, R. Mitha, L. Johansson and H. Welinder, *Occup. Environ. Med.*, 1994, **51**, 205.

2. http://www.youthxchange.net/main/b164_ff1_textile-fibres-intro_b.asp (retrieved 12.08.2008).
3. I. Lee, B. R. Evans and J. Woodward, *Ultramicroscopy*, 2000, **82**, 231.
4. A. Peterlin and P. Ingham, *Textile Res. J.*, 1970, **40**, 345.
5. J. W. S. Hearle, in *Ullmann's Fibers*, Wiley-VCH Verlag GmbH & Co. KGaA, Weinheim, 2008, vol. 1, ch. 2, p. 39.
6. A. C. O'Sullivan, *Cellulose*, 1997, **4**, 173.
7. A. Sarko, in *Cellulose-structure, Modification and Hydrolysis*, ed. R. A. Young and R. M. Rowell, John Wiley & Sons, New York, 1986, p. 29.
8. J. Sugiyama, R. Vuong and H. Chanzy, *Macromolecules*, 1991, **24**, 4168.
9. A. A. Baker, W. Helbert, J. Sugiyama and M. J. Miles, *Biophys. J.*, 2000, **79**, 1139.
10. J. Hayashi, A. Sufoka, J. Okita and S. Watanabe, *J. Polymer Sci. Polymer Letters Edition*, 1975, **13**, 23.
11. A. Sarko, in *Wood and Cellulosics: Industrial Utilization, Biotechnology, Structure and Properties*, ed. J. F. Kennedy, Ellis Horwood, Chichester, UK, 1987, p. 55.
12. E. S. Gardiner and A. Sarko, *Can. J. Chem.*, 1985, **63**, 173.
13. K. Mühlethaler, *J. Polym. Sci.*, 1969, **C28**, 305.
14. P. Zugenmeier, *Prog. Polym. Sci.*, 2001, **26**, 1341.
15. H. Zahn, K. Schaeffer and C. Popescu, in *Biopolymers*, ed. A. Steinbüchel and S. R. Fahnestock, Wiley-VCH, Weinheim, 2003, vol. 8, ch. 7, p. 155.
16. C. Popescu and H. Hoecker, *Chem. Soc. Rev.*, 2007, **36**, 1282.
17. H. Ito and Y. Muraoka, *Textile Res. J.*, 1995, **65**, 755.
18. W. T. Astbury and A. Street, *Phil. Trans. R. London Soc.*, 1931, **A230**, 75.
19. W. T. Astbury and H. J. Woods, *Phil. Trans. R. London Soc.*, 1933, **A232**, 333.
20. L. Pauling, R. B. Corey and H. R. Branson, *Proc. Nat. Acad. Sci. USA*, 1951, **37**, 205.
21. F. H. C. Crick, *Nature*, 1952, **170**, 882.
22. F. H. C. Crick and J. C. Kendrew, *Adv. Protein Chem.*, 1957, **12**, 133.
23. L. Pauling and R. B. Corey, *Proc. Nat. Acad. Sci. USA*, 1951, **37**, 251.
24. W. T. Astbury, *Trans. Faraday Soc.*, 1933, **29**, 193.
25. L. Y. Yatsu, E. Espelie Karl and P. E. Kolattukudy, *Plant Physiol.*, 1983, **73**, 521.
26. E. Lord, *J. Textile Inst.*, 1956, **47**, T209.
27. United States Department of Agriculture, *Better Cottons*, Government Printing Office, Washington, 1957.
28. P. Kassenbeck, *Textile Res. J.*, 1970, **40**, 330.
29. World Wide Fund for Nature (WWF) International Report, *The impact of cotton on fresh water resources and ecosystems*, May 1999.
30. S. K. Batra, in *Handbook of Fibre Chemistry*, ed. M. Lewin and E. M. Pearce, Marcel Dekker, NY, 1998, p. 530.
31. B. Wahle, W. Becker and C. Hartschen, EP 1 177 341, 2000 (Cognis).
32. J. Knott, M. Grandmaire and J. Thelen, *J. Textile Inst.*, 1981, **72**, 19.
33. R. Höfer and J. Bigorra, *Green Chem. Lett. Rev.*, 2008, **1**, 79.

34. R. D. Swisher, in *Surfactant Biodegradation*, Marcel Dekker, New York, Basel, 1987, 2nd edn, revised and expanded; CEFIC, *Anaerobic Biodegradation of Surfactants*, Summary and Position Paper (February 1999), www.erasm.org/position/Anaerobic_Bio1.pdf (retrieved 06.11.2007); J. L. Berna, N. Battersby, L. Cavalli, R. Fletcher, A. Guldner, D. Schowanek and J. Steber, *Anaerobic Biodegradation of Surfactants*, Scientific Review, ERASM (1999), www.lasinfo.org/pos_papers/pos_paper_bioanaerobia.pdf (retrieved 06.11.2007).
35. B. Wahle and J. Falkowski, *Rev. Progr. Col.*, 2002, **32**, 118.
36. Deutsche Haut- und Allergiehilfe e.V., *Skin and Textiles – Protection for Sensitive Skin* (in German: *Haut und Textilien – Schutz für empfindliche Haut*), Informationsreihe für Patienten und Verbraucher, MedCom Publ., Bonn, 2005.
37. R. Tyagi, V. K. Tyagi and R. K. Khanna, *J. Oleo. Sci.*, 2006, **55**, 337; S. Mishra and V. K. Tyagi, *J. Oleo. Sci.*, 2007, **56**, 269; A. Scheidgen, Development and Trends for Softeners (in German: *Entwicklung und Trends bei Weichspülern*), Multiplikatorenseminar 2007, http://www.haushaltstechnik.uni-bonn.de/PDF2007/SCHEIDGEN_Weichspueler.pdf (retrieved 05.01.2009).
38. W. Ushakowa and B. van Roy, *External Validation of Silicone Technologies for Fabric Care*, Technical Literature, Dow Corning, 2003, http://www.dowcorning.com/content/publishedlit/27-1114-01.pdf (retrieved 06.01.2009).
39. R. Christie, in *Environmental Aspects of Textile Dyeing*, Woodhead Textiles Series No. 66, Woodhead Publ., Abington, Cambridge, 2007.
40. K. von Allwörden, *Z. Angew. Chem.*, 1916, **29**, 77.
41. E. Heine and H. Hocker, *Rev. Prog. Coloration*, 1995, **25**, 57.
42. T. Wakida, T. Tokuyama, M. Lee, J. Hun Jeon, H. Kuriyama and S. Ishida, *Textile Res. J.*, 2004, **60**, 656.
43. A. Hesse, H. Thomas and H. Höcker, *Textile Res. J.*, 1995, **65**, 355.

CHAPTER 9.7

Plant-based Biologically Active Ingredients for Cosmetics

CHARLOTTE D'ERCEVILLE, FLORENCE HENRY, PATRICE LAGO AND ANDREAS RATHJENS

Laboratoires Sérobiologiques, Division of Cognis France, 3 rue de Seichamps, CS 71040 Pulnoy, 54272 Essey Les Nancy Cedex, France

9.7.1 Introduction

Plant-based biologically active ingredients are used in different kinds of industries and have achieved a particularly high importance in the development of cosmetic actives with a proven biological effect. However, what about other aspects of modern economies? Is it sufficient that the products are just derived from plant origin? The simple answer is no. In today's world, we have to find the right balance between economic interests and the ecological limits imposed by the capacity of the biosphere to provide society with all of its resources for present and future generations. This is not a contradiction at all; instead, society and industry are forced to respect nature, to set up and follow clear rules on what sustainable development should look like and to accept their Corporate Social Responsibility (CSR). These aspects are discussed using a concrete example, cosmetic active ingredients from the Argan tree.

RSC Green Chemistry No. 4
Sustainable Solutions for Modern Economies
Edited by Rainer Höfer
© The Royal Society of Chemistry 2009
Published by the Royal Society of Chemistry, www.rsc.org

9.7.2 Active Ingredients and their Functionality in Cosmetic Applications

One definition of active ingredients comes from the pharmaceutical area.[1] An active ingredient, also active pharmaceutical ingredient or API, is a drug that is pharmaceutically active. The traditional word for API is pharmacon, adopted from the Greek "pharmacos", which originally denoted a magical substance. However, the term "active ingredient" is now also common and applied in other areas, for example nutrition and personal care, especially in the field of cosmetics. In all cases, we speak about a component that is considered to fulfil the intended activity of a final product, described by its claim, and gives the desired physiological effect. The main distinction lies in the delivery mode, which is oral intake for nutritional application and topical application for cosmetic formulations. Going a step deeper into the cosmetic domain and focusing on skin care applications, we can therefore describe an active ingredient as a substance or a group of molecules with the ability to improve the structure and appearance of skin when applied topically. To give some examples, knowing that the list of claims is much longer, we might speak about skin moisturization, reduction of wrinkles and protection against environmental stress, such as sun light or pollution or the wish to change one's skin colour either by whitening or by tanning. Figure 9.7.1 gives a representative overview of biological targets for skin care active ingredients.

Cosmetics in general are more than simple externally applied formulations to change or enhance the beauty of skin, hair, nails, lips and eyes. The key issues of today are related to well-being and convenience. Consumers are looking for solutions that reduce the complexity of their daily life, support their demand of

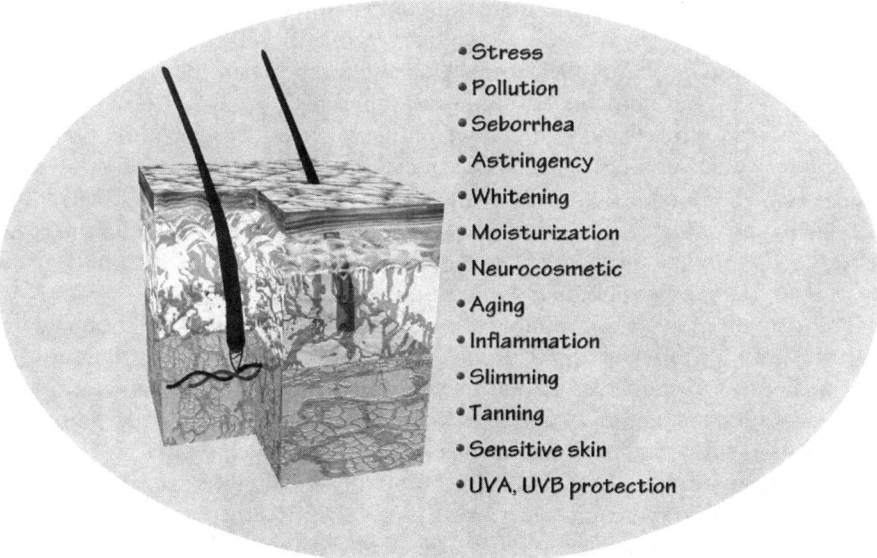

Figure 9.7.1 Biological targets for skin care active ingredients.

excellent performance and are in line with their "green" consciousness. As a result, cosmetic products are intended to deliver not only the functional benefits they promise, such as an anti-aging effect or protection against harmful environmental influences *etc.* (see Figure 9.7.1), but also to give the user a sense of well-being. Thus, consumers are looking more and more for a holistic product experience that appeals to all their senses.

9.7.3 Plant-based Raw Materials

Plant-based raw materials are important for the development of active ingredients. It is not mandatory that the source for a cosmetic active substance is a plant; mineral and animal materials or synthetic molecules are also widely used in the industry in order to achieve or improve performance and benefits of a cosmetic formulation. Although the limited fossil resources are still the main basis for the chemical industry, they are more and more regulated, have the disadvantage of not being "renewable" and, in addition, prices are continuously increasing due to the growing worldwide demand, predominantly for energy generation. Animal-based raw materials have received negative reports since the 1990s, regarding bovine spongiform encephalopathy, which have left a feeling of discomfort with consumers when coming into contact with them. The request for plant extracts has continuously increased since that time in order to replace animal-derived ingredients.[2,3] Moreover, taking into account the modern consumer demand for "green" and environmentally friendly products, supporting the wellness and sustainability trend, natural renewable sources are the raw material of choice for extraction and chemical synthesis. Thanks to the huge potential and diversity of the plant kingdom with more than 287,000 species described on Earth, plant-based active ingredients are today the driver of innovation and most probably their importance will increase in the coming years.[4]

Aside from flora biodiversity, plants contain a high variety of phytochemical compounds with performance on a wide range of cosmetics functionalities. For instance, proteins and polysaccharides act as film formers and tightening active ingredients. Seed oils, rich in fatty acids and triglycerides, are used for their emollient and moisturizing properties and, finally, secondary metabolites are increasingly applied in cosmetics to reduce the effects of different environmental stress, *e.g.* flavonoids for UV photo-protection.[2]

The use of plant-based biologically active ingredients in all branches of industry and in different applications is the subject of intensive discussions in the media and in the wider public. Historically, natural products from plants have provided an endless source of medicine and remain an undiminished starting point for the development of new pharmaceuticals.[5]

The conservation of biological diversity, the sustainable use of its components, and the fair and equitable sharing of benefits arising out of the utilization of genetic resources is promoted by the Convention of Biological Diversity (CBD), a legally-binding instrument with 191 parties to date, which was negotiated under the auspices of the United Nations Environment Program

(UNEP) at the Earth Summit in Rio de Janeiro and entered into force in 1993.[7] The 9th Conference of the Parties (COP) recently held in Bonn has specifically discussed the economic importance of genetic resources and the direct and indirect relationship of companies with biodiversity, as well as potential pathways for integrating biodiversity into their strategies and daily activities ("Business & Biodiversity Initiative"). It was mentioned that 35,000 plant species worldwide are used and 2000 species are marketed in Europe for medicinal and aromatic purposes. Finally, 70% of modern pharmaceuticals are based on plants.

The worldwide market size for medicinal plants, exotic fruits and seeds has been estimated to be 18,000 Mio US$. In parallel, it seems clear that there is an increasing number of consumer goods using plant derived raw materials, such as phytopharmaceuticals, cosmetics and dietary products.[6]

From here, we would like to focus on the cosmetic industry and explain in more detail the modern principles of sustainability and Corporate Social Responsibility with a concrete example: the Argan tree story.

9.7.4 Sustainability Concept and Corporate Social Responsibility (CSR)

Behind the term "sustainability" lies the quest for future-viable solutions for the environment, technology, business and our social structures. The drivers of the sustainability trend are society's changing values. According to a study published in 2005 by the Natural Marketing Institute (NMI), a new group of consumers is emerging all across the world. It is referred to as LOHAS (Lifestyle Of Health And Sustainability), and already one-third of US consumers belong to this group.[8] They are not searching for a compromise in terms of excellent product performance, but want the complete package including high efficacy, environmentally friendly manufacturing processes and no risk on product safety.

According to the European Commission, CSR is "a concept whereby companies integrate social and environmental concerns in their business operations and in their interactions with their stakeholders on a voluntary basis".[9]

Stakeholders exist both within and outside a firm. CSR is concerned with treating the stakeholders of the firm ethically and in a socially responsible manner. The aim of social responsibility is to create higher and higher standards of living, while preserving the profitability of the corporation for all its stakeholders.[10]

A company operates sustainably when its actions do not impair the planet's ability to provide the necessary resources in both the present and the future and when it integrates, on a voluntary basis, social and environmental concerns in its business operations and its interactions with its stakeholders. This is a commitment to find a healthy balance between financial, environmental and social imperatives, without compromising the development opportunities of

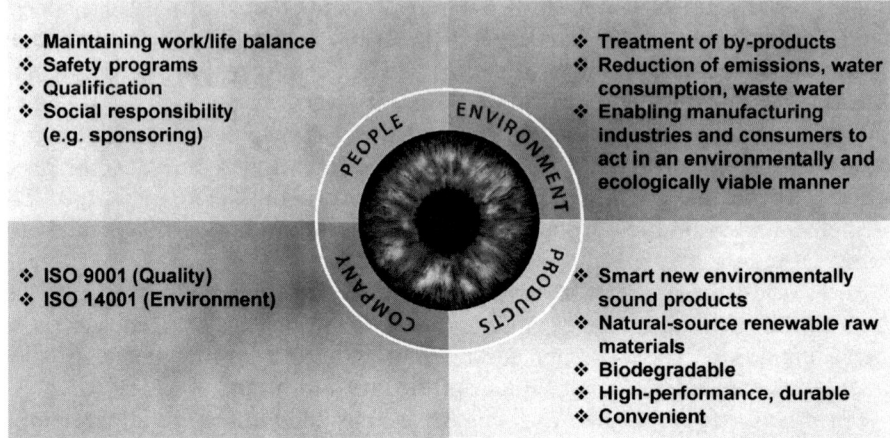

Figure 9.7.2 Four dimensions of sustainability.

future generations. Based on the principles of Agenda 21,[11] we are focusing on four key aspects (Figure 9.7.2):

- **People:** the business is built around people; we aim to be a responsible and attractive employer, develop and improve continuously the skills and expertise of our employees, and to be a good neighbour.
- **Environment:** the use of natural renewable raw materials is a key. At the same time, we offer to manufacturing industries and consumers solutions that are both cost-effective and environmentally friendly.
- **Company:** strict quality, environmental, health and safety standards are part of our integrated management system. Working in partnership with the communities, either in industrialized regions or developing countries, is essential to our philosophy.
- **Products:** to develop innovative performance products that meet the highest standards in terms of occupational and consumer health, safety and environmental protection.

9.7.5 From the Botanical Raw Material towards the Final Product

The development of a novel plant-based biologically active ingredient generally follows well-defined steps correlated to a logic process flow, which is less dependent on the part of the plant used, *e.g.* fruits, flowers, leaves, by-products such as oilcakes, or the intended final performance. All work starts with the identification of a suitable raw material and corresponding suppliers. The aforementioned principles of sustainability have to be applied already at this stage, and all available information must be analyzed in a critical manner. Key elements of our sourcing process take into account a responsible and environmentally friendly approach by evaluating:

- Bioavailability of the natural resource;
- Plant status, *e.g.* protected, vulnerable, threatened or endangered species, *etc.*;
- Preference for value creation of by-products, *e.g.* oilcakes or fruit hull from other industrial applications;
- Social impact of controlled collection and farming;
- Cultivation and collection following organic principles;
- Traceability, GMO (genetically modified organisms) risk, pesticide level, *etc.*;
- Implementation of environmental impact studies;
- Intellectual property rights of third parties;
- Toxicological profile of the plant or its extracts based on literature information.

Concerning the natural raw material supplier, it is fundamental to have reliable partners. This is in practice only possible not by multiplying the number of suppliers, but by capitalizing on long-term partnerships with selected ones, who share the same targets on social and environmental commitment. They are responsible for a more extensive group of raw materials and have therefore the critical mass to build up a network for cultivation and harvesting. Moreover, the implementation of local added value in the form of technical or operational empowerment is easier for a larger organization.

Once the source and potential supplier are selected, the product development phase begins. A number of extracts are prepared, by using a different solvent system, preferably water, and purifying protocols. These prototypes are extensively tested *in vitro* in order to evaluate their efficacy and to identify the one performing best with regard to the desired claim. The final version can then be focused upon. Product safety must be assessed by performing a series of toxicological tests, the product stability itself and in the final application form, for instance cosmetic frame formulations, has to be confirmed. Finally, the biological activity is validated by a clinical study involving healthy human volunteers. The transfer to industrial manufacturing is then prepared. Questions of the environmental impact of industrial production are taken into consideration, and the most appropriate processes using novel technologies to reduce waste, energy and water consumption are selected.

9.7.6 Sustainable Development and CSR for the Supply of Natural Products Derived from the Argan Tree

The Argan tree (*Argania spinosa*), Figure 9.7.3, is an endemic tree of South Morocco, growing in a region that runs from Safi in the north to the edge of the Sahara in the south. The main zone extends south-east from Essaouira to the

Figure 9.7.3 Environmental impact study: experimental area, localization of Argan forest, Argan.

Souss plain, covers about 800,000 ha containing more than 2 million trees and acts as a green belt against desert advancement. Argan is the second most common species in Morocco, behind the evergreen oak (*Quercus ilex*) and ahead of the thuya (*Tetraclinis articulata*). The tree is particularly resistant to the dry and arid conditions of this region, can tolerate temperatures ranging from 3 to 50°C and grows at altitudes of up to 1500 metres. Its roots, which extend over a large area and are very deep, can search out water at more than 30 metres under the ground, which helps it to survive the dry periods that can last for several months each year. The tree reaches heights of 8 to 10 metres and has a shape similar to an olive tree. Its life is quite long and it is not uncommon for it to reach ages of 150 to 200 years; even some 250-year-old trees have been recorded.[12]

Argan leaves are small, dark green and grow on branches that can be spiky at their ends. It is usually an evergreen plant but may lose its leaves in a persistent dry period. The tree has small oval greenish yellow fruits that become brown when they ripen and contain a very hard shell enclosing one to three almond-like kernels. The ripening period is quite long taking 2 years.[13]

The Argan forest plays a crucial social and economical role as it ensures the subsidence of over 3 million people of the rural Moroccan population. Unfortunately, due to the demographic pressure, Argan woodland suffers from continuous degradation. Over-exploitation, soil erosion and advancing desertification are amongst the threats to this unique heritage. The natural growth of the population leads to an increase in wood collection, which is used both for construction and as firewood. Excessive grazing makes it more difficult for the Argan trees to recover. The climate, which is always hard in these regions, is accompanied by erosion that makes the soil even more arid.[14]

UNESCO, under the program "Man and the Biosphere" (MAB), took this up and classed the Moroccan Argan forest as a Biosphere Reserve (RBA) in 1998.[15] Biosphere Reserves are delimited regions in land-based ecosystems where one tries to reconcile economic development with conserving the biodiversity. Biosphere Reserves serve in some ways as "living laboratories" for testing out and demonstrating integrated management of land, water and biodiversity, which is the principle of the ecosystem approach that has resulted from the biological diversity convention (UNESCO 1996).[16] Biosphere Reserves have to fulfil three main tasks that are complementary and synergistic:

- **Conservation role**: to contribute to the conservation of the whole range of biodiversity elements especially the landscapes, the ecosystems, the species and the genetic variation;
- **Development role**: to foster economic and human development, which is socio-culturally and ecologically sustainable;
- **Logistic role**: to provide support for research, monitoring, education and information exchange related to local, national and global issues of conservation and development.

9.7.6.1   Targanine Network

In this context, the Targanine network was founded in 1996 by Professor Zoubida Charrouf with the support of Mohamed V University in Rabat and the association Ibn Al Baytar with the aim to:

- educate stakeholders of the Argan forest to protect their ecological patrimony;
- enhance economic valorization of Argan fruits;
- improve the social status of women: increase their incomes and access to literacy programs.

Today, 15 cooperatives for crushing and 6 cooperatives for pressing, involving more than 2000 women, are involved in the production of organic Argan oil used in food and cosmetics. The EIG Targanine (Economic Interest Group) was established in 2003 in Agadir with the goal of assisting cooperatives in marketing, promotion and value creation at the national and international level.[17] Figure 9.7.4 illustrates the principles of sustainability applied by the cooperatives of EIG Targanine.

9.7.6.2   Partnership Between EIG Targanine and Cognis

For nearly six years, the Targanine network of cooperatives and Cognis have been working together in a close partnership. Initialized and supported by Professor Zoubida Charrouf, this collaboration focuses on new products derived from the Argan tree with the condition that these activities must

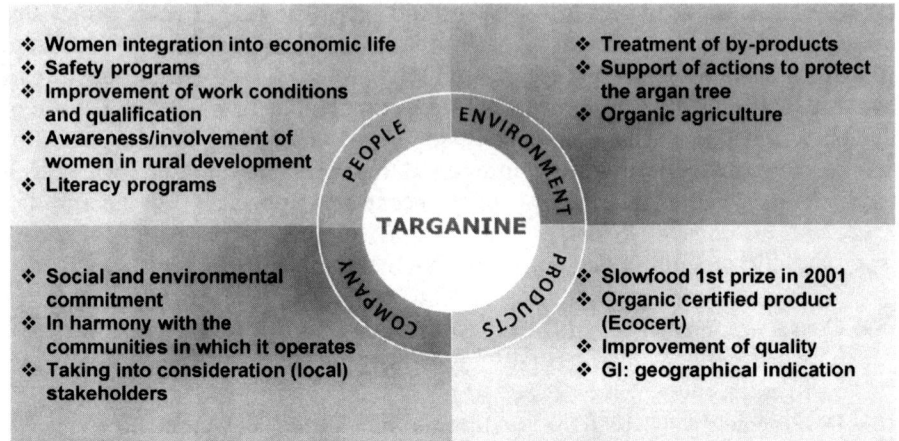

Figure 9.7.4   Sustainable approach of EIG Targanine.

Table 9.7.1   Composition of Argan oil according to literature[18,19]

	Benzaria et al.	*Khallouki et al.*
Fatty acid (%)		
Myristic acid (C14:0)	0.42 ± 0.01	
Palmitic acid (C16:0)	13.49 ± 0.05	16.5
Stearic acid (C18:0)	5.40 ± 0.08	3.7
Oleic acid (C18:1 ω-9)	41.19 ± 0.15	45
Linoleic acid (C18:2 ω-6)	38.9 ± 0.22	
Eicosadienoic acid (C20:2 ω-6)	0.4 ± 0.02	
Dihomo-γ-linoleic acid (C20:3 ω-6)	0.21 ± 0.01	
Saturated fatty acid	19.31 ± 0.11	
Polyunsaturated fatty acid	39.5 ± 0.22	
Tocopherols		629 ± 28
Squalene		311 ± 2
Sterols		272 ± 10

support and encourage the protection of the Moroccan Argan forest and in parallel add local value. The following points were defined as main targets:

- Identification of specific parts of the Argan tree (leaves and oil cake) that could be valuable in cosmetic applications and provide additional incomes to the Targanine members;
- Evaluation of the environmental impact of harvesting Argan leaves and definition of the most appropriate and respectful way to do this;
- Involvement of local populations in the protection of the Argan forest by, for example, creation of a tree nursery.

What is the situation now after a couple of years? Whilst the Argan is mainly known for its seed kernel oil (see Table 9.7.1), the joint research program focused on adding value to its by-product, the oil cake rich in proteins, and as a second target on the leaves. After intensive phytochemical studies and a comprehensive performance-testing program including *in vitro* as well as clinical assays, we were able to finalize two new active ingredients for application in cosmetic products. The first, Argatensyl™ is a high-molecular-weight protein complex extracted from the oil cake, which possesses skin-tightening properties and a biological effect on existing wrinkles. The second one, Arganyl™, is derived from the leaves and contains valuable polyphenols that protect the skin from premature aging and strengthens the cutaneous structure.

Furthermore, we decided together also to commercialize a cosmetic grade of the precious oil Lipofructyl™ Argan, rich in polyunsaturated fatty acids (including linoleic acid (omega-6) and oleic acid) and natural tocopherol, for its nourishing and protective properties. Figure 9.7.5 gives an overview on the development activities and the outcome.

Since 2004, Targanine supplies Cognis directly *via* its local cooperative network with the Argan oil, the oil cake and the dried leaves. The principles of sustainable trade, direct with the cooperative, paying fair prices with prepayment conditions (50% payment at the time of order of the goods), are all part of this business. These guarantee Targanine an additional and diversified income for the network's members and ensures a sustainable activity.

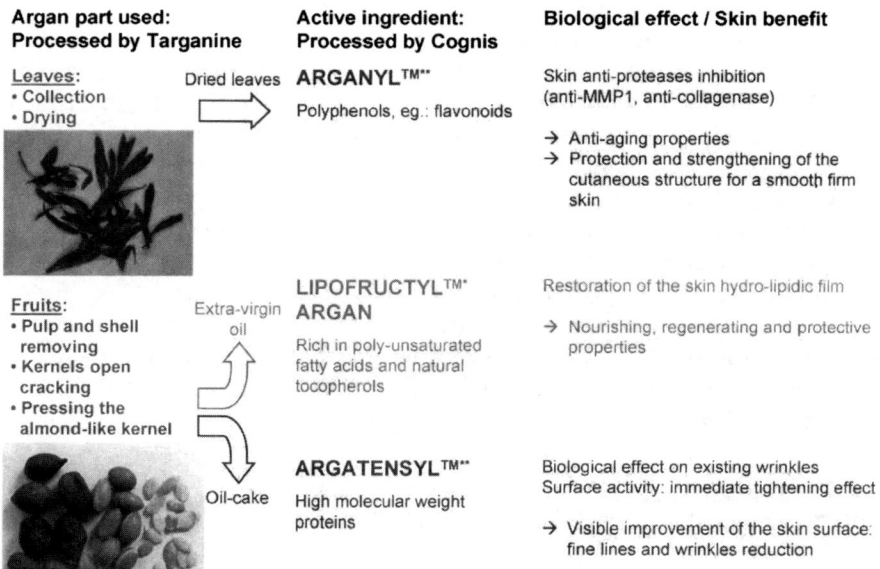

Argan part used:
Processed by Targanine

Active ingredient:
Processed by Cognis

Biological effect / Skin benefit

Leaves:
• Collection
• Drying

Dried leaves **ARGANYL™****

Polyphenols, eg.: flavonoids

Skin anti-proteases inhibition
(anti-MMP1, anti-collagenase)

→ Anti-aging properties
→ Protection and strengthening of the
cutaneous structure for a smooth firm
skin

LIPOFRUCTYL™*
ARGAN

Extra-virgin oil

Rich in poly-unsaturated
fatty acids and natural
tocopherols

Restoration of the skin hydro-lipidic film

→ Nourishing, regenerating and protective
properties

Fruits:
• Pulp and shell
removing
• Kernels open
cracking
• Pressing the
almond-like kernel

ARGATENSYL™**

Oil-cake High molecular weight
proteins

Biological effect on existing wrinkles
Surface activity: immediate tightening effect

→ Visible improvement of the skin surface:
fine lines and wrinkles reduction

* *Raw material certified by Ecocert SAS BP 47 32600 L'Isle Jourdain according to the standard of Natural and Organic Cosmetics: 100% of the ingredients are coming from Organic Farming*
** *Raw material conforms to Ecocert standard of natural and organic cosmetics*

Figure 9.7.5 Argan-based products.

The environmental effect of collecting Argan leaves for cosmetic applications, target 2, depends on the tree's ecology, the dynamics of the ecosystems and the socio-economic circumstances. An impact study looked at these different factors during a three-month mission to Morocco in 2004 by monitoring an experimental area in the village of Amelne, Tiznit province, in the South of Agadir. The experimental area, established on a plot of the Argan forest, involved the follow-up of growth, flowering and fruit production of trees, which were subject to a range of sampling protocols. This work was done after having received approval from the provincial governor and the local authorities.

There are several options for collecting leaves:

• Collection of yellowing leaves on the ground, or on the trees;
• Picking green leaves on the trees;
• Recovering leaves from sylvicultural operations (maintenance) on the Argan forest: coppicing.

After evaluating several possibilities, the practice of coppicing was selected. This operation consists of thinning the tree in order to rationalize its growth and enhance fruiting. The procedure is recommended by the Water and Forest Government Agency and supports the perpetual improvement of the Argan tree in the forest. Thanks to the cosmetic value of Argan leaves, an incentive for the local population to perform this often neglected, but important, maintenance task was established.

Finally, a technology transfer program to extend the existing tree nursery was implemented in the Targanine network cooperatives. The project has been submitted to the provincial governor, the Water and Forest Government Agency and local authorities who have given their approval and offered their support. Cognis has financed this project with the aim to extend and increase reforestation of the Argan forest, and it has developed a protocol for germinating Argan seeds and planting new trees. A forestry engineer has trained the staff to manage and monitor the nursery, and each woman in the cooperatives is involved in this reforestation program.[20,21]

The cooperation between IEG Targanine and Cognis is based on a sustainable partnership including all elements mentioned before and by sharing the same principles of CSR, although translated to their specific environment (see Figure 9.7.6).

9.7.7 Conclusion

The use of natural renewable sources for the development of active ingredients is only one of many alternatives for modern economies. Taking up the trends of modern consumer demand and their growing environmental consciousness forces society and industry to search for solutions, and plant-based raw materials fit perfectly. There is an increasing need to facilitate, demonstrate and

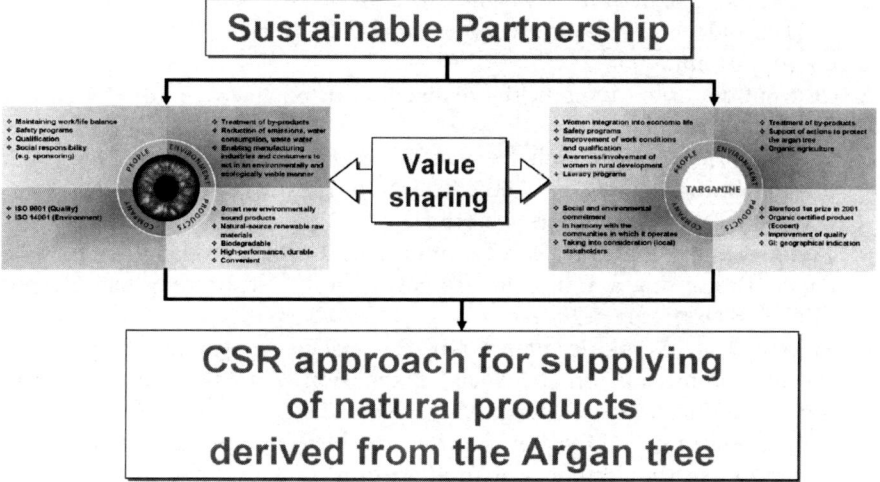

Figure 9.7.6 Shared CSR approach of IEG Targanine and Cognis.

promote Corporate Social Responsibility to enhance long-term sustainable economic developments. The example of products from the Argan tree shows that economic interests can be in alignment with environmentally friendly sourcing which helps to preserve our ecosystem, in this case the fragile Argan forest. All stakeholders have to be involved in this process (authorities, local partners, manufacturers, *etc.*) and their commitment is necessary to ensure a sustainable social impact and to support the local economical development. Looking to the future, we speak about an ongoing process, and further improvements must be generated proactively from within, driven by individuals as well as by corporate conscience.

References

1. Food and Drug Administration guidance, http://www.fda.gov/cder/guidance/index.htm.
2. T. Aburjai and F. M. Natshem, *Phytother. Res.*, 2003, **17**, 987.
3. P. Hövelmann, *Perspektiven nachwachsender Rohstoffe in chemischen Anwendungen* in 7. Symposium Nachwachsende Rohstoffe für die Chemie, Schriftenreihe Nachwachsende Rohstoffe, Bd. 18 Landwirtschaftsverlag, Münster (2001).
4. International Union for the Conservation of Nature and Natural Resources (IUCN), Table 1: Numbers of threatened species by major groups of organisms (1996–2004), http://www.iucnredlist.org/info/tables/table1.
5. B. M. Schmidt, D. M. Ribnicky, P. E. Lipsky and I. Raskin, *Nat. Chem. Biol.*, 2007, **7**, 360.

6. A. Drews, The requirements in the convention on Biological Diversity to Using Genetic Resources, *International Conference "Business & Biodiversity"*, Bonn, 2008.
7. Convention on biological diversity, http://www.cbd.int (retrieved 12.07.2008).
8. LOHAS Market Research Review: Marketplace Opportunities Abound, S. French and G. Rogers, 2005, The Natural Marketing Institute, http://www.lohas.com/journal/trends.html.
9. Commission of the European Communities, COM (2001), 366 final, Green Paper: Promoting a European framework for Corporate Social Responsibility, Brussels, 18.7.2001.
10. M. Hopkins, Working Paper No. 27, Policy Integration Department, World Commission on the Social Dimension of Globalization, International Labour Office, Geneva, May 2004.
11. United Nations – Division for Sustainable Development, http://www.un.org/esa/sustdev/publications/publications.htm.
12. J. F. Morton and G. L. Voss, *Econ. Bot.*, 1987, **41**, 221.
13. O. M'Hirit, M. Benzyane, F. Benchekroun, S. M. El Yousfi and M. Bendeenoun, in *L'arganier : une espèce fruitière-forestière à usages multiples*, ed. Mardaga, Belgique, 1998, p. 150.
14. Z. Charrouf and D. Guillaume, *J. Ethnopharmacol.*, 1999, **67**, 7.
15. Biosphere Reserve Information – Morocco – Arganeraie, http://www.unesco.org/mabdb/br/brdir/directory/biores.asp?code=MOR+01&mode=all.
16. UNESCO, Réserve de Biosphère : la stratégie de Séville et le cadre statuaire du réseau mondial, Paris, UNESCO, 1996.
17. 20 ans de recherche-action pour le développement durable de l'Arganier – Association Ibn Al Baytar, Professeur Z. Charrouf, 2007.
18. A. Benzaria, N. Meskini, M. Dubois, M. Croset, G. Nemoz, M. Lagarde and A. F. Prigent, *Nutrition*, 2006, **22**, 628.
19. F. Khallouki, C. Younos, R. Soulimani, T. Oster, Z. Charrouf, B. Spiegelhalder, H. Bartsch and R. W. Owen, *Eur. J. Canc. Prev.*, 2003, **12**, 67.
20. L. Pumadera, F. Henry, Z. Charrouf and G. Pauly, *Environmental impact of the cosmetic valorization of the leaves of Argania spinosa (L.) Skeels, Morocco, Beyond wood*, Cornwall, 2004.
21. L. Pumadera, F. Henry, Z. Charrouf, G. Pauly and G. Falconnet, *Bois & Forêts des Tropiques*, 2006, **287**, 35.

CHAPTER 10

Sustainable Solutions – Green Solvents for Chemistry

CARLES ESTÉVEZ

IUCT, Alvarez de Castro, 63, E-08100 Mollet del Valles, Barcelona, Spain

10.1 Introduction

Solvents as a class of chemicals appear late in the history of mankind. Indeed, Nature does not need any solvents apart from water. Ancient caveman paintings, the Chinese and Japanese art of lacquering (based on urushiol, the latex of the Chinese lacquer tree),[1] the Mediterranean fresco painting and medieval oil paintings did not use solvents (except eventually spirit of turpentine obtained by the distillation of resin from terebinth or pine trees).The notion of solvents appeared (in the tradition of the Arabian *al-kimiya*) with the European alchemists who believed in the possibility of "transmuting" inferior metals, more particularly lead and mercury, into gold. The English friar and alchemist Roger Bacon believed that gold dissolved in *aqua regia* was the elixir of life. The Swiss alchemist Philippus Paracelsus believed in the existence of one undiscovered element common to all, of which the four elements (Fire, Earth, Air, Water) posited by Empedokles (*c.* 430 BC) were merely derivative forms. Linking alchemy with medicine, Paracelsus maintained that this prime element of creation termed *alkahest*, if it were found, would prove to be the philosopher's stone, the universal medicine and the irresistible solvent.[2] When in the twelfth century alembics with efficient cooling were invented, distillation of wine yielding concentrated alcohol, *aqua vitae*, was possible.[3] Extraction of aromatic substances (essential oils, from Latin *quinta essentia*, quintessence)

RSC Green Chemistry No. 4
Sustainable Solutions for Modern Economies
Edited by Rainer Höfer
© The Royal Society of Chemistry 2009
Published by the Royal Society of Chemistry, www.rsc.org

with alcohol followed by distillation would eventually become the oldest application of a solvent as such.

The ability of non-water-based solvents to remove soil and stains from fabrics was accidentally discovered in 1825 by French dye-works owner Jean Baptiste Jolly after his maid knocked over a lamp, spilling spirits of turpentine. Jolly had noticed that when the oil evaporated the area of the cloth was cleaner and he developed a service for cleaning people's clothes in this manner, which he termed *nettoyage à sec* (dry cleaning in English). In the early days, garment scourers and dryers found several fluids that could be used as dry-cleaning solvents, including camphene, benzene, kerosene and gasoline. These fluids are all dangerously flammable, so dry cleaning was a hazardous business until the introduction of carbon tetrachloride in 1910, and perchloroethylene in the 1930s.

The history of industrial chemistry cannot be dissociated from the development of organic solvents. On the contrary, both went hand in hand. Between 1826 and 1845 August Wilhelm Hofmann's research on coal tar led to the development of synthetic coal tar dyes as well as to the synthesis of aromatic solvents like benzene, xylene and toluene based on coal tar.

Solvents play vital roles in modern economies.[4] Oxygenated, hydrocarbon and chlorinated solvents, to mention only a few examples of the existing chemical families, make possible the formulation and manufacture of adhesives, paints, household products, active pharmaceutical ingredients, agrochemical compositions and edible oils. These and many other important products and processes demand a massive solvent usage. According to a study by the Freedonia Group[5] the overall world demand for solvents, including hydrocarbon and chlorinated types, is on the order of 20 Mio mto annually (2007), the lion's share of 58% being used in the coatings, inks and adhesives industries. Demand for hydrocarbon and chlorinated solvents, though, will continue its downward trend as a result of environmental regulations, with oxygenated and green solvents replacing them and growing at an average of 6–7%.

Because of the solvent properties of existing solvents and the engineering design of process technologies, a significant fraction of the solvents used in processes or formulations is released to the environment or has to be managed as a liquid organic waste and ultimately incinerated. Only in the fine chemicals sector, solvents contribute to as much as 50% of the total mass of waste.[6] The impact on small and medium enterprises (SMEs) is higher because SMEs often lack efficient means to develop environmentally benign second generation processes. In 2006, the European Commission alerted that SMEs are responsible for 60–70% of all industrial pollution in the European Union.[7]

10.2 The Design of Safer Chemicals and Solvent Innovation

Historically, solvents have been designed to maximize technical performance and minimize cost. To achieve a high efficacy of function, molecular properties of solvents (*e.g.* polarity, solubility, viscosity, vapour pressure, density and surface tension) and, hence, their molecular structures have been tailored to fit in the very narrow windows of specifications characteristic of each particular industrial application. Therefore, the characterization of physico-chemical properties and the discovery of the chemical laws governing solvent effects in chemical systems has been a major research goal over the past decades.[8] Unfortunately, a major characteristic of many traditional solvents is their potential to pose health and environmental risks due to their intrinsic (eco)-toxicity and potential for environmental damage. Note that both characteristics are also manifestations of molecular structure. Table 10.1 indicates the health hazard categories used to characterize chemical substances. Figure 10.1 provides evidence of the health hazard categories of the 64 most used traditional solvents. A large portion falls into the toxic and very toxic classes and only 3% can be considered completely safe for human health.

Table 10.1 Health hazard categories and their corresponding health effects.

Level	Hazard description	Health effects
0	Negligible	None
1	Low	Irritation
2	Moderate	Serious damage if in contact with skin, eyes or respiratory tract. Accumulation in breast milk. Toxic to reproduction (Cat. 3). Impaired fertility.
3	High	Toxic if in contact with skin, eyes or respiratory tract. Toxic to reproduction (Cat. 1 or 2). Carcinogenic and mutagenic (Cat. 3). Cumulative effects.
4	Very high	Very toxic by inhalation, in contact with skin and if swallowed. Carcinogenic and mutagenic (Cat. 1 or 2). Cancer and heritable genetic damage.

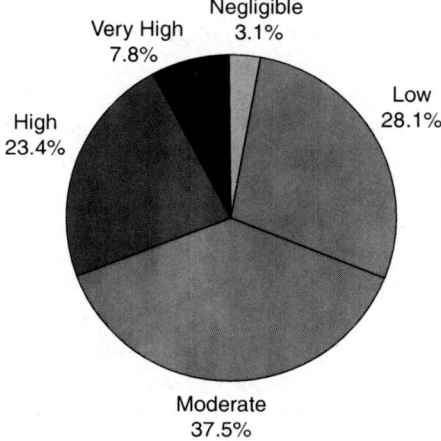

Figure 10.1 Distribution of the 64 most used traditional solvents in 5 health hazard categories ranging from negligible hazard to very high hazard.

Table 10.2 Hazardous chemicals, their Environment Health & Safety effects and major protocols, regulations and programmes.

Chemical	EHS Effects	Protocols, Regulations and Programmes	Year
CFCs, CCl$_4$, CH$_3$Cl, CH$_3$Br, Halons, HBFCs	• Stratospheric ozone depletion	• Montreal Protocol	1987
Fossil fuels	• Greenhouse effect (CO$_2$) • Global Earth warming	• Kyoto Protocol	1997
Dangerous/toxic substances	• Water contamination	• Framework Directive: 2000/60/CE • Related directives: 80/68/CE, 76/464/CE	2000
Persistent Organic Pollutants (POPs)	• Bioaccumulation • Toxicity • Persistent	• UNEP Chemicals Programme • Rotterdam Convention • Stockholm Convention	1997
Solvents	• VOCs emission • Tropospheric ozone formation • Occupational health	• Directives: 99/13/CE, 2001/81/CE • CAFE Programme • Occupational Health	1999
Commercial chemicals	• Health hazard	• REACH	2003

Table 10.2 summarizes the different regulations affecting most of the currently used solvents.

Basically, there are two principal approaches available to delineate the risk profile and address the concerns created by the usage of solvents. First, *control of exposure* minimizes the exposure to the hazardous solvent by using engineering designs such as closed systems with 100% recovery and zero emissions. An obvious disadvantage of these systems is that they do not eliminate the intrinsic hazard of the chemical substance. The second approach is based on *control of hazard*. It implies the substitution of the hazardous solvent by safer alternatives which may in certain cases involve process modifications to accommodate the new solvent. Replacement of hazardous solvents by bulk or full solids technologies without the need of solvents, water-based systems or the use of green solvents are examples of hazard control strategies.

Since the identification by Anastas and Warner of the replacement of undesired hazardous substances by safer alternatives as a key green chemistry research area,[9] many researchers in industry and academy have united in what

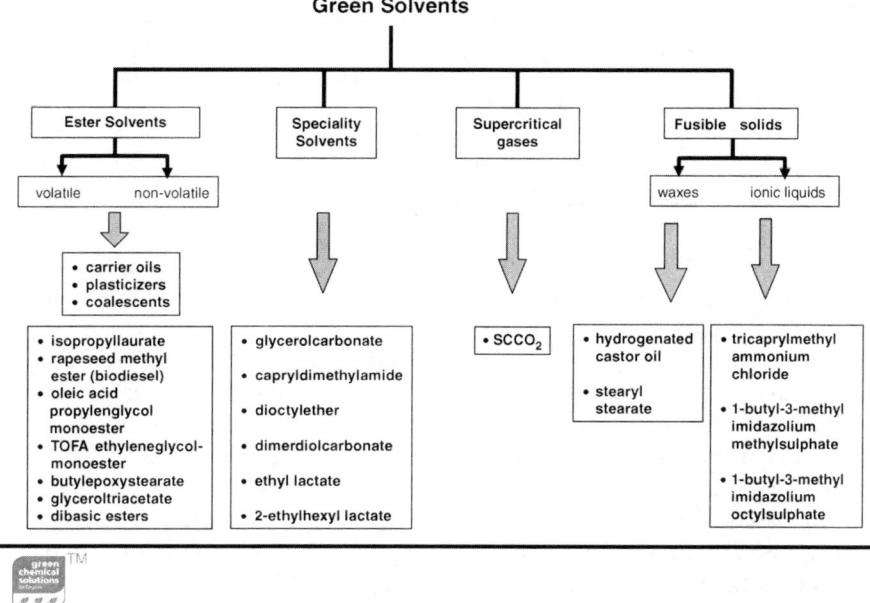

Figure 10.2   Examples of commercially available green solvents (proprietary design; reproduction with the permission of Cognis GmbH, Monheim/D; *ref.* R. Höfer, J. Bigorra, Comunicaciones, *Jorn. Com. Esp. Deterg.*, 2006, **36**, 43–64).

we envision as an incipient global effort to systematically design safer chemicals.[10] Solvents are no exception. Over the past decade, supercritical fluids,[11] ionic liquids,[12] fluorous phases[13] and switchable or expanded solvents[14] have been proposed by the academic research community as alternative solvents and reaction media. The growing experimental evidence suggests that these innovative solvents can not only replace traditional solvents in performing a wide variety of chemical processes and transformations but also have led to the discovery of new applications. Figure 10.2 shows examples of existing green solvents commercially available in industrial quantities.[15]

Despite these promising advances, many solvents with undesirable health and environmental profiles are still being used in commercial processes and formulations because satisfactory safer replacements have yet to be developed.

10.3   SOLVSAFE: A Roadmap for the Design and Application of Safer Functional Organic Solvents

10.3.1   Background and Sustainability Goals

The general lack of knowledge and strategies to reconcile functional performance with environmentally benign properties, reduced human risk and

Table 10.3 SOLVSAFE EHS objectives.

1	Reduction in the number and quantity of hazardous solvents utilized.
2	Reduction of Volatile Organic Compounds (VOC) emissions to the atmosphere.
3	Reduction of CO_2 emissions.
4	Increase in the quantity of renewable materials utilized.
5	Reduction of health risks on production sites by decreasing the Occupational Exposure Level (OEL) due to hazardous solvents.

economic viability prompted IUCT chemists to develop new methodologies to discover safer and functional replacements of existing solvents with undesired properties. In January 2003, the research team felt that a useful way to test the discovery tools was to create a research consortium where molecular designers and industrial researchers could collaboratively design effective green solvents. A primary focus of the project was to achieve simultaneously both a maximum efficacy of function for each specialized technical application and a benign environmental, health and safety profile. The projected research activities encompassed design, synthesis, physico-chemical characterization, performance tests, toxicology characterization and economic viability calculation. The project, SOLVSAFE, was sent to the EU FP6 NMP Programme for evaluation and was granted by the European Commission. In 2005 the Consortium started R&D activities with 19 partners from 6 different European countries.[16]

Sustainability being a central concept in SOLVSAFE, the project was designed to fulfil five environment, health and safety (EHS) working objectives aimed at reducing human health hazard, eco-toxicity, fossil carbon utilization and waste generation (Table 10.3).

10.3.2 Design Strategy

Molecular control over EHS as well as performance properties were attained through a design strategy that started with the creation of a chemical library composed of 239 potential solvent candidates. Relevant platform molecules were selected in order to fulfil a diversity of design criteria including a high renewable atom index, moderate to low vapour pressure and a cost-effective synthetic accessibility (see Table 10.4).

The ensemble was composed of 11 chemical families:

- A: Carboxylic acid dialkylamides
- B: Carboxylic acid esters
- C: Monopropyleneglycol esters
- D: Glycerol acetals and ketals
- E: Glycerol esters
- F: Glycerol diethers
- G: Glycerol triethers
- H: Fluorinated glycerol diethers

Table 10.4 Design criteria for platform molecules used as sources of SOLV-SAFE solvents.

	Design criteria	Category	Parameter
1	Molecular structure derived from renewable platform molecules (*e.g.* glycerol, carboxylic acids and derivatives thereof).	Environmental	Renewable Atom Index
2	Low VOC emission.	Environmental & Safety	Vapour pressure
3	Low flammability.	Safety	flash point
4	Efficient and green synthetic route.	Technical	Number of synthetic steps
		Environmental	CO_2 utilization
5	Functional group compatibility.	Technical	Reactivity

- I: Fluorinated glycerol triethers
- J: Organic carbonates
- K: Hydrocarbons

Figure 10.3 compares the distributions of a dataset containing the 108 most used existing solvents and a dataset of 239 SOLVSAFE solvent candidates in two principal components which account for the structural diversity of both datasets. One of the defining features of chemical spaces is that molecular structures can be represented as points whose coordinates depend on the values of relevant descriptors or variables. To characterize each molecular structure, SOLVSAFE used 52 structural descriptors. The principal component statistical analysis projects the data contained in the 52-dimensional chemical space into a two-dimensional space (plot in Figure 10.3). This approximation provides an overview of the systematic variation and distribution of the structural information and reveals how significant is the dissimilarity of the SOLVSAFE dataset when compared with the traditional solvents dataset.

In a second design stage, the chemical library was evaluated with *in silico* health hazard estimation methods developed by IUCT and state of the art eco-tox expert systems. Prediction of the health hazard and eco-toxicological profiles and physico-chemical properties with a sufficient degree of confidence allowed selection of those molecular structures exhibiting lower levels of intrinsic hazard. Figure 10.4 shows the calculated health hazard distribution of the SOLVSAFE dataset.

Note that the distribution of traditional solvents is peaked on health hazard level 2 (Figure 10.1 and Table 10.1) while the SOLVSAFE virtual library distribution is peaked on hazard level 0 (Figure 10.4). Furthermore, the fraction of SOLVSAFE solvents in the lowest hazard levels 0 and 1 is 77%, which is significantly higher than the corresponding fraction of traditional solvents which amounts to 31%. In addition, no solvent candidate was predicted in the higher hazard level 4 in the SOLVSAFE dataset. Figure 10.5 shows the percentage of solvents in the low or negligible hazard levels according to their effects in five

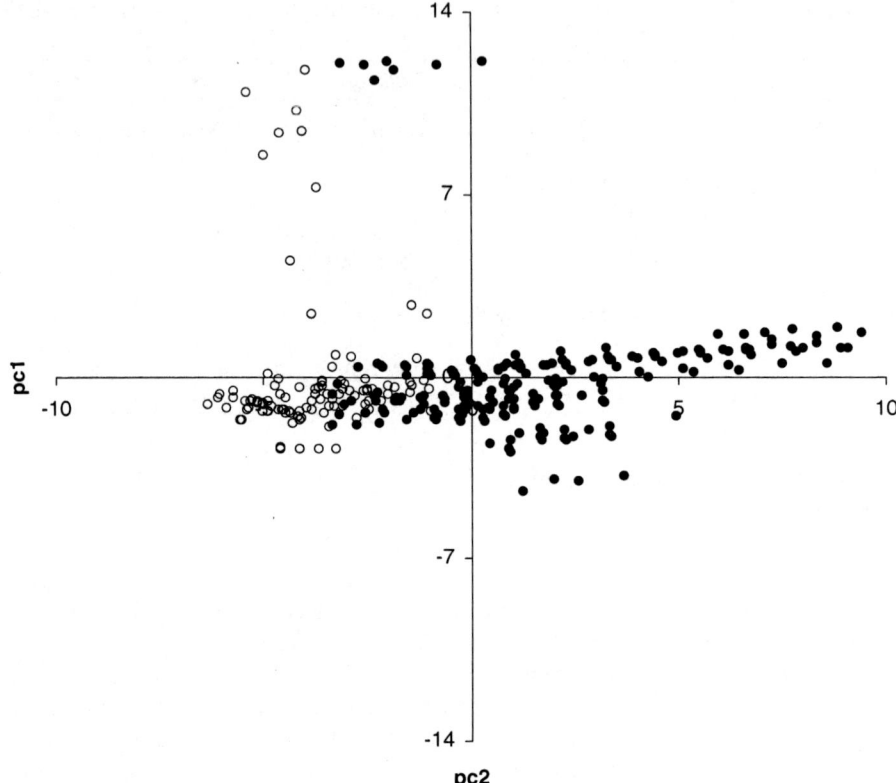

Figure 10.3 Principal components analysis of the structural characteristics of SOLVSAFE (filled circles) and traditional (circles) solvents.

environmental, health and safety categories, namely health hazard, aquatic toxicity, volatile organic compound emission, fire hazard and occupational exposure potential. The fraction of safer solvents is higher than 70% in all cases.

Most of the SOLVSAFE solvent candidates are characterized by a relatively low aquatic toxicity for the three trophic levels. This is a key property in those applications where the environmental impact should be minimized. The vapour pressures of SOLVSAFE solvents ranged from 0.001 to 200 hPa. The average vapour pressure of 1.35 hPa was significantly lower than the average value of the 64 most used solvents, 87 hPa. Interestingly, the dataset was composed of 70% non VOC solvents, an invaluable feature for those applications in open systems. The solvents in the dataset exhibit high flashpoints (87% of the solvents have flash points higher than 55 °C) assuring a safer alternative for those uses in which fire hazard needs to be minimized. The occupational exposure hazard of the SOLVSAFE dataset is negligible according to current risk evaluation procedures.[17] The average renewable mass index, a measure of the renewable mass content, exceeds 60%. The average CO_2 emission potential, a

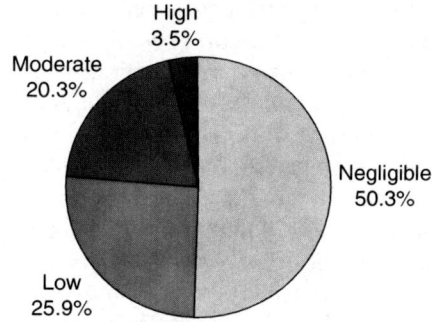

Figure 10.4   Distribution of 143 SOLVSAFE solvent candidates according to their calculated health hazard.

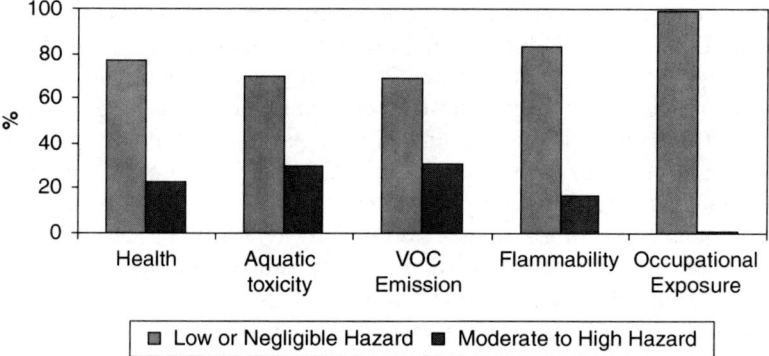

Figure 10.5   Classification of SOLVSAFE solvent candidates according to their effects in five environmental, health and safety categories. Health adverse effects according to Table 10.1. Toxicity for aquatic organisms based on LC50 96 h (fish) and ready biodegradability. VOC emission potential is considered when the vapour pressure at normal temperature is higher than 0.1 hPa. Low fire hazard is assigned whenever the flashpoint is higher than 55 °C. A low exposure potential in workplaces is considered when the vapour pressure is less than 10 hPa.

parameter that quantifies the CO_2 emission upon complete oxidation, is 47% lower than the average value of 68 traditional solvents. Note that the high renewable mass index of SOLVSAFE solvents may contribute to greenhouse gas savings provided that sustainable practices are used in the manufacture of the platform molecules.

Once a subset of potentially safer solvents is available, the next step involves the synthesis of the actual solvents. Green synthetic methodologies are selected in order to provide economically viable manufacturing processes leading to high quality solvents with the lowest cost per kilogram. In a subsequent stage, technical performance in different industrial applications follows a two-stage evaluation. First, efficacy of function for a specific industrial application is estimated using a predictive model developed by IUCT. Subsequently, an experimental investigation of the relevant solvents is carried out and the results compared with the predicted values. In the final step, the best performing solvents are subjected to experimental toxicological evaluation using protocols that minimize the use of laboratory animals.

We use the term *safer functional organic solvent* (*SFOS*) to refer generically to the green solvents designed under the SOLVSAFE design strategy. For the purpose of our investigation, we defined an SFOS as *an organic solvent displaying such molecular properties that enables it to achieve maximum functional performance for a particular application while minimizing its intrinsic potential for environmental or human health hazard*. It is worth clarifying our understanding of the term *green solvent* in view of the different contributions to this subject.[18] If we consider a green solvent as an evolutionary product of chemical innovation, then a green solvent has to bear a number of environment, health and safety (EHS) characteristics that differentiate it from existing hazardous solvents. Toxicologists, environmental scientists and industrial safety experts have identified quantifiable EHS parameters, metrics (*e.g.* LD50, LC50, flammability, ozone depleting potential, global warming potential, *etc.*) and threshold values above which the property poses a serious hazard to human health or the environment. Therefore, each parameter can be in the safe or unsafe domain depending ultimately on the specific molecular structure which determines this property. Any chemical substance with efficient solvency characteristics displaying a higher number of green traits (EHS parameters in the safe domain) when compared to a reference substance, typically the current industrial standard, can be considered a green or safer solvent. Alternatively, we could define a green solvent in absolute terms imposing the more restrictive condition that it possesses the maximum number of green traits (*i.e.* all of the EHS parameters in the safe domain). In the absence of consensus, each author will choose the relative or absolute definition according to specific circumstances. In any case, a full specification of the EHS parameters is necessary to describe and evaluate the "green solvent" under consideration.

SOLVSAFE is aimed at developing SFOS for their innovative application in key products and processes within six different sectors ranging from metal degreasing, paints and varnishes, crop protection formulations to fine chemicals manufacture and extraction of vegetable oils.

10.4   Industrial Application of SOLVSAFE Solvents: Results and Perspectives

10.4.1   Fine Chemicals

The manufacture of fine chemicals is one of the greatest challenges in synthetic chemistry. With a variety of commercial products ranging from organic pigments, explosives and surfactants to more sophisticated molecules such as pharmaceutical active ingredients, the sector is committed to delivering high purity substances to a market that is highly regulated in terms of product quality and safety and process environmental performance. Solvents are of extraordinary importance to the sector since their role is to achieve the maximum amount of synthetic construction in the shortest amount of time, with the highest isolated yield for the lowest cost per kilogram. Their impact actually goes beyond the product quality, affecting the safety of operators during processing and during transportation as well as the environmental burden of the process (spills, emissions to the atmosphere, waste water streams and disposal of wastes).[19] Following the design strategy, a subset of 22 SOLVSAFE solvent candidates were investigated for the replacement of methyl isobutyl ketone in the synthesis of Pimozide, an antipsychotic drug for the treatment of schizophrenia, chronic psychosis, the Gilles de la Tourette syndrome and resistant tics. The synthetic route to Pimozide has five steps (Scheme 10.1). In the final step, the one leading to the active principle, an alkyl chloride (1) reacts with a secondary amine (2) in methyl isobutyl ketone at 120 °C to yield Pimozide (3).

Methyl isobutyl ketone has a boiling point of 117 °C, a vapour pressure of 15 mmHg and a TLV-TWA exposure limit of 50 ppm. Glycerol formal and glycerol isobutyral were selected because of their expected superior technical performance. The comparison of relevant EHS properties for the two candidates is summarized in Table 10.5. The lower CO_2 emission potential and higher renewable mass index of glycerol formal prompted us to select this candidate. Glycerol formal is a non toxic glycerol acetal, less flammable than MEK and exhibits superior technical performance as evidenced by the

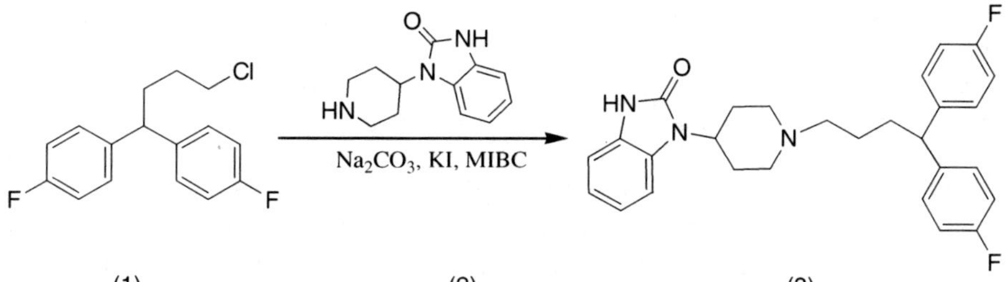

Scheme 10.1

Table 10.5 Comparison of solvent properties and EHS data of two solvent candidates for methyl isobutyl replacement in an S_N2 reaction.

Properties	Glycerol formal	Glycerol isobutyral
Melting point	−59 °C	−49 °C
Boiling point	191 °C	195 °C
CO_2 emission potential	0.42 Tm CO_2/Tm	0.90 Tm CO_2/Tm
Renewable mass index	86%	60%
Health Hazard Level	0	Under evaluation

Table 10.6 Comparison of selected reaction parameters between traditional synthesis using methyl isobutyl ketone and the new glycerol formal route.

Reaction parameters	MEK	Glycerol formal
Molar ratio (1)/(2)	1.2	1.1
Temperature	120 °C	90 °C
Time	6 h	6 h
Yield	80%	97%

improved stoichiometric and volumetric yields and reduced temperature regime[20] (see Table 10.6).

The new process was scaled-up successfully. The yield obtained was similar to that of the classical route and the new process proved to be cost effective. Only considering raw materials is the classical route less economic. The 15% reduction in the overall process cost is due to a higher efficiency of the glycerol formal route and a smaller waste stream generated during the manufacture.

10.4.2 Metal Degreasing

Over the past few years, one of the major challenges in the area of metal degreasing has been the transition from fully emissive open-top systems based on the use of chlorinated solvents to closed-loop metal-degreasing systems based on low-VOC-emission, low-toxicity solvents. Alternative chlorinated solvents such as trichloroethanol, chloroform, methyl chloride, CFC-113, HFCs, HCFCs, CO_2 jets, $scCO_2$, semi-aqueous solvents, alkaline cleaning agents, emulsifying detergent-based cleaners, aliphatic-hydrocarbon-based solvents and azeotropic mixtures have been proposed to replace trichloroethylene, one of the most widely used metal degreasers. However, none of the proposed alternatives fully satisfy the key industrial needs of the metal finishing sector. The ideal process would allow the carrying out of metal degreasing operations in highly variable settings, with metal parts of different sizes and shapes, while minimizing diffuse emission, release of contaminated air during loading and unloading, and solvent release from cleaned metal parts. The process would not generate large waste streams, allowing an easy and cost effective recycling of solvent and rinsing water, and ultimately deliver adequately conditioned parts for immediate use in subsequent steps of the metal finishing process. The SOLVSAFE metal degreasing procedure involves a two-step process in which the metal parts protected with grease are put in contact with a solvent of low volatility and toxicity at room temperature. In a second phase, the solvent is removed by rinsing the metal parts with water. Both solvent and water can be separated and reutilized in the process (Figure 10.6).

Functional efficacy prediction led to the discovery of two solvent candidates with excellent technical performance. Laboratory tests confirmed the *in silico* estimation and found an excellent efficiency when metal parts were degreased by immersion and subsequently rinsed with water.

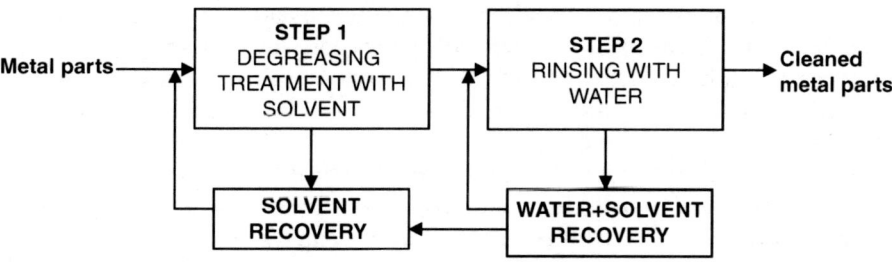

Figure 10.6   Metal degreasing process with SOLVSAFE solvents.

Table 10.7   Degreasing efficacies and EHS parameters for three fatty acid dimethylamides as compared with the current industrial standard trichloroethylene.

Degreasing solvent	*Degreasing efficacy (%)*	*Health hazard[a]*	*Aquatic hazard[a]*	*Vapour pressure at 20°C (hPa)*	*CO_2 emission potential (Tm CO_2/ Tm)*	*Renewable mass index (%)*
CO-01	100	2[b]	0	0.13	0.61	69.2
CO-02	98	2[b]	0	0.06	0.51	74.3
CO-03	81	0	0	0.02	0.44	77.9
Trichloro-ethylene	100	4	4	78	0.67	0
Average value (68 traditional solvents)	–	–	–	87	1.96	0

[a]Level 0 corresponds to negligible hazard. Level 4 indicates maximum hazard.
[b]Predicted values. Health properties under evaluation.

Table 10.7 compares the degreasing efficacy and various EHS properties of three dimethylamides of natural fatty acids with different chain lengths, trichloroethylene, and the average value for traditional solvents. Because of the low vapour pressure, the VOC emission potential is very low and the occupational exposure and flash point reach safer values. As they are based on natural vegetable oils, the alternative solvents exhibit a high renewable mass index.[21]

10.4.3   Paints and Varnishes

The paints and varnishes sector still consumes large volumes of industrial solvents despite the enormous progress made over the past decade to reduce the solvent content in formulations. The SOLVSAFE goal is to evaluate and find new medium- and high-boiling-point solvents or film forming agents (coalescents), with very low environmental impact and very high solvent or coalescence efficiency, for high solid 2-pack PUR coating systems (based on high solid polyols and high solid aliphatic isocyanates) on the one hand, and for waterborne coatings (*e.g.* 2-pack waterborne PUR systems) on the other. The

ultimate goal is to replace existing solvents and coalescents and obtain formulations with higher solid content at application viscosity (high solid coatings), and with less coalescent content (waterborne coatings), in such a way as
to maintain or even enhance their efficacy of function while improving their
environmental profile.

A new series of SOLVSAFE solvents with reduced VOC potential have been
obtained and tested in paint formulations. Coalescent products are employed in
aqueous paints in order to improve film formation and levelling. There are well-
established products on the market that are mainly based on non-renewable
feedstock. Any greener alternative should contribute to impart the desired
properties of the final decorative paint quality, for example the hiding power,
gloss, drying, durability and cost. Laboratory research has found that certain
dialkyl carbonates display excellent viscosity and coalescence properties when
compared with the current industrial standard butyl glycol. Propyleneglycol
monoleate[22] and glycerol acetals and, more particularly, glycerol isobutyral
have been evaluated as greener alternatives to existing coalescent products with
good results. Glycerol isobutyral exhibits similar film-forming properties when
compared with existing well-established industrial solvents and coalescents
such as butyl glycol and 2,2,4-trimethyl-1,3-pentanediol monoisobutyrate
(TMB).[23]

10.4.4 Crop Protection Formulations

The ideal composition of an agrochemical formulation would enable full
exploitation of the biological potential of each active ingredient as a crop
protection or nutrition agent (*i.e.* improving uptake and leaf coverage), with
full respect to the environment, users (mainly farmers) and consumers of
treated crops or processed food obtained from treated crops. The design of
safer organic solvents (SFOS) having the potential to increase plant protection,
bioavailability and reduce environmental impact of existing crop protection
formulations is a primary focus of the SOLVSAFE project. A significant
number of emulsifiable concentrate (EC) formulations have been investigated
using greener alternatives to existing hazardous solvents. Emulsifiable concentrates are liquid compositions containing an agricultural active ingredient in
a liquid form. They usually consist of a single phase. ECs are to be diluted in
water, in order to form a direct emulsion in which a liquid hydrophobic phase
containing the active ingredient is dispersed throughout the water. For example, a farmer would dilute an emulsifiable concentrate comprising a hydrophobic agricultural active ingredient in water, and readily obtain an emulsion
to be applied on a field. This procedure, where the farmer prepares from a
concentrated composition the final product to be applied on a field, is usually
referred to as a "tank mix" procedure. An emulsifiable concentrate is also
referred to as a "tank mix" composition. One of the main advantages of EC
formulations is that the solubilized active ingredient can display a higher
bioavailability which results in an increased efficacy. Insecticides are generally
activated using this kind of formulation. The active ingredients delthametrin,

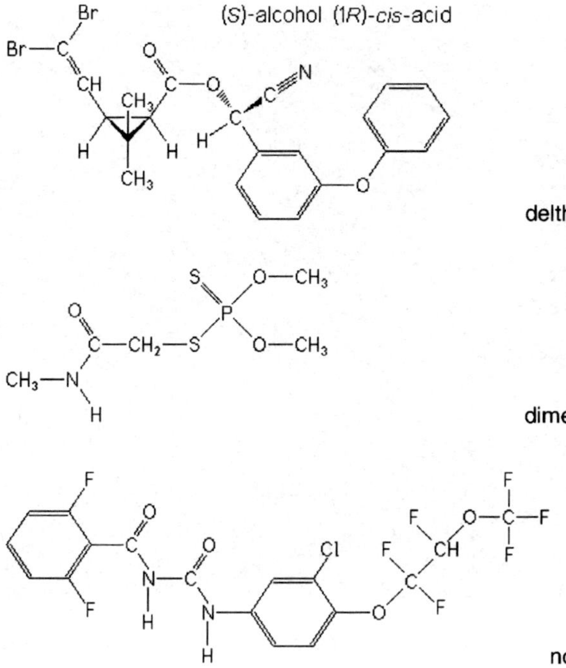

(S)-alcohol (1R)-cis-acid

delthametrin

dimethoate

novaluron

Scheme 10.2

dimethoate and novaluron (Scheme 10.2) are, respectively, representative examples of three important insecticidal chemical classes: pyrethroids, organophosphates and benzoylureas.

Esters, dialkylamides and carbonates were investigated as components of agrochemical EC formulations, obtaining promising preliminary data. Prototypes with acceptable stability were evaluated for their biological efficacy and selectivity under field conditions.

In addition, EC formulations were prepared using a few proprietary experimental herbicides, belonging to the protoporphyrinogen oxidase (PPO) inhibitors (Figure 10.7). Several safer solvents were tested and most formulations resulted in a generally increased efficacy, but only a select number of long-chained lactic acid esters were found to provide a significantly increased selectivity toward crops.[24]

The SOLVSAFE strategy also incorporated the research of innovative water-based micro-emulsions containing green solvents as the lipophilic component. Micro-emulsions are environmentally friendly formulations as water is the main component. The addition of lipophilic substances such as long-chain alkanes (*i.e.* mineral oil), lipophilic esters (preferably, SFOS) or their compositions was shown to increase the magnitude of the expected biological effects. The functional performance of SFOS was tested when micro-emulsions were successfully applied to a number of triazole and strobilurin fungicides.[25] The

Figure 10.7   Post-emergence control of *Gallium aparine* treated with PPO inhibitors.

bioavailability of each active ingredient was magnified when compared to a traditional formulation, as demonstrated in greenhouse tests and field trials. The novel micro-emulsion technology is expected to benefit fungicides, herbicides and insecticides.

10.5   Conclusions

The development of less hazardous and commercially viable solvents, the shift away from inefficient, solvent-consuming chemical routes towards bio-based synthesis and the replacement of oil-based feedstock by renewable starting materials are only a few examples of necessary strategic goals for the twenty-first-century chemical industry. Research efforts over the past decades have evidenced the formidable complexity of designing commercially viable and benign solvents and other functional products. Reconciling non-toxicity with efficacy of function, or environmental performance with industrial operability, to mention just two examples of simultaneously desired and often conflicting properties, are major challenges that need to be addressed in the upfront design stage.

SOLVSAFE, an industrial research consortium under the EU Sixth Framework Programme, is a first attempt at collaborative research between industry, specialized research centres, and academia towards a common research agenda for the development of safer organic industrial solvents.

The innovative SOLVSAFE design strategy, based on novel basic science linking molecular structure with environment, health hazard and chemical functionality, has been very successful in surmounting current barriers to bringing greener solvents to commercial manufacturing facilities and industrial research laboratories. The new solvents display not only good functional characteristics but also fundamental sustainability traits, making them highly competitive in the marketplace.

References

1. M. Kopplin in *Museum für Lackkunst*, BASF Lacke + Farben AG, ed., Münster, *s.a.*
2. R. Höfer, P. Birnbrich, S. Busch and J. Bigorra, *Chimica & L'Industria,*, Marzo 2007, **2**, 118–121.
3. W. H. Brock, *The Fontana History of Chemistry*, Fontana Press, London, 1992.
4. H. Kittel, *Lehrbuch der Lacke und Beschichtungen*, 2. völlig neu bearb. Aufl., Bd. 4, Lösemittel, Weichmacher, Additive, M. Ortelt, Ed. S. Hirzel Verl, Stuttgart, 2007; E. W. Flick, ed., *Industrial Solvents Handbook*, William Andrew Publishing/Noyes, Westwood, 5th edn, 1998; D. Stoye and W. Freitag, *Paints, Coatings and Solvents*, Wiley-VCH, Weinheim, 1998; H. Gnamm and W. Sommer, *Die Lösungsmittel und Weichmachungsmittel*, Wissenschaftl, Verlagsges, Stuttgart, 1958.
5. Study #:1715 (09/2003), http://www.freedoniagroup.com/World-Solvents.html.
6. Spanish Ministry of the Environment, *Report on the Spanish Reference Document for Best Available Techniques in the Organic Fine Chemicals Sector*, 2003.
7. A. Vettori, *European sustainability policy: key elements and drivers*, EBC Ann. Congress, Edinburgh, 2006, http://www.eubuilders.org/DOC/Misc/1%20-%20Andrea%20Vettori.ppt (retrieved 25.11.2008).
8. C. Reichardt, *Solvents and Solvent Effects in Organic Chemistry*, Wiley-VCH, Weinheim, 2003.
9. P. T. Anastas and T. C. Williamson, in *Green Chemistry: Designing Chemistry for the Environment*, ed. P. T. Anastas and T. C. Williamson, American Chemical Society Symposium Series, No. 626, ACS, Washington D.C., 1996, pp. 1–17; P. T. Anastas and J. C. Warner, *Green Chemistry: Theory and Practice*, OUP, Oxford, New York, 2000.
10. http://www.epa.gov/greenchemistry/pubs/pgcc/past.html; P. T. Anastas, W. Leitner, P. G. Jessop, C. Li, P. Wasserscheid, A. Stark, ed., Handbook of Green Chemistry - Green Solvents, Wiley-VCH, Weinheim 2009.
11. P. G. Jessop and W. Leitner, ed., *Chemical Synthesis Using Supercritical Fluids*, Wiley-VCH, Weinheim, 1999.
12. K. R. J. Seddon, *Chem. Technol. Biotechnol.*, 1997, **68**, 351.
13. A. Ogawa and D. P. Curran, *J. Org. Chem.*, 1997, **62**, 2917.

14. N. Jiang, D. Vinci, C. L. Liotta, C. A. Eckert and A. J. Ragauskas, *Ind. Eng. Chem. Res.*, 2008, **47**(3), 627–631; L. Phan, H. Brown, J. White, A. Hodgson and P. G. Jessop, *Green Chem.*, 2009, **11**, 53–59.

15. R. Höfer and J. Bigorra, *Green Chem.*, 2007, **9**, 203.

16. http://www.solvsafe.org.

17. W. Coenen, *Expositionsermittlung und -begrenzung bei Gefahrstoffen : eine Standortbestimmung des Berufsgenossenschaftlichen Instituts für Arbeitssicherheit-BIA = Determination and limitation of exposure to hazardous substances. Definitions of the approach adopted by the Berufgenossenschaftliches Institut für Arbeitssicherheit-BIA*, Münchener Gefahrstoff-Tage, München, 1992, Staub. Reinhaltung der Luft, 1993, **53**(5), 171–176.

18. W. M. Nelson, *Green Solvents for Chemistry: Perspectives and Practice* (Green Chemistry Series), OUP, Oxford, New York, 2003; C. Capello, U. Fischer and K. Hungerbühler, *Green Chem.*, 2007, **9**, 927–934; E. J. Beckman, in *Book of Abstracts, DECHEMA Symposium, Green Solvents for Processes*, Lake Constance, ed. B. Feißt, DECHEMA, Frankfurt, 2006, p. 14; R. A. Sheldon, *Green Chem.*, 2005, **7**, 267–278.

19. D. J. C. Constable, C. Jiménez-González and R. K. Henderson, *Org. Process Res. Dev.*, 2007, **11**, 133.

20. C. Estévez, N. Bayarri and J. Castells, PCT/EP 2007/011420, 2006 (IUCT).

21. J. Bigorra, J. Raya, R. Valls, C. Estévez, L. Galià and J. Castells, EP 08 007 673.0, 2008 (Cognis/IUCT); C. Estévez and J. Bigorra, Comunicaciones, *Jorn. Com. Esp. Deterg.*, 2009, **39**, 93–103.

22. S. Shah, S. Singhal, A. Khan and V. Shah, *Asia Pac. Coatings J.*, June 2005, **18**(3), 31.

23. J. Bigorra, S. Sato, B. Ramiro and E. Graupera, EP 07020568.7, 2007 (Cognis).

24. M. Vanzulli, EP-A 08151742.7, 2007 (Isagro Ricerca).

25. Patent application pending.

CHAPTER 11

Sustainable Solutions for Adhesives and Sealants[1,2]

JÜRGEN O. WEGNER

The ChemQuest Group Inc., Bilker Strasse 27, D-40213 Düsseldorf, Germany

11.1 Introduction

"Sustainability is changing the way products are being designed, produced, and discarded. It will certainly change the way we think about the raw materials that go into these products, and it will certainly have a major effect on the adhesives and sealants industry."[3]

Traditionally adhesives and sealants are typical cross-sectional technologies with myriad applications across all industries. In daily life they play an important active role to enable sustainable solutions in the first place. Passively, through intelligent formulations, they also contribute to minimizing any negative kind of footprints, and maximizing material, energy and process efficiency. This chapter will concentrate on the sustainable formulation of adhesives and sealants themselves, and will only briefly touch on the enormous impact the bonding technology as such has on many sustainability improvements in construction, woodworking, paper and packaging, transportation, aerospace, in labelling, electronics and medical applications. Often enough the adhesive bonding in itself is from a sustainability point of view advantageous over other bonding techniques, be it welding and soldering, be it "classic" mechanical assembly like bolting, screwing, nailing, the impact, however, depends on actual conditions. Energy generation from renewable sources like

RSC Green Chemistry No. 4
Sustainable Solutions for Modern Economies
Edited by Rainer Höfer
© The Royal Society of Chemistry 2009
Published by the Royal Society of Chemistry, www.rsc.org

wind and sun radiation would simply be impossible without appropriate adhesives, adhesive resins and sealants, as are any desired weight reductions of airplanes and cars for gaining better fuel efficiencies, as is triple glazing and building insulation in construction for reducing energy transfer in residential living – just to give a few illustrative examples.

The use of adhesives and sealants is almost as old as mankind. For many millennia adhesives have been natural, based on rubbers, plant and animal glues, and in Mesopotamia 2500 years ago even some boat sealants were already in use based on naturally occurring crude oil tars – an early example of the petrochemical base of today's modern chemistry in general, and the formulation of state-of-the-art adhesives and sealants in particular. Before the invention of macromolecular or polymer chemistry in the 1920s and 1930s all adhesives came from nature, and returned back to it through decomposition at the end of service life. Definitely polymer chemistry disrupted the circle of natural recycling. The use and performance of today's adhesives and sealants would even be unthinkable without the latest insights in polymer science and technology. Progress in macromolecular science and advancements in adhesives and sealants depend upon each other – adhesives are applied polymers. The more demanding a bonding is to take place, the more precise the chemistry at given physical frame conditions has to be adjusted for achieving a decent and reliable job. The adhesives industry constantly strives to lowering the environmental impact of its products. Often sustainability and natural base of materials are used interchangeably, and this chapter will shed some light on the question to what extent, based on current knowledge, modern design of adhesives and sealants can be achieved with technologies from renewable sources. How dependent are adhesive and sealant formulators upon petrochemical raw materials? What prospects and limitations are to be seen in this field? What do we learn from Mother Nature with its abundant examples of bonding techniques? Do we see those techniques one day replacing petrochemical-based, depletable and thus non-sustainable adhesive and sealant alternatives?

11.2 Features and Requirements of Adhesives and Sealants

Adhesives are designed to bond different or similar materials with each other. Adhesives function through well-understood adhesion and cohesion phenomena. They all interact with the surface to be bonded (adhesion), and need to develop sufficient internal strength for bearing the load applied onto bonded parts (cohesion).[4] Adhesives must be liquid at the stage of application for perfectly wetting the surface to be bonded. They must solidify for bearing the load no matter if through chemical reaction and/or through physical means. The only exemption is pressure sensitive adhesives (PSA), which are highly viscous liquids usually spread on a carrier substrate, *i.e.* an adhesive tape or an adhesive label. Despite often-heard marketing claims there is no such thing as a universal adhesive. The contrary is true: the more demanding an adhesive bond relative to dissimilarity of materials, force to be applied onto the bond, chemical and physical compatibility with surrounding conditions during assembly and bond life, assembly speed, initial tack required, the more likely the ultimate formulation requires a very specific adhesive for meeting all requirements. Structural adhesives are one of the fastest growing global market segments.[5] On top of all the other requirements they have to meet very specific norms, safety codes and/or customer specifications that call for a narrow and consistent quality. Fine-tuned precision is paramount, as is specialization. Today two- and multi-component adhesives with well-defined curing characteristics dominate in structural adhesion, as does light, radiation, induction and/or microwave curing that constantly moves the chemical prerequisites to new frontiers.

The development of sealants takes a similar direction from a linseed-oil-based caulk of yesterday to a hi-tech, silicone-based structural sealant of today to a molecular designed, silane-terminated technology of tomorrow. Sealants by definition form a permanently flexible joint between moving parts, they adhere well to the flanks of both parts and fill the ever changing gap dimension at any time during service life. The deformation can be plastic or elastic in nature, with the percentage of elastic recovery in most cases directly correlating to sealant quality. It is an everyday experience that a rubber band perfectly resembles the performance of a sealant: when pulled only slightly, the band fully recovers its shape upon release. However, the more stress is applied, the more the shape changes irreversibly until at a certain point a final rupture occurs, a phenomenon which also depends on age, exposure and temperature. Sealants in construction, transportation and aerospace are permanently exposed to thermal expansion and shrinkage, to vibration, UV and other radiation, to chemicals, oils, greases and cleaning agents, but nevertheless have to endure the lifetime of its host product, or at least a full service period. Preferably, without primering sealants should perfectly adhere to all materials to be sealed. This may be glass, metals, construction elements, wooden planks, tiles and sanitary elements. These materials play an important role in automotive assembly, in marine and aerospace applications. As with adhesives, the quality requirements for sealants are rapidly increasing. They have to meet very high quality hurdles defined by norms and specifications, and in fact we even see a tendency of gradually blurring boundaries between adhesives and sealants caused by an increasing overlap of functionalities. Insulating glazing and photovoltaic elements are just two examples where huge amounts of edge sealants have to assure proper functioning for at least 15 years of guaranteed service life despite heavy exposure to UV radiation and atmospherics plus permanent temperature changes.

As a consequence adhesives and sealants of today and tomorrow truly are hi-tech products designed and optimized for the very specific and narrow area of their use. Their chemical composition must be well defined to meet all technical requirements, and must on top of that fulfil very strict safety, health and environmental prerequisites as usually set forth by local and government authorities.

11.3 Chemical Composition of Adhesives and Sealants Over Time

Until about 1910 all adhesives and sealants were based on "natural" components, most of them resulting from renewable animal and plant sources, the rest representing early usage of chemical by-products like hemicelluloses from paper and cellulose manufacturing, coal tars from gas production, plus naturally occurring bitumen. This picture dramatically changed in the 1920s with the then booming rubber industry, new discoveries in macromolecular science and technology, dramatic progress in the chemical industry as a whole and finally the intensive use of crude oil, natural gas and coal as a rising feedstock for a vast variety of petrochemical technologies. Figure 11.1 shows the relative share of adhesives and sealants based on renewable resources over the past 80 years including a projection into 2020.[6] It shows that especially in the 1950s and 1960s a sharp decline occurred coupled with a steep rise in adhesives and sealants tonnage: whereas the absolute volume of "natural"-based renewable adhesives remained fairly constant over most of the time, it was almost all growth in tonnage (and even more dominant in value) clearly associated with petrochemical-based raw materials.

Today all formulated adhesives and sealants combined represent a global tonnage of 8.5 Mio mto (dry, annually), and are projected to grow to 10.2 Mio mto by 2010.[7] These figures do not include typical binders for paper manufacture and for use in the woodworking industry. From this tonnage a mere 15.5% (2006), respectively 15.0% (2010), represent adhesives and sealants based on natural origin with or without additional chemical modification.

With the volume ratio of adhesives and sealants remaining fairly constant over time in the 80-85 : 15–20% range[2,8] the amount of crude-oil-based

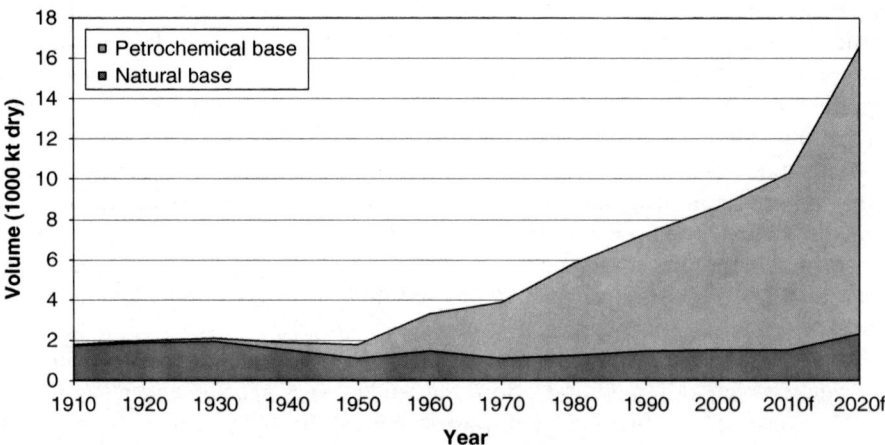

Figure 11.1 Total worldwide production of adhesive and sealants including share of natural-based components.

formulations may look quite immense with 7.2 Mio mto in 2006 and 8.5 Mio mto in 2010.

However, this represents just 0.006% of all oil consumed per annum even if this amount has been doubled by factoring in the material and energy needs for processing the base ingredients for adhesives and sealants, and also taking into account the difference between "dry" and "wet" adhesives as representing on average a factor of 3.

Considering a global plastic production of 245 Mio mto (2006),[9] petro-chemical-based adhesives and sealants represent just 3% of this. "Dry" adhesive in this sense is the solid portion of an adhesive governing its overall performance properties, whereas "wet" adhesives describe the ready-to-use product in its average proportion of solid (34%) and liquid (66%) ingredients. The liquid phase commonly and increasingly is water, but 0.71 Mio dry mto (2006), respectively 0.78 Mio dry mto (2010), remain solvent- and thus crude-oil-based,[10] a sizable factor that has to be regarded as well.

The following graphs (Figures 11.2 to 11.5) inform about the global usage of adhesives relative to area of use, base chemistry, regional spread and adhesives technology.

11.4 Ongoing Sustainability Evolution

Considerations of safety, health and environmental impact always played a governing role in formulating adhesives and sealants, and one of the greatest challenges was and remains to be the replacement of questionable ingredients against those with a proven record without sacrifice in technical performance, or even by improving both simultaneously, the sustainability and technical profile as well. Adhesive formulators invest heavily in R&D to prove that benign sustainability does not equal inferior performance. In fact proactive anticipation and compliance with ever more demanding regulations is one of

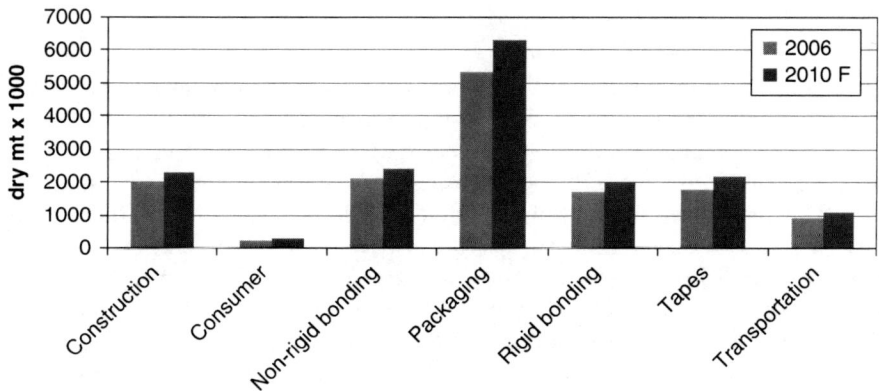

Figure 11.2 Global adhesive volume per use area.

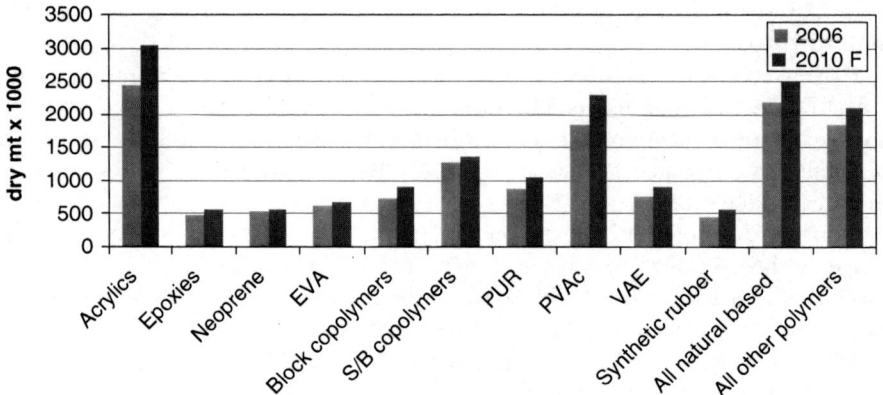

Figure 11.3   Global adhesive volume per base chemistry.

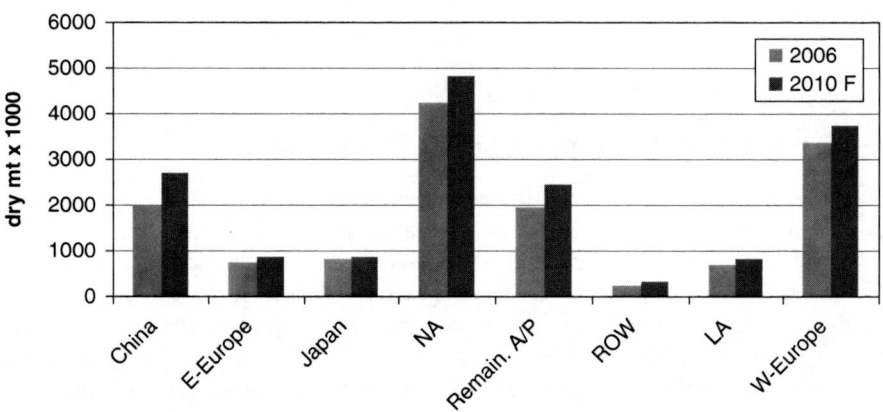

Figure 11.4   Adhesive usage per region.

the most important sources for innovations in this industry, and ranks second next to solving or improving so far unmet customer needs.

The replacement of solvent-based adhesives by either water-based or full solid systems is just one of those examples. Solvent-based adhesives usually are proven technologies with a robust usage profile, whereas their rather sophisticated, water-based counterparts are sensitive to low-temperature and high-humidity applications, require film-forming (often VOC-based) agents themselves, and are usually 20–30% more expensive per adhesive bond. In any case the switch from solvent to water requires drastic usage changes at higher costs, one of the reasons why solvent-based adhesives show negative growth in Western Europe and North America, but nevertheless grow globally with 2.2% relative to 4.2% for adhesives in general[11] due to local production in emerging

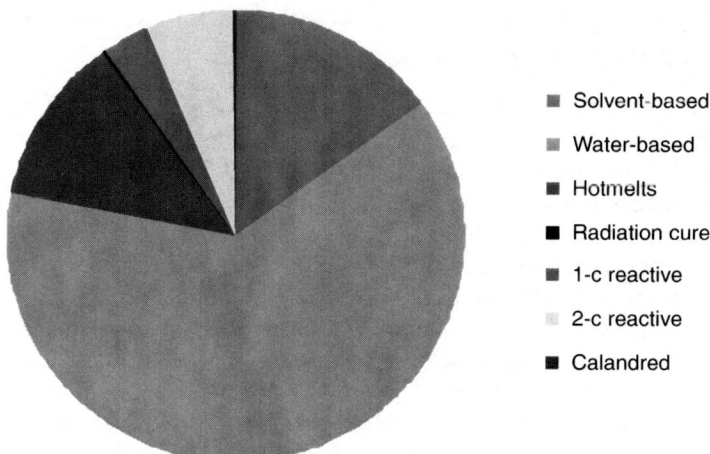

Figure 11.5 Volume spread per adhesives technology (in 2006).

economies. No doubt this is an area where the enforcement of government-imposed restrictions could help to drastically reduce the use of solvents in adhesives on a global scale, and thus to cut back the relevant environmental footprint. The aim of the adhesives and sealants industry should be to limit the use of solvents to the utmost necessary, *i.e.* where in weld bonding of plastic pipes the solvent plays the role of the adhesive.

Hand in hand with the replacement of solvents goes the development of full-solid adhesives and sealants, and so called hotmelt adhesives are just one of those technologies progressing well over the last two decades. Usually hotmelts are completely emission-free, and represent an almost ideal sustainability profile. They are applied above melting temperature, and develop their bonding strength immediately with cooling. Thermoplastic by nature they de-bond with elevated temperature which might be desirable or disadvantageous. Reactive hotmelts overcome those difficulties as they eventually after application become thermoset through cross-linking reactions, and represent one of the fastest growing adhesive specialities with numerous fields of usage in woodworking and packaging, in electronics, bookbinding and transportation.

In general the quality evolution of adhesives and sealants over the past decades is characterized by numerous examples of technological steps that all have in common the relative reduction of ecological footprint coupled with meeting ever higher safety standards at affordable costs. Be it the drastic reduction of monomers, the ban of heavy metal catalysts and tin-organic compounds, the replacement of formaldehyde-containing binders against other polymer systems, the substitution of solvents as described above, the responsible use of plastification technologies or the drastic reduction of VOC, all these examples prove full commitment of the adhesives industry towards a sustainability level as high as possible.

11.5   Quality Features and Gaps with Natural-based Adhesives and Sealants

Adhesives and sealants based on renewable sources show a number of advantages, but also major disadvantages that effectively limit their broader use at present. On the positive side they usually show an excellent bio-compatibility given that their chemical modification did not completely alter this. Natural-based products in general show a positive Greenhouse Gas (GHG) balance, and thus do not directly contribute to climate change. They show a fairly good price/performance ratio, usually are available in sufficient quantity and show no or low toxicity.

However, natural-based materials like starches and soybean oils are used in direct competition to animal or human food consumption. Technically, they are not at all suitable for high modulus adhesive and sealant joints. They usually are sensitive to hydrolysis and oxidization. Exposure to atmospheric conditions or chemical attack can drastically limit their service life expectancy. They are sensitive to elevated temperatures and need to be protected against microbiological decay with more or less toxic preservatives. In light of those features and based on today's knowledge some important conclusions can be drawn:

- For the time being and in the foreseeable future high performance and so-called reactive adhesives and sealants will rely on petrochemical key components;
- There is on the other hand room for stopping the downturn trend or a further decline of adhesives and sealants based on renewable resources;
- Structural adhesion and structural sealants are important growth markets in themselves and through increasing replacement of physical joining techniques like bolting, welding and soldering, they often require chemical or physical means for the desired curing, and can by no means be formulated based on natural key components except functional moieties like polyurethanes based on oleochemical polyols, dimer acid based polyesters or blends of natural and synthetic rubber in the tyre industry, but the cross-linking reaction in all these cases is all but "natural";
- In low to medium modulus applications of sealant caulks and adhesives, however, where curing takes place through physical solidification, *i.e.* due to cooling and/or drying, the use of raw materials based on renewable sources is quite common, and can certainly be expanded with focused research for overcoming inherit shortcomings.

11.6   Current Use of Renewable Raw Materials in Adhesives and Sealants

Table 11.1 shows an overview reflecting the actual state-of-the art.[11] We see that "natural"-based key components often have a direct impact on product and quality performance criteria.

Table 11.1 Typical quality features when using natural-based components in adhesives, foams and sealants.

Base material	Typical use area	Special technical feature
Starches & Dextrines	Paper & packaging, labelling, wallpaper	Remoistenable, low-to-no toxicity, bonding speed
Cellulose ethers	Tiling, construction, wallpaper	Quick hydration, long open time, flow, workability
Tall resins & colophonium derivatives	Hotmelts, flooring adhesives	Extender, tackifier
Caseines	Labelling, levelling compounds	Cold water resistance, flow
Oleochemical	PU foams	Homogene & uniform foam lamelles
Dimeric fatty acids	PE and PA-based hotmelts, epoxies	Flexibilizing extenders

Source: The ChemQuest Group Inc.; own market survey.

Table 11.2 Global status for use of natural-based adhesives (wet mto, average solids 50%).

Natural resin base	2006	2010	AGR (%)
Natural rubber	455	502	2.5
Starches/dextrins	1668	1892	3.2
Natural tree resins	17	16	–0.1
Resin emulsions	15	15	1.1
Cellulose ethers	40	44	2.4
Total natural based adhesives	2195	2469	3.0
Total wet adhesives natural and petrochemical based	14120	16608	4.1
Total share of natural-based adhesives	15.50%	15%	

Source: The ChemQuest Group Inc.; proprietary database.

11.7 Major Use Areas for Adhesives Based on Natural Resources

Table 11.2 gives a survey about major use areas for adhesives based on renewable sources. The share of renewable raw materials in the sealants market is well below 5%. Numbers in Table 11.2 do not include polyurethanes containing oleochemical-based polyols, vegetable oil-based polyesters and polyamides based on dimer fatty acids in total estimated at 35 mto in 2006, 42 mto in 2010. Projection into 2010 shows that without disruptive innovations the share of natural-based adhesives will further decline to a mere 15% in total.

11.8 Outlook and Conclusion

Research teams around the world are currently studying adhesive strategies and bonding techniques developed in the biosphere by plants and animals. The way molluscs attach themselves to the hull of a ship, the way geckos and spiders walk through bonding and de-bonding on-command, even upside down against gravity on low-energy surfaces, are striking examples where in nature very sophisticated bonding techniques have emerged through evolution. Although quite fascinating from a scientific point of view, the practical value of biological insight translated into everyday adhesives and sealants is very limited so far. We are particularly optimistic about new discoveries in protein chemistries, and also believe in new technological routes of producing adhesive raw materials *via* bio-engineering techniques. This will help to produce them by using "green" chemistries for better efficiency with a lower carbon footprint at a much improved sustainability profile.

But undoubtedly the major sustainability impact of adhesives and sealants lies today and in the future in their direct usage as an enabling bonding technology. Relative to their indirect impact, estimates range from 100 to several 1000 times. Adhesives and sealants contribute through their sheer existence to sustainable solutions in all walks of life, ranging from medical to transportation to construction, electrical and electronic, packaging and woodworking applications. Through higher performance the specific amount of an adhesive to form a reliable bond consistently declines and this also has to do with ever more precise dosing of application machinery. Adhesives are made to adhere, and sealants to seal, with no compromise whatsoever in their expected reliability. At present bio-based or natural-based adhesives in their specific performance are at best second class especially under demanding or critical conditions, and clearly are outperformed by petro-chemical-based alternatives. One day this may change, however, but for the foreseeable future we will have to live with them, and benefit from the much larger sustainability impact they indirectly offer through their almost limitless usage.

References

1. Contains part of a presentation titled *Adhesives Based on Renewable Sources*, held by the author at the World Adhesives Conference in Miami, April 2008.
2. Daniel S. Murad, *Adhesives & Sealants Industry*, July 1, 2008.
3. E. M. Petrie, *Adhesives and Sealants*, Editorial, Oct 1, 2008.
4. G. Gierenz and W. Karmann, *Adhesives and Adhesive Tapes*, Wiley-VCH, Weinheim, 2001.
5. *World Adhesives Database* of The ChemQuest Group Inc., partly accessible *via* www.chemquest.com.
6. ASC, FEICA, IVK data, compiled by The ChemQuest Group Inc.
7. Global Market Forecast conducted by The ChemQuest Group Inc., Cincinnati.
8. Estimation of the author.

9. S. Marcinowski, *Renewable Raw Materials – a Novel Approach in Polymers*, Bio & Polymers, Biannual Meeting of the GdCh-Division of "Macromolecular Chemistry", Aachen, 2008, http://www.gdch.de/vas/tagungen/tg/5325__e.htm; Anonymous, ...
10. ChemQuest Survey.
11. From a presentation the author held at the European Coatings Show 2007, Nuremberg/Germany titled *Drivers of growth for the worldwide adhesives business*, partly published in Adhäsion, 2007, **718**, 14–19 and in *European Coatings Journal*, 2007, **6**, 34–46.

CHAPTER 12

White Biotechnology

THOMAS HAAS,[a] MANFRED KIRCHER,[a] TIM KÖHLER,[b] GÜNTER WICH,[c] ULRICH SCHÖRKEN[d] AND RAINER HAGEN[e]

[a] Evonik Degussa GmbH, Creavis Technologies & Innovation, Paul-Baumann-Straße 1, D-45772 Marl, Germany; [b] Evonik Goldschmidt GmbH, Goldschmidtstraße 100, D-45127 Essen, Germany; [c] Wacker Chemie AG, Consortium für elektrochemische Industrie, Zielstattstraße 20, D-81378 München, Germany; [d] Cognis GmbH, Henkelstraße 67, D-40589 Düsseldorf, Germany; [e] Uhde Inventa-Fischer GmbH, Holzhauser Straße 157–159, D-13509 Berlin, Germany

12.1 The Status of White or Industrial Biotechnology

12.1.1 Introduction

White Biotechnology and industrial biotechnology are synonymous. According to EuropaBio – the European Association for Bioindustries – "industrial biotechnology is the application of biotechnology for processing and production of chemicals, materials and energy. It uses enzymes and micro-organisms to make products in sectors such as chemicals, food and feed, paper and pulp, textiles and energy. By using biotech processes, all these sectors can make significant contributions to mitigate climate change. Industrial biotechnology can help prevent pollution and offers new ways to produce goods and services. Environmental benefits are often combined with increased economic efficiency, leading to cost savings in the production process while the production output and quality remains the same or increases. In summary, industrial

RSC Green Chemistry No. 4
Sustainable Solutions for Modern Economies
Edited by Rainer Höfer
© The Royal Society of Chemistry 2009
Published by the Royal Society of Chemistry, www.rsc.org

biotechnology often delivers environmental and economic benefits at the same time."

The relevance of White Biotechnology is well documented. OECD, the Royal Belgian Academy Council of Applied Science, Dechema as well as US-NREL and US-DOE, outlined White Biotechnology's innovative power and future potential.[1–6] The worldwide volume of white biotech production sums up to an estimated €50 billion – close to biotech's pharma product volume of $55 billion.[7]

All biotechnological processes depend on enzymes – either in whole-cell living systems or isolated out of their biological context. In fact, purified enzymes are established in processing food and textiles and are supplements in feed and detergents – all products of everybody's daily life. Whole-cell systems particularly in the field of specialty chemicals provide a broad range of methods and processes. For many years, enantiomerically pure amino acids for food, feed and pharma industries have been produced by microbial fermentation. Ambitious R&D resulted in enzymes especially modified and optimized for a desired chemical reaction such as biocatalysis of optically active amines, alcohols, epoxides and more.[8]

12.1.2 Relevant Market Segments

12.1.2.1 *Enzymes in Food, Feed and Households*

The most powerful driver of White Biotechnology is its ability to perform specific biochemical reactions synthetic chemistry is not able to provide. An example is White Biotechnology's first product – a protease – which was launched in 1909 by Röhm & Haas in Darmstadt. Its application is digesting tissue residues from animal skin when starting the tanning process. Still today enzymes are an inevitable tool in the tanning industry. Enzymes captured more and more industrial fields – either as a processing tool like this example from tanning or as the active agent itself. Processing of food today depends strongly on enzymes: 8 Mio mto of liquid sugar are produced annually mainly for the beverage industry by splitting corn starch by amylases. Pectinases clear fruit juices and the production of cheese depends on chymosin and more enzymes. In detergents a mix of especially selected and optimized proteases, lipases and amylases not only remove protein, fat and starch stains but also allow for washing laundry effectively at a temperature of only 30 °C. Current R&D efforts target even room temperature by further lowering the laundry enzyme mix's optimum temperature while keeping its activity. This example demonstrates how biotechnological processes deliver products of high functional quality and at the same time save energy significantly. Like this one, many biotechnological processes and products help in minimizing the ecological footprint.

Another example from a significant but often overlooked market is the feed additive industry. Phytase is an enzyme added to feed enabling swine and poultry to utilize the typical plant phosphate storage compound phytic acid because these animals do not express the digestive means of splitting phytic acid. Therefore providing phytase with feed has an economical and especially an ecological impact: The farmer saves adding phosphate to the feed and – even more important – the phosphate burden of liquid manure from undigested phytic acid is significantly reduced. More applications of enzymes in different markets are shown in Table 12.1.

Table 12.1   Use of technical enzymes in diverse applications.

	Products from Enzymatic Catalysis			
Product/Process	Annual Production mto a^{-1}	Price EUR/ kg	Application	Producer
Chemicals				
Acrylamid	100,000	1,4	Polymers	Mitsubishi Rayon
Amino Acids				
L-asparaginic acid	13,000		Aspartam	Tanabe Seiyaku, DSM
L-methionine	400	20	Pharma	
L-Dopa	300		Pharma	Ajinomoto
L-alanine	500		Pharma	Tanabe Seiyaku
D-, L-valine	50		Pharma	DSM
L-tert. Leucine	10	500	Pharma	Evonik Industries
L-Carnitin	200			Lonza
ß-phenylalanine	1			Dow Pharma, Evonik Industries
Food				
Glucose	20,000,000	0,3	Liquid sugar	
Isoglucose	8,000,000	0,8	Liquid sugar	
L-malic acid	100	20	Acidification	Tanabe Seiyaku
Isomalt	70,000		Sweetener	Südzucker, Cerestar
Aspartam	10,000		Sweetener	HSC
Fructooligosaccharides	10,500	2,5	Prebiotic	Beghin Meji, Orafti, Sensus, Cosucra
Antibiotic-derivatives				
6-APA	10,000			Kaneka, DSM, Novartis
7-ACA	4,000			Novo, Novartis
D-4-Hydroxyphenylglycin	7,000			Kaneka, Recordati
Intermediates, chiral products				
(S)-2-cloropropionic acid	2,000			Herbicides, Avecia
D-pantolacton	2,000			Pharma, Fuji Pharma
(S)-Methoxyisopropylamin	1,000			Herbicides, BASF
Mandelic acid	200	20	Pharma	Mitsubishi Rayon, BASF
Ethyl (S)-4-clor-3-hydro-xibutyric acid	150		Pharma	Daicel Chemical Industries, Rütgers
Ephedrin	1,500	90	Pharma	BASF

Source: Dechema Positionspapier, Frankfurt, 2004; personal communication[3]

12.1.2.2   Cosmetic Ingredients

Also in the field of cosmetics White Biotechnology unfolds its ecological and economical benefits. A successful example is Evonik Industry's enzymatic process to emollients. Emollients are functional esters in skin care products giving skin a soft and comfortable feeling. Conventionally they are produced by chemical synthesis through a multi-step process which needs a lot of energy input and produces volatile, aqueous and solid waste. In contrast the new enzymatic catalysis based on lipase needs only one step. It saves energy, avoids waste and leads to superior product quality (Figure 12.1).

Obviously the enzymatic process is beneficial concerning the ecological balance. Energy input, side products and emissions are reduced or avoided completely. By simplifying the production plant to only a few steps fixed and variable costs drop down.

More cosmetic ingredients or precursors thereof are produced by fermentation. The example of sphingolipids is discussed in more detail in Chapter 12.2.1.

12.1.2.3   Fine and Specialty Chemistry: Amino Acids

Another innovation was added to White Biotechnology when in 1956 Kyowa Hakko selected *Corynebacterium glutamicum* for large-scale fermention of

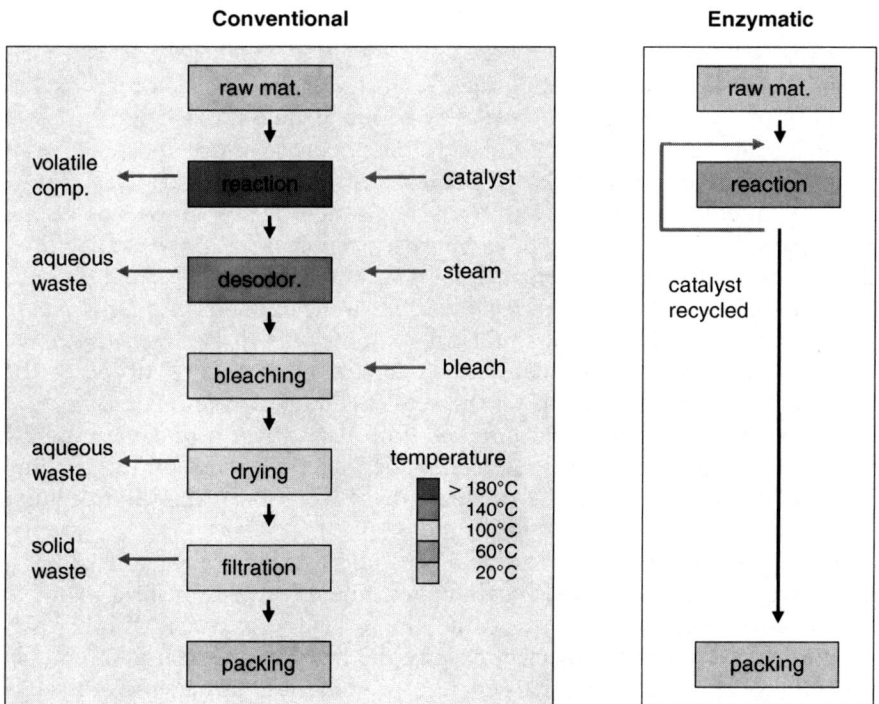

Figure 12.1   Conventional and enzymatic production of emollient esters.

L-glutamic acid. Today 1.5 Mio mto a^{-1} of this amino acid are produced worldwide and predominantly used in food as a flavour-enhancing additive. The essential amino acids L-lysine, L-threonine and L-tryptophan are produced by fermentation for the feed additive market. In 2004 worldwide 770,000 mto a^{-1} L-lysin and 65,000 mto a^{-1} L-threonine were produced by bacterial fermentation. According to Ajinomoto sales of feed additive amino acids grow up to 9% per year. Today Evonik Industries produces feed grade L-lysine, L-threonine and L-tryptophan by large-scale fermentation[9] and produces pharma-grade L-alanine, L-aspartic acid, L-isoleucine, L-proline, L-valine and L-tryptophan for applications in the pharma markets. More products from fermentation are shown in Table 12.2.

Amino acid fermentation is predominantly based on *Corynebacterium glutamicum* and *Escherichia coli*. *Corynebacterium*'s decisive advantage is the relatively easy to manipulate anabolic network. *E.g.* the metabolic regulation of *Corynebacterium*'s lysine biosynthesis is in contrast to other species quite simple. By manipulating only a few key steps the precursor flow is directed to lysine (Figure 12.2). Aspartate kinase is the key enzyme to enhance and deregulate the intermediate flow towards lysine and threonine. Furthermore reducing homoserine dehydrogenase pushes the carbon flow towards lysine. However, developing competitive high-performance strains requires optimizing the whole complex metabolic network and de-bottlenecking of lysine biosynthesis itself including all precursors starting from the carbon source uptake.

Corynebacterium actively excretes amino acids through its cell wall membrane and does not degrade L-lysine due to the lack of lysine-decarboxylase. For 60 years all these characteristics have made this microbe the species of choice in L-lysine production. In addition it demonstrates the potential of natural biosynthesis pathways for commercial purposes. In contrast *Escherichia coli* entered the field of industrial amino acid fermentation not because of comparable advantages provided by nature but because of the availability of effective tools for genetic engineering. In the early 1980s such methods were state of the art for *Escherichia coli* but were only on an infant level for *Corynebacterium*. Developing industrial strains based on *Escherichia coli*, which at that time was not broadly covered by intellectual property (IP), provided room to build new IP in the field of amino acid fermentation.

This example of amino acid fermentation demonstrates that the establishment of a species as an industrial production system depends on three factors: (i) suitability given by nature, (ii) availability of R&D tools and (iii) freedom to operate.

As already described for phytic acid, also adding amino acids to feed has an economical as well as an ecological impact. The most relevant feed plants are soy and corn. Their profile of essential amino acids – and that of other plant materials as well – does not meet exactly the livestock's essential amino acid demand profile. Since the animals eat feed-protein until the demand of the most limited amino acid is filled the remaining overflow of other amino acids cannot be transformed into body mass. It is excreted. Consequently liquid manure of

Table 12.2 Products produced by fermentation.

	Fermentative Products			
Product/Process	Annual Production mto a^{-1}	Price EUR/kg	Application	Producer
Amino Acids				
L-glutamate	1,500,000	1,2	Flavor Enhancer	Ajinomoto
L-lysine	700,000	2	Feed	Evonik Industries, Ajinomoto, ADM
L-threonine	30,000	6	Feed	Evonik Industries, ADM
L-phenylalanine	10,000	10	Sweetener, Pharma	DSM
L-tryptophane	1,200	20	Feed-, Food	Ajinomoto
L-arginine	1,000	20	Pharma, Cosmetic	Kyowa Hakko
L-Cystein	500	20	Food, Pharma	Wacker
Acids				
Lactic acid	150,000	1,8	Food, Leather, Textiles	
Gluconic acid	100,000	1,5	Food, Textiles, Metall, Construction	Jungblunzlauer
Citric acid	1,000,000	0,8	Pharma, Food, Metal, Laudry	
Acetic acid	190,000	0,5	Food	
Itaconic acid	4,000		Comonomer	Cargill
Enzymes				
Total	3,000,000			Novozymes, Genencor
Laundry				Novozymes, Genencor, Henkel
Food processing				Novozymes, Genencor
Textile-, leather-processing				Novozymes, Genencor
Feed additive				Novozymes, Genencor, BASF
Antibiotics				
Bacitracin A	4	3.000,00	Organ transplatation	
Cyclosporin	3	5.200,00	Pharma, Feed additive	
Penicilline	45,000	300		
Biopolymers				
Polylactid	140,000	2,25	Food, Oilproduction	Nature Works
Xanthan	40,000	8,4	Blood substitute	Evonik Industries
Dextrane	2,600	200		Novozymes
Hyaluronic acid	500			
Vitamins			**Feed**	
Riboflavin (B2)	30,000		Feed	BASF, DSM
Cyanocobalamine (B12)	20	25.500,00	Feed	
Vitamin C	80,000	8		
Lipids				
Phytosphingosine			Cosmetics	Evonik Industries
Energy				
Ethanol	39,000,000	0,35	Energy	ADM, Brazil Sugar Refineries

Source: Dechema Positionspapier, Frankfurt, 2004; personal communication[3]

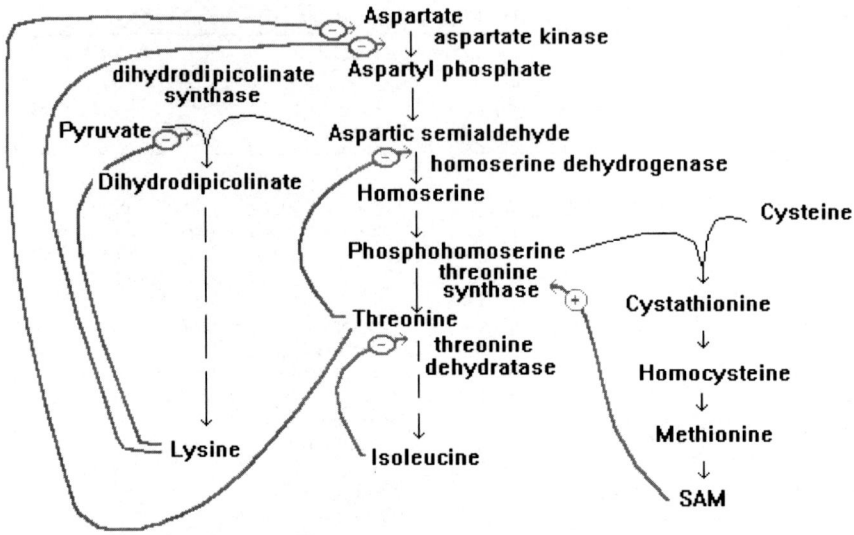

Lysine plus threonine severely inhibits growth of maize in a synergistic manner. Growth inhibition could result from combined effects of lysine on aspartate kinase and threonine on homoserine dehydrogenase, resulting in starvation for methionine. Growth inhibition by lysine + threonine can be overcome by supplying methionine. Bryan (1980) & Miflin (1977).

Figure 12.2   Metabolic scheme and key steps in lysine biosynthesis.

cattle bred on pure plant feed carries a high burden of nitrogen. Adapting the feed's amino acid profile to the animal's need by adding essential amino acids therefore reduces the manure's nitrogen burden significantly.

L-Cysteine – another amino acid produced by fermentation will be described in more detail in Chapter 12.2.2.

More essential amino acids are produced by fermentation for the pharmaceutical markets – as ingredients in infusion solutions or as a precursor for pharmaceutical amino acid derivatives. The latter requires extremely ambitious bio-catalytic processes including co-factor regeneration. Evonik Industries developed a process to the pharma-building block L-tert.leucine based on formate-dehydrogenase and NAD+-regeneration (Figure 12.3). Today this process runs on a tons scale.

12.1.2.4   Monomers

12.1.2.4.1   Diamines and Diacids.   In addition to the diverse applications of amino acids even products of their degradation are also of industrial interest. 1,5-diaminpentane or cadaverine results from decarboxylation of L-lysine.

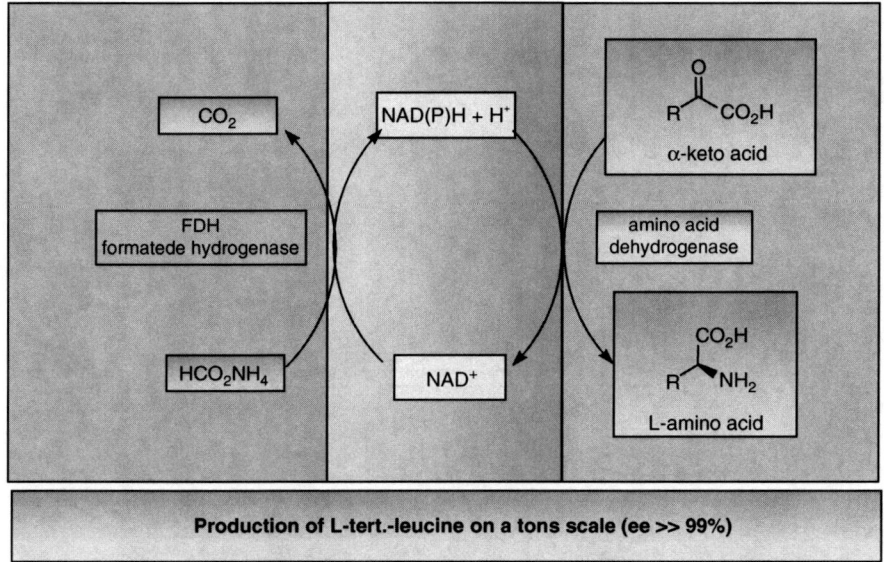

Production of L-tert.-leucine on a tons scale (ee >> 99%)

Figure 12.3 1-Step bio-catalysis including co-factor regeneration.

Cadaverine is only a representative example of diamines and related diacids which are the building blocks of polyamides – polymers currently produced synthetically from fossil carbon sources. Based on industrial lysine producing strains of *Corynebacterium glutamicum* a fermentative process to cadaverine starting from sugar has been developed.[10] A comparable biotechnological alternative might become reality for dicarboxylic acids providing the possibility to produce polyamides completely from renewable feedstock.

Lactic acid is, besides cadaverine, another example of easily accessible microbial intermediates which provide the potential for monomers leading to polymers and plastic materials. It is also an example of biotechnology's disruptive power penetrating and modifying traditional chemical processes: in contrast to pure biotechnological production of amino acids described earlier these new processes combine biotechnological and chemical steps. The monomer or its precursor is produced by fermentation followed by chemical synthesis and polymerization. Poly-lactic acid (PLA) is a prototypical example of such type of new polymers based on sugar. In 2001 Cargill (NatureWorks) set a 140,000 mto a^{-1} PLA plant on stream. Today PLA is applied in quite different markets from packaging, benches and clothing. It will be discussed in more detail in Chapter 12.2.4.

More bio-based polymers are in the chemical industry's pipeline for all performance specifications from commodities and engineering up to high-performance polymers. Biotechnology restarted the innovation cycle which stopped for petrochemical polymers 15 years ago (Figure 12.4).

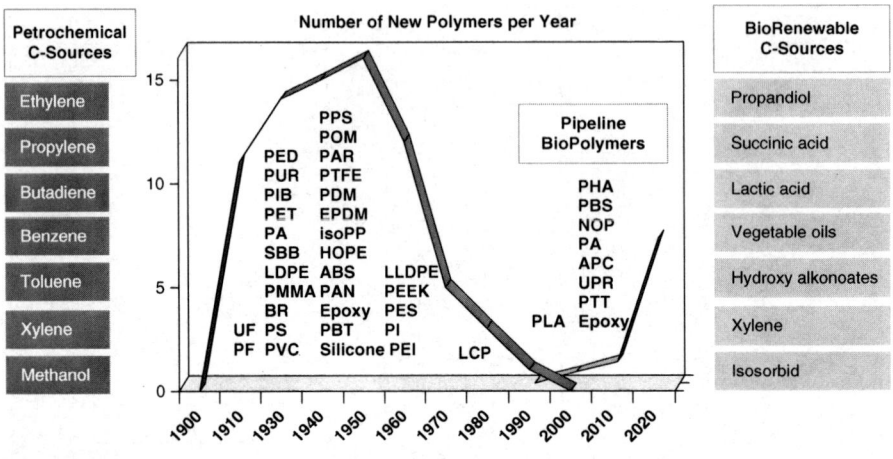

Source: DSM

Figure 12.4 Innovation in monomers and polymers.

12.1.2.4.2 1,3-Propanediol, PDO. Whereas all examples mentioned so far are based on natural biochemical pathways, biotechnology really enters a new level in industrial application when synthetic pathways engineered towards a specific product are tolerated and expressed by living systems. In White Biotechnology recently the first successful example in microbial production of 1,3 PDO has been reported. 1,3 PDO is the bio-based building block for producing DuPont's polymers Sorona® (Figure 12.5) and Cerenol® (Figure 12.6), both new polymers with a promising market potential. DuPont claims a broad range of applications for PDO-formulations and its PDO-based polymer Sorona® and PDO-based polyols Cerenol® in automotive refinishing, inkjet inks, personal care, lube and greases, elastic fibres and a variety of articles for automotive and sporting end uses (Figure 12.7).

The future market success depends decisively on the polymers' performance where *e.g.* Cerenol® is superior to PLA and starch-based polymers concerning the strength versus stiffness index (Figure 12.8).

Also according to DuPont Cerenol® shows value-adding characteristics which give it a competitive advantage in comparison to conventional

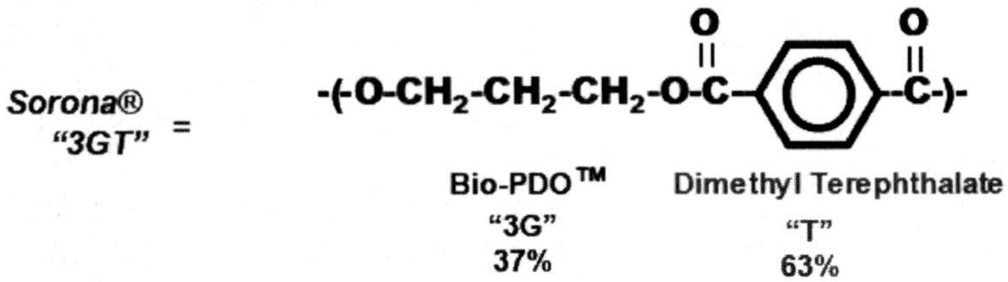

Sorona® = -(-O-CH$_2$-CH$_2$-CH$_2$-O-C-⬡-C-)-
"3GT"

Bio-PDO™   Dimethyl Terephthalate
"3G"   "T"
37%   63%

Figure 12.5 Sorona®.

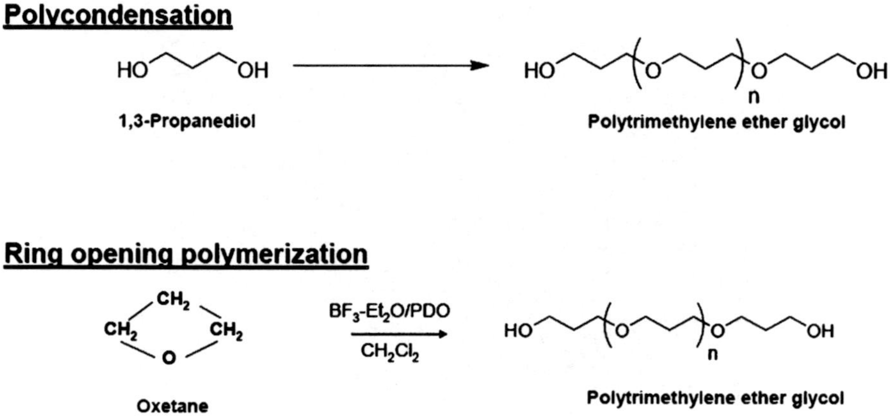
Polycondensation
HO OH
1,3-Propanediol
HO O O OH
n
Polytrimethylene ether glycol
Ring opening polymerization
CH2
CH2 CH2
O
Oxetane
BF3-Et2O/PDO
CH2Cl2
HO O O OH
n
Polytrimethylene ether glycol

Figure 12.6 Cerenol®.

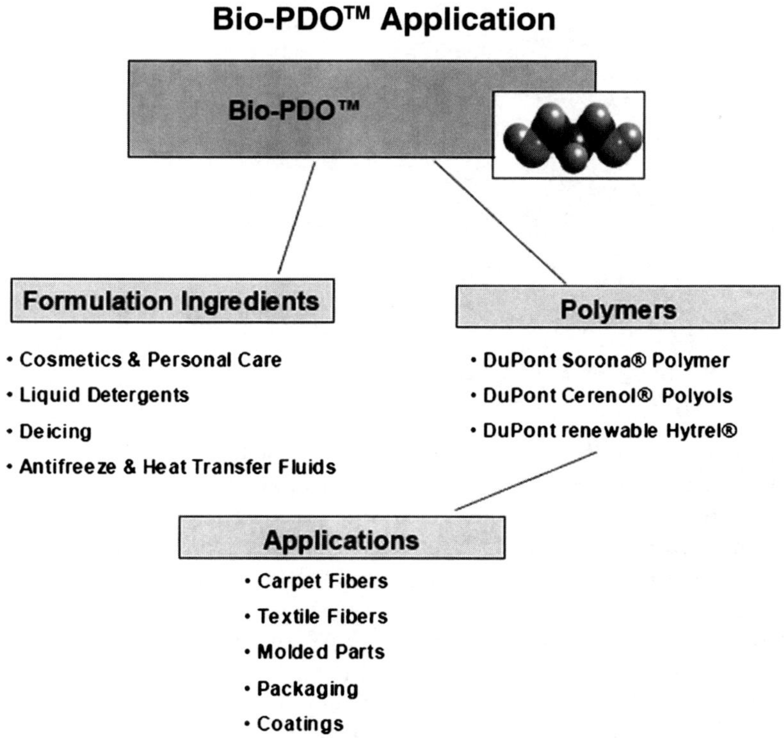
Bio-PDO™ Application
Bio-PDO™
Formulation Ingredients
• Cosmetics & Personal Care
• Liquid Detergents
• Deicing
• Antifreeze & Heat Transfer Fluids
Polymers
• DuPont Sorona® Polymer
• DuPont Cerenol® Polyols
• DuPont renewable Hytrel®
Applications
• Carpet Fibers
• Textile Fibers
• Molded Parts
• Packaging
• Coatings

Figure 12.7 Applications for PDO-formulations and its PDO-based polymer Sorona®.

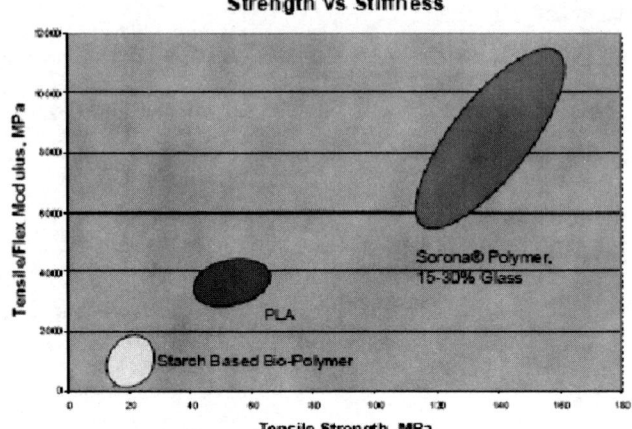

Figure 12.8 Performance profile of Sorona[R].[11]

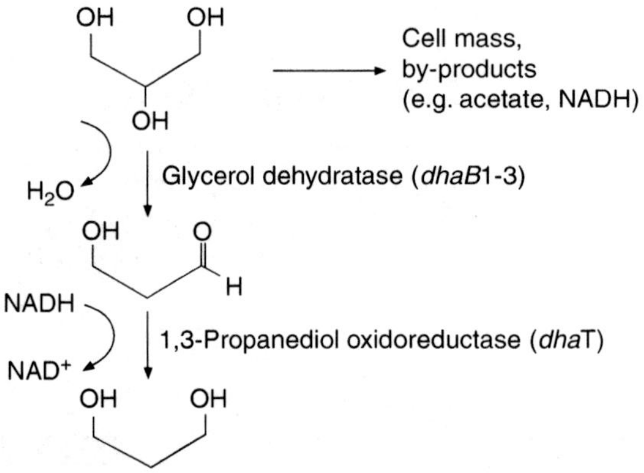

Figure 12.9 Biosynthesis of 1,3 PDO from glycerol.

polyols in low volatility, non-flammability, high flash-point, less reactivity, low toxicity, high oxidative stability and overall easier handling, processing and transporting.[12]

The new synthetic pathway to produce 1,3-PDO from glucose is based on intracellularly formed glycerol. This glycerol is converted to the product (Figure 12.9).

Citrobacter, Chlostridium, Enterobacter, Klebsiella and Lactobacillus species use this pathway when growing anaerobically on glycerol. However, it runs with low yield and does not allow producing 1,3-PDO when growing aerobically on D-glucose. To overcome these limitations pathways from several species have been combined in one host strain. In a first step in *Escherichia coli* glyceraldeyde 3-phosphate dehydrogenase (gap) has been down-regulated resulting in an increased flow of precursors towards 1,3-PDO. Furthermore D-glucose uptake was increased by manipulating the phosphotransferase system (PTS).[13] However, the link between the aerobic glucose metabolism and the PDO pathway starting from glycerol was still missing. Therefore the genetic pathway to glycerol from yeast was introduced. Finally a strain combining the benefits of three different species has been constructed which directs 60% of its carbon uptake towards 1,3-PDO and only 11% into biomass whereas the original *E. coli* uses 63% of its uptake for building biomass.[11] In this strain the procaryote *Escherichia coli* is the host providing a biological system well established in industrial processes and delivering an efficient sugar uptake system, the eukaryote *Saccharomyces cerevisiae* contributes the pathway genes to glycerol and the genes for the final catalytic steps to 1,3-PDO originate from the eukaryote *klebsiella pneumoniae* (Figure 12.10).[11]

1,3-PDO is the first example of a trend-setting scientific and technical breakthrough in engineering a synthetic pathway targeting a biotechnological intermediate well known to chemists but new to the engineered organism. The chemical industry turns out to be the driver in biotechnological innovation.

12.1.3 The Drivers of White Biotechnology

The products mentioned above are just examples for the most relevant market segments of the chemical industry: monomers and polymers, fine and specialty chemistry, pharmaceuticals as well as detergents and body care. With a sales volume of €98 bn in 2007 these market segments represent 75% of the German chemical industry. In addition these markets are those with the highest annual growth. In 2007 inorganics and petrochemicals grew by only 1% whereas the annual growth rate of polymers has been 3% and fine- and specialty chemistry, pharmaceuticals as well as detergents and body care even reported growth rates between 5 and 9% (Figure 12.11).

White Biotechnology is generally well established especially in those fields where it is the only option. There is no economical alternative to biotechnological production of functional proteins such as enzymes or enantiomerically pure amino acids. However, new ecological and economical drivers move the chemical industry's focus on products so far produced by chemistry starting from fossil carbon sources. The highly visible climate change and unpredictable economics of fossil oil have pushed the chemical industry towards bio-renewable feedstocks for many years. However, bio-renewable carbon sources come with a significant burden. In contrast to fossil carbon sources 50% of their molecular weight is oxygen. Comparing fossil and bio-renewable carbon sources therefore needs to standardize on the cost of carbon (Figure 12.12).

On this basis bio-renewables have become quite competitive in the last few years (Figure 12.13).

In summary, changing feedstock markets in combination with the enormous progress in biotechnologies are the boosters of White Biotechnology.

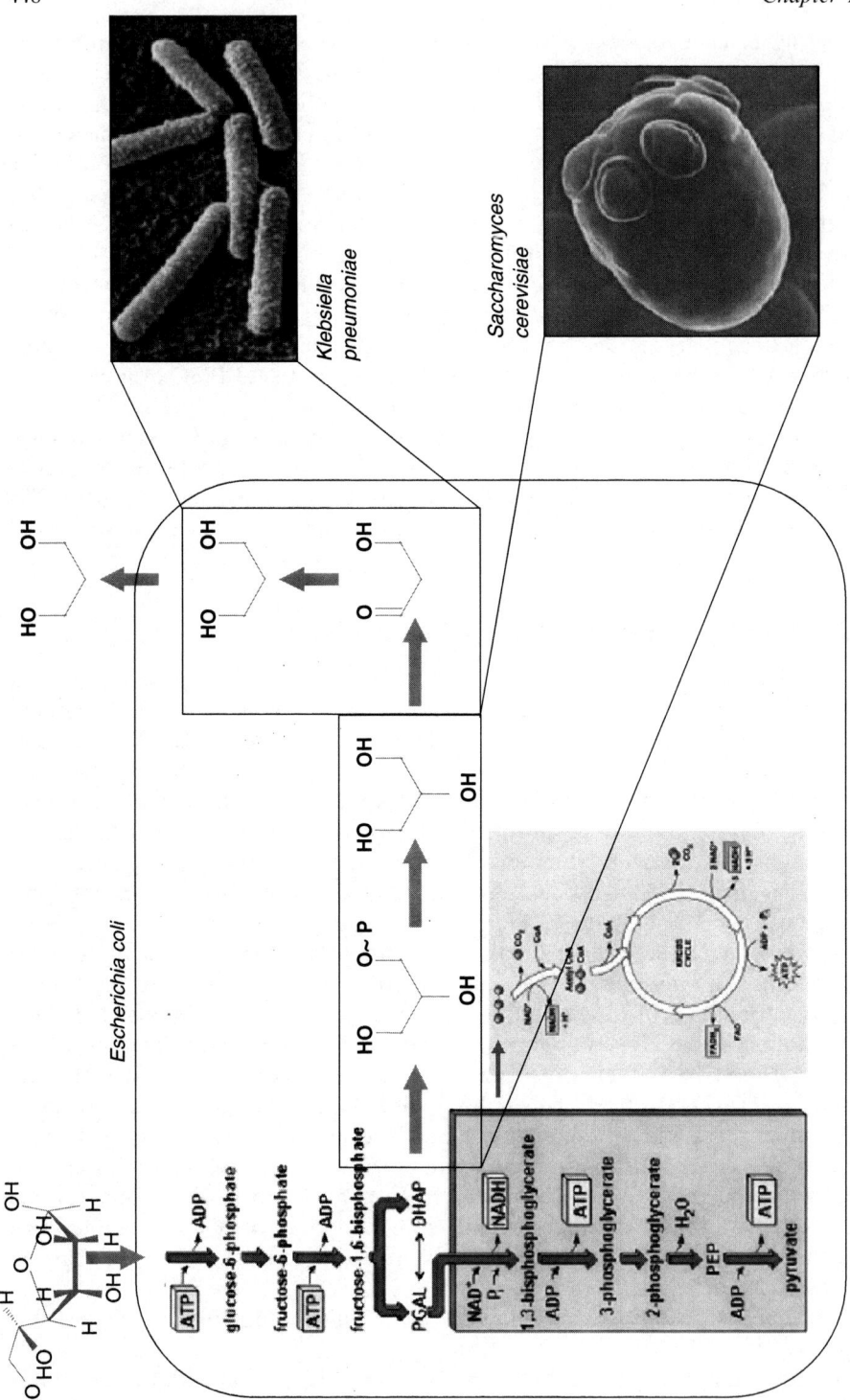

Figure 12.10   Genetic pathways to industrial production of 1,3 PDO.

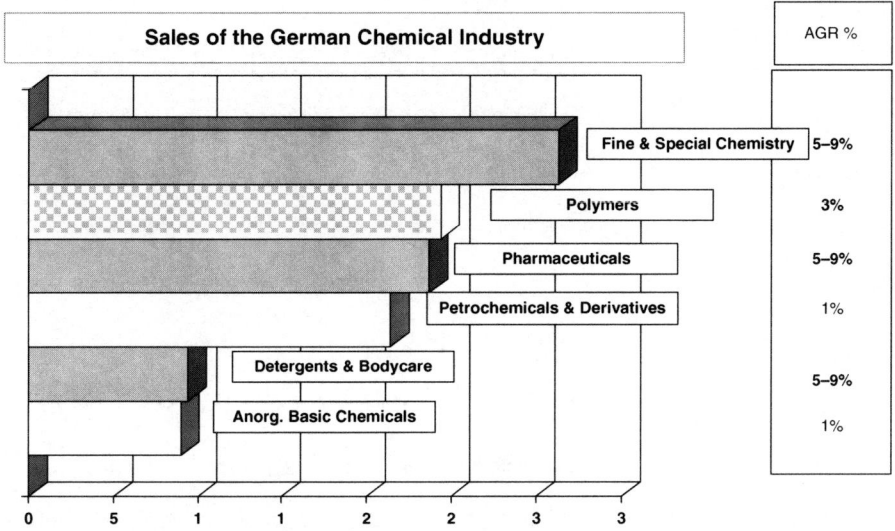

Annal Sales of the German Chemical Industry (bn €: 2006)

Figure 12.11 Sales volume in the market segments of the German chemical industry (VCI). In the grey segments industrial biotechnology is already established or is starting (monomers and polymers).

Biotechnological plants of industrial scale have been announced all over the world (Figure 12.14).

Consequently chemical industries are in urgent need of clean technologies, processes and products. Targeting on alternative biotech processes leading to well-established compounds is just the beginning not the end. With growing familiarity with biotechnological processes the chemical industry more and more realizes the potential of modern biotechnology to deliver new high-performance products available by pure biotechnology or combined with chemistry.

12.2 Recent Examples

12.2.1 Sphingolipids

12.2.1.1 Sphingolipid Functions and Applications

Sphingolipids along with sterols and glycerophospholipids are the most abundant class of lipids constituting eukaryotic cellular membranes.

Their functions have been best studied in the model yeast *Saccharomyces cerevisiae.* Here, apart from structural functions, they play important roles as second messengers in signal transduction pathways, regulating diverse cellular processes like growth, nutrient uptake, heat stress response, cell wall synthesis

Petrochemicals

		Production	Price	C	H	O	Carbon
Ethylene	C_2H_4	110 mio mto a^{-1}	1228 €/mto	86%	14%		**1428 €/mto C**
Propylene	C_3H_6	75 mio mto a^{-1}	1015 €/mto	86%	14%		**1180 €/mto C**
Benzene	C_6H_6	45 mio mto a^{-1}	871 €/mto	92%	8%		**947 €/mto C**
Naphtha	C_nH_{2n}		605 €/mto	86%	14%		**703 €/mto C**

BioRenewables

		Production	Price	C	H	O	Carbon
Sucrose	$C_{12}H_{22}O_{11}$	172 mio mto a^{-1}	180 €/mto	42%	6%	52%	**429 €/mto C**
Bio-Ethanol	C_2H_6O	39 mio mto a^{-1}	357 €/mto	52%	13%	35%	**687 €/mto C**
Corn-Glucose	$C_6H_{12}O_6$	330 mio mto a^{-1}	223 €/mto	40%	7%	53%	**558 €/mto C**

Prices as of September, 2008, 1€ = 1.48 US$,
Sources: Naphtha Europe - CMAI; **Propylene** Europe Contract–CMAI; **Benzene** Europe –ICIS;
Sugar No. 11$^®$ – ICE; **Ethanol** (Hydrous) Brazil – Cepea; Glucose in **Corn** (65%) - CBOT.

Figure 12.12   Cost of carbon from fossil and biorenewable feedstocks.

and repair, endocytosis, cellular aging, protein transport and degradation, and others.[14–16] To a large extent, sphingolipid biosynthesis pathways as well as sphingolipid functions are conserved from yeast to man.

In mammals, a specific class of sphingolipids termed ceramides serves an additional important function related to formation and maintenance of the epidermal permeability barrier, protecting skin from desiccation and preventing chemicals from penetrating the skin into underlying tissues.[17,18] Ceramides are amide-linked fatty acids containing a long-chain amino alcohol (sphingoid base).

Skin barrier functions can be mainly attributed to the outermost layer of mammalian epidermis, the *stratum corneum* (SC). This region consists of dead, cornified cells embedded in a matrix of extracellular lipids, containing mainly ceramides, cholesterol, cholesterol sulphate and free fatty acids. The SC lipids are organized in a specific lamellar, quasi-crystalline structure, which is critical for optimal skin barrier homeostasis. However, formation of the lamellar structure requires presence of the lipid subclasses in a defined ratio, and ceramides by weight comprise roughly 50% of total SC lipids. The structure and formation of the epidermal lipid barrier is explained in Figure 12.15.

Mammalian SC ceramides contain one of four sphingoid bases, namely sphingosine, 6-hydroxysphingosine, phytosphingosine and dihydrosphingosine

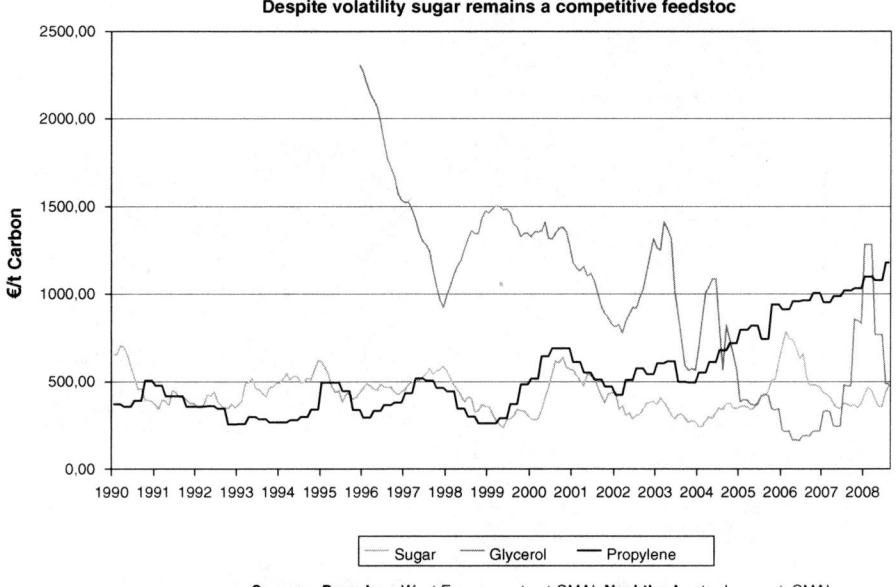

Figure 12.13 Prices of fossil and renewable carbon are coupled.

Recent Press Releases: Bio-Monomers & Polymers

Supplier	Monomer Production	Polymer Production	Marketing	Market
Tate&Lyle	JV 1,3-PDO	DuPont	Mohawk TheHomeDepot	Polyester 2007 -45 kta
ADM	JV Telles Metabolix/ADM	Telles PHB	Telles	PHB 2008 -50 kta
Tianjin Green Bio-Science	JV	JV PHB	DSM	PHB 2009 -10 kta
Odebrecht	JV Ethene	Braskem	Braskem	Polyethylene 2009 -200 kta
Crystallsev	JV Ethene	Dow	Dow	Polyethylene 2011 -350 kta
Roquette	JV Succinic acid	DSM	???	PBS 2011 -? kta
Cargill	Carg./Novozy. 3-HP	???	???	Polyacrylate 2011 -? kta

Figure 12.14 Biotechnological plants are announced worldwide.

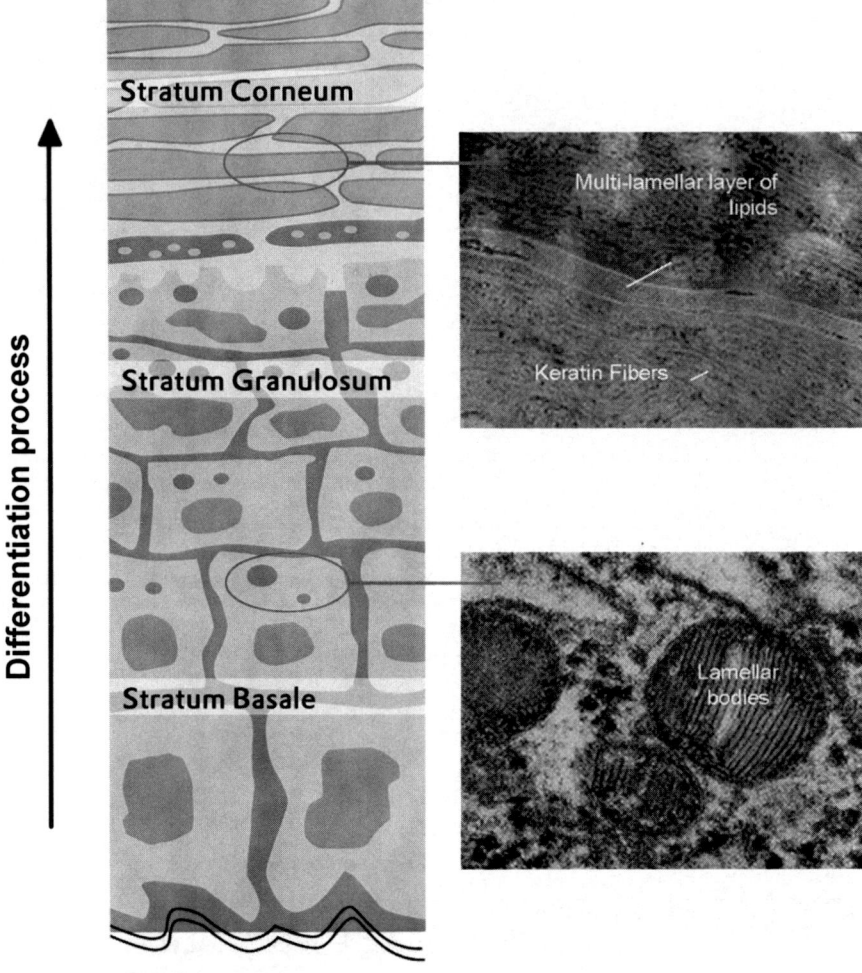

Epidermis

Figure 12.15   Formation of the lipid barrier of human skin. The top layer of the epidermis called *stratum corneum* is a hornified and inert barrier. Its primary functions are regulation of the skin's moisture content and protection of the underlying tissues against external influences. Due to its structure it is often compared to a brick wall in which the non-viable keratin-filled corneocytes are embedded like bricks in a matrix of intercellular lipids. Synthesis of the *stratum corneum* lipids starts in deeper skin layers, where lipids (mainly glucosylceramides and sphingomyelin) are produced and packaged in so-called "lamellar bodies". During differentiation and maturation, these lipids are enzymatically converted to ceramides and finally assembled into densely packed lamellar structures surrounding the corneocytes and filling the intercellular spaces of the *stratum corneum*.

(sphinganine). To date, 11 different ceramide classes, with more than 300 molecular variants, have been identified in human skin. Their classification is based on the sphingoid bases contained and the fatty acids they are linked to (straight-chain fatty acids, α-hydroxylated fatty acids, ω-hydroxylated and esterified fatty acids). Depending on the sphingoid base present in a specific ceramide, two or three asymmetric C-atoms occur, meaning that up to eight different stereoisomers are conceivable. However, the sphingoid bases in natural skin ceramides have only one conformation for each base. Conformation has proven crucial for SC skin barrier function, because interactions with other lipids, and as a consequence formation of the lamellar structure of the SC lipid layer, depend on proper ceramide stereochemistry.

Topical application of compositions comprising sphingolipids, such as ceramides, improves the barrier function and moisture-retaining properties of the skin.[19–23] Moreover, sphingoid bases exhibit anti-inflammatory and anti-microbial activities, making sphingolipids valuable active ingredients in cosmetic formulations.[24,25]

12.2.1.2 Biotechnological Production of Phytosphingosine by Fermentation of Pichia ciferrii

Sphingoid bases as core building blocks for ceramide production can be produced by extraction from different natural sources, such as brain or chicken eggs. However, the amount of sphingoid bases gained from those resources is fairly low. Moreover, due to concerns raised by the BSE crisis, products of animal origin lack customer acceptance. Therefore, many companies producing cosmetics generally banned those compounds from their portfolios.

Chemical production of sphingoid bases could represent an alternative, and several routes have been described. However, due to the presence of several stereocentres, chemical synthesis results in the production of racemic mixtures rather than defined enantiomers. Moreover, extensive protection chemistry is required due to the presence of several functional groups within the molecules, making those processes very laborious and expensive.

For the sphingoid base phytosphingosine, an alternative production method relies on fermentation of the non-conventional yeast *Pichia ciferrrii*. Originally isolated in 1932, it was discovered in 1960 that certain isolates of this yeast secrete substantial amounts of acetylated sphingoid bases, mainly phytosphingosine, into the growth medium.[26] The major form detected was tetraacetyl phytosphingosine (TAPS), but other, not fully acetylated forms were detected as well.[27] It is assumed that acetylation of phytosphingosine in this yeast is a prerequisite for secretion.[28] The physiological significance of sphingolipid secretion remains obscure. Growth inhibition of competing microbial strains is one possibility. However, acetylated sphingoid bases precipitate in the medium, and are likely inactive in this form. For the moment, acetylation of sphingoid bases and subsequent secretion into the medium is a unique trait of certain *P. ciferrii* strains.

   Soon after discovery, the production behaviour of *P. ciferrii* for TAPS was investigated in large-scale fermentation.[29] During the past decades, *P. ciferrii* TAPS producers have been continuously improved by re-isolation and mutagenesis/screening programs, and product titres in fermentation processes today yield multigram dimensions. Several companies, such as Evonik Industries and Doosan Corporation, produce TAPS by fermentation. After extraction and purification from the fermentration broth, chemical deacetylation yields the sphingoid base, which is then chemically acylated to yield the desired ceramides. Alternatively, phytosphingosine may be directly incorporated as an active ingredient into cosmetic formulations.

12.2.1.3   *Sphingolipid Biosynthesis in the Yeast* Pichia ciferrii

Sphingolipid biosynthesis in *P. ciferrii* has been extensively studied for many years. An overview of the biosynthetic pathway is given in Figure 12.16. For a more detailed description, the reader is referred to the excellent work of

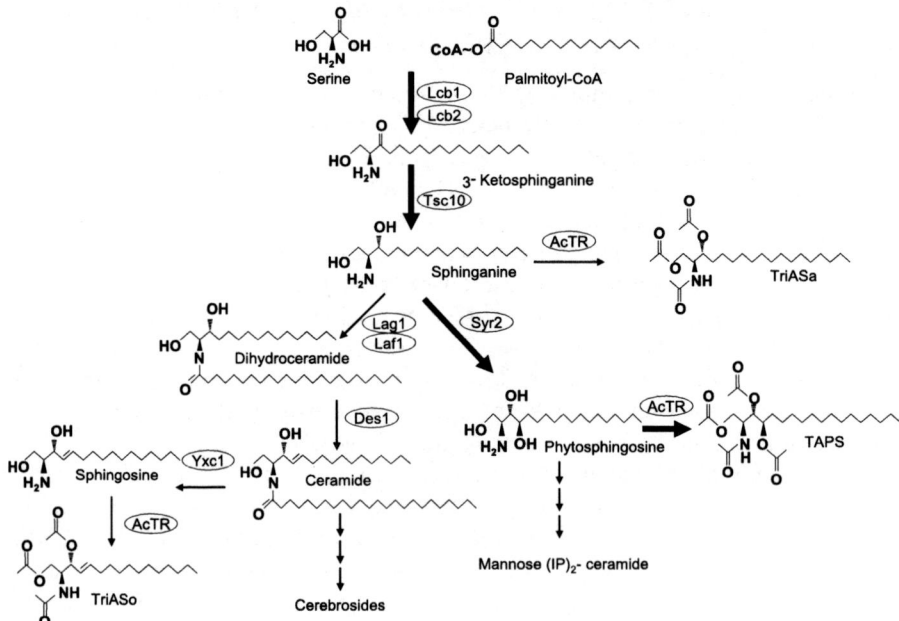

Figure   12.16   Sphingolipid biosynthesis in *Pichia ciferrii*. Enzymes are shown as ovals and named according to the corresponding homologues from *Saccharomyces cerevisiae*. Abbreviations: AcTR, acetyl-CoA:long-chain base acetyltransferase; TAPS, tetraacetyl phytosphingo-sine; TriASa, triacetyl sphinganine; TriASo, triacetyl sphingosine; IP, inositol phosphoryl. TAPS is by far the most abundant sphingo-lipid formed in wild-type cells, as reflected by the strength of the arrows.

Boergel.[30] Biosynthesis starts with condensation of serine and palmitoyl-CoA to form 3-ketosphinganine. The reaction is catalyzed by heterodimeric serine palmitoyltransferase (Lcb1p, Lcb2p) and is considered rate-limiting in the closely related yeast *Saccharomyces cerevisiae*. The gene encoding one subunit (*LCB2*) was cloned and over-expressed, yielding two-fold increased TAPS levels.[31] Reduction of 3-ketosphinganine by NADPH-dependent 3-ketosphinganine reductase (Tsc10p) leads to formation of sphinganine (dihydrosphingosine), the direct precursor of phytosphingosine.[32] Hydroxylation of sphinganine at the C2-position to give phytosphingosine is catalyzed by sphinganine hydroxylase (Syr2p/Sur2p). The corresponding gene was also cloned and characterized by Bae *et al.*[33]

Acetylation of phytosphingosine (and to a minor extent also sphinganine) is catalyzed by acetyl-CoA:long-chain base acetyltransferase.[34,35] The enzyme was partially purified and characterized from enriched membrane fractions. It accepts various sphingoid bases as substrates, and experimental data suggests that a single enzyme catalyzes both N- and O-acetylation.[36]

Besides free and acetylated sphingoid bases, *P. ciferrii* produces a range of ceramides with long or very long acyl side chains attached to the base *via* amide linkage.[32] Those ceramides may comprise phytosphingosine, sphinganine, sphingosine or sphingadienine as sphingoid base. They can be further converted to complex glucosylceramides (glucocerebrosides) by attachment of a glucose residue to the hydroxy group in C1-position of the sphingoid base.[37] Whether or not *P. ciferrii* like *e.g. S. cerevisiae* produces phytoceramides containing myo-inositol is currently not known. Cerebroside biosynthesis starts by transfer of a long acyl residue from acyl-CoA to the amino group of sphinganine, catalyzed by the dimeric dihydroceramide synthase Lag1p/Laf1p.[38] Next, Des1p (dihydroceramide-Δ4-desaturase) introduces a double bond at Δ4 position in the sphingoid base backbone, leading to formation of sphingosine-based ceramide from sphinganine-based dihydroceramide.[39] Obviously, Des1p does not accept the free sphingoid base sphinganine as substrate, but is only active on the corresponding dihydroceramide. Ceramide represents a branching point in *P. ciferrii* sphingolipid biosynthesis. It can be further converted to complex cerebrosides (see above), or alternatively be hydrolyzed by action of the ceramidase Yxc1p to give the free sphingoid base and fatty acid. So far, Yxc1p represents the only ceramidase identified in *P. ciferrii* and, based on homology studies and knowledge about closely related yeast species, it is likely that Yxc1p from *P. ciferrii* represents a ceramidase hydrolyzing sphingosine- as well as sphinganine- and phytosphingosine-based ceramides.[38]

Although *P. ciferrii* contains the complete genetic information required for conversion of sphinganine to sphingosine, the pathway seems to be largely inactive. Only traces of sphingosine are produced, acetylated and secreted by *P. ciferrii* wild-type strains, whereas large amounts of acetylated phytosphingosine (along with some acetylated sphinganine) are produced.

12.2.1.4   Strain Improvement of Pichia ciferrii for Efficient Production of Sphingosine

As sphingosine is one of the major sphingoid base components of human skin ceramides, it is of considerable commercial interest to produce the corresponding ceramides for cosmetic applications. Sphingosine can be produced chemically from phytosphingosine, but the process is laborious and expensive, adding significant costs to the final ceramide prices and preventing their broad application in cosmetics so far. Therefore, Evonik Industries started a broad research program several years ago, and since then has undertaken great research efforts to redirect sphingolipid biosynthesis in *P. ciferrii* from phytosphingosine to sphingosine production. The program included classical strain improvement methods (random mutagenesis/screening) as well as rational elements (metabolic engineering).

As a first step, *P. ciferrii* wild-type strains were subjected to treatment with the antifungal drug syringomycin E, which is known to be toxic for yeast strains carrying an intact *SYR2* gene.[33,40] A number of *SYR2* strains with a complete block of phytosphingosine production could be isolated.[39] However, initial expectations that a block in phytosphingosine production would directly translate into elevated titres of acetylated sphingosine (triacetyl sphingosine, TriASo) were not fulfilled. Rather, *SYR2* strains secreted greatly elevated levels of acetylated sphinganine (triacetyl sphinganine, TriASa), pointing to one or more bottlenecks in the three-step pathway leading from sphinganine to sphingosine formation. This finding suggested that a purely classical strain improvement program probably would not be successful, making a rational approach more promising. The toolbox for metabolic engineering of *P. ciferrii* has so far been strictly limited. For instance, no autonomously replicating plasmids are available, so gene over-expression can only be achieved by stable chromosomal integration of the corresponding expression cassettes. However, a transformation system based on a point-mutated allele of a small ribosomal protein (*PcL41**) conferring cycloheximide resistance has been described.[31] Together with a repetitive rDNA element for targeted integration of heterologous DNA, the system has proven useful for gene over-expression purposes, as it favours multi-copy integration into the chromosome.

To identify and overcome metabolic bottleneck(s), a large number of dihydroceramide synthase, dihydroceramide-Δ4-desaturase and ceramidase genes were functionally expressed in *P. ciferrii*, either alone or in different combinations.[30] The genes evaluated were taken from a diverse set of organisms, including yeasts, mammals and viruses. Finally, it turned out that concerted over-expression of all three enzyme types, covering the whole pathway from sphinganine to sphingosine, was required for efficient sphingosine formation. Moreover, large differences were observed concerning the impact of homologous genes from various organisms on sphingosine titres, and many different genes and gene combinations had to be screened prior to isolation of a strain efficiently converting sphinganine to sphingosine.

Initially, over-expression of the different sphingosine biosynthesis genes had only minimal effects, although expression in all cases was driven by strong promoters driving expression of glycolytic genes. A closer inspection of the known *P. ciferrii* open reading frames revealed that *P. ciferrii* coding DNA has a GC content below 35%, accompanied by a very narrow codon usage with a high bias towards A or T in third-base codon positions. Therefore, a number of codons occur very rarely in *P. ciferrii*, and foreign genes seem to be only poorly expressed unless their sequence is adapted to *P. ciferrii* codon usage. After the foreign genes had been adapted accordingly, sphingosine productivity was dramatically improved, resulting in titres several fold higher compared to the starting strains.

Research work to further increase sphingosine production titres in recombinant *P. ciferrii* strains is still ongoing. Usually, production titres of biomolecules are constrained by regulatory mechanisms like *e.g.* feedback inhibition, a common biological principle applying to many biosynthetic pathways in nature. Future work will hopefully help to elucidate the underlying regulatory principles, and lead to a better understanding of sphingolipid biosynthesis in the unique yeast *Pichia ciferrii*.

12.2.2 L-Cysteine

12.2.2.1 *Biotechnical Production of L-Cysteine*

12.2.2.1.1 L-Cysteine – Relevance in Nature and Industry. L-cysteine is an important sulphur-containing compound in nature. Due to the accumulation of several functional groups in close proximity (thiol, amino and carboxyl group) this amino acid is predestined to participate in a wide variety of different biochemical reactions and processes. For living organisms L-cysteine has an outstanding relevance since it is involved in protein biosynthesis as one of the 21 canonical building blocks. In proteins or peptides L-cysteine residues often constitute structural functions by the formation of either intra- or inter-molecular disulfide bonds thus stabilizing the protein's overall structure or by the constitution of Fe/S clusters in the catalytic domain of enzymes. Apart from its requirement in proteins, L-cysteine functions also as major donor of reduced sulphur atoms in the biosynthesis of essential low molecular weight sulphur-containing compounds of the cell, like L-methionine, lipoic acid, biotin, thiamine, coenzyme A and others.

Due to its extraordinary reactivity L-cysteine plays not only a crucial role for the cell's metabolism but it has gained importance in several fields of humans' daily lives as well.

Currently the world market of L-cysteine represents approximately 5,000 mto a^{-1}, approximately 55 Mio US$ sales and an annual growth rate of 4%, whereby its main fields of application are the pharmaceutical, food and cosmetic industries. About 30% of the annual L-cysteine requirement is converted into the derivatives N-acetylcysteine and S-carboxymethylcysteine as mucolytic pharmaceuticals for bronchial diseases. Also due to its strong reductive ability L-cysteine is applied in the food industry; as baking additive L-cysteine breaks up the disulfide bonds of gluten thus facilitating the kneading of the dough and helping to increase its volume. Another significant portion of L-cysteine is used as a flavour enhancer and for the production of meat flavour in pet food. In cosmetics L-cysteine is used for the generation of permanent waves, for anti-aging products and even for whitening creams.

12.2.2.2   Manufacturing of L-cysteine

12.2.2.2.1   Extraction from Natural Sources.   Today, L-cysteine is one of the few amino acids still to be produced by extraction from natural material in industrial scale.[41] Keratin-rich waste products like pig bristles, poultry feathers and human hair are used as raw material for the isolation process. After acidic hydrolysis of the keratin at elevated temperatures the low-solubility L-cysteine-dimer L,L-cystine is precipitated easily upon pH neutralization. Finally, by electrolytic reduction L-cysteine is formed from L,L-cystine. However, during the extraction process approximately 27 kg of hydrochloric acid are needed to receive 1 kg of L-cysteine, which means an immense challenge for waste water treatment.[42]

12.2.2.2.2   Bioconversion   of   DL-2-Amino-Thiazoline-4-Carboxylic   Acid. Another process makes use of the ability of some bacteria belonging to the genus *Pseudomonas* to convert DL-2-amino-thiazoline-4-carboxylic acid (DL-ATC) into L-cysteine.[43,44] DL-ATC is synthesized chemically from 2-chloroacrylic acid and thiourea. Formation of L-cysteine from DL-ATC is a three-step process catalyzed by different enzymes inside the cells: 1) racemization of DL-ATC to L-ATC by ATC racemase, 2) formation of N-carbamoyl-L-cysteine (L-NCC) from L-ATC by L-ATC hydrolase and 3) hydrolysis of L-NCC to L-cysteine, NH_3 and CO_2 by L-NCC-amino-hydrolase.

12.2.2.2.3   Microbial Fermentation – a Powerful and Sustainable Process. However, in order to face the increasing needs for L-cysteine and to provide the market with a product meeting the highest quality standards a new process for the large-scale production of L-cysteine has been established by Wacker Chemie in 2001. This process is based upon microbial fermentation using a special strain of *Escherichia coli* which has been generated by means of metabolic engineering and produces large amounts of L-cysteine when grown on glucose as carbon source. Already in 2004 the fermentative production process shared more than 12% of the L-cysteine market representing a volume of 500 mto and an annual growth rate of 10%.[45] One important aspect for the success of the Wacker process is the high quality of the product which is needed for different applications. Especially for the pharmaceutical industry fermentation based L-cysteine is desirable because the danger of product contamination with agents like BSE or SARS can be excluded.

Another benefit of this new process is the environmental sustainability since in contrast to the extraction process only 1 kg of hydrochloric acid is needed to receive 1 kg of L-cysteine, which corresponds to an acid reduction of more than 96%.[42] This issue was honoured by the *Federation of German Industries* (BDI) in June 2008 when Wacker won the environmental prize in the *"Environmentally Compatible Technology"* category.

12.2.2.2.4 L-Cysteine Biosynthesis in *Escherichia coli*. In *E. coli* the amino acid L-serine serves as a precurser for L-cysteine biosynthesis. L-serine in turn is derived from 3-phosphoglycerate, an intermediate of the cell's central metabolism.[46] The initial step of the specific L-serine biosynthetic pathway is the oxidation of 3-phosphoglycerate to 3-phosphohydroxypyruvate catalyzed by 3-phosphoglycerate dehydrogenase (SerA) (see Figure 12.17). For control of the cell's L-serine pool the SerA activity is subjected to an efficient feed-back inhibition by L-serine ($K_i = 40\,\mu\text{M}$). In the second step an amino group is inserted by phosphoserine transaminase (SerC) to give 3-phosphoserine. Finally the phosphate is cleaved off by 3-phosphoserine phosphatase (SerB) to end up with L-serine.

As in many other organisms in *E. coli* L-cysteine synthesis represents the sole mechanism for incorporation of inorganic reduced sulphur atoms into organic matter. Thus, due to the central position of L-cysteine in sulphur metabolism the enzymatic activities of the different biosynthetic branches finally leading to the formation of L-cysteine are stringently regulated and coordinated on the expression level or on the enzyme activity level, respectively (see Figure 12.17).[47] The final step in L-cysteine biosynthesis is the incorporation of a reduced sulphur atom into O-acetylserine (OAS) to give L-cysteine. In the case

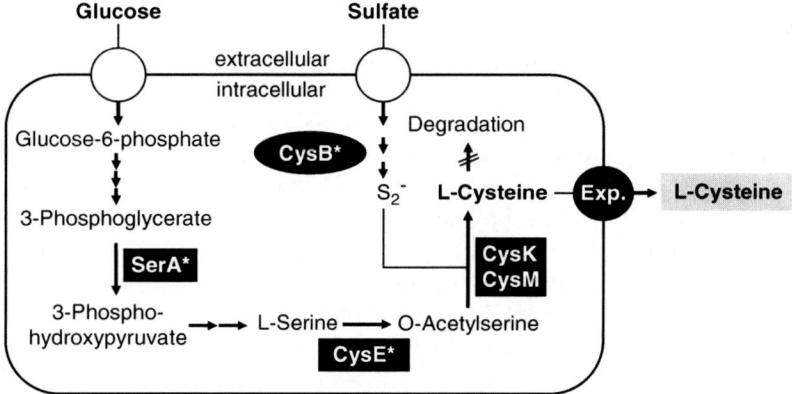

Figure 12.17 Important targets for the generation of an L-cysteine production strain. Mutant proteins are marked with an asterisk. Exp.: cysteine export proteins. SerA: 3-phosphoglycerate dehydrogenase. CysE: serine transacetylase. CysB: transcriptional activator of the cystein regulon. CysK: O-acetyl-L-serine (thiol)-lyase A. CysM: O-acetyl-L-serine (thiol)-lyase B.

of sulphate being the only available sulphur source the sulphur atom has to be reduced to sulphide prior to incorporation via an energy-dependent multi-stage process. However, when the cell is provided directly with a reduced sulphur compound like sulphide or thiosulphate, respectively, L-cysteine biosynthesis is a relatively simple two-stage process in which the hydroxyl group of L-serine is replaced by a thiol group. The first step consists of the O-acetylation of L-serine catalyzed by the L-serine transacetylase, the product of the *cysE* gene (see Figure 12.17). Subsequently, the activated acetyl group of OAS is replaced by reduced sulphur to give L-cysteine and acetate. This reaction is managed by either of the O-acetylserine (thiol)-lyases, isoenzymes with different substrate preferences, which are encoded by the *cysK* or the *cysM* gene, respectively. The CysK enzyme (isoenzyme A) acts with sulphide as substrate and yields L-cysteine directly as the end-product, whereas the CysM enzyme (isoenzyme B) has a broader substrate specificity and is able to accept thiosulphate as a sulphur donor as well.[48] In the latter case S-sulphocysteine is formed as an intermediate which requires hydrolysis or reduction to deliver L-cysteine.

Most of the genes coding for the enzymes involved in L-cysteine biosynthesis (*e.g. cysK, cysM* and genes for uptake and reduction of sulphur compounds) are integrated in a regulatory unit, the so-called cysteine regulon. The expression of these genes (except *cysE*) is under the control of the transcriptional activator CysB which requires N-acetylserine (NAS) as a co-inducer molecule as well as a sulphur limitation, because sulphide and thiosulphate act as so-called anti-inducers. NAS is formed from OAS in an irreversible and purely chemical reaction preferentially occurring in neutral solutions or even better at an alkaline pH value.[49] Thus NAS accumulates in the cell rather quickly under conditions when OAS is formed by the CysE catalyzed reaction but cannot be consumed due to the lack of reduced sulphur compounds.

The L-cysteine biosynthesis is also kinetically controlled by the end-product L-cysteine, which efficiently inhibits the activity of serine transacetylase.[50] By this means, the inhibition of the metabolic flux from L-serine to L-cysteine is integrated into the regulation of the expression of all genes necessary for sulphate assimilation. Since CysB and its positive effector NAS are required for transcriptional activation of the cysteine regulon the feedback inhibition of OAS synthesis by L-cysteine (resulting in a reduced intracellular NAS accumulation) elegantly counterbalances gene expression when the L-cysteine pool of the cell is replenished. Stringent balance of L-cysteine biosynthesis is essential, as it is the major sulphur donor for the synthesis of other organo-sulphur compounds and also because higher levels of L-cysteine have been reported to be inhibitory or even toxic.[51–54]

12.2.2.2.5 Generation of an L-Cysteine Production Strain by Metabolic Engineering. In order to overcome the natural regulation and the limitations of L-cysteine biosynthesis several targets have to be engineered to finally end up with an effective overproduction of the metabolite (see Figure 12.17). One major task is the generation of feedback-resistant enzymes which are no longer inhibited by intermediates or by the end-product of the L-cysteine biosynthetic pathway. For example, variants of the 3-phosphoglycerate dehydrogenase with a tremendously increased resistance towards L-serine were isolated by means of a sophisticated mutagenesis/screening strategy. The most effective SerA mutant protein described has a 3750-fold higher resistance against L-serine compared to the wild-type SerA.[55] Besides point mutations, which are described in the European patent EP1496111, a complete removal of the regulatory domain of the 3-phosphoglycerate dehydrogenase could be an interesting starting point for further optimization of the cysteine production process.[56] The proof of this concept was demonstrated in an engineered strain of *Corynebacterium glutamicum* for an improved L-serine production.[57]

The second important feedback mechanism which has to be eliminated regulates the penultimate reaction in L-cysteine biosynthesis. The L-serine transacetylase CysE is normally inhibited effectively by the end-product L-cysteine ($K_i \sim 1 \, \mu M$). Also for this enzyme various muteins were generated showing a very efficient decoupling of feedback regulation by the end-product. Most powerful is the mutant allel named cysE23 encoding a protein with a K_i value of 2.3 mM.[58]

An alternative issue for avoidance of feedback inhibition by L-cysteine is the use of heterologous L-serine transacetylases. Very promising proteins can be found in plants which are summarized in the review by Wada and Takagi.[59]

Besides elimination of feedback regulation mechanisms a very fundamental subject is the enhancement of end-product excretion for an effective L-cysteine overproduction. Recently four different export systems for L-cysteine belonging to various classes of membrane proteins could be identified in *E. coli*, which implies the importance of such proteins for the cell. The first exporter identified was the YdeD (EamA) protein which belongs to the PecM family of transport proteins.[60] Even in a wild-type strain over-expression of the *ydeD* gene leads to the excretion of both L-cysteine and its precursor OAS. YfiK (EamB) was detected by the same group and belongs to the RhtB type of transporters. Like YdeD YfiK also excretes both L-cysteine and OAS but only in strains already containing a deregulated CysE enzyme.[61] Over-expression of a third transporter confers resistance to high extracellular L-cysteine concentrations. This L-cysteine transport system, encoded by the operon *cydDC*, belongs to the group of ABC transport proteins and is required for cytochrome assembly.[62] The L-cysteine export protein identified most recently can be assigned to the large group of multidrug resistance transporters. The discovery of the Bcr protein as an L-cysteine exporter displays that global screening procedures are valuable tools for metabolic engineering.[63] For effective L-cysteine production a highly sophisticated balance is needed when several export proteins are combined in the production process.

Since even a slightly elevated L-cysteine concentration is inhibitory or possibly toxic for the cells *E. coli* possesses another mechanism for detoxification of this compound in addition to excretion: degradation of L-cysteine. Five enzymes with L-cysteine desulfhydrase activity have been identified so far in this organism: L-tryptophanase (TnaA), L-cystathionine β-lyase (MetC),

O-acetyl-L-serine (thiol)-lyase A (CysK), O-acetyl-L-serine (thiol)-lyase B (CysM) and MalY.[64] Thus in order to engineer a potent L-cysteine over-producer excessive degradation of the amino acid must be prevented by inactivation of the genes encoding the major L-cysteine desulfhydrase activities.

Finally, an efficient and elegant tool for the deregulation of L-cysteine biosynthesis in general is the expression of a constitutively active variant of the transcriptional activator CysB no longer being dependent on the co-inducer molecule NAS. Thus expression of the genes belonging to the L-cysteine regulon is ensured under all physiological conditions.

Although good progress has been achieved so far by application of conventional metabolic engineering tools (*e.g.* mutagenesis and screening, targeted genetic engineering), new comprehensive approaches like genome sequencing and metabolic flux analysis of production strains are necessary to identify new genetic targets for further manipulation which will help to improve the process for future demands.

12.2.3   Lipid Biotechnology

The microbial production and the biotechnological transformation of lipids and lipid soluble compounds are summarized under the term "Lipid Biotechnology". Target substances are triacylglycerols including their different fatty acid types and to a minor extent phospholipids, sphingolipids, glycolipids, sterols, carotenoids and other lipid soluble compounds (Figure 12.18). Excellent overviews about the different lipid classes, a lipid glossary and fatty acid structures of seed oils can be accessed online.[65–68] Several classes of biocatalysts including lipases, esterases or phospholipases may be utilized for the

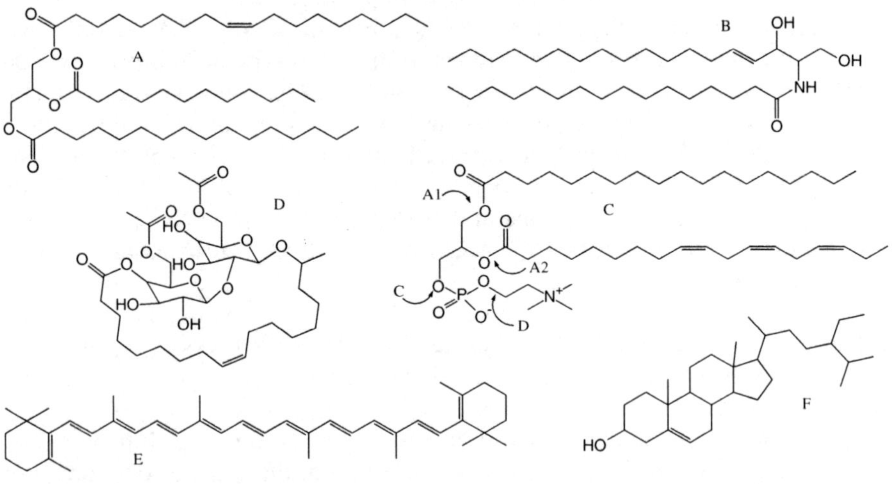

Figure 12.18   Examples of lipid structures and oil soluble compounds: A) triacylglycerol; B) ceramide; C) phospholipid (with enzymatic modification sites); D) sophorose lipid; E) carotene; F) sitosterol.

Table 12.3 Typical examples for Lipid Biotechnology. Commercialized lipid (including glycerol) based products manufactured biotechnologically.

Product	Process (typical)
Microbial oils (polyunsaturated)	Fermentation of mainly marine micro-organisms
"Designer Lipids"/Structured lipids/ Margarine	Enzymatic modification with regioselective lipases
Carotenoids (Carotene/Astaxanthin)[80,81]	Fermentation of *Dunaliella* or *Haematococcus*
Cosmetic esters[82]	Lipase catalyzed synthesis
Sophorose lipids[83]	Fermentative production with *Candida bombicola*
Sphingolipids[84]	Fermentative production with *Pichia ciferrii*
Steroid hormones (sterol based)[85]	Transformations with several microorganisms
Decalactone (aroma compound)[86,87]	Biocatalytic production with *Yarrowia lipolytica*
Hexenal (+ other grass aroma compounds)[86,87]	Lipoxygenase and Hydroperoxide Lyase
Dicarboxylic acids (mainly from alkanes)[88,89]	Biocatalytic production with *Candida tropicalis*
Biodiesel (in development)[90,91]	Lipase catalyzed transesterification
Dihydroxyacetone (glycerol based)[92]	Production with *Gluconobacter oxydans*
1,3-Propanediol (glycerol based)[93]	Production with *Clostridium sp.* or *Klebsiella sp.*

modification of lipids, fats and oils. Several of these enzymes from microbial, plant and animal origin are available commercially from different suppliers.[69] An excellent book supplying a broad overview about biocatalysts for lipid modification is available.[70]

Lipases are the most versatile catalysts in the field of Lipid Biotechnology with a variety of industrial applications *e.g.* in the food, cosmetic, pharmaceutical, leather, pulp and paper or detergent industries (Table 12.3).[70,71] Most of them display certain fatty acid selectivity; some possess regioselective properties and show enantioselective catalytic behaviour. Fatty acid selectivity and regioselectivity are especially interesting properties in the field of lipid biotechnology. The lipases are not only able to modify ester bonds of lipids but may also catalyze non-natural reactions like the modification of hydrophilic polyol compounds, the peroxidation of fatty acids or the transformation of amine based compounds.[70–73] Additionally the lipase-catalyzed synthesis of several polyester compounds has been shown.[74,75] The availability of high-throughput screening methods for lipases makes them ideal targets for modern biotechnology including *e.g.* metagenome screening approaches for the identification of new catalysts or genetic engineering for alteration or optimization of the catalyst properties.[76–79]

Microbial production of oils and the enzymatic synthesis of "Designer Lipids" will be exemplified in this chapter in more detail.

12.2.3.1 Production of Microbial Lipids

Microbial lipid production has been known for more than a century; however, the economics of microbial oil production utilizing yeasts and fungi could not compete with that of plant-derived oils for commodity products. Therefore the concept of commodity oil production by fermentation was generally abandoned.[94,95]

12.2.3.1.1 Microbial PUFA Oils. Micro-organisms, mainly of marine origin, are able to synthesize polyunsaturated fatty acids (PUFAs). In contrast to plants, micro-organisms and animals are able to build up fatty acids with chain lengths of >C18 via sequential elongase and desaturase cycles. Recently a completely separate pathway for the synthesis of polyunsaturated long-chain fatty acids was detected in marine bacteria and some marine algae, which uses a polyketide synthase system.[96,97]

Several species are of commercial interest for the fish feed industry and as natural sources of PUFAs for health food and infant nutrition formulas. Health food applications focus on products containing eicosapentaenoic (EPA) and docosahexaenoic acid (DHA) while infant formulas contain arachidonic acid (ARA) and DHA as PUFA components. EPA enriched products are mainly based on purified fish oils, whereas ARA and DHA are also produced microbially in an industrial scale.[96–98]

Mortierella alpina was identified to be an excellent production organism with up to 70% ARA in the storage lipid fraction and it is the only commercially used micro-organism for ARA manufacturing.[99,100] With the application of molecular technology, increases in the total ARA content of *Mortierella* have been achieved.[101]

DHA-rich oils are produced industrially with micro-organisms of either the *Thraustochytrid* genus or *Crypthecodinium cohnii*. These micro-organisms are able to grow phototrophically or heterotrophically. For commercial production, heterotrophic conditions are applied for cost reasons.[96] Several other micro-organisms are able to produce PUFAs including marine bacteria; however, they are not of commercial importance for the time being.[102] A promising approach is the introduction of gene clusters encoding for PUFA synthesis into lipid-accumulating yeasts. The engineering of a *Yarrowia lipolytica* strain with an EPA content of 39% of the total fatty acids was reported recently.[96]

12.2.3.2 Enzymatic Synthesis of "Designer Lipids"

The term "Designer Lipids" comprises artificial glycerides, synthesized either chemically or enzymatically, with a specific fatty acid composition and/or a specific regioselective pattern, the latter ones called structured lipids. Enzymatic

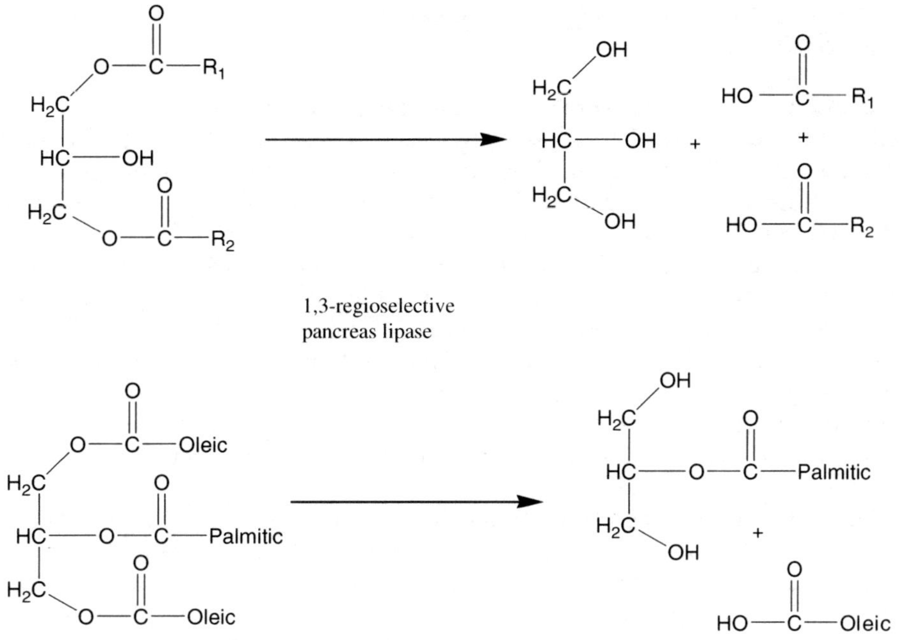

Figure 12.19 Simplified scheme of the pancreatic hydrolysis of structured lipids (top: 1,3-diacylglycerols; bottom: human milk fat replacers).

processes have several advantages over chemical processes: the regioselectivity of enzymes can be exploited to obtain specific structures and fewer by-products are formed due to lower reaction temperatures. Several structured lipids have been synthesized enzymatically and concepts for reactor designs for the production of these lipids have been elucidated.[103–105] Structured lipids are interesting products for infant and medical nutrition as well as for dietetic products. The structure of the lipid influences e.g. the gastric emptying, the fecal loss of lipids and the lipid profiles in tissues after resynthesis.[106] The pancreatic lipase is a 1,3-regioselective enzyme, thus the structure of a lipid determines the composition of fatty acids and glyceride species absorbed in the intestine (Figure 12.19).

12.2.3.2.1 Human Milk Fat Replacers. Human milk fat is a structured lipid with an amount of more than 70% of saturated fatty acids, mainly palmitic acid, in the sn-2 position of the triglyceride backbone. This structure, in comparison to a randomized glyceride with human milk fat composition, guarantees a high absorption of energy-rich palmitic acid and a low loss of calcium in the faeces.[107] Human milk fat substitutes are produced enzymatically on an industrial scale. The process is based on the regioselective exchange of oleic acid with palmitic acid at the sn-1 and sn-3 positions of tripalmitin.[108] Alternatively the regioselective alcoholysis of tripalmitate yields

2-monoacylglycerols, which are then re-esterified at the 1- and 3-positions with oleic acid.

12.2.3.2.2 1,3-Diacylglycerols. 1,3-diacylglycerols are produced enzymatically on a commercial scale as dietetic cooking oil and have been a great success especially in Japan. Long-term clinical studies including a recent 1-year human study suggest that the exchange of standard cooking oils with diacylglycerol oil results in a moderate loss of body weight.[109,110] The working principle of the oil is a decreased formation of 2-monoglyceride after hydrolytic action of pancreatic lipase, resulting in a lower re-synthesis and storage of triglycerides and in an increase of metabolic fat oxidation.[109] The oil is produced enzymatically in a 1,3-regioselective synthesis of fatty acid with glycerol.[111] Acyl migration from the sn-1 to the sn-2 position decreases the content of 1,3-diacylglycerol, thus reaction conditions must be controlled carefully to obtain a high concentration of the desired product.

12.2.3.2.3 Triglyceride Concentrates. Triglycerides of concentrated PUFAs and conjugated linoleic acid (CLA) are produced enzymatically by several companies starting from either free fatty acids or the corresponding ethyl esters. The triglycerides are advantageous for food applications, because free fatty acids and ethyl esters cannot be formulated into food products directly. Due to the oxidative instability of the products, enzymatic processes are favoured over chemical routes. The enzymatic synthesis is possible without the use of solvents under a high vacuum at moderate temperatures.[112,113] The removal of water or ethanol from the reaction is a critical step; improvements in the synthesis reaction were achieved with nitrogen as strip gas.[114]

12.2.3.2.4 Margarine Production. Enzymatic interesterification at low reaction temperatures for the production of *trans* fat free margarine and confectionary fats is a recently developed process, which is now running industrially at several manufacturers.[115] The ban of *trans* fats, which were identified as causing a variety of negative health effects in humans,[116] has given the enzymatic process a competitive advantage over classical margarine production.

12.2.4 PLA (Polylactic Acid)

12.2.4.1 Introduction

Polylactide or Polylactic Acid (PLA) is a synthetic, aliphatic polyester from lactic acid (Figures 12.20 and 12.21).

For industrial applications, such as fibres, films and bottles, the chain length n should be between 700 and 1400. This is significantly higher than with partially aromatic polyesters like PET and PBT, where n is between 100 and 200. Therefore, the requirements on both raw material purity and technical effort are much higher.

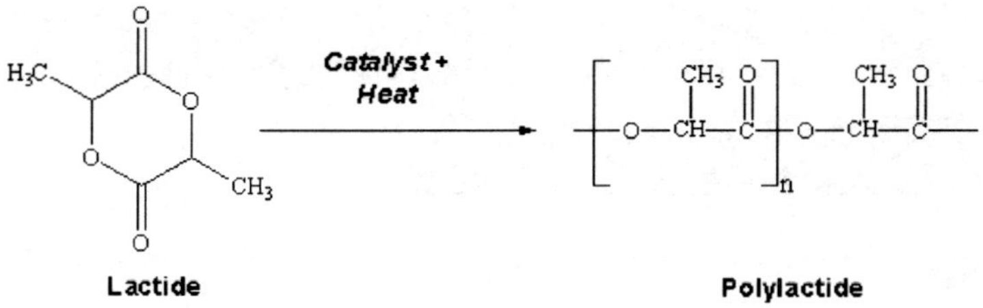

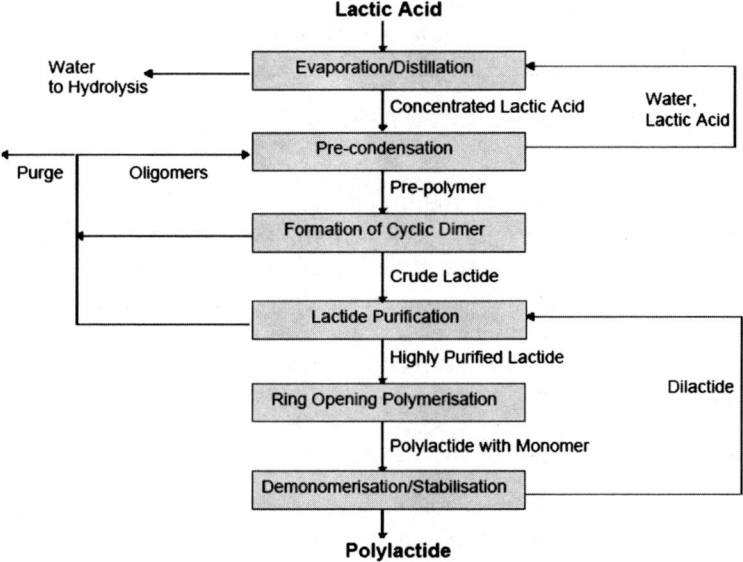

Figure 12.20 Ring opening polymerization of lactic acid dimer yielding polylactide (reproduced from Wikipedia: Polylactic Acid).

Figure 12.21 Synthesis of polylactic acid (PLA).

At temperatures below its glass transition point (*e.g.* 55 °C, depending on comonomer content) PLA is as stable as PET or PBT. Only in an industrial composting facility, the high temperature (60 °C) and humidity required for the hydrolysis are achieved. After hydrolysis, PLA is biologically degradable by common micro-organisms. Lactic acid, the monomer building block of PLA, can frequently be found in plants and animals as a by-product or intermediate product of metabolism. Lactic acid is non-toxic by its nature.

12.2.4.2 Non-depleting Properties of PLA

Lactic acid can be industrially produced from a number of starch- or sugar-containing agricultural products. Competition between human food, industrial lactic acid and PLA production is not to be expected: for example, using PLA

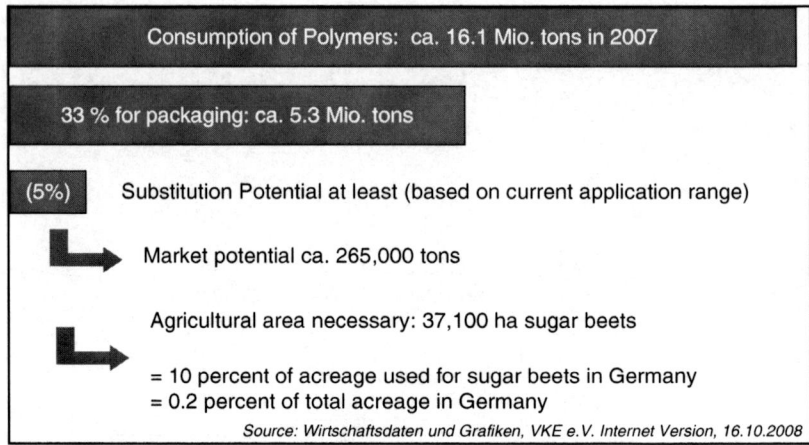

Figure 12.22 Packaging and acreage.

as substitute for 5% of the German packaging plastics consumption requires only 0.5% (sugar beet) to 1.25% (wheat) of the agricultural area available (Figure 12.22). At the same time, approximately 30% of the available area lies fallow mainly for economic reasons. Research is in progress on processes and micro-organisms that produce lactic acid from cellulose coming from agricultural residues such as maize stalks or straw.

Several recycling methods can be applied to waste PLA:

- Industrial composting

  — Most attractive method of disposal based on public acceptance
  — No recovery of material and energy

- Mechanical recycling

  — Loss of product properties cannot be recovered
  — "Downcycling"

- Burning (energy recycling)

  — Recovers "green energy"

- Chemical recycling

  — Back into polymerisation
  — Collecting and sorting yet to be solved

Composting allows only moderate benefits. In future, sorting, purification of PLA waste and re-feeding into the polymerization plant seems to be the most attractive way of recovery.

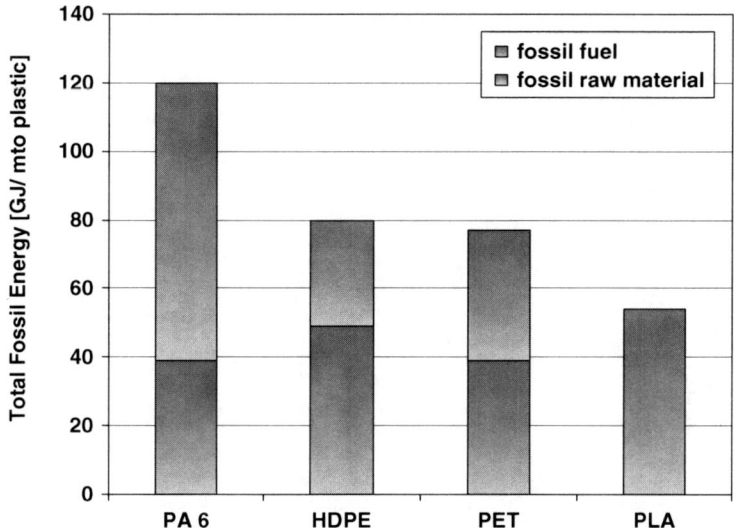

Source: M. Patel, R. Narayan, in Natural Fibers, Biopolymers and Biocomposites, A. Mohanty, M. Misra, L. Drzal, Taylor & Francis Group, 2005, Boca Raton.

Figure 12.23 Consumption of fossil resources by PLA *vs.* polymers from fossil feedstock, "cradle to gate".

PLA – like other biopolymers – is often criticized for the need of process energy from fossil resources. Even if this is the case at present, 1 kg of PLA represents fewer energy equivalents than 1 kg of polymers from petrochemical feedstock (Figure 12.23). Consequently, PLA producers can also reap financial benefits by trading CO_2 emission certificates (Figure 12.24).

If process energy is supplied by biomass, *e.g.* biogas, the fossil energy required for 1 kg PLA can be cut by half, thus duplicating the benefits from trading CO_2 emission certificates. Additionally, significant potential exists for saving process energy by improving lactic acid and polymerization technologies.

12.2.4.3 Market Potential of PLA

The first marketplace for PLA is Europe, with Italy, Germany, The Netherlands, the United Kingdom and France in the frontline, followed by Japan, USA and South Korea. In Germany and France the governments have issued specific regulations to give preference to renewable and biodegradable polymers.

As PLA is a rather young polymer, many companies are developing applications. These are mainly focused on the field of packaging fresh food, but other applications are emerging. Processing PLA is possible on standard polymer equipment for the various process techniques.

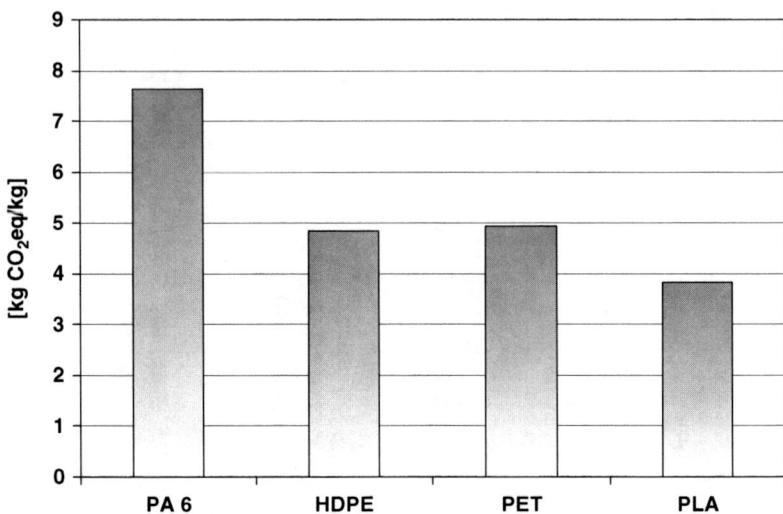

Source: M. Patel, R. Narayan, in Natural Fibers, Biopolymers and Biocomposites, A.Mohanty, M. Misra, L. Drzal, Taylor & Francis Group, 2005, Boca Raton.

Figure 12.24 CO_2 emissions by PLA *vs.* polymers from fossil feedstock, "cradle to gate".

Table 12.4 PLA producers, production capacity. State: 09/2008; no claim to completeness.

Producer	Trade Name	Route	Capacity t/y	Country
Nature Works	Ingeo	ROP	70,000	USA
Mitsubshi Chemicals	Lacea	DP-S	500	Japan
Toyota	Eco Plastic	ROP	1,000	Japan
Hisun	Hisun	ROP	5,000	PR China
Toray	Ecodea	ROP	5,000	RO Korea
Shimadzu	Lacty	ROP	1,000	Japan

Besides some smaller suppliers (Table 12.4) NatureWorks LLC is the only producer of PLA based on a 70,000 mto a^{-1} facility which has been brought to full capacity over the last years and intends to increase capacity to 140,000 mto a^{-1} in 2009.

12.2.4.4 Process Routes to PLA

Several process routes have been developed or are practised on industrial scale: Ring Opening Polymerization (ROP), Direct Polycondensation in high boiling solvents (DP-S) and Direct Polymerization in bulk followed by chain extension with reactive additives.

ROP (Figures 12.20 and 12.21) is the route which delivers by far the highest proportion of PLA chips available on the market. The other routes produce only minor amounts or did not get past the pilot scale. Figure 12.21 depicts the steps of an ROP process, starting from lactic acid. In the first part lactide is formed, which – after fine purification – is converted by ROP to PLA.

12.2.4.5 Processing of PLA

A major advantage of PLA is the possibility to process the polymer on common process equipment. Especially the converters of polyolefins do not require a change to other process equipment. They only need to change the handling of granulate. It is very important to dry the polymer before processing otherwise it will degrade. Water and high temperatures (up to 240 °C) facilitate fast degradation.

PLA is a polymer which can be processed by:

- injection moulding;
- sheet extrusion;
- extrusion blow moulding;
- thermoforming;
- injection stretch blow moulding;
- fibre spinning;
- non woven spinning, spun bonding.

12.2.4.6 Properties of PLA

PLA is a crystal clear, transparent material when amorphous that becomes hazier the higher the crystallinity. Crystallized material is opaque. When producing lactide, meso-lactide is formed as a by-product. It is difficult to separate the meso-lactide from the L-lactide in the purification step. When polymerizing L-lactide with small contents of meso-lactide a co-polymer is formed. Increasing meso-lactide leads to decreasing crystallinity. With more than 10–15% meso-lactide the polymer is amorphous.

By varying the amount of meso-lactide the properties of the polymer can be adjusted for specific applications.

One of the reasons for the limited consumption of PLA up to now is the low thermal resistance. The T_g (glass transition temperature) is about 55 °C depending on comonomer content to a small extent (Table 12.5).

Methods of improving thermal resistance are to prepare a stereo complex (sc PLA) or a stereo-block copolymer (sbc PLA). Melting point and heat distortion temperature (HDT) will increase significantly.

Improving the thermal properties can extend the applications of PLA considerably in the future.

There are also various additives that improve the properties of PLA with respect to impact strength, melt viscosity, HDT, crystallinity *etc.*

Table 12.5 Properties of PLA type.

Type	T_m	T_g	σ_n	E_b
PLLA	160–180 °C	55–65 °C	45–55 MPa	3–5 %
PL/DLA	–	55 °C		50–200 %
sc PLA	220–230 °C	60 °C		3–5 %
sbc PLA	185–195 °C	55 °C		5–10 %

T_m-melting temperature E_b-elongation at break
T_g-glass transition temperature σ_n-tensile strength at break

12.2.4.7 Perspective

PLA combines all prerequisites of sustainability with important properties of well-established polymers. Applications have already been found in many niches of packaging and textile products. Within those niches fast growth of consumption is expected to continue depending on the availability of PLA polymer.

High research activity is dedicated to overcome typical weaknesses of PLA – low impact strength and low heat distortion temperature – and to develop tailor-made PLA grades in order to serve special applications. These activities will conquer new niches for PLA and will help to increase PLA consumption at high velocity.

Other growth factors are the availability and prices of raw oil, agricultural products and production plants and technology.

Within the foreseeable future PLA will not become a commodity polymer like PE, PP and PS – this is considered to be an advantage for both PLA producers and converters. However, this could change in the long term.

12.3 Outlook of White Biotechnology

White Biotechnology has an enormous future potential, simply because today enzymes and micro-organisms can be rationally designed. In addition the biotechnologic processes are characterized by water-based, harmless and less toxic conditions.

At present the unique performance of biotechnological products is the main driver of this technology, especially in the pharma and cosmetic markets. API's, active pharma ingredients, are often protein based, like antibodies. These are one of the backbones of red biotechnology. But also other API's and nutrition ingredients, which can only be synthesized in a complex series of chemical steps are nowadays synthesized in microbes. Examples are insulin or vitamins, examples of White Biotechnology. Many of these are described in the beginning of this chapter.

Nevertheless the share of White Biotechnology for all chemical processes and products today is only 5%. This share will grow steadily. But the growth will take time, on the one hand because of the high complexity of the biological systems and on the other hand because of missing solutions for the interface of biology and chemistry. One of the important interfaces is the downstream processing of the aqueous biotechnological process solutions. To work up these solutions energy efficiently, achieving chemical products in a high purity, more biochemical engineering breakthroughs are needed. Therefore in the near future White Biotechnology will still mainly grow in chemical niches, but as soon as solutions for the biochemical interfaces are provided, White Biotechnology will also grow into chemical bulk product markets.

Another important factor for the development of White Biotechnology will be simply the oil price. Bulk chemicals derived from renewables will be unlikely at low prices.

But with the next economic recovery in addition to the climate change derived CO_2-debate new, white biotech derived processes will be in place and take their share of the new economic growth.

A large amount of effort in research and development is necessary for White Biotechnology to become a significant part of a green future.

References

1. OECD, *The Application of Biotechnology to Industrial Sustainability*, 2001.
2. Royal Belgian Academy Council of Applied Science, *Industrial Biotechnology and Sustainable Chemistry*, 2004.
3. Dechema, *Weiße Biotechnologie, Chancen für Deutschland*, Frankfurt, 2004.
4. NREL, *Lignocellulosic Biomass to Ethanol Process Design and Economics Utilizing Co Current Dilute Acid Prehydrolysis and -enzymatic Hydrolysis Current and Futuristic Scenarios*, 1999.
5. T. Werpy and G. Petersen, (Ed.), *Top Value Added Chemicals from Biomass*, US DOE, Oak Ridge, **vol. 1**, 2004.
6. OECD , *Task Force Industrial Biotechnology*, 2008.
7. EuropaBio, *White Biotechnology: Gateway to a More Sustainable Future*, 2004, 13.
8. M. Breuer, K. Ditrich, T. Habicher, B. Hauer, M. Keßeler, R. Stürmer and T. h. Zelinski, *Angew. Chem. Int. Ed.*, 2004, **43**(7), 788.
9. M. Kircher, *Biotechnol. J.*, 2006, **787–794**.
10. O. Zelder, W. K. Jeong, C. Klopprogge, A. Herold and H. Schröder, WO2007/113127, 2007 (BASF).
11. R. W. Miller, *Opening Degussa Science-to-Business Center Bio*, Marl, 15.3.2007.
12. H. Sunkara and H. Ng, *Biobased Polyols*, Sarnia, 19.10.2006.
13. C. E. Nakamura and G. M. Whited, *Curr. Opin. Biotechnol.*, 2003, **14**, 454–459.

14. L. A. Cowart and Y. A. Hannun, *Top. Curr. Gen.*, 2004, **6**, 383.
15. L. A. Cowart and L. M. Obeid, *Biochim. Biophys. Acta*, 2007, **1771**(3), 421.
16. R. C. Dickson, C. Sumanasekera, L. Chiranthani and L. Robert, *Prog. Lipid Res.*, 2006, **45**(6), 447.
17. C. R. Harding, A. Watkinson, A. V. Rawlings and I. R. Scott, *Int. J. Cosmet. Sci.*, 2000, **22**(1), 21.
18. W. M. Holleran, Y. Takagi and Y. Uchida, *FEBS Lett.*, 2006, **580**(23), 5456.
19. K. De Paepe, D. Roseeuw and V. Rogiers, *J. Eur. Acad. Dermatol. Venereol.*, 2002, **16**(6), 587–94.
20. M. Kucharekova, J. Schalkwijk, P. C. M. Van De Kerkhof and P. G. M. Van De Valk, *Contact Dermatitis*, 2002, **46**(6), 331–338.
21. V. Rawlings, *Int. J. Cosmet. Sci.*, 2003, **25**(1/2), 63–95.
22. U. Schick, U. Wollenweber, K. Korevaar and A. V. Rawlings, *Fragrance Journal*, 2004, **32**(11), 17–22.
23. U. Wollenweber, K. Korevaar, A. V. Rawlings and U. Schick, *SOFW Journal*, 2004, 130(9), 12–14, 16–18.
24. D. J. Bibel, R. Aly and H. R. Shinefield, *J. Invest. Dermatol.*, 1992, **98**(3), 269–273.
25. T. Pavicic, U. Wollenweber, M. Farwick and H. C. Korting, *Int. J. Cosmet. Sci.*, 2007, **29**(3), 181–190.
26. L. J. Wickerham and F. H. Stodola, *J. Bac.*, 1960, **80**, 484.
27. M. L. Greene, T. Kaneshiro and J. H. Law, *Biochim. Biophys. Acta*, 1965, **98**(3), 582.
28. Y. Barenholz, N. Gadot, E. Valk and S. Gatt, *Biochim. Biophys. Acta*, 1973, **306**(2), 341.
29. H. G. Maister, S. P. Rogovin, F. H. Stodola and L. J. Wickerham, *Appl. Microbiol.*, 1962, **10**, 401.
30. D. Boergel, Metadata Internet Doc. [Ger. Diss.], 2007, (D0313-3).
31. J. H. Bae, J. H. Sohn, C. S. Park, J. S. Rhee and E. S. Choi, *Appl. Env. Microbiol.*, 2003, **69**(2), 812.
32. W. Stoffel, G. Sticht and K. D. Le, *Hoppe-Seyler's Zeitschrift fuer Physiologische Chemie*, 1968.
33. J. H. Bae, J. H. Sohn, C. S. Park, J. S. Rhee and E. S. Choi, *Yeast*, 2004, **21**(5), 437.
34. Y. Barenholz, I. Edelman and S. Gatt, *Biochim. Biophys. Acta*, 1971, **248**(3), 458.
35. Y. Barenholz and S. Gatt, *Biochem. Biophys. Res. Comm.*, 1969, **35**(5), 676.
36. Y. Barenholz and S. Gatt, *J. Biol. Chem.*, 1972, **247**(21), 6827.
37. B. Kaufman, S. Basu and S. Roseman, *J. Biol. Chem.*, 1971, **246**(13), 4264.
38. S. Schaffer, M. A. Van den Berg, D. Boergel and T. Hueller, WO2007/131720, 2007 (Evonik Degussa).

39. M. A. Van den Berg and S. Schaffer, WO2006/048458, 2005 (Cosmoferm).
40. F. C. Adetuyi, A. Isogai, D. Di Giorgio, A. Ballio and J. Y. Takemoto, *FEMS Microbiol. Lett.*, 1995.
41. M. Ikeda, in *Advances in Biochemical Engineering/Biotechnology*, ed. T. Scheper, R. Faurie and J. Thommel, Springer, Berlin, Heidelberg, New York, 2003, **vol. 79**, pp. 1–35.
42. Wacker Chemie AG, Press release 20, 23.06.2008.
43. K. Sano and K. Mitsugi, *Agric. Biol. Chem.*, 1978, **42**, 2315–2321.
44. K. Sano, C. Eguchi, N. Yasuda and K. Mitsugi, *Agric. Biol. Chem.*, 1979, **43**, 2373–2374.
45. C. Winterhalter, in *Biotechnologie für Einsteiger*, ed. R. Renneberg, Elsevier GmbH, München, 2006, p. 92.
46. G. V. Stauffer, in *Escherichia coli and Salomnella typhimurium: Cellular and Molecular Biology*, ed. F. C. Neidhardt, R. Curtiss III, J. L. Ingraham, E. C. C. Lin, K. B. Low, B. Magasanik, *et al.*, ASM Press, Washington, 1996.
47. N. M. Kredich, in *Escherichia coli and Salomnella typhimurium: Cellular and Molecular Biology*, ed. F. C. Neidhardt, R. Curtiss III, J. L. Ingraham, E. C. C. Lin, K. B. Low, B. Magasanik, *et al.*, ASM Press, Washington, 1996.
48. T. H. P. Maier, *Nat. Biotechnol.*, 2003, **21**, 422–427.
49. M. Flavin and C. Slaughter, *Biochemistry*, 1965, **4**, 1370–1375.
50. D. Denk and A. Böck, *J. Gen. Microbiol.*, 1987, **133**, 515–525.
51. P. Datta, *Proc. Natl. Acad. Sci. USA*, 1967, **58**, 635–641.
52. C. L. Harris, *J. Bacteriol.*, 1981, **145**, 1031–1035.
53. M. A. Sørensen and S. Pedersen, *J. Bacteriol.*, 1991, **173**, 5244–5246.
54. J. M. Delaney, D. Ang and C. Georgopoulos, *J. Bacteriol.*, 1992, **174**, 1240–1247.
55. T. Maier and R. Flinspach, EP1496111, 2004 (Consortium für elektrochemische Industrie).
56. J. K. Bell, P. J. Pease, J. E. Bell, G. A. Grant and L. J. Banaszak, *Eur. J. Biochem.*, 2002, **269**, 4176–4184.
57. R. Netzer, R. Faurie, P. Peters-Wendisch, L. Eggeling, H. Sahm, R. Faurie and B. Klaßen, DE10231297, 2002 (Forschungszentrum Jülich/Amino GmbH).
58. W. Leinfelder and P. Heinrich, EP0858510, 1995 (Consortium für elektrochemische Industrie).
59. M. Wada and H. Takagi, *Appl. Microbiol. Biotechnol.*, 2006, **73**, 48–54.
60. T. Daßler, T. Maier, C. Winterhalter and A. Böck, *Mol. Microbiol.*, 2000, **36**, 1101–1112.
61. I. Franke, A. Resch, T. Daßler, T. Maier and A. Böck, *J. Bacteriol*, 2003, **185**, 1161–1166.
62. M. S. Pittman, H. Corker, G. Wu, M. B. Binet, A. J. G. Moir and R. K. Poole, *J. Biol. Chem.*, 2002, **277**, 49841–49849.

63. S. Yamada, N. Awano, K. Inubushi, E. Maeda, S. Nakamori, K. Nishino, A. Yamaguchi and H. Takagi, *Appl. Environm. Microbiol.*, 2006, **72**, 4735–4742.
64. N. Awano, M. Wada, H. Mori, S. Nakamori and H. Takagi, *Appl. Environm. Microbiol.*, 2005, **71**, 4149–4152.
65. Cyberlipid Centre, www.cyberlipid.org/index.htm (retrieved 24.07. 2008).
66. Lipid Library, www.lipidlibrary.co.uk/index.html (retrieved 24.07.2008).
67. F. D. Gunstone and B. G. Herslöf, Lipid Glossary 2, The Oily Press, Bridgwater, 2000, accessible via www.pjbarnes.co.uk/op/lg2pdf.htm (retrieved 24.07.2008).
68. Seed oil fatty acids databank; http://sofa.bfel.de/ (retrieved 24.07.2008); details in K. Aitzetmüller, B. Mattäus and H. Friedrich, *Eur. J. Lipid Sci. Technol.*, 2003, 105, 92.
69. Association of Manufacturers and Formulators of Enzyme Products, accessible via http://www.amfep.org/index.html (retrieved 24.07.2008).
70. U. T. Bornscheuer, (Ed.), *Enzymes in Lipid Modification*, Wiley VCH, Weinheim, 2000.
71. D. G. Hayes, *JAOCS*, 2004, **81**, 1077.
72. F. Hasan, A. Shah and A. Hameed, *Enz. Microb. Technol.*, 2006, **39**, 235.
73. P. Villeneuve, *Biotech. Advances*, 2007, **25**, 515.
74. R. A. Gross, A. Kumar and B. Kalra, *Chem. Rev.*, 2001, **101**, 2097.
75. H. Uyama and S. Kobayashi, *Adv. Polym. Sci.*, 2006, **194**, 133.
76. M. Schmidt and U. T. Bornscheuer, *Biomol. Eng.*, 2005, **22**, 51.
77. P. Lorenz, K. Liebeton, F. Niehaus and J. Eck, *Curr. Opin. Biotech.*, 2002, **13**, 572.
78. K.-E. Jaeger and T. Eggert, *Curr. Opin. Biotech.*, 2004, **15**, 305.
79. M. Bertram, P. Hildebrandt, D. P. Weiner, J. S. Patel, F. Bartnek, T. S. Hitchman and U.-T. Bornscheuer, *J. Am. Oil Chem. Soc.*, 2008, **1**, 47.
80. L. Dufosse, P. Galaup, A. Yaron, S. M. Arad, P. Blanc, K. N. C. Murthy and G. A. Ravishankar, *Trends Food Sci. Technol.*, 2005, **16**, 389.
81. M. A. Borowitzka, *J. Biotech.*, 1999, **70**, 313.
82. O. Thum, *Inform*, 2005, **16**, 186.
83. S. Mukherjee, P. Das and R. Sen, *TIBS*, 2006, **24**, 509.
84. T. Pavicic, U. Wollenweber, M. Farwick and H. C. Korting, *Int. J. Cosmet. Sci.*, 2007, **29**, 181.
85. H. L. Holland, in *Lipid Biotechnology*, ed. T. M. Kuo and H. W. Gardner, Marcel Dekker, New York, 2002, p. 575.
86. J. Schrader, M. M. W. Etschmann, D. Sell, J.-M. Hilmer and J. Rabenhorst, *Biotechnol. Lett.*, 2004, **26**, 436.
87. M. Aguedo, M. H. Ly, I. Belo, J. A. Teixeiro, J.-M. Belin and Y. Wache, *Food Technol. Biotechnol.*, 2004, **42**, 327.
88. K. Kroha, *Inform*, 2004, **15**, 568.
89. A. Weiss, in *Modern Biooxidation*, ed. R. D. Schmid and V. L. Urlacher, Wiley VCH, Weinheim, 2007, p. 193.

90. M. J. Haas, G. J. Piazza and T. A. Foglia, in *Lipid Biotechnology*, ed. T. M. Kuo and H. W. Gardner, Marcel Dekker, New York, 2002, p. 587.
91. Y. Shimada, T. Nagao and Y. Watanabe, in *Handbook of Industrial Biocatalysis*, ed., C. T. Hou, CRC press, Boca Raton, 2005, Chap. 8.
92. D. Hekmat, R. Bauer and J. Fricke, *Bioprocess Biosyst. Eng.*, 2003, **26**, 109.
93. S. Hirschmann, K. Baganz, I. Koschik and K.-D. Vorlop, *Landbauforschung Völkenrode*, 2005, **55**, 261.
94. C. Ratledge, in *Industrial Applications of Single Cell Oils*, ed. D. J. Kyle and C. Ratledge, AOCS Press, Champaign, Illinois, 1992, p. 1.
95. C. Ratledge, *TIBTECH*, 1993, **11**, 278.
96. J. P. Wynn and C. Ratledge, in *Long Chain Omega-3 Specialty Oils*, ed. H. Breivik, The Oily Press, Bridgwater, 2007, p. 43.
97. C. Ratledge, *Biochimie*, 2004, **86**, 807.
98. O. P. Ward and A. Singh, *Process Biochem.*, 2005, **40**, 3627.
99. Y. Shinmen, S. Shimizu, K. Akimoto, H. Kawashima and H. Yamada, *Appl. Microbiol. Biotechnol.*, 1989, **31**, 11.
100. N. Totani, K. Someya and K. Oba, in *Industrial Applications of Single Cell Oils*, ed. D. J. Kyle and C. Ratledge, AOCS Press, Champaign, Illinois, 1992, p. 52.
101. S. Takeno, A. Sakuranadi, A. Tomi, M. Inohara-Ochiai, H. Kawashima and S. Shimizu, *J. Biosci. Bioeng.*, 2005, **100**, 617.
102. J.-P. Berge and G. Barnathan, *Adv. Biochem Eng./Biotechnol.*, 2005, **96**, 49.
103. X. Xu, *Eur. J. Lipid Sci. Biotechnol.*, 2000, **102**, 287.
104. C. C. Akoh, in *Food Lipids*, ed. C. C. Akoh and D. B. Min, Marcel Dekker, New York, 2002, p. 877.
105. X. Xu, *Eur. J. Lipid Sci. Biotechnol.*, 2003, **105**, 289.
106. M. Mu and T. Porsgaard, *Prog. Lipid Res.*, 2005, **44**, 430.
107. R. G. Jensen, in *Structured and Modified Lipids*, ed. F. D. Gunstone, Marcel Dekker, New York, 2001, p. 455.
108. C. C. Akoh and X. Xu, in *Lipid Biotechnology*, ed. T. M. Kuo and H. W. Gardner, Marcel Dekker, New York, 2002, p. 461.
109. Y. Katsuragi, T. Yasukawa, N. Matsuo, B. D. Flickinger, I. Tokimitsu and M. G. Matlock, (Ed.), *Diacylglycerol Oil*, AOCS Press, Champaign, Illinois, 2004.
110. H. Kawashima, H. Takase, K. Yasunaga, Y. Wakaki, Y. Katsuragi, K. Mori, T. Yamaguchi, T. Hase, N. Matsuo, T. Yasukawa, I. Tokimitsu and W. Koyama, *J. Am. Diet. Ass.*, 2008, **108**, 57.
111. T. Watanabe, H. Yamaguchi, N. Yamada and I. Lee, in *Diacylglycerol Oil*, ed. Y. Katsuragi, T. Yasukawa, N. Matsuo, B. D. Flickinger, I. Tokimitsu and M. G. Matlock, AOCS Press, Champaign, Illinois, 2004, p. 253.
112. G. G. Haraldsson, B. Ö. Gudmundsson and Ö. Almarsson, *Tetrahedron*, 1995, **51**, 941.

113. J. A. Arcos, C. Otero and C. G. Hill, *Biotech. Lett.*, 1998, **20**, 617.
114. P. Borg, M. Girardin, B. Rovel and D. Barth, *Biotech. Lett.*, 2000, **22**, 777.
115. M. W. Christensen, L. Andersen, O. Kirk and H. C. Holm, *Lipid Technology Newsletter*, 2001, **33**.
116. M. Ledoux, P. Juaneda and J.-L. Sebedio, *Eur. J. Lipid Sci. Biotechnol.*, 2007, **109**, 891.

Subject Index

acarbose precursor 247–8
acetate/mevalonate pathway 347–9
acetylation 330
acetylserine 459–60
active ingredients 247
  in cosmetics 395–6
activity-based criteria, exclusion 17–18
acylglycerols 179, 462, 465, 466
adenosine triphosphate 279
adhesion 426
adhesives 425–34
aesculin 317, 318
AFNOR reference model 39
Agenda 21 6, 398
agility, logistic 45–6
agricultural land
  increased biomass use 116, 121
  polymer consumption 468
agriculture
  changes in usage 171, 172
  dairy farming 71–80
  greenhouse gas emissions 56
  product utilization 8
  production 69
  see also cereals; crops; pesticides
alchemy 407
alcohol see ethanol
algae 157, 182, 212–13
algal oil 212–13
alkaline neutralization 177
alkyl glucosides 225–6
alkyl polyglycosides 60, 226, 252–3
allergies 387

Alwörden's reaction 390
amino acids
  α-amino acids 375–80
  in white biotechnology 438, 439–42
amphoteric 376
  surfactants 387–8
amylopectin 239–41, 244
amylose 239–41, 244, 281, 288–9
amylosucrase 288
anaerobic biodegradability 386
anaerobic digestion 110–13
  biogas from 115
  methane from 142
  process scheme 111
animal fats and oils 167
  fatty acids in 182
  global production/demand 169
  production 178–9
  see also fats; milk
animal feed 30–3, 187
  additives 437–8, 440
  cattle feed 74–80
animal fibres 368, 374–80
  sourcing 383–4
  structure 379
anti-parallel β-sheet 379
antibiotics 72, 249, 438, 441
antioxidants 202, 203, 207, 208
APG® surfactants 60, 226, 252–3
aquatic toxicity 62, 414, 415, 419
arachidonic acid 464
ARBOFORM® 312, 328–9
Argan oil 402

Argan tree 399–404, 405
Arganyl™ 403
Argatensyl™ 403
asbestos fibres 369
L-ascorbic acid 251–2
ash 306, 318–19
asset management 16–22
asset value 13–14
automatic fed combustion technologies 99
automotive industry 391
autothermal process 152

babassu oil 195
bacteria
    amino acid fermentation 439–40
    anaerobic digestion 110–13
    bioethanol fermentation 134–5
    in biopolymer production 446–7, 448
    fuel-producing 157–8
    *see also Corynebacterium
        glutamicum; Escherichia coli*
bacterial inulin 287–8
bagasse 266, 269
banking 12–23
BASIL process 33–4
bast fibres 381–3, 385
best available technology 320
best-in-class 18
betaines 387–8
betulin 317, 318
beverages 246–7
bio-SNG *see* synthetic natural gas
biobutanol 144
bio-chemical conversion
    biomass 90, 109–13
    biomethane 114, 115
biodegradability
    plastics 249–51
    surfactants 62, 384, 386
    test methods 62
biodiesel
    flow diagram 143
    glycerine market 222
    HTU diesel 145
    jet-based 142
    from palm 193

    property comparisons 141
    from rapeseed 198
    from vegetable oils 138–40
    worldwide capacity 8
    see also biofuels; diesel fuel; synthetic
        biofuels
bioethanol 109, 114, 127–37
    bio-ethylene from 272
    in Brazil 121, 127, 136
    plant materials for 127–8
    production processes 128–37
    as transportation fuel 136
bio-ethylene 272
biofuels 125–58
    carbon costs 450, 451
    combustion technologies 99
    EU directives 125–6, 138
    first-generation 126–37
    global trade 173
    lipid-based 137–42, 143
    major forms of 89–90
    market expansion 114–15
    second-generation 142, 144–57
    syngas-based 145–57
    third-generation 157–8
    usage 86–8
    from vegetable oils 107–9
    see also biodiesel; bioethanol;
        synthetic biofuels
biogas
    production 110–13, 114–15
    yield targets 111
bioliq®-process 156–7
biological diversity 6, 396–7, 401
biomass 86–123
    bioethanol from 128–32
    biofuels from 126
    competition, effects on 118–19
    competition areas 116–18
    configuration approaches 119–22
    definition 164
    demand 116
    different options for use 120
    electricity from 88, 114, 115
    energy content and elements in 91
    environment/climate advantages 88

gasification 96–7, 98, 104–7, 144, 150–1, 152
   for green chemistry 164–6
   heat, power and fuels from 89
   market expansion 119
   potential 93–5, 113–15
   properties 90–2
   thermo-chemical conversion 89–90, 95–107
   uses 113–15, 122, 301
   see also feedstock; renewable resources; wood
biomass conversion
   bio-chemical 90, 109–13
   new technologies 121
   physico-chemical 90, 107–9
   thermo-chemical 89–90, 95–107
biomethane 112, 114–15, 142
   see also synthetic natural gas
biopolymers 443–7, 448, 451
   fats and oils as precursors 214–16
   from fermentation 441
Biosphere Reserves 401
biosyncrude 156
biotechnology *see* white biotechnology
bleaching 320–1, 389
blow-in combustion units 101–2
Botryococcus braunii 157
Brundtland, Mrs Gro Harlem 4
BtL fuels *see* synthetic biofuels
BTL process 155–6
bubbling fluidized bed (BFB) 152–3
building materials 305–6, 327
Business and Biodiversity Initiative 397
butter 178, 204–5

cadaverine 443
calcium oxide 267
calendula oil 208
Camelina oil 207–8
Cannabis sativa 381
CANOLA 197
caoutchouc 339
capryl dimethylamide 220
carbohydrates 343, 346, 355, 357–8
   see also starch; sucrose; sugar

carbon
   budget 344–5
   costs 450, 451
   in plants 343
carbon dioxide emissions
   from cattle 80
   from polylactic acid and polymers 470
   potential 414–15, 417, 419
carbon monoxide 96, 97
carbonization 384
carrier oils 218–21
castor oil 209–11, 216
caterpillars 384–5
cationic surfactants 387
cattle feed 74–80
cavitation 344
cedar trees 3
cell membrane 213
cellulose 130, 131, 132, 133, 307–9
   lignocellulose 129, 130–2, 142
   see also hemicellulose
cellulose esters 324
cellulose fibres
   chemistry and structure 372–4
   regenerated 324
ceramides 450, 452–7
cereals
   bioethanol production 128, 129, 130, 133
   in starch production 239, 240, 241
   see also individual cereals
Cerenol® 444–5
cetane 140, 141
channel relationships 39
charcoal 102
cheeses 80
chemical fibres 370, 391
Chemical Industry Responsible Care® 168
chemical pulping 319–20, 321, 322
   applications 323
chemical refining 175–7
chemically processed biofuels 89–90
China wood oil 208
Chlorella vulgaris 157
Choren CarboV® 155–6

CHRIGAS project 155
circulating fluidized bed 153, 154–5
cis-fatty acids 179, 180
cleaning, fabrics 388–9
Clostridia 134, 144
coal 91
  briquettes 256
  gasification 106
coalescents 419–20
coalification 102
coating 258, 259
  fabrics 389–91
cobalt catalysts 149–50
cocoa butter 204
coconut oil 59, 60, 193–5
Cognis 401–4, 405
cohesion 426
Collaborative Planning, Forecasting
    and Replenishment 44
colouring, fabrics 390
combined heat and power plants 114
combustion 95–6
  direct combustion 98–102
  technologies 99
compaction, detergents 64, 65
competitive advantage 41
composting 109, 468
confectioneries 246
construction industry 305–6, 327
consumer products 53–65, 68
  washing process life-cycle 57–62
consumer purchasing behaviour 54–6
contamination, toleration levels 151
Convention of Biological Diversity 6,
    396, 401
cooperatives, dairy 72–4
coppicing 404
copra 193
corn 75
corn oil 200–1, 239
corn syrup, high fructose 246–7
corporate social responsibility 12, 394,
    397–8
  and Argan tree products 399–404, 405
corrugated board 259
cortex 377, 379

Corynebacterium glutamicum 439, 440,
    443
cosmetics 193, 195, 203, 205, 224
  from ceramides 456
  plant-based ingredients for 394–405
  from white biotechnology 439
cossettes 267
costs
  logistic 44–5
  see also prices
cotton 370, 373, 380–1, 382
  properties 385
Cotton Belt 381
cottonseed 381, 382
cottonseed oil 200, 381
cradle-to-grave analysis 26
crops
  energy crops 93, 94, 95, 113, 121, 212
  protection 185, 228, 420–2
  silage 75
  see also cereals
crude oil
  extraction 226
  prices 174, 249, 261
  see also petrochemicals; petroleum oil
crude palm oil 189, 191, 192
crystallization 268, 269, 270
  strain-induced 360
Cuphea seed oil 195, 196
curing systems 29–30, 31, 34, 431, 432
cuticle 377, 379
L-cysteine 457–62
cysteine regulon 460
cytosol 356

dairy farming
  cattle feed 74–80
  milk procurement 71–4
deforestation 172, 188, 400
degumming 175–6
dehydrated castor oil 211
dehydration 135–6
delthametrin 421
demographic dynamics 54–7, 58
deodorization 177
derivatized starches 242

design strategy, SOLVSAFE 412–16
designer lipids 464–6
detergents 57–62
  from fatty alcohols 222
  sustainability profiles 62–5
  using white biotechnology 437
developed countries
  biofuel usage 87
  consumer purchasing behaviour 54–6
developing countries, biofuel usage 87, 88
development megatrends 54–7, 58
dextran 281, 289–90
dextransucrase 289
dextrins 243
dextrose 244, 247, 253
dextrose equivalent value 243, 244
diacids 442–3
1,3-diacylglycerols 465, 466
diamines 442–3
diesel fuel 137–8
  selected properties 108
  specifications 139–40
  see also biodiesel; gasoline
dilute acid hydrolysis 132
dimethoate 421
dimethyl isosorbide 260
dimethylether 147, 148
dimethylol dihydroxy ethylene urea 330–1
direct combustion 98–102
direct supply chain 39
disasters, natural and ecological 3
Disponil® 227
disposal phase 62
dissolving grade pulp 323–4
distillation 135–6
DL-2-amino-thiazoline-4-carboxylic
  acid 458
DMAPP 347, 349, 350
DMDHEU 330–1
docosahexaenoic acid 464
Domini 400 Social Index 15
double-metal cyanide catalysis 211
Dow Jones Sustainability Indexes 15, 16
downstream processing 136, 173, 473
drugs *see* pharmaceuticals
dry acid degumming 175

dry adhesives 429
dry cleaning 408
dry fractionation 191
dry-milling 130, 201
dry rubber 354, 355, 358
DVFA framework 13, 14
DXP-MEP pathway 347–9
dyes 390
Dynamin 127

e-commerce 40
easy-care finishing 390
ECN, Netherlands 155
ecobalance 26
eco-efficiency analysis 26–8
  beneficial uses 34–5
  conducting 27–8
  curing systems 29–30, 31
  ionic liquids 33–4
  vitamin B_2 production 30–3
ecological cleansers 63, 64
ecological disasters, first 3
ecological risk assessment 62
eco-logistics 46–8
economic circuit 13
Eco-packaging program 46
ecosystems, forests 303
edible rendering 179
effectiveness 43
efficiency, logistic 43–5
EHS Guidelines *see* Environmental
  Health and Safety Guidelines
eicosapentaenoic acid 464
EIG Targanine 401–4, 405
elasticity 360
elastomer market 340, 341
electricity
  from biomass 88, 114, 115
  generation 98
elemental chlorine-free bleaching 320
elongation phase 348, 349–50
emollients 439
emulsifiable concentrates 420–1
emulsifiers 224, 226–7, 277, 421
emulsion polymerization 226–7
(end-)energy sources 117–18

energy consumption
  Germany 86, 87
  pulp and paper production 321
  rubber 342
  sugar production 268
  washing process 61, 62
energy content, biomass and coal 91
energy crops 93, 94, 95, 113, 121, 212
energy generation, from biomass 86–123
energy prices 118
engagement 19
Enhanced Analytics Initiative 18, 21
entrained-flow gasifiers 153, 154
environment impacts
  Argan tree 400, 404
  cotton 381
  fabric production 390
  natural fibres 370
  pulp and paper processing 320–1
  solvents 409, 414–15
  surfactants 386, 387
  wool 384
  see also recycling
Environmental Health and Safety
  Guidelines 174, 175, 410, 412, 416
enzymes
  in cosmetics 439
  degumming 175, 176
  designer lipid synthesis 464–6
  feedback-resistant 460–1
  in food, feed and households 437–8
  hydrolysis 132, 133
  production 441
  reactions 279–80, 281–4
  see also fermentation
epichlorohydrin 222
epoxidized oils 214
epoxidized soybean oil 187
erucamide 198
erucic acid 196, 197, 198
Escherichia coli 157–8, 440, 446–7, 448
  in L-cysteine biosynthesis 458–60
ESG (Environmental, Social,
  Governance) 13, 14, 16–22
essential fatty acids 76, 77, 78
essential oils 182, 224

ester solvents 411
esterification 219, 220, 253, 277–8
  see also transesterification
Esterquats 387
ethanol
  fermentation 109–10
  selected properties 109, 110
  from syngas 147
  see also bioethanol
ethyl-tertiary-butyl-ether 127, 136
European Sustainable Development
  Strategy (2006) 48
European Union
  biofuels directives 125–6, 138
  climate objectives 302
  REACH standard 342
  RENEW project 155
  SOLVSAFE project 412
evening primrose oil 205, 206
Evian mineral water 80–4
evolution, human 2
ex ante screening 17
ex post monitoring 17
excess air ratio 97, 98
exclusion 17–18
expanded polystyrene 250
exposure, control of 410
extended supply chain 39
external lubrication 219
extra-financial criteria 16–17, 22
extra virgin olive oil 206
extractives 306, 307, 308, 313–18
  examples 315
  solubility 316

fabric conditioners 386–8
fabrics 388–91
FAME *see* fatty acid methyl ester
FASAL® 312, 328–9
fat replacers 278
fats 76
  biopolymers precursors 214–16
  chemical composition 179–213
  fatty acid compositions 183
  processing 217
  production, use and trade, changes to
    168–74

for synthetic biofuels 140–1
value chain 213–28
see also animal fats and oils; fatty
acids; lipids
fatty acid dimethylamides 419
fatty acid esters 218, 219–21
hydrogenation 223
sucrose 278
fatty acid methyl ester 90, 107–9
hydrogenolysis 140–1
selected properties 108
synthesis of 138–9
fatty acids 76–80, 137, 194
in animal fats and oils 182
biodiesel specifications and 140
chemical composition 179–80
in neutral lipids 358–9
nut oils and olive oil 209
in oils and fats 183
as oleochemicals 216–18
in vegetable oils 182
see also omega-3 fatty acids; omega-6
fatty acids; *trans*-fatty acids;
unsaturated fatty acids
fatty alcohols 222–4
Fischer glycosidation 253
synthesis 140–1
feedstock
biodiesel 138–40
biogas production 112, 113
see also biomass; renewable resources
fermentation 31, 32, 60
bacteria 134–5, 439–40
ethanol 109–10
glucose syrups used in 248–9
main products from 249
microbial 458–9
new molecules 261–2
products produced by 441
and saccharification 132
sucrose in 264
vitamin C production 252
yeast 132–4
see also enzymes; white
biotechnology
fertilizers, mineral 266, 267

fibres
global production 371
properties of glass and natural 385
see also natural fibres; *individual
fibres*
fibrils 308, 372, 373, 376, 377–8, 379
fibroin 374–5, 376, 379–80
FICFB gasifier 155
finance, sustainability 12–23
fine chemicals 416–18, 439–42
first-generation biofuels 126–37
Fischer glycosidation 253
Fischer–Tropsch process 149–50, 155
fish meal 179, 187
fish oil 179, 182, 213
fixed-bed gasifiers 105, 152, 153, 154
flammability 414, 415
flash pyrolysis 103, 145
flax 75, 381–3
flax oil 206–7
flexible fuel vehicles 126, 136
fluidized-bed gasifiers 105, 152–3, 154–5
food
eating habits 56
enzymes in 437
functional 208, 287
prices 60
processed 246
safety 70
sustainability in 69–84
food applications, starch 240, 241, 245–7
Food-Based Dietary Guidelines 71
forest residues 93, 94
forests
deforestation 172, 188, 400
as ecosystem and resource 303
historical sustainability 4
pressure on 60–1
reforestation 404
see also wood
formaldehyde-based resins 251
fossil fuels 88, 116, 126, 447
alternatives to 142, 301
carbon costs 450, 451
environmental health and safety 410
polymer and polylactic acid
consumption 469

see also coal; crude oil;
    petrochemicals; petroleum oil
fractionation 191
Freshfields Bruckhaus Deringer study 20
Frey-Wyssling particles 356
fructans 286–8
fructose 133, 266, 273, 280
    high fructose corn syrup 246–7
    oligofructose 287
fructosyltransferase 287–8
FTSE4Good Index Series 15
fuel-ethanol 135
fuels
    from biomass 89, 114, 120
    from producer gas 106–7
    selected properties 108
    see also bioethanol; biofuels; diesel
        fuel; transportation fuels
full oxidation 97
full-solid adhesives and sealants 431
functional foods 208, 287
functional unit 26, 57
fungal diseases 185
furan fatty acid 358–9
furanose ring 287
furfural 275
furfurylation 330, 331
fusible solids 411

gamma-linolenic acid 205
gas cleaning 105, 106, 151
gasification 96–7, 98, 104–7, 144,
    150–1, 152
gasifiers
    current developments 154–6
    types 105, 152–3
gasoline 110, 147–9, 148
    see also diesel fuel
gel phase 353
gelatinization 240, 242
genetic modification
    cotton 381
    organisms 136
    rapeseed 197
    soya bean 184–5
Gingko biloba 317

glass fibres 385
Global Commerce Initiative 49–50
Global Ethic 2
Global Food Safety Initiative 70
Global Product Strategy 6
global production/demand
    adhesives and sealants 428–9, 430, 433
    biodegradable polymers 251
    fibres 371
    oils and fats 169, 170, 171
    papermaking 325
    pulp 322, 324
    solvents 408
global trends, demographics 54–7, 58
global warming 2, 56
glucoheptonic acid 254, 255
gluconic acid 254, 255
glucono delta-lactone 256
D-glucopyranose 307, 372
glucose 132, 133, 266
    D-glucose 239, 243, 244, 245, 255, 280
glucose syrups 243–4, 246–7, 261
    in fermentation 248–9
glucosinolates 196–7
glues 325
L-glutamic acid 440
glycerine 139, 222, 248
glycerol 446–7
glycerol formal 417–18
glycerol isobutyral 417, 420
glycerolipids 213
glycolipids 358
glycosidation 253
glyphosphate 184–5
good governance 69–70
Gossypium hisutum 380
grasslands 171
green chemistry 6–7
    biomass for 164–6
    principles 334
    starch-related 247–56
    24 principles of 7, 8
green diesel *see* biodiesel; biofuels;
    synthetic biofuels
green gas *see* biomethane
green solvents 220, 407–23
    commercially available 411

green strength 361
greenhouse gases 37–8, 342
    agriculture 56
    from cattle 80
    transportation 41, 50
    see also carbon dioxide emissions
Grenelle Environment meeting (2007) 41
gross calorific value 92
groundnut oil 200
groundwaters 81–2
GSP gasifier 154
gypsum boards 257

hardening 219
hardwoods 306, 307, 308
    extractives contents 314
    lignin in 311
    polyoses in 310
hay 75
hazard, control of 410
health hazards 419
    categories 409
    SOLVSAFE 413–15
HEAR (high erucic acid rape seed)
    197, 198
heartwood 314
heat build-up 360
heat production, from biomass 89, 90,
    113–14, 120
heating and drying 95, 96
α-helices 376, 377
hemicellulose 95, 130, 132, 307, 308,
    309–10
hemp 381–3
HERA project 62
herbaceous residues 93, 94
herbicides 184–5, 266
Hevea brasiliensis see rubber tree
hexane 175
High Environment Quality method 46–7
high fructose corn syrup 246–7
high oleic sunflower oil 199, 200
higher heating value 92
horse chestnut 318
hot gas cleaning 106, 151
hot rotating wheel 103

hotmelts 431
HTU diesel 145
hubs, logistic 40, 50
human milk fat replacers 465–6
humectant 248
hunger 69, 70
hybrid products 45
hydrochloric acid scavengers 33
hydrodeoxygenation step 145
hydroformylation 224
hydrogen, from biomass gasification 144
hydrogen bridges, cellulose 308, 309,
    372–3
hydrogen peroxide 389
hydrogenated castor oil 211
hydrogenation 177–8, 180, 219, 223,
    275, 285
hydrolysate 244, 249
hydrolysis
    cellulose 132, 133
    starch 243–4
hydrophilic extractives 314
hydrophilization 191
hydroxyethyl starches 247
7-hydroxymatairesinol 317–18
5-(hydroxymethyl)furfural 272–5

immobilized cells 283–4
import, biomass 117
impregnation modification 330–1
indigenous communities 172, 188, 205
industrial binders 240, 256–7
industrial biotechnology *see* white
    biotechnology
industrial cleaning applications 225–6
inerts 220, 228
infectious diseases 56–7, 58
infiltrating 389–91
inorganic chemicals 59, 306, 318–19
    in natural rubber 355, 359
inositols 357–8
insect fibres 374–5
insecticides *see* pesticides
intensive farming 74–5
intercompany collaboration 44–5
intermediate filaments 376, 378

internal lubrication 219
internal optimization 44
inulin 281, 286–8
invert sugar 270, 280
investments 12–13
   see also socially responsible
      investment
ion exchange process 280
ionic liquids 33–4
iron catalysts 149–50
ISO 14000 standards 26, 27, 46
isomalt 282, 285–6
isomaltulose 280–6
isopentenyl diphosphate 347–50
isosorbide 260

Jatropha oil 211–12
jet fuels 141–2
'Jockey' 78, 79, 80
jojoba oil 181
Just-in-Time 40

Karlsruhe Institute of Technology 156–7
keratin 374–7, 383, 458
key performance indicators 13, 14
Klebsiella pneumoniae 447, 448
knotwood extractives 317–18
Kraft pulping 312, 319–20, 321, 322
Kyoto Protocol 37

lactic acid 443, 466–7
   see also polylactic acid
lactide 471
lamellae 380
lamellar structures 450, 452–3
land area, biomass 116
landfill gas 110
latex yield 344–5, 346
laticifer vessels 344, 345
laurics 182, 192, 193
lead-times, reduction 45
lecithin 175
legislation
   hazardous chemicals 410
   pension funds 20
   see also European Union
Leuconostoc mesenteroides 289

levan 281, 286–7, 288
levansucrase 288
levulinic acid 274–5
LHV *see* lower heating value
life-cycle, washing process 57–62
life-cycle assessment 26–8, 335
   curing systems 30
   washing process 57–8, 60
life-cycle impact assessment 26
life-cycle interpretation 26
life-cycle inventory 26
lignan 207, 317
lignin 130, 132, 307, 308, 311–13
   basic units of 312
   in pulping 319–20
lignocellulosic feedstock 129, 130–2, 142
Limits to Growth (Club of Rome) 3
Linium usitatissimum 381
Linola™ 207
linoleic acid 198–9, 200, 203, 208, 209
linseed 75, 78–80
linseed oil 206–7
linters 381, 382
lipases 463, 465, 466
lipids 76–80
   biofuels based on 137–42, 143
   biotechnology 462–6
   designer lipids 464–6
   jet fuels based on 141–2
   in natural rubber 355, 358–9
   sphingolipids 449–57
   see also fats; fatty acids; oils
Lipofructyl™ 403
lipophilic extractives 314, 316
liquefied petrol gas 147
localization, rubber biosynthesis 351
logistic agility 45–6
logistic efficiency 43–5
logistic reliability 42–3
logistics 37–51
   current situation 40–1
   definition and role 38–40
   detergent life-cycle 61
   employment in 47
   four drivers 41–8
   sustainable 48–50
   see also transportation

LOHAS (Lifestyle Of Health And
  Sustainability) 54–5, 397
low-density lipoprotein-cholesterol 180
lower heating value 90–2
  ethanol 109, 110
lubricants 218–21
lumen 380
lutoid particles 356
lysine 440, 442, 443

macroalgae 212
macrofibrils 308, 373, 377, 379
macrogel 353, 354
macrostructure, rubber 351, 352
maize oil 200–1
maltodextrins 243–4, 246, 247, 261
maltotriose 248
mannans 310
manually fed combustion technologies 99
Marco Polo programme 48–9
margarine production 466
mass customization 45
Mater-bi® 250
mechanical extraction 175
mechanical pulping 319–20, 321, 322
mechanically processed biofuels 89
medicines *see* pharmaceuticals
Mediterranean diet 71, 205
medium-chain fatty acids 194
melt process 278
mercaptodextran 290
mercerization 373, 374, 389
mesostructure, rubber 352–5
metabolic engineering 460–2
metabolites best 62
metal degreasing 418–19
methacrylic acid 272, 273
methane 56, 80, 97
  in biogas 111
  via anaerobic digestion 142
  see also biomethane
methanol
  contamination levels tolerated 151
  from syngas 146–7, 148
methionine 442
methyl esters 193, 218, 219, 221, 223
  see also fatty acid methyl ester

methyl isobutyl ketone 417–18
methyl soyate 187
metrics, sustainability 25–35
mevalonic acid 347–9
microalgae 157, 182, 212–13
microbial fermentation 458–9
microbial lipids 464
micro-credit instrument 69
micro-emulsions 421
microfibrils 308, 372, 373, 376, 377, 379
microgel 353, 354–5
microorganisms *see* bacteria;
  *Escherichia coli; Pichia ciferrii;
  Saccharomyces cerevisiae*
microstructure, rubber 351, 352
MILENA gasifier 155
milk 178
  chemical composition 182
  human milk fat replacers 465–6
  procurement 71–4
  quotas 171
milk production, cattle feed for 74–80
modified solid wood products 329–34
modified starches 242–3, 261
MOGD process 148
molar mass distribution 353
molasses 269, 270
molecular sieves 135
monomers 442–7, 451
Mortierella alpina 464
mother liquor 268–9
MTG process 147–9
MTS process 148–9
mulesing 383–4
multi-modal transport 46, 48–9
MVA pathway 347–9

natural fibres 368–9
  classification 370
  processing 385–91
  sourcing 380–5
  structure 371–80
  sustainable 369–71
natural gas grid 114–15, 142
nature 2–3
Neo-amylose™ 281, 288–9

net calorific value 90–2
nettle cloth 368
neutral lipids 358–9
neutral oil 177
NExBTL 141
NF X 50-600 standard 38
non-ionic surfactants 388
non-isoprene components, natural
    rubber 355–9
non-timber-forest-products 301–2
norm-based screening 18
novaluron 421
NuLin™ 207
NuSun 199–200
nut oils 209
nutrition 68–85
    definition 70–1
    see also food

'Observatory' 84
occupational exposure 414, 415
octane index 136
oil extraction 175
oil palm *see* palm oil
oil seeds 167
    global production 169, 170, 171
    global trade 174
    production 174–5
oils
    biopolymers precursors 214–16
    carrier oils 218–21
    chemical composition 179–213
    essential oils 182, 224
    fatty acid compositions 183
    processing 217
    production, use and trade 168–74
    for synthetic biofuels 140–1
    value chain 213–28
    see also vegetable oils
oilseed rape *see* rapeseed oil
oleic acid 216
oleics 182, 203
oleochemicals
    chemical composition 180
    fatty acids as 216–18
    global consumption 170

polyols 214, 215
    utilization 8
    value chain 218, 220
    see also biodiesel; fats; fatty acids; oils
Olestra 200
oligofructose 287
olive oil 205–6, 209
omega-3 fatty acids 76–80, 182, 207,
    208, 213
omega-6 fatty acids 76–80, 213
organic chemicals 59–60
organic solvents 408
    safer functional 411–16, 420, 421
organosolv pulping 320
Our Common Future (Brundtland
    Commission) 4, 5
oven, wood chips 100–1
over-fishing 179
oxidation 96–8
oxidative bleaching 320
oxidized starches 242–3
Oxo-process 224

packaging materials 250–1, 468
    recycling 46, 80–1
paints 419–20
Palatinose™ 282–6
palm-kernel oil
    composition and uses 192–3
    production 189–92
palm oil 59, 60–1, 172, 187–93
    composition and uses 192
    production 189–92
    sustainable cultivation 61, 188–9
palm olein 191, 193
palm stearin 191, 193
palmitic acid 78
pancreatic lipase 465, 466
paper grade pulp 323
papermaking 258–9, 319–25
pasture 75
peanut oil 200, 201
pellet combustor 99–100
pension funds 20
People for the Ethical Treatment of
    Animals 384

per capita consumption, raw materials 55
persistant organic pollutants 410
pesticides 220, 228, 381, 383, 420–1
  see also herbicides
petrochemicals 164–5, 369–70, 428–9
  see also crude oil; fossil fuels;
    petroleum oil
petroleum diesel *see* diesel fuel
petroleum oil 8, 173
Phakopsora pachyrhizi 185
pharmaceuticals 205, 208, 210, 213
  from dextran 290
  formulations 247
  green solvents in 417
  plant-based 396, 397
  from starches 240, 247–8, 260
  using white biotechnology 438, 442
  from wood extractives 317–18
  see also antibiotics
phosphatides 175, 177
phospholipids 358
photosynthesis 127, 305, 343–4
phthalates 260
physical refining 177
physico-chemical conversion, biomass
  90, 107–9
phytase 437–8
phytochemicals 71, 396, 403
phytoplankton 212–13
phytosphingosine 453–4, 455
Pichia ciferrii 453–7
pigment printing 390
Pimozide 417
plant fibres 368, 372–4
  sourcing 380–3
plant inulin 287
plants
  bioethanol production 127–8
  carbon and water in 343
  for cosmetics 394–405
  essential oils 182
  feed from 440, 442
  see also forests; phytochemicals;
    wood
plasma expanders 247, 289
plasterboards 257

plasticizers 214, 220, 251, 260
plastics
  biodegradable 249–51
  wood-plastic-composites 326–8
platform molecules 412–13
PLATO process 332–3
plugging index 346
pollution *see* environment impacts;
  health hazards
polyamides 443
poly(*cis*-1,4-isoprene) 347–51, 352
polydimethylsiloxane 388
polyether polyols 254, 255
polyethylene 272
polyethyleneterephthalate 260
polyhydric alcohols 275–6
polyisoprenoids 347–9
polylactic acid 443, 466–72
  market potential 469–70
  non-depleting properties 467–9
  process routes 470–1
  processing 471
  properties 471–2
polylactic fibres 369
polymers
  biodegradable plastics 249–51
  carbon dioxide emissions 470
  fossil fuel consumption 469
  packaging and acreage in
    consumption 468
  see also biopolymers
polymorphs, cellulose 373–4
polyoses *see* hemicellulose
polysorbates 253–4
polyunsaturated fatty acids 206–9, 213
  microbial oils 464
polyurethane foams 214, 215, 254, 279
polyurethanes 210–11, 278–9
polyvinyl chloride 272
postponement 45
pot marigold 208
potatoes 239
poverty, and hunger 69, 70
power, from biomass 89, 114, 120
practice-based criteria, exclusion 18
pre-treatment, cellulose 130, 131, 132

predicted environmental
    concentration 62
predicted no-effect concentration 62
prenyltransferases 349–50
pressure sensitive adhesives 426
prices
    crude oil 174, 249, 261
    energy market 118
    food 60
    fossil fuels and renewable carbon
        450, 451
    increased biomass use 116, 117, 118,
        120, 121
    natural rubber 342
    sugar 264, 265
primary extractives 314, 316
principal component statistical
    analysis 413, 414
Principles for Responsible Investment
    21
printing inks 220, 221
probiotics 71
processed foods 246
producer gas 98, 104, 106–7
product-specific requirement 26
Project Village Farm 72
1,3-propanediol 444–7, 448
1,2-propylene glycol 275–6
Protaminobacter rubrum 283, 284
protein–fatty acid condensates 224,
    225
proteins
    fibres 374–80
    in natural rubber 355, 357
    small rubber particle 350, 357
    in soya beans 185–7
    see also amino acids; enzymes
protoporphyrinogen oxidase 421, 422
pulp oils 167, 169, 174
pulping 312, 314, 319–25
    global production 322, 324
    trade flow 323
pure plant oils 139
pyrolysis 97–8, 102–4
    decomposition 95–6
pyrolysis oil 103–4, 145

quality management
    adhesives and sealants 432, 433
    Evian water 84
    high environment quality method
        46–7
    milk 72
    olive oil 206
    Total Quality Management 44
quebrachitol 357–8

radiation curing 431, 432
rainforests *see* forests
rapeseed methyl ester 139–40
rapeseed oil 171, 195–8
    '00-rape' 197, 198
    selected properties 108
raw materials 59–61
    bioethanol production 127–8
    developmental megatrends 55, 58
    plant-based 396–7, 398–9
    price increases 116, 117, 118
    production 55
    for synthetic biofuels 140–1
    see also feedstock; renewable
        resources; *individual raw
        materials*
raw sugar 270
REACH standard 342
reactivity 45
recharge area 84
rectification 135–6
Rectisol process 151
recycling 426
    bioethanol production 136
    packaging 46, 80–1
    polylactic acid 468
    wood-plastic-composites 328
red palm oil 193
reductive bleaching 320
refining 175–7
reforestation 404
regenerated cellulose fibres 324
regeneration 373, 374
Reichskraftstoff 127
reliability, logistic 42–3
rendering 178–9

RENEW project 155
renewable mass index 414–15, 417, 419
renewable resources
  in adhesives and sealants 432–3
  consumption 165
  energy 86, 88
  fibres 369, 370
  historical depletion 4
  renaissance of 7–9
  sucrose 264–5
  wood 304
  yielding chemical intermediates 166
  see also biomass; feedstock; recycling
resins, formaldehyde-based 251
response surface methodology 282
Responsible Care® 6–7
Responsible Care Global Charter 6
responsible investment *see* socially
    responsible investment
retrogradation 241, 243
retting 382
reverse logistics 46
rewetting 191
rice bran oil 202
ricin 209
ricinoleic acid 216
Ricinus communis agglutinin 209–10
ring opening polymerization 470–1
Rio Earth Summit (1992) 5–6, 397, 398
risk management 84
Roundtable on Sustainable Palm Oil
    61, 188–9
Roundup Ready® soya beans 184
rubber, natural 339–40
  see also rubber tree; synthetic rubber
  non-isoprene components 355–9
  poly(*cis*-1,4-isoprene) biosynthesis
    347–51
  prices 342
  production 341
  specific properties 359–61
  structure 351–5
  supply chain 340–2
rubber clonal typology 345
rubber cream 356
rubber elongation factor 350, 357

rubber tree 340, 341
  photosynthesis and water in 343–4
  tapping, latex yield and carbon
    budget 344–5
  tapping and water budget 345–6
Rubisco 344
RuBP 344

saccharides 286
saccharification 130, 132
Saccharomyces cerevisiae 127–8,
    132–4, 136, 280
  in biopolymer production 447, 448
  sphingolipids in 449
safer organic solvents 411–16, 420, 421
safety
  food 70
  see also Environmental Health and
      Safety Guidelines; health hazards
safflower oil 204
sanforizing 389
sanitary defence perimeter 84
saponification 217
sapwood 314
saturated fatty acids 76, 80, 140, 209
saw mill industry 326
'schmoo' 226
scouring 384, 388
scutching 382
sealants 425–34
seaweed 212
second-generation biofuels 142, 144–57
secondary extractives 314, 316
SEEBALANCE 35
seed oils *see* oil seeds
semi-chemical pulping 319, 322
Sephadex® 290
sericin 376, 379–80
L-serine 459–60, 461
serum 356
sesame oil 202–4
shareholder advocacy 19
shea butter 204–5
sheep 383–4
Shikimate biosynthesis 311
short-chain fatty acids 194
silk 374–5, 376, 379–80, 384–5

simultanaeous saccharification and
    fermentation 132
Single-Minute Exchange of Die 45
sizing 258
skin barrier functions 450, 452–3
skin care active ingredients 395
slenderness ratio 326
small and medium enterprises 408
    logistics 40, 41, 48, 50
small rubber particle protein 350, 357
smallholders 188, 204, 340, 341
soaps, manufacture 217
socially responsible investment 15–19
    best-in-class 18
    engagement 19
    exclusion 17–18
    mainstreaming of 19–22
    principles for 21
softwoods 306, 307, 308
    extractives contents 314
    lignin in 311, 313
    polyoses in 310
solar energy 88
Sollbruchstelle 215
solvent-based adhesives 430, 431
solvents
    environment impacts 409
    environmental health and safety 410
    extraction 175
    fractionation 191
    green solvents 220, 407–23
    history 407–8
    safer organic 411–16, 420, 421
SOLVSAFE 411–16
    industrial applications 416–22
soot 87
sorbitan esters 253–4
sorbitol 244–5, 248, 252, 253–5, 260
Sorona® 444–6
soya bean 172, 175, 184–5
    composition and uses 185–7
soya bean oil 184–7
soya polyols 187
space heating 87
Span® 253
SPC model, Sustainability Master®
    63, 64

speciality chemicals 439–42
speciality solvents 411
sperm oil 168, 181
sphinganine 455, 456
sphingolipids 449–57
sphingosine 456–7
Splenda® 276
splitting 220
    fats 217, 219
stabilizing, fabrics 389
stakeholders 397, 405
    logistics reliability 43
    natural rubber 340, 341
    palm oil 189
standards
    biodiesel 138
    environmental management 26, 27, 46
    logistics 38
    REACH standard 342
starch
    in bioethanol production 128, 129, 130
    food applications 240, 241, 245–7
    industrial binder applications 240,
        256–7
    markets 238–9, 240, 241
    modified 242–3, 261
    outlook for 259–62
    paper and board applications 240,
        257–9
    pharmaceutical and chemical
        applications 240, 247–56, 260
    product family tree 261
    refinery products 243–5
    structure and properties 239–42
steam explosion 132
storage organs, plant 127–8
strain-induced crystallization 360
stratum corneum 450, 452
structural adhesives 427, 432
substitution, biomass 117
sucralose 276–7
sucrose 127–8, 133, 264–5
    chemistry 270–90
    outlook 290–1
    production 266–70
    in the rubber tree 344–5, 357
    see also sugar

sucrose acetate isobutyrate 278
sucrose derivates
  degradation of framework 271, 272–6
  maintaining carbohydrate structure
      271, 279–90
  maintaining sucrose skeleton 271,
      276–9
sucrose esters 277–8
sucrose isomerases 280, 282
sucrose polyesters 278
sugar 109
  bioethanol production 128, 129, 133
  as carbon feedstock 450, 451
  invert sugar 270, 280
  prices 264, 265
  replacer 282, 285–6
  see also sucrose
sugar beet/cane 264, 265, 266
  sugar manufacture from 267–70
sulphite pulping 311, 312, 314, 319–22
sunflower oil 171, 198–200
supercritical gases 411
supply chain 38–9, 51
  eco-logistics 47
  efficiency 44–5
  natural rubber 340–2
  problems with 40
  reliability 43
  sustainable logistics 49–50
Supply Chain Event Management 43
Supply Chain Management 39, 50
Supply Chain Wheel 47–8
supramolecular structure, rubber 352
surfactants 59–60, 62, 63–4, 193, 384
  from fatty alcohols 222
  in fibre processing 386–8
  fossil vs renewable base stock 223
  green 224–8, 252–3
  from sucrose 290–1
sustainability 302, 369
  adhesives and sealants 425–34
  Argan tree products 399–404, 405
  concept 397–8
  fibres 369–71
  finance 12–23
  in food and nutrition 69–84

four dimensions 398
green solvents 407–23
history of 1–9
metrics 25–35
palm oil cultivation 188–9
Sustainability Master® 63, 64
sustainability profiles, detergents 62–5
sustainability zone 63
sustainable chemistry *see* green
  chemistry
sustainable development, logistics
  and 48–50
SVZ gasifiers 154
sweeteners 283, 286
syngas 106, 144
  based biofuels 145–57
  based products 146–50
  current developments 154–7
  generation 150–4
synthetic biofuels 115, 120
  fats and oils as raw material 140–1
  flow diagram 143
  property comparisons 141
  see also biofuels
synthetic natural gas 114–15, 120, 144
synthetic rubber 217, 341, 360
syrup 268–9

tack 361
tall oil 180, 181, 217
'tank mix' 420
tanning industry 437
tapping 344–6
Targanine network 401–4, 405
technical biomass potential 93–5, 113–15
temperature
  anaerobic digestion 110, 112
  Fischer–Tropsch process 149–50
  heat build-up 360
  and photosynthesis 343–4
tetraacetyl phytosphingosine 453–4
textile manufacturing 369
textile softeners 386–8
thermal energy, conversion 98
thermal modification, wood 331–4
thermal properties, polylactic acid 471–2

thermo-chemical conversion
  biomass 89–90, 95–107
  biomethane 114, 115
  wood 95, 96
thermophilic clostridia 134
thermoplastic starch 250, 261
'thick juice' 268, 270
'thin juice' 268, 270
thinned starches 243
third-generation biofuels 157–8
threonine 442
TIGAS process 148
timber 306
TOE (tonnes of oil equivalent) 342
toothpaste 248
Total Quality Management 44
toxins, castor oil 209–10
trace elements, in biomass 90
trans-fatty acids 179, 180, 182, 199
transesterification 107, 108, 138–9,
    217, 218, 219, 220
transpiration 343
transportation
  capacity utilization 40
  drawbacks to 37, 50
  eco-logistics 46
  forecast rates 37, 49
  greenhouse gas emissions 41
  sustainable logistics 48–9
  see also logistics
transportation fuels
  bioethanol as 136
  jet fuels 141–2
  vegetable oils as 137–8
  see also bioethanol; biofuels; diesel
      fuel; synthetic biofuels
trees *see* forests; wood
trehalulose 282–6
triacylglycerols 462
trichloroethylene 418, 419
triglycerides 179, 466
tung oil 208
Tween® 253
tyres 341, 360

ultimate supply chain 39

UN Conference on Environment and
    Development (1992) 5–6
UN Stockholm Conference on the
    Human Environment (1972) 3, 6
UN World Commission on Environment
    and Development 4, 5
UNEP FI 16, 20
unsaponifiables 359
unsaturated fatty acids 76, 80, 209
  chemical composition 179–80
  polyunsaturated 206, 207, 208, 209, 213
use phase 61
UV curing 29–30, 31, 34

value chain, fats and oils 213–28
Värnamo demonstration plant 154–5
varnishes 419–20
vegetable fats 177–8, 204–5
vegetable oils
  as biodiesel feedstock 138–40
  from biomass 90, 107
  in cattle feed 76–80
  chemical composition 182–213
  classification 167
  global production 169
  major uses 167–8
  polymer building blocks 216
  production 174–8
  raw materials 59–61
  as transportation fuels 137–8
  uncertain market 115
  see also oils; *individual vegetable oils*
Vendor-Managed Inventory 44
vernonia oil 181
viscosity, starch 241, 243
vitamin B_2 production 30–3
vitamin C 251–2
volatile organic chemicals 226, 317,
    389, 414, 415, 419, 420
vulcanization 361

walnut oil 209
washed bottom fraction particles 350
washing process, life-cycle 57–62
washing temperature 61
waste paper 321

water
  budget 345–6
  degumming 175
  in plants 343
  retting 382
  in the rubber tree 343–4
  stress 344
water-based adhesives 430, 431
water-based emulsion 226–7
water resources
  developmental megatrends 56–7, 58
  Evian mineral water 80–4
water use efficiency 343
waxes, natural 182
wet adhesives 429
wet-end 258
wet gas cleaning 106, 151
wet-milling 130, 201
whaling, banning of 168
white biotechnology 165, 436–7
  L-cysteine 457–62
  drivers 447–9
  lipids 462–6
  market segments 437–47
  outlook 472–3
  plant announcements 451
  polylactic acid 466–72
  sphingolipids 449–57
  vitamin B$_2$ production 31–2
  see also enzymes; fermentation
Who Cares Wins initiative 20–1
whole-cell systems 437
winterization 191

wood
  chemical composition 307
  chemistry 306–19
  coalification 102
  commercial species 306, 308
  increasing demand 116
  lower heating value 92
  modification 329–34
  from resources to products 303–6
  sustainability, perspectives 302
  thermo-chemical conversion 95, 96
  uses 301
  see also forests
wood-based composites 325–9
wood chips, oven 100–1
wood gas 126
wood pellet combustor 99–100
wood-plastic-composites 326–8
wood residues 93, 94, 304
wool 370, 371, 374–5, 376, 383–4, 385
wool fabric 389, 390

xylans 310
xylose 310

yarns 388
yeast *see Pichia ciferrii;*
    Saccharomyces cerevisiae

zeolite ZSM-5 catalyst 147, 148
zero-waste operations 128, 168, 192
Ziegler technology 224
Zymomonas mobilis 134–5